T0184361

Grundlehren der mathematischen Wissenschaften 249

A Series of Comprehensive Studies in Mathematics

Kai Lai Chung and John B. Walsh

Markov Processes, Brownian Motion, and Time Symmetry

Second Edition

Springer

Berlin Heidelberg New York
Hong Kong London
Milan Paris Tokyo

Kai Lai Chung
Department of Mathematics
Stanford University
Stanford, CA 94305
USA

John B. Walsh
Mathematics Department
University of British Columbia
Vancouver, BC V6T 1Z2
Canada

AMS Subject Classifications (2000): 60Jxx, 60J65

Library of Congress Cataloging in Publication Data
Chung, Kai Lai
Markov processes, Brownian motion, and time symmetry / Kai Lai Chung, John B. Walsh.
p. cm.
Includes bibliographical references and index.

1. Markov processes. 2. Brownian motion processes. I. Walsh, John B. II. Title.
QA274.7.C52 2004
519.2′33- -dc22 2004048547

Printed on acid-free paper.
ISBN-13: 978-1-4419-1960-1 e-ISBN-13: 978-0-387-28696-9

Printed in the United States of America. (TS/SB)

9 8 7 6 5 4 3 2 1

Springer is a part of Springer Science+Business Media
springeronline.com

Contents

Preface to the New Edition

This book consists of two parts, to be called Part I and Part II. Part I, Chapters 1 through 5, is essentially a new edition of Kai Lai Chung's Lectures from Markov Processes to Brownian Motion (1982). He has corrected a number of misprints in the original edition, and has inserted a few references and remarks, of which he says, "The latter must be regarded as randomly selected since twenty-some years is a long time to retrace steps ..." This part introduces strong Markov processes and their potential theory. In particular, it studies Brownian motion, and shows how it generates classical potential theory.

Part II, Chapters 6 through 15, began life as a set of notes for a series of lectures on time reversal and duality given at the University of Paris. I originally planned to add the essential parts of these notes to this edition to show how the reversal of time—the retracing of steps—explained so much about Markov processes and their potential theory. But like many others, I learned that the inessential parts of a cherished manuscript form at most a fuzzy empty set, while the essential parts include everything that should have been in the original, even if it wasn't. In short, this, like Topsy, just grow'd.

Indeed, reversal and duality are best understood in light of Ray processes and the Ray–Knight compactification. But it is fitting that a study of symmetry be symmetrical itself, so once I had included the Ray compactification, I had to include its mirror image, the Martin boundary. This was followed by a host of examples, remarks and theorems to show how these new ideas influence the theory and practice developed in the first part. The result was the present Part II.

In a sense, Part II deals with the same subjects as Part I, but more narrowly: using Part I for a general understanding, we are free to focus on the effects of time reversal, duality, and time-symmetry on potential theory. Certain theorems in Part I are re-proved in Part II under slightly weaker hypotheses. This is not because I want to generalize the theorems, but because I want to show them in a different light: the proofs in Part II are quite different from those of Part I.

The class of Markov processes in Part I is slightly less general than it first appears—it does not include all Markov chains, for example, nor is it closed under time-reversal. Thus, after setting the stage with preliminary sections on general theory and Markov chains, I introduce Ray processes, which are in a sense the Platonic ideals of Markov processes: any nice Markov process with stationary transition probabilities has its ideal Ray process, and the two are connected by a procedure called Ray–Knight compactification. Ray processes introduce branching points. These are customarily ignored in the Markov process theory, but they arise naturally when one considers boundary conditions. Although they may initially seem to complicate the theory, they clarify some important things, such as the classification of stopping times and path discontinuities, and they help to explain some otherwise-mysterious behavior. I consider them an essential part of the subject.

The key tools in this study are the fundamental theorem of time-reversal in §10 and Doob's theory of h-transforms in §11. These figure systematically in the proofs.

We study probabilistic potential theory under duality hypotheses. In one sense, duality and time-reversal are two aspects of the same phenomenon: the reversed process is an h-transform of the dual. In another, deeper sense, discussed in §15, duality simply implies the existence of a cofine topology. In any case, duality is a natural setting for a study of time-symmetry.

One of the themes of Part II is that the strength and elegance of the potential theory increase as the process and its dual become more alike, i.e. as the process becomes more symmetric. However, some say that near-symmetry is beautiful but perfect symmetry is bland, and I feel that the most interesting theory arises when the process and its dual are nearly alike, but when there is still some tension between symmetry and assymmetry.

Another theme is the importance of left limits, particularly at the lifetime. (The whole structure of the Martin boundary can be viewed as an attempt to understand this quantity in depth.) This leads to an interplay between the strong Markov property, the moderate Markov property, and time-reversal. Watch for it.

I have included several chapters dubbed "Fireside chats" on subjects which are relevant and illuminating, but not strictly necessary for the rest of the material. They contain careful statements of results, but no rigorous proofs. The treatment is informal and is more concerned with why the results should be true rather than why they are true. The reader can treat these as extended remarks, or simply ignore them entirely.

Finally, Part II uses the same notation and terminology as Part I, with a few notable exceptions: for instance, "optional times" become "stopping times" and "superaveraging functions" become "supermedian functions." There is no deep reason for this, other than a misjudgement of the difficulty of changing these in the final computer file by a simple search-and-replace. It should not cause the reader any problems.

John B. Walsh
Vancouver, B.C., Canada
July 4, 2004

Preface to the First Edition

This book evolved from several stacks of lecture notes written over a decade and given in classes at slightly varying levels. In transforming the overlapping material into a book, I aimed at presenting some of the best features of the subject with a minimum of prerequisites and technicalities. (Needless to say, one man's technicality is another's professionalism.) But a text frozen in print does not allow for the latitude of the classroom; and the tendency to expand becomes harder to curb without the constraints of time and audience. The result is that this volume contains more topics and details than I had intended, but I hope the forest is still visible with the trees.

The book begins at the beginning with the Markov property, followed quickly by the introduction of optional times and martingales. These three topics in the discrete parameter setting are fully discussed in my book *A Course In Probability Theory* (second edition, Academic Press, 1974).* The latter will be referred to throughout this book as the *Course*, and may be considered as a general background; its specific use is limited to the material on discrete parameter martingale theory cited in §1.4. Apart from this and some dispensable references to Markov chains as examples, the book is self-contained. However, there are a very few results which are explained and used, but not proved here, the first instance being the theorem on projection in §1.6. The fundamental regularity properties of a Markov process having a Feller transition semigroup are established in Chapter 2, together with certain measurability questions which must be faced. Chapter 3 contains the basic theory as formulated by Hunt, including some special topics in the last three sections. Elements of a potential theory accompany the development, but a proper treatment would require the setting up of dual structures. Instead, the relevant circle of ideas is given a new departure in Chapter 5. Chapter 4 grew out of a short compendium as a particularly telling example, and Chapter 5 is a splinter from unincorporated sections of Chapter 4. The venerable theory of Brownian motion is so well embellished and ramified that once begun it is hard to know where to stop. In the end I have let my own propensity and capability make the choice. Thus the last three sections of the book treat several recent developments which have engaged me lately. They are included here with the hope of inducing further work in such fascinating old-and-new themes as equilibrium, energy, and reversibility.

*Third edition, 2001.

I used both the Notes and Exercises as proper non-trivial extensions of the text. In the Notes a number of regretably omitted topics are mentioned, and related to the text as a sort of guide to supplementary reading. In the Exercises there are many alternative proofs, important corollaries and examples that the reader will do well not to overlook.

The manuscript was prepared over a span of time apparently too long for me to maintain a uniform style and consistent notation. For instance, who knows whether "semipolar" should be spelled with or without a hyphen? And if both $|x|$ and $\|x\|$ are used to denote the same thing, does it really matter? Certain casual remarks and repetitions are also left in place, as they are permissible, indeed desirable, in lectures. Despite considerable pains on the part of several readers, it is perhaps too much to hope that no blunders remain undetected, especially among the exercises. I have often made a point, when assigning homework problems in class, to say that the correction of any inaccurate statement should be regarded as part of the exercise. This is of course not a defense for mistakes but merely offered as prior consolation.

Many people helped me with the task. To begin with, my first formal set of notes, contained in five folio-size, lined, students' copybooks, was prepared for a semester course given at the Eidgenössiche Technische Hochschule in the spring of 1970. My family has kept fond memories of a pleasant sojourn in a Swiss house in the great city of Zürich, and I should like to take this belated occasion to thank our hospitable hosts. Another set of notes (including the lectures given by Doob mentioned in §4.5) was taken during 1971–2 by Harry Guess, who was kind enough to send me a copy. Wu Rong, a visiting scholar from China, read the draft and the galley proofs, and checked out many exercises. The comments by R. Getoor, N. Falkner, and Liao Ming led to some final alterations. Most of the manuscript was typed by Mrs. Gail Stein, who also typed some of my other books. Mrs. Charlotte Crabtree, Mrs. Priscilla Feigen, and my daughter Marilda did some of the revisions. I am grateful to the National Science Foundation for its support of my research, some of which went into this book.

August 1981 Kai Lai Chung

Chapter 1

Markov Process

1.1. Markov Property

We begin by describing a general Markov process running on continuous time and living in a topological space. The time parameter is the set of positive numbers, considered at first as just a linearly ordered set of indices. In the discrete case this is the set of positive integers and the corresponding discussion is given in Chapter 9 of the *Course*. Thus some of the proofs below are the same as for the discrete case. Only later when properties of sample functions are introduced will the continuity of time play an essential role. As for the living space we deal with a general one because topological properties of sets such as "open" and "compact" will be much used while specific Euclidean notions such as "interval" and "sphere" do not come into question until much later.

We must introduce some new terminology and notation, but we will do this gradually as the need arises. Mathematical terms which have been defined in the *Course* will be taken for granted, together with the usual symbols to denote them. The reader can locate these through the Index of the *Course*. But we will repeat certain basic definitions with perhaps slight modifications.

Let $(\Omega, \mathscr{F}, P)$ be a probability space. Let

$$\mathbf{T} = [0, \infty).$$

Let $\mathbf{E}$ be a locally compact separable metric space; and let $\mathscr{E}$ be the minimal Borel field in $\mathbf{E}$ containing all the open sets. The reader is referred to any standard text on real analysis for simple topological notions. Since the Euclidean space $\mathbf{R}^d$ of any dimension d is a well known particular case of an $\mathbf{E}$, the reader may content himself with thinking of $\mathbf{R}^d$ while reading about $\mathbf{E}$, which is not a bad practice in the learning process.

For each $t \in \mathbf{T}$, let

$$X_t(\omega) = X(t, \omega)$$

be a function from Ω to $\mathbf{E}$ such that

$$X_t^{-1}(\mathscr{E}) \subset \mathscr{F}.$$

This will be written as

$$X_t \in \mathscr{F}/\mathscr{E};$$

and we say that X_t is a random variable taking values in $(\mathbf{E}, \mathscr{E})$. For $\mathbf{E} = \mathbf{R}^1$, $\mathscr{E} = \mathscr{B}^1$, this reduces to the familiar notion of a real random variable. Now any family $\{X_t, t \in \mathbf{T}\}$ is called a *stochastic process*. In this generality the notion is of course not very interesting. Special classes of stochastic processes are defined by imposing certain conditions on the random variables X_t, through their joint or conditional distributions. Such conditions have been formulated by pure and applied mathematicians on a variety of grounds. By far the most important and developed is the class of Markov processes that we are going to study.

Borel field is also called σ-field or σ-algebra. As a general notation, for any family of random variables $\{Z_\alpha, \alpha \in A\}$, we will denote the σ-field generated by it by $\sigma(Z_\alpha, \alpha \in A)$. Now we put specifically

$$\mathscr{F}_t^0 = \sigma(X_s, s \in [0, t]); \qquad \mathscr{F}_t' = \sigma(X_s, s \in [t, \infty)).$$

Intuitively, an event in $\mathscr{F}_t^0$ is determined by the behavior of the process $\{X_s\}$ up to the time t; an event in $\mathscr{F}_t'$ by its behavior after t. Thus they represent respectively the "past" and "future" relative to the "present" instant t. For technical reasons, it is convenient to enlarge the past, as follows.

Let $\{\mathscr{F}_t, t \in \mathbf{T}\}$ be a family of σ-fields of sets in $\mathscr{F}$, such that

(a) if $s < t$, then $\mathscr{F}_s \subset \mathscr{F}_t$;
(b) for each t, $X_t \in \mathscr{F}_t$.

Property (a) is expressed by saying that "$\{\mathscr{F}_t\}$ is increasing"; property (b) by saying that "$\{X_t\}$ is adapted to $\{\mathscr{F}_t\}$". Clearly the family $\{\mathscr{F}_t^0\}$ satisfies both conditions and is the minimal such family in the obvious sense. Other instances of $\{\mathscr{F}_t\}$ will appear soon. The general definition of a Markov process involves $\{\mathscr{F}_t\}$ as well as $\{X_t\}$.

Definition. $\{X_t, \mathscr{F}_t, t \in \mathbf{T}\}$ is a Markov process iff one of the following equivalent conditions is satisfied:

(i) $\forall t \in \mathbf{T}, A \in \mathscr{F}_t, B \in \mathscr{F}_t'$:

$$P(A \cap B \mid X_t) = P(A \mid X_t) P(B \mid X_t).$$

(ii) $\forall t \in \mathbf{T}, B \in \mathscr{F}_t'$:

$$P(B \mid \mathscr{F}_t) = P(B \mid X_t).$$

(iii) $\forall t \in \mathbf{T}, A \in \mathscr{F}_t$:

$$P(A \mid \mathscr{F}_t') = P(A \mid X_t).$$

The reader is reminded that a conditional probability or expectation is an equivalence class of random variables with respect to the measure P. The equations above are all to be taken in this sense.

We shall use the two basic properties of conditional expectations, for arbitrary σ-fields $\mathscr{G}, \mathscr{G}_1, \mathscr{G}_2$ and integrable random variables Y and Z:

(a) If $Y \in \mathscr{G}$, then $E\{YZ|\mathscr{G}\} = YE\{Z|\mathscr{G}\}$;

(b) If $\mathscr{G}_1 \subset \mathscr{G}_2$, then

$$E\{E(Y|\mathscr{G}_1)|\mathscr{G}_2\} = E\{E(Y|\mathscr{G}_2)|\mathscr{G}_1\} = E\{Y|\mathscr{G}_1\}.$$

See Chapter 9 of *Course*.

Let us prove the equivalence of (i), (ii) and (iii). Assume that (i) holds, we will deduce (ii) in the following form. For each $A \in \mathscr{F}_t$ and $B \in \mathscr{F}'_t$ we have

$$E\{1_A P(B|X_t)\} = P(A \cap B). \tag{1}$$

Now the left member of (1) is equal to

$$\begin{aligned} E\{E[1_A P(B|X_t)]|X_t\} &= E\{P(A|X_t)P(B|X_t)\} \\ &= E\{P(A \cap B|X_t)\} = P(A \cap B). \end{aligned}$$

Symmetrically, we have

$$E\{1_B P(A|X_t)\} = P(A \cap B),$$

which implies (iii).

Conversely, to show for instance that (ii) implies (i), we have

$$\begin{aligned} P(A \cap B|X_t) &= E\{E(1_A \cdot 1_B|\mathscr{F}_t)|X_t\} \\ &= E\{1_A P(B|\mathscr{F}_t)|X_t\} = E\{1_A P(B|X_t)|X_t\} \\ &= P(B|X_t)E\{1_A|X_t\} = P(B|X_t)P(A|X_t). \end{aligned}$$

From here on we shall often omit such qualifying phrases as "$\forall t \in \mathbf{T}$". As a general notation, we denote by $b\mathscr{G}$ the class of bounded real-valued $\mathscr{G}$-measurable functions; by C_c the class of continuous functions on $\mathbf{E}$ with compact supports.

Form (ii) of the Markov property is the most useful one and it is equivalent to any of the following:

(iia) $\forall Y \in b\mathscr{F}'_t$:

$$E\{Y|\mathscr{F}_t\} = E\{Y|X_t\}.$$

(iib) $\forall u \geq t, f \in b\mathscr{E}$:

$$E\{f(X_u)|\mathscr{F}_t\} = E\{f(X_u)|X_t\}.$$

(iic) $\forall u \geq t, f \in C_c(E)$:

$$E\{f(X_u)|\mathscr{F}_t\} = E\{f(X_u)|X_t\}.$$

It is obvious that each of these conditions is weaker than the preceding one. To show the reverse implications we state two lemmas. As a rule, a self-evident qualifier "nonempty" for a set will be omitted as in the following proposition.

Lemma 1. *For each open set G, there exists a sequence of functions* $\{f_n\}$ *in* C_c *such that*

$$\lim_n \uparrow f_n = 1_G.$$

This is an easy consequence of our topological assumption on **E**, and gives the reader a good opportunity to review his knowledge of such things.

Lemma 2. *Let S be an arbitrary space and* $\mathbb{D}$ *a class of subsets of S.* $\mathbb{D}$ *is closed under finite intersections. Let* $\mathbb{C}$ *be a class of subsets of S such that* $S \in \mathbb{C}$ *and* $\mathbb{D} \subset \mathbb{C}$. *Furthermore suppose that* $\mathbb{C}$ *has the following closure properties:*

(a) *if* $A_n \in \mathbb{C}$ *and* $A_n \subset A_{n+1}$ *for* $n \geq 1$, *then* $\bigcup_{n=1}^{\infty} A_n \in \mathbb{C}$;
(b) *if* $A \subset B$ *and* $A \in \mathbb{C}$, $B \in \mathbb{C}$, *then* $B - A \in \mathbb{C}$.

Then $\mathbb{C} \supset \sigma(\mathbb{D})$.

Here as a general notation $\sigma(\mathbb{D})$ is the σ-field generated by the class of sets $\mathbb{D}$. Lemma 2 is Dynkin's form of the "monotone class theorem"; the proof is similar to Theorem 2.1.2 of *Course*, and is given as an exercise there. The reader should also figure out why the cited theorem cannot be applied directly in what follows.

Let us now prove that (iic) implies (iib). Using the notation of Lemma 1, we have by (iic)

$$E\{f_n(X_u) \mid \mathscr{F}_t\} = E\{f_n(X_u) \mid X_t\}.$$

Letting $n \to \infty$ we obtain by monotone convergence

$$P\{X_u \in G \mid \mathscr{F}_t\} = P\{X_u \in G \mid X_t\}. \tag{2}$$

Now apply Lemma 2 to the space **E**. Let $\mathbb{D}$ be the class of open sets, $\mathbb{C}$ the class of sets A satisfying

$$P\{X_u \in A \mid \mathscr{F}_t\} = P\{X_u \in A \mid X_t\}. \tag{3}$$

Of course $\mathbb{D}$ is closed under finite intersections, and $\mathbb{D} \subset \mathbb{C}$ by (2). The other properties required of $\mathbb{C}$ are simple consequences of the fact that each member of (3), as function of A, acts like a probability measure; see p. 301 of *Course* for a discussion. Hence we have $\mathbb{C} \supset \mathscr{E}$ by Lemma 2, which means that (3) is true for each A in $\mathscr{E}$, or again that (iib) is true for $f = 1_A$, $A \in \mathscr{E}$. The class of f for which (iib) is true is closed under addition, multiplication by a constant, and monotone convergence. Hence it includes $b\mathscr{E}$ by a standard approximation. [We may invoke here Problem 11 in §2.1 of *Course*.]

To prove that (iib) implies (iia), we consider first

$$Y = f_1(X_{u_1}) \cdots f_n(X_{u_n})$$

where $t \le u_1 < \cdots < u_n$, and $f_j \in b\mathscr{E}$ for $1 \le j \le n$. For such a Y with $n = 1$, (iia) is just (iib). To make induction from $n - 1$ to n, we write

$$\begin{aligned} E\left\{\prod_{j=1}^{n} f_j(X_{u_j}) \,\middle|\, \mathscr{F}_t\right\} &= E\left\{E\left[\prod_{j=1}^{n} f_j(X_{u_j}) \,\middle|\, \mathscr{F}_{u_{n-1}}\right] \middle|\, \mathscr{F}_t\right\} \\ &= E\left\{\prod_{j=1}^{n-1} f_j(X_{u_j}) E[f_n(X_{u_n}) \mid \mathscr{F}_{u_{n-1}}] \,\middle|\, \mathscr{F}_t\right\}. \end{aligned} \tag{4}$$

Now we have by (iib)

$$E[f_n(X_{u_n}) \mid \mathscr{F}_{u_{n-1}}] = E[f_n(X_{u_n}) \mid X_{u_{n-1}}] = g(X_{u_{n-1}})$$

for some $g \in b\mathscr{E}$. Substituting this into the above and using the induction hypothesis with $f_{n-1} \cdot g$ taking the place of f_{n-1}, we see that the last term in (4) is equal to

$$\begin{aligned} E\left\{\prod_{j=1}^{n-1} f_j(X_{u_j}) E[f_n(X_{u_n}) \mid \mathscr{F}_{u_{n-1}}] \,\middle|\, X_t\right\} &= E\left\{E\left[\prod_{j=1}^{n} f_j(X_{u_j}) \,\middle|\, \mathscr{F}_{u_{n-1}}\right] \middle|\, X_t\right\} \\ &= E\left\{\prod_{j=1}^{n} f_j(X_{u_j}) \,\middle|\, X_t\right\} \end{aligned}$$

since $X_t \in \mathscr{F}_{u_{n-1}}$. This completes the induction.

Now let $\mathbb{D}$ be the class of subsets of Ω of the form $\bigcap_{j=1}^{n} \{X_{u_j} \in B_j\}$ with the u_j's as before and $B_j \in \mathscr{E}$. Then $\mathbb{D}$ is closed under finite intersections. Let $\mathbb{C}$ be the class of subsets Λ of Ω such that

$$P\{\Lambda \mid \mathscr{F}_t\} = P\{\Lambda \mid X_t\}.$$

Then $\Omega \in \mathbb{C}$, $\mathbb{D} \subset \mathbb{C}$ and $\mathbb{C}$ has the properties (a) and (b) in Lemma 2. Hence by Lemma 2, $\mathbb{C} \supset \sigma(\mathbb{D})$ which is just $\mathscr{F}'_t$. Thus (iia) is true for any indicator $Y \in \mathscr{F}'_t$ [that is (ii)], and so also for any $Y \in b\mathscr{F}'_t$ by approximations. The equivalence of (iia), (iib), (iic), and (ii), is completely proved.

Finally, (iic) is equivalent to the following: for arbitrary integers $n \ge 1$ and $0 \le t_1 < \cdots t_n < t < u$, and $f \in C_c(\mathbf{E})$ we have

$$E\{f(X_u) \mid X_t, X_{t_n}, \ldots, X_{t_1}\} = E\{f(X_u) \mid X_t\}. \tag{5}$$

This is the oldest form of the Markov property. It will not be needed below and its proof is left as an exercise.

1.2. Transition Function

The probability structure of a Markov process will be specified.

Definition. The collection $\{P_{s,t}(\cdot,\cdot), 0 \le s < t < \infty\}$ is a *Markov transition function* on $(\mathbf{E}, \mathscr{E})$ iff $\forall s < t < u$ we have

(a) $\forall x \in \mathbf{E}: A \to P_{s,t}(x, A)$ is a probability measure on $\mathscr{E}$,
(b) $\forall A \in \mathscr{E}: x \to P_{s,t}(x, A)$ is $\mathscr{E}$-measurable;
(c) $\forall x \in \mathbf{E}, \forall A \in \mathscr{E}$:

$$P_{s,u}(x, A) = \int_{\mathbf{E}} P_{s,t}(x, dy) P_{t,u}(y, A).$$

This function is called [*temporally*] *homogeneous* iff there exists a collection $\{P_t(\cdot,\cdot), 0 < t\}$ such that $\forall s < t, x \in \mathbf{E}, A \in \mathscr{E}$ we have

$$P_{s,t}(x, A) = P_{t-s}(x, A).$$

In this case (a) and (b) hold with $P_{s,t}$ replaced by P_t, and (c) may be rewritten as follows (*Chapman-Kolmogorov equation*):

$$P_{s+t}(x, A) = \int_{\mathbf{E}} P_s(x, dy) P_t(y, A). \tag{1}$$

For $f \in b\mathscr{E}$, we shall write

$$P_t f(x) = P_t(x, f) = \int_{\mathbf{E}} P_t(x, dy) f(y).$$

Then (b) implies that $P_t f \in b\mathscr{E}$. For each t, the operator P_t maps $b\mathscr{E}$ into $b\mathscr{E}$, also $\mathscr{E}_+$ into $\mathscr{E}_+$, where $\mathscr{E}_+$ denotes the class of positive (extended-valued) $\mathscr{E}$-measurable functions. The family $\{P_t, t > 0\}$ forms a semigroup by (1) which is expressed symbolically by

$$P_{s+t} = P_s P_t.$$

As a function of x and A, $P_t(x, A)$ is also called a "kernel" on $(\mathbf{E}, \mathscr{E})$.

Definition. $\{X_t, \mathscr{F}_t, t \in \mathbf{T}\}$ is a homogeneous Markov process with (P_t) as its transition function (or semigroup) iff for $t \ge 0$, $s > 0$ and $f \in b\mathscr{E}$ we have

$$E\{f(X_{t+s}) | \mathscr{F}_t\} = P_s(X_t, f). \tag{2}$$

Observe that the left side in (2) is defined as a function of ω (not shown!) only up to a set of P-measure zero, whereas the right side is a completely determined function of ω since X_t is such a function. Such a relation should be understood to mean that one version of the conditional expectation on the left side is given by the right side.

Henceforth a homogeneous Markov process will simply be called a Markov process.

The distribution μ of X_0 is called the *initial distribution* of the process. If $0 \le t_1 < \cdots < t_n$ and $f \in b\mathscr{E}^n$, we have

$$E\{f(X_{t_1}, \ldots, X_{t_n})\}$$
$$= \int \mu(dx_0) \int P_{t_1}(x_0, dx_1) \cdots \int P_{t_n - t_{n-1}}(x_{n-1}, dx_n) f(x_1, \ldots, x_n) \quad (3)$$

where the integrations are over $\mathbf{E}$. In particular if f is the indicator of $A_1 \times \cdots \times A_n$, where $n \ge 1$ and each $A_j \in \mathscr{E}$, this gives the finite-dimensional joint distributions of the process.

If Ω is the space of all functions from $\mathbf{T}$ to $\mathbf{E}$, it is possible by Kolmogorov's extension theorem (see e.g. Doob [1]) to construct a process with the joint distributions given in (3). *We shall always assume that such a process exists in our probability space.* If x is any point in $\mathbf{E}$, and $\mu = \varepsilon_x$ (the point mass at x), the corresponding process is said to start at x. The probability measure on the σ-field $\mathscr{F}^0$ generated by this process will be denoted by P^x, and the corresponding expectation by E^x. For example if $Y \in b\mathscr{F}^0$:

$$E^x(Y) = \int_\Omega Y(\omega) P^x(d\omega),$$

and if $Y = 1_A(X_t)$, where $A \in \mathscr{E}$, then the quantity above reduces to

$$P^x(X_t \in A) = P_t(x, A). \quad (4)$$

Furthermore for each $\Lambda \in \mathscr{F}^0$, the function $x \to P^x(\Lambda)$ is $\mathscr{E}$-measurable. For $\Lambda = X_t^{-1}(A)$ this follows from (4) and property (b) of the transition function. The general case then follows by a monotone class argument as in the proof that (iib) implies (iia) in §1.1.

The Markov property (2) can now be written as

$$P\{X_{s+t} \in A \mid \mathscr{F}_t\} = P^{X_t}\{X_s \in A\} = P_s(X_t, A) \quad (5)$$

where $t \ge 0$, $s > 0$, $A \in \mathscr{E}$. Beware of the peculiar symbolism which allows the substitution of X_t for the generic x in $P^x(A)$. For instance, if $s = t$ in the second member of (5), the two occurrences of X_t do not have the same significance. [There is, of course, no such confusion in the third member of (5).] Nevertheless the system of notation using the superscript will be found workable and efficient.

We want to extend the equation to sets more general than $\{X_{s+t} \in A\} = X_{s+t}^{-1}(A)$. This can be done expeditiously by introducing a "shift" $\{\theta_t, t \geq 0\}$ in the following manner. For each t, let θ_t map Ω into Ω such that

$$\forall t\colon (X_s \circ \theta_t)(\omega) = X_s(\theta_t\omega) = X_{s+t}(\omega). \tag{6}$$

With this notation we have

$$X_{s+t}^{-1} = \theta_t^{-1}X_s^{-1},$$

so that (5) becomes

$$P\{\theta_t^{-1}(X_s^{-1}(A))\,|\,\mathscr{F}_t\} = P^{X_t}\{X_s^{-1}(A)\}. \tag{7}$$

In general if $\Lambda \in \mathscr{F}^0$, then $\theta_t^{-1}\Lambda \in \mathscr{F}_t'$ (proof?), and we have

$$P\{\theta_t^{-1}\Lambda\,|\,\mathscr{F}_t\} = P^{X_t}\{\Lambda\}. \tag{8}$$

More generally, if $Y \in b\mathscr{F}^0$, we have

$$E\{Y \circ \theta_t\,|\,\mathscr{F}_t\} = E^{X_t}\{Y\}. \tag{9}$$

The relations (8) and (9) follow from (7) by Lemma 2 of §1.1.

Does a shift exist as defined by (6)? If Ω is the space of all functions on $\mathbf{T}$ to $E\colon\Omega = E^{\mathbf{T}}$, as in the construction by Kolmogorov's theorem mentioned above, then an obvious shift exists. In fact, in this case each ω in Ω is just the sample function $X(\cdot,\omega)$ with domain $\mathbf{T}$, and we may set

$$\theta_t\omega = X(t + \cdot, \omega),$$

which is another such function. Since $X_s(\omega) = X(s,\omega)$ the equation (6) is a triviality. The same is true if Ω is the space of all right continuous (or continuous) functions, and such a space will serve for our later developments. For an arbitrary Ω, a shift need not exist but it is always possible to construct a shift by enlarging Ω without affecting the probability structure. We will not detail this but rather postulate the existence of a shift as part of our basic machinery for a Markov process.

For an arbitrary probability measure μ on $\mathscr{E}$, we put

$$P^\mu(\Lambda) = \int P^x(\Lambda)\mu(dx), \qquad \Lambda \in \mathscr{F}^0. \tag{10}$$

This is the probability measure determined by the process with initial distribution μ. For instance, equation (9) remains true if the E there is replaced by E^μ. Note that P^μ (in particular P^x) is defined so far only on $\mathscr{F}^0$, in contrast to P which is given on $\mathscr{F} \supset \mathscr{F}^0$. Later we shall extend P^μ to a larger σ-field by completion.

The transition function $P_t(\cdot, \cdot)$ has been assumed to be a strict probability kernel, namely $P_t(x, \mathbf{E}) = 1$ for every $t \in \mathbf{T}$ and $x \in \mathbf{E}$. We will extend this by allowing

$$P_t(x, \mathbf{E}) \leq 1, \qquad \forall t \in \mathbf{T}, x \in \mathbf{E}. \tag{11}$$

Such a transition function is called *submarkovian*, and the case where equality holds in (11) [*strictly*] *Markovian*. A simple device converts the former to the latter as follows. We introduce a new $\partial \notin \mathbf{E}$ and put

$$\mathbf{E}_\partial = \mathbf{E} \cup \{\partial\}, \qquad \mathscr{E}_\partial = \sigma\{\mathscr{E}, \{\partial\}\}.$$

The new point ∂ may be considered as the "point at infinity" in the one-point compactification of $\mathbf{E}$. If $\mathbf{E}$ is itself compact, ∂ is nevertheless adjoined as an isolated point. We now define P'_t as follows for $t > 0$ and $A \in \mathscr{E}$:

$$\begin{aligned} &P'_t(x, A) = P_t(x, A), \\ &P'_t(x, \partial) = 1 - P_t(x, E), \qquad \text{if } x \neq \partial; \\ &P'_t(\partial, \mathbf{E}) = 0, \qquad P'_t(\partial, \{\partial\}) = 1. \end{aligned} \tag{12}$$

It is clear that $P'_t(\cdot, \cdot)$ on $(\mathbf{E}_\partial, \mathscr{E}_\partial)$ is Markovian. Let $(X_t, \mathscr{F}_t)$ be a Markov process on $(\mathbf{E}_\partial, \mathscr{E}_\partial)$ with (P'_t) as transition function. Notwithstanding the last two relations in (12), it does not follow that ∂ will behave like an "absorbing state" (or "trap"). However it can be shown that this may be arranged by an unessential modification of the probability space. We shall assume that this has already been done, so that ∂ is absorbing in the sense that

$$\forall \omega, \forall s \geq 0\colon \{X_s(\omega) = \partial\} \subset \{X_t(\omega) = \partial \text{ for all } t \geq s\}. \tag{13}$$

Now we define the function ζ from Ω to $[0, \infty]$ as follows:

$$\zeta(\omega) = \inf\{t \in \mathbf{T}\colon X_t(\omega) = \partial\} \tag{14}$$

where, as a standard convention, $\inf \varnothing = \infty$ for the empty set $\varnothing$. Thus $\zeta(\omega) = \infty$ if and only if $X_t(\omega) \neq \partial$ for all $t \in \mathbf{T}$, in other words $X_t(\omega) \in \mathbf{E}$ for all $t \in \mathbf{T}$. The random variable ζ is called the *lifetime* of the process X.

The observant reader may remark that so far we have not defined $P_0(\cdot, \cdot)$. There are interesting cases where $P_0(x, \cdot)$ need not be the point mass $\varepsilon_x(\cdot)$, then x is called a "branching point". There are also cases where $P_0(\cdot, \cdot)$ should be left undefined. However, we shall assume until further notice that we are in the "normal" case where

$$\forall x \in \mathbf{E}_\partial\colon P_0(x, \cdot) = \varepsilon_x(\cdot),$$

namely P_0 is the identity operator. Equivalently, we assume

$$\forall x \in \mathbf{E}_\partial \colon P^x\{X_0 = x\} = 1. \tag{15}$$

Before proceeding further let us give a few simple examples of (homogeneous) Markov processes.

EXAMPLE 1 (Markov chain).

$\mathbf{E}$ = any countable set, for example the set of positive integers.

$\mathscr{E}$ = the σ-field of all subsets of $\mathbf{E}$.

We may write $p_{ij}(t) = P_t(i, \{j\})$ for $i \in \mathbf{E}$, $j \in \mathbf{E}$. Then for any $A \in \mathbf{E}$, we have

$$P_t(i, A) = \sum_{j \in A} p_{ij}(t).$$

The conditions (a) and (c) in the definition of transition function become in this case:

(a) $\forall i \in \mathbf{E} \colon \sum_{j \in \mathbf{E}} p_{ij}(t) = 1$;

(c) $\forall i \in \mathbf{E}, k \in \mathbf{E} \colon p_{ik}(s + t) = \sum_{j \in \mathbf{E}} p_{ij}(s) p_{jk}(t)$;

while (b) on p. 6 is trivially true. For the submarkovian case, the "$=$" in (a) is replaced by "$\leq$". If we add the condition

(d) $\forall i \in \mathbf{E}, j \in \mathbf{E} \colon \lim_{t \downarrow 0} p_{ij}(t) = \delta_{ij}$;

then the matrix of transition function

$$\Pi(t) = (p_{ij}(t))$$

is called a "standard transition matrix". In this case each $p_{ij}(\cdot)$ is a continuous function on $\mathbf{T}$. See Chung [2] for the theory of this special case of Markov process.

EXAMPLE 2 (uniform motion).

$\mathbf{E} = R^1 = (-\infty, +\infty)$; $\mathscr{E}$ = the classical Borel field on R^1.

For $x \in R^1$, $t \geq 0$, we put

$$P_t(x, \cdot) = \varepsilon_{x+t}(\cdot).$$

Starting from any point x, the process moves deterministically to the right with uniform speed. This trivial example turns out to be the source of many counterexamples to facile generalities. A slight modification yields an example for which (15) is false. Let $\mathbf{E} = \{0\} \cup (-\infty, -1] \cup [1, \infty)$,

$$\begin{aligned} &P_0(0, \{1\}) = \tfrac{1}{2}, \qquad P_0(0, \{-1\}) = \tfrac{1}{2}; \\ &P_t(x, \cdot) = \varepsilon_{x+t}(\cdot), \quad \text{if } x \geq 1;\, t \geq 0; \\ &P_t(x, \cdot) = \varepsilon_{x-t}(\cdot), \quad \text{if } x \leq -1;\, t \geq 0; \\ &P_t(0, \cdot) = \tfrac{1}{2}\{\varepsilon_{1+t}(\cdot) + \varepsilon_{-1-t}(\cdot)\}. \end{aligned}$$

Note that although P_0 is not the identity, we have $P_0P_t = P_tP_0 = P_t$ for $t \geq 0$.

Example 3 (Poisson process). $\mathbf{E} = \mathbf{N}$ = the set of positive (≥ 0) integers or the set of all integers. For $n \in \mathbf{N}$, $m \in \mathbf{N}$, $t \geq 0$:

$$P_t(n, \{m\}) = \begin{cases} 0 & \text{if } m < n, \\ \dfrac{e^{-\lambda t}(\lambda t)^{m-n}}{(m-n)!}, & \text{if } m \geq n. \end{cases}$$

Note that in this case there is spatial homogeneity, namely: the function of the pair (n, m) exhibited above is a function of $m - n$ only.

Example 4 (Brownian motion in R^1).

$$\mathbf{E} = R^1, \mathscr{E} = \mathscr{B}^1.$$

For real x and y and $t > 0$, put

$$p_t(x, y) = \frac{1}{\sqrt{2\pi t}} \exp\left[-\frac{(y-x)^2}{2t}\right]$$

and define the transition function as follows:

$$\begin{aligned} &P_t(x, A) = \int_A p_t(x, y)\, dy, \qquad t > 0; \\ &P_0(x, A) = \varepsilon_x(A). \end{aligned}$$

The function $p_t(\cdot, \cdot)$ is a *transition probability density*. In this case it is the Gaussian density function with mean zero and variance t. As in Example 3 there is again spatial homogeneity, indeed p_t is a function of $|x - y|$ only. This example, and its extention to R^d, will be the subject matter of Chapter 4.

1.3. Optional Times

A major device to "tame" the continuum of time is the use of random times. The idea is very useful even for discrete time problems and has its origin in considering "the first time when a given event occurs". In continuous time it becomes necessary to formalize the intuitive notions in terms of various σ-fields.

We complete the increasing family $\{\mathscr{F}_t, t \in \mathbf{T}\}$ by setting

$$\mathscr{F}_\infty = \bigvee_{t \in \mathbf{T}} \mathscr{F}_t.$$

Recall that the notation on the right side above means the minimal σ-field including all $\mathscr{F}_t$, which is not the same as $\bigcup_{t \in \mathbf{T}} \mathscr{F}_t$. Although we shall apply the considerations below to a Markov process, we need not specify the family $\{\mathscr{F}_t\}$ to begin with.

Definition. The function $T:\Omega \to [0, \infty]$ is called optional relative to $\{\mathscr{F}_t\}$ iff

$$\forall t \in [0, \infty): \{T \le t\} \in \mathscr{F}_t.$$

The preceding relation then holds also for $t = \infty$ by letting $t \uparrow \infty$ through a sequence. Define

$$\forall t \in (0, \infty): \mathscr{F}_{t-} = \bigvee_{s \in [0,t)} \mathscr{F}_s;$$

$$\forall t \in [0, \infty): \mathscr{F}_{t+} = \bigwedge_{s \in (t,\infty)} \mathscr{F}_s.$$

We have clearly for each t: $\mathscr{F}_{t-} \subset \mathscr{F}_t \subset \mathscr{F}_{t+}$.

Definition. The family $\{\mathscr{F}_t\}$ is called right continuous iff

$$\forall t \in [0, \infty): \mathscr{F}_t = \mathscr{F}_{t+}.$$

It follows from the definition that the family $\{\mathscr{F}_{t+}\}$ is right continuous. Note the analogy with a real-valued increasing function $t \to f(t)$.

Proposition 1. *T is optional relative to $\{\mathscr{F}_{t+}\}$ if and only if*

$$\forall t \in [0, \infty): \{T < t\} \in \mathscr{F}_t. \tag{1}$$

Proof. If T is optional relative to $\{\mathscr{F}_{t+}\}$, then by definition

$$\forall t \in (0, \infty): \{T \le t\} \in \mathscr{F}_{t+}.$$

Hence we have for $1/n < t$:

$$\left\{T \le t - \frac{1}{n}\right\} \in \mathscr{F}_{(t-1/n)+} \subset \mathscr{F}_t$$

and consequently

$$\{T < t\} = \bigcup_{n=1}^{\infty} \left\{T \le t - \frac{1}{n}\right\} \in \mathscr{F}_t.$$

Conversely if (1) is true, then for $t \in [0, \infty)$:

$$\left\{T < t + \frac{1}{n}\right\} \in \mathscr{F}_{t+1/n}$$

and consequently

$$\{T \le t\} = \bigcap_{n=1}^{\infty} \left\{T < t + \frac{1}{n}\right\} \in \bigwedge_{n=1}^{\infty} \mathscr{F}_{t+1/n} = \mathscr{F}_{t+}. \qquad \square$$

Examples.

1. The lifetime ζ defined in (14) of §1.2 is optional relative to $\{\mathscr{F}^0_{t+}\}$, because ∂ is absorbing which implies (Q is the set of rational numbers)
$$\{\zeta < t\} = \bigcup_{r \in Q \cap [0,t)} \{X_r = \partial\} \in \mathscr{F}^0_t.$$
2. Suppose $\mathscr{F}_0$ contains all P-null sets and $\forall t$: $P\{T = t\} = 0$. Then we have
$$\forall t\colon \{T \le t\} - \{T < t\} \in \mathscr{F}_t$$
so that optionality relative to $\{\mathscr{F}_t\}$ is the same as relative to $\{\mathscr{F}_{t+}\}$.
3. Consider a Poisson process with left continuous paths and T = first jump time. Then T is optional relative to $\{\mathscr{F}^0_{t+}\}$ but not to $\{\mathscr{F}^0_t\}$, because
$$\{T = t\} \notin \mathscr{F}^0_t [= \mathscr{F}^0_{t-}].$$

The reader is supposed to be familiar with elementary properties of the Poisson process which are "intuitively obvious", albeit sometimes tedious to establish rigorously.

From now on we fix the family $\{\mathscr{F}_t, t \in [0, \infty]\}$ and call T an "optional time" or "strictly optional time" according as it is optional relative to $\{\mathscr{F}_{t+}\}$ or $\{\mathscr{F}_t\}$. We shall prove most results for the former case but some of them extend to the latter.

Proposition 2. *If S and T are optional, then so are*

$$S \wedge T, \qquad S \vee T, \qquad S + T.$$

Proof. The first two are easy; to treat the third we consider for each t:

$$\{S + T < t\} = \bigcup_{r \in Q_t} \{S < r; T < t - r\}$$

where $Q_t = Q \cap (0, t)$. Clearly each member of the union is in $\mathscr{F}_t$; hence so is the union. □

Proposition 3. T_n *is optional for each* $n \geq 1$, *then so are*

$$\sup_n T_n, \qquad \inf_n T_n, \qquad \limsup_n T_n, \qquad \liminf_n T_n. \tag{2}$$

Proof. This follows from Proposition 1 and

$$\left\{\sup_n T_n \leq t\right\} = \bigcap_n \{T_n \leq t\} \in \mathscr{F}_{t+};$$

$$\left\{\inf_n T_n < t\right\} = \bigcup_n \{T_n < t\} \in \mathscr{F}_t;$$

$$\limsup_n T_n = \inf_{m \geq 1} \sup_{n \geq m} T_n;$$

$$\liminf_n T_n = \sup_{m \geq 1} \inf_{n \geq m} T_n.$$

□

For strictly optional T_n only the first item in (2) is strictly optional. In Example (3) above, if we set $T_n = T + 1/n$ then each T_n is strictly optional but $T = \inf_n T_n$ is not.

Associated with an optional time T, there is a σ-field defined as follows.

Definition. If T is strictly optional, $\mathscr{F}_T$ is the class of subsets of $\mathscr{F}_\infty$ such that

$$\forall t \in [0, \infty): \Lambda \cap \{T \leq t\} \in \mathscr{F}_t.$$

If T is optional, $\mathscr{F}_{T+}$ is the class of subsets of $\mathscr{F}_\infty$ such that

$$\forall t \in [0, \infty): \Lambda \cap \{T \leq t\} \in \mathscr{F}_{t+}. \tag{3}$$

It follows from the proof of Proposition 1 that (3) is equivalent to

$$\forall t \in (0, \infty]: \Lambda \cap \{T < t\} \in \mathscr{F}_t.$$

Let us verify, for example, that $\mathscr{F}_{T+}$ is a σ-field. We see from (3) that $\Omega \in \mathscr{F}_{T+}$ because T is optional. It then follows that $\mathscr{F}_{T+}$ is closed under complementation. It is also closed under countable union; hence it is a

σ-field. For $T = a$ constant t, $\mathscr{F}_T$ reduces to $\mathscr{F}_t$, and $\mathscr{F}_{T+}$ reduces to $\mathscr{F}_{t+}$. Thus our definition and notation make sense and it is a leading idea below to extend properties concerning constant times to their analogues for optional times.

It is easy to see that if $\{\mathscr{F}_t\}$ is right continuous, then any optional T is strictly optional, and furthermore we have

$$\mathscr{F}_T = \mathscr{F}_{T+}.$$

Proposition 4. *If T is optional and $\Lambda \in \mathscr{F}_{T+}$, we define*

$$T_\Lambda = \begin{cases} T, & \text{on } \Lambda, \\ \infty, & \text{on } \Lambda^c. \end{cases} \tag{4}$$

Then T is also optional.

Proof. For $t < \infty$, we have

$$\{T_\Lambda < t\} = \Lambda \cap \{T < t\} \in \mathscr{F}_t.$$

This implies $\{T_\Lambda = \infty\} \in \mathscr{F}_\infty$ by the remark after Proposition 1. □

Theorem 5.

(a) *If T is optional, then $T \in \mathscr{F}_{T+}$;*
(b) *If S and T are both optional and $S \leq T$, then $\mathscr{F}_{S+} \subset \mathscr{F}_{T+}$;*
(c) *If T_n is optional, $T_{n+1} \leq T_n$ for each $n \geq 1$, and $\lim_n T_n = T$, then*

$$\mathscr{F}_{T+} = \bigwedge_{n=1}^{\infty} \mathscr{F}_{T_n+}. \tag{5}$$

In case each T_n is strictly optional, we may replace $\mathscr{F}_{T_n+}$ by $\mathscr{F}_{T_n}$ in (5).

Proof. We leave (a) and (b) as exercises. To prove (c), we remark that T is optional by Proposition 3. Hence we have by (b):

$$\mathscr{F}_{T+} \subset \bigwedge_{n=1}^{\infty} \mathscr{F}_{T_n+}. \tag{6}$$

Now let $\Lambda \in \bigwedge_{n=1}^{\infty} \mathscr{F}_{T_n+}$; then for each t:

$$\Lambda \cap \{T < t\} = \Lambda \cap \left[\bigcup_{n=1}^{\infty} \{T_n < t\}\right] = \bigcup_{n=1}^{\infty} [\Lambda \cap \{T_n < t\}] \in \mathscr{F}_t,$$

so that $\Lambda \in \mathscr{F}_{T+}$. Hence the inclusion in (6) may be reversed. □

Theorem 6. *If S and T are both optional, then the sets*

$$\{S < T\}, \qquad \{S \leq T\}, \qquad \{S = T\}$$

all belong to $\mathscr{F}_{S+} \wedge \mathscr{F}_{T+}$.

Proof. We have for each t:

$$\{S < T\} \cap \{T < t\} = \bigcup_{r \in Q_t} \{S < r < T < t\},$$

where $Q_t = Q \cap (0, t)$. For each r, $\{r < T < t\} \in \mathscr{F}_t$, $\{S < r\} \in \mathscr{F}_r \subset \mathscr{F}_t$. Hence the union above belongs to $\mathscr{F}_t$ and this shows $\{S < T\} \in \mathscr{F}_{T+}$. Next we have

$$\begin{aligned}\{S < T\} \cap \{S < t\} = \{S < T \wedge t\} &= \bigcup_{r \in Q} \{S < r < T \wedge t\} \\ &= \bigcup_{r \in Q_t} \{S < r < T\} \in \mathscr{F}_t,\end{aligned}$$

since for each $r < t$, $\{r < T\} = \{T \leq r\}^c \in \mathscr{F}_{r+} \subset \mathscr{F}_t$. Hence, $\{S < T\} \in \mathscr{F}_{S+}$. Combining the two results we obtain $\{S < T\} \in \mathscr{F}_{S+} \wedge \mathscr{F}_{T+}$. Since

$$\begin{aligned}\{S \leq T\} &= \{T < S\}^c, \\ \{S = T\} &= \{T \leq S\} - \{T < S\},\end{aligned}$$

the rest of the theorem follows. □

We will now extend the notion of the σ-field $\mathscr{F}_{t-}$ to a random time T. It turns out that we can do so with an arbitrary positive function T, but $\mathscr{F}_{T-}$ will have nice properties when T is an optional time.

Definition. For any function T from Ω to $[0, \infty]$, $\mathscr{F}_{T-}$ is the σ-field generated by $\mathscr{F}_{0+}$ and the class of sets:

$$\{t < T\} \cap \Lambda, \quad \text{where } t \in [0, \infty) \text{ and } \Lambda \in \mathscr{F}_t. \tag{7}$$

It is easy to see that $\mathscr{F}_{T-}$ contains also sets of the form

$$\{T = \infty\} \cap \Lambda, \quad \text{where } \Lambda \in \mathscr{F}_\infty. \tag{8}$$

As defined above, $\mathscr{F}_{T-}$ need not be included in $\mathscr{F}$; but this is the case if $T \in \mathscr{F}$, as we shall assume in what follows. It is obvious that $T \in \mathscr{F}_{T-}$.

If T is optional, then $\mathscr{F}_{0+} \subset \mathscr{F}_{T+}$ by Theorem 5, (b). Next, for each $u \geq 0$, the set

$$[\{t < T\} \cap \Lambda] \cap \{T < u\}$$

is empty for $u \le t$, and belongs to $\mathscr{F}_u$ if $t < u$ and $\Lambda \in \mathscr{F}_t$. Hence each generating set of $\mathscr{F}_{T-}$ in (7) belongs to $\mathscr{F}_{T+}$, by definition of the latter. In other words, we have for an optional T:

$$\mathscr{F}_{T-} \subset \mathscr{F}_{T+}. \tag{9}$$

Similarly if T is strictly optional, then $\mathscr{F}_{T-} \subset \mathscr{F}_T$ provided $\mathscr{F}_{0+} = \mathscr{F}_0$. The next proposition is sharper; note that it reduces to $\mathscr{F}_{0+} \subset \mathscr{F}_{T-}$ for $S \equiv 0$, which is true by definition of $\mathscr{F}_{T-}$.

Proposition 7. *If S and T are both optional and $S \le T$, then $\mathscr{F}_{S-} \subset \mathscr{F}_{T-}$. If moreover $S < T$ on $\{S < \infty\} \cap \{T > 0\}$, then*

$$\mathscr{F}_{S+} \subset \mathscr{F}_{T-}. \tag{10}$$

Proof. The first assertion is easy. Under the hypothesis of the second assertion, we have for any Λ:

$$\Lambda = \bigcup_{r \in Q} [\Lambda \cap \{S < r < T\}] \cup [\Lambda \cap \{S = \infty\}] \cup [\Lambda \cap \{T = 0\}].$$

If $\Lambda \in \mathscr{F}_{S+}$, then $\Lambda \cap \{S < r\} \in \mathscr{F}_r$ for each r, and so

$$[\Lambda \cap \{S < r\}] \cap \{r < T\} \in \mathscr{F}_{T-}.$$

Next, $\Lambda \cap \{S = \infty\} \in \mathscr{F}_\infty$ and so

$$\Lambda \cap \{S = \infty\} = [\Lambda \cap \{S = \infty\}] \cap \{T = \infty\} \in \mathscr{F}_{T-}.$$

Finally,

$$\Lambda \cap \{T = 0\} = [\Lambda \cap \{S = 0\}] \cap \{T = 0\} \in \mathscr{F}_{T-},$$

because $\Lambda \cap \{S = 0\} \in \mathscr{F}_{0+} \subset \mathscr{F}_{T-}$ and $\{T = 0\} \in \mathscr{F}_{T-}$. Combining these results we obtain $\Lambda \in \mathscr{F}_{T-}$. □

Next, we introduce a special class of optional times which plays an essential role in the general theory of stochastic processes. Although its use will be limited in this introductory text, the discussion below will serve the purpose of further strengthening the analogy between optional and constant times.

Definition. T is a *predictable time* iff there exists a sequence of optional times $\{T_n\}$ such that

$$\forall n\colon T_n < T \quad \text{on } \{T > 0\}; \quad \text{and} \quad T_n \uparrow T. \tag{11}$$

Here "$T_n \uparrow T$" is a standard notation which means: "$\forall n: T_n \le T_{n+1}$ and $\lim_n T_n = T$." The limit T is optional by Proposition 3. Observe also that $T_n < \infty$ for each n. The sequence $\{T_n\}$ is said to *announce* T.

Theorem 8. *If T_n is optional and $T_n \uparrow T$, then*

$$\mathscr{F}_{T-} = \bigvee_{n=1}^{\infty} \mathscr{F}_{T_n-}. \tag{12}$$

If T is predictable and $\{T_n\}$ announces T, then we have also

$$\mathscr{F}_{T-} = \bigvee_{n=1}^{\infty} \mathscr{F}_{T_n+}. \tag{13}$$

Proof. We have "$\supset$" in (12) by Proposition 7. To prove the reverse inclusion, we have if $\Lambda \in \mathscr{F}_t$:

$$\{t < T\} \cap \Lambda = \bigcup_{n=1}^{\infty} [\{t < T_n\} \cap \Lambda] \in \bigvee_{n=1}^{\infty} \mathscr{F}_{T_n-}.$$

Also $\mathscr{F}_{0+} \subset \mathscr{F}_{T_n-}$ for each n. Hence each generating set of $\mathscr{F}_{T-}$ belongs to the right member of (12), and (12) is proved. If $\{T_n\}$ announces the predictable T, then $\mathscr{F}_{T-} \supset \mathscr{F}_{T_n+}$ by Proposition 7, and so we have "$\supset$" in (13). Since $\mathscr{F}_{T_n+} \supset \mathscr{F}_{T_n-}$ for each n, the reverse inclusion follows from (12). □

We now return to the Markov process $X = \{X_t, \mathscr{F}_t, t \in \mathbf{T}\}$.

Definition. X is said to be Borel measurable iff

$$(t, \omega) \to X(t, \omega)$$

is in $\mathscr{B} \times \mathscr{F}$, where $\mathscr{B}$ is the Euclidean Borel field on $\mathbf{T}$. X is said to have right limits iff for each ω, the sample function $t \to X(t, \omega)$ has right limits everywhere in $\mathbf{T}$; X is said to be right continuous iff for each ω, the sample function $t \to X(t, \omega)$ is right continuous in $\mathbf{T}$. Similarly, "left" versions of these properties are defined in $(0, \infty]$.

Proposition 9. *If X is right [or left] continuous, then it is Borel measurable.*

Proof. For each $n \ge 0$, define

$$X_n(t, \omega) = X\left(\frac{[2^n t] + 1}{2^n}, \omega\right). \tag{14}$$

It is clear that

$$(t, \omega) \to X_n(t, \omega) = \sum_{k=0}^{\infty} 1_{[k/2^n, k+1/2^n)}(t) X\left(\frac{k+1}{2^n}, \omega\right)$$

is Borel measurable. If X is right continuous, then

$$X(t, \omega) = \lim_n X_n(t, \omega)$$

for $(t, \omega) \in \mathbf{T} \times \Omega$. Hence X is Borel measurable; see the Lemma in §1.5 below for a proof. If X is left continuous we need only modify the definition in (14) by omitting the "+1" in the right member. □

Definition. Let X be Borel measurable and T an $\mathscr{F}$-measurable function from Ω to $[0, \infty]$ (namely an extended-valued random variable). We define X_T, or $X(T)$, on the set $\{T < \infty\}$ by

$$X_T(\omega) = X(T(\omega), \omega). \tag{15}$$

If $X_\infty(\omega)$ is defined for all $\omega \in \Omega$, then X_T is defined on Ω by (15).

The next result extends the hypothesis "$X_t \in \mathscr{F}_t$".

Theorem 10. *Let X have right limits. If T is optional, then*

$$X_{T+} 1_{\{T < \infty\}} \in \mathscr{F}_{T+}. \tag{16}$$

If X has left limits as well, and T is predictable, then

$$X_{T-} 1_{\{T < \infty\}} \in \mathscr{F}_{T-} \tag{17}$$

where $X_{0-} = X_0$.

Note. If $T(\omega) = \infty$, the left members of (16) and (17) are defined to be zero by convention, even when $X_{\infty+}$ or $X_{\infty-}$ are not defined.

Proof. Define for $n \geq 0$:

$$T_n = \frac{[2^n T] + 1}{2^n}. \tag{18}$$

This is just the usual dyadic approximation of T from above. For each n, T_n is strictly optional (proof ?) and $T_n > T$, $\lim_n \downarrow T_n = T$. For convenience let us introduce the "dyadic set"

$$D = \left\{\frac{k}{2^n} \,\middle|\, k \geq 0, n \geq 0\right\}. \tag{19}$$

For each $d \in D$ and $B \in \mathscr{E}$ we have

$$\{T_n = d;\ X(T_n) \in B\} = \{T_n = d;\ X(d) \in B\} \in \mathscr{F}_d$$

because T_n is strictly optional. Hence for each t,

$$\{X(T_n) \in B\} \cap \{T_n < t\} = \bigcup_{d \in D_t} \{T_n = d;\ X(T_n) \in B\} \in \mathscr{F}_t$$

where $D_t = D \cap [0, t)$. This shows that $\{X(T_n) \in B;\ T_n < \infty\}$ belongs to $\mathscr{F}_{T_n+}$ (actually to $\mathscr{F}_{T_n}$); since B is arbitrary we have $X(T_n)1_{\{T_n < \infty\}} \in \mathscr{F}_{T_n+}$. Since $X(T+) = \lim_n X(T_n)$ on $\{T < \infty\}$ which belongs to $\mathscr{F}_{T+}$, and $\mathscr{F}_{T_n+} \downarrow \mathscr{F}_{T+}$, it follows that

$$X(T+)1_{\{T < \infty\}} \in \bigwedge_{n=1}^{\infty} \mathscr{F}_{T_n+} = \mathscr{F}_{T+}$$

by Theorem 5.

If T is predictable, let $\{T_n\}$ announce T. Then

$$\lim_n X(T_n) = X(T-)$$

on $\{0 < T < \infty\}$ provided X has left limits in $(0, \infty)$, and trivially on $\{T = 0\}$ provided we define $X(0-) = X(0)$. Now for each n, we have just shown that $X(T_n) \in \mathscr{F}_{T_n+}$; hence

$$X(T-)1_{\{T < \infty\}} \in \bigvee_n \mathscr{F}_{T_n+} = \mathscr{F}_{T-}$$

by Theorem 8, if we note that $T \in \mathscr{F}_{T-}$. If perchance X has left limits in $(0, \infty]$ so that $X_{\infty-}$ is defined, then we have $X(T-) \in \mathscr{F}_{T-}$. □

We can extend the shift θ_t to θ_T for any function T from Ω to $[0, \infty]$. Observe that the defining equations for the shift, (6) of §1.2, is an identity in s and t for each ω. It follows that on the set $\{T < \infty\}$ we have

$$X_t(\theta_T(\omega)) = X_{t+T(\omega)}(\omega) = X_{T+t}(\omega). \tag{20}$$

Note that $T + t$ is another function of the same type as T. Thus we have

$$X_t \circ \theta_T = X_{T+t}$$

as functions on $\{T < \infty\}$. If T is optional, then so is $T + t$; indeed $T + t$ is strictly optional for $t > 0$. Hence by Theorem 10, if X is right continuous:

$$X_t \circ \theta_T 1_{\{T < \infty\}} = X_{T+t} 1_{\{T < \infty\}} \in \mathscr{F}_{T+t+}. \tag{21}$$

So far so good, but the next step is to consider the inverse mapping θ_T^{-1}. This would be awkward if θ_T were to be defined on $\{T < \infty\}$ only. Here is the device to extend it to Ω. Put

$$\forall \omega \in \Omega: X_\infty(\omega) = \partial.$$

Postulate the existence of a point ω_∂ in Ω for which

$$\forall t \in \mathbf{T}: X_t(\omega_\partial) = \partial;$$

this then holds for $t \in [0, \infty]$. Finally, define

$$\forall \omega \in \Omega: \theta_\infty(\omega) = \omega_\partial.$$

After this fixing all three terms in (20) reduce to ∂ when $t = \infty$ or $T = \infty$ or both. Hence (20) now holds for each $t \in [0, \infty]$ and any T from Ω to $[0, \infty]$. The existence of ω_∂ is of course trivial in the case where $\Omega = \mathbf{E}_\partial^\mathbf{T}$, and can in general be arranged by enlarging the given Ω.

With this amendment, we can rewrite (16) simply as

$$X_T = X_{T+} \in \mathscr{F}_{T+}.$$

In equation (21) we can now omit the factor $1_{\{T<\infty\}}$, and the result is equivalent to

$$\theta_T^{-1}\{\sigma(X_t)\} \subset \mathscr{F}_{T+t+}$$

where $\sigma(X_t)$ is the σ-field generated by X_t. Since $\mathscr{F}_{T+t+}$ increases with t it follows by the usual extension that

$$\theta_T^{-1}\{\mathscr{F}_t^0\} \subset \mathscr{F}_{T+t+} \subset \mathscr{F}_\infty. \tag{22}$$

Let us give a useful application at once.

Theorem 11. *Let S be optional relative to $\{\mathscr{F}_t^0\}$ and T be optional relative to $\{\mathscr{F}_t\}$, then*

$$T + S \circ \theta_T \tag{23}$$

is optional relative to $\{\mathscr{F}_t\}$.

Proof. For each t, we have by (22)

$$\{S \circ \theta_T < t\} = \theta_T^{-1}\{S < t\} \in \mathscr{F}_{T+t+}. \tag{24}$$

Hence for each $r < t$:

$$\{S \circ \theta_T < t - r\} \cap \{T + t - r < t\} \in \mathscr{F}_t$$

by (24) with $t - r$ replacing t and the definition of $\mathscr{F}_{T+t-r}$. Now we have

$$\{T + S \circ \theta_T < t\} = \bigcup_{r \in Q_t} \{T < r; S \circ \theta_T < t - r\}$$

and each member of the union above belongs to $\mathscr{F}_t$. Hence the set above belongs to $\mathscr{F}_t$ and this proves the theorem. □

Remark. Later we will extend the result to an S which is optional relative to $\{\mathscr{F}_t^\mu\}$ or $\{\mathscr{F}_t^\sim\}$; see §2.3.

What does the random variable in (23) mean? Let A be a set in $\mathscr{E}$ and define

$$D_A(\omega) = \inf\{t \geq 0 \mid X(t, \omega) \in A\}. \tag{25}$$

This is called the "first entrance time into A", and is a prime example of an optional time. Indeed, it was also historically the first such time to be considered. The reader may well recall instances of its use in the classical theory of probability. However for a continuous time Markov process, it is not a trivial matter to verify that D_A is optional for a general A. This is easy only for an open set A, and we shall treat other sets in §3.3. Suppose this has been shown; we can identify the random variable in (23) when $S = D_A$ as follows:

$$T(\omega) + D_A \circ \theta_T(\omega) = \inf\{t \geq T(\omega) \mid X(t, \omega) \in A\}. \tag{26}$$

In particular if $T(\omega) \equiv s$, $s + D_A \circ \theta_s$ is "the first entrance time into A on or after time s"; if $T(\omega) = D_B(\omega)$ where $B \in \mathscr{E}$, then $D_B + D_A \circ \theta_{D_B}$ is "the first entrance time into A on or after the first entrance time into B." Of course, a similar interpretation is meaningful for any optional time T. We shall see many applications to Markov processes involving such random times.

The definitions and results of this section have been given without mention of probability. It is usually obvious how to modify them when exceptional sets are allowed. For instance, if each T_n is defined only P-a.e., and $\lim_n T_n$ exists only P-a.e., then there is a set Ω_0 with $P(\Omega_0) = 1$ on which all T_n's are defined and $\lim_n T_n$ exists for all ω in Ω_0. Furthermore if $\{T_n \leq t\}$ differs from a set in $\mathscr{F}_t$ by a P-null sets, then $\Omega_0 \cap \{T_n \leq t\} \in \mathscr{F}_t$ provided that $(\mathscr{F}, P)$ is complete and $\mathscr{F}_0 (\subset \mathscr{F}_t)$ is augmented (by all P-null sets). Thus T_n is optional considered on the trace of $(\Omega, \mathscr{F})$ on Ω_0. In this way we can apply the preceding discussions to a reduced probability space, and we shall often do so tacitly. On the other hand, there are deeper results concerning optionality in which probability considerations are expedient; see for example Problem 12 below.

Exercises

1. If T is optional and $S = \varphi(T) > T$ where φ is a Borel function, then S is strictly optional, indeed predictable. In general a strictly positive predictable time is strictly optional.

2. $\mathscr{F}_{T-}$ is also generated by $\mathscr{F}_{0+}$ and one of the two classes of sets below:
(i) $\{t \leq T\} \cap \Lambda$, where $0 \leq t < \infty$ and $\Lambda \in \mathscr{F}_{t-}$;
(ii) $\{t < T\} \cap \Lambda$, where $0 \leq t < \infty$ and $\Lambda \in \mathscr{F}_{t+}$.

3. If S and T are functions from Ω to $[0, \infty]$ such that $S \leq T$, then $\mathscr{F}_{S-} \subset \mathscr{F}_{T-}$ if and only if $S \in \mathscr{F}_{T-}$. This is the case if S is optional. On the other hand if S is arbitrary and T is optional, $\mathscr{F}_{S-} \subset \mathscr{F}_{T+}$ may be false.

4. Define for any function T from Ω to $[0, \infty]$ the σ-field
$$\mathscr{F}_{T+} = \bigwedge_{n=1}^{\infty} \mathscr{F}_{(T+1/n)-}.$$
Prove that if T is optional, this coincides with the definition given in (3).

5. If T is optional then
$$\mathscr{F}_{T+} = \bigvee_{n=1}^{\infty} \mathscr{F}_{(T \wedge n)+}.$$

6. Let S be optional relative to $\{\mathscr{F}_t\}$ and define $\mathscr{G}_t = \mathscr{F}_{S+t}$, where $\mathscr{G}_0 = \mathscr{F}_{S+}$. Prove that if $T \geq S$, T is optional relative to $\{\mathscr{F}_t\}$ if and only if $T - S$ is optional relative to $\{\mathscr{G}_t\}$.

7. If T is predictable and $\Lambda \in \mathscr{F}_{T-}$, then the T_Λ defined in (4) is predictable.

8. Let $\{T^{(k)}\}$ be a sequence of predictable times. If $T^{(k)}$ increases to T, then T is predictable. If $T^{(k)}$ "settles down" to T, namely $T^{(k)}$ decreases and for each ω, there exists an integer $N(\omega)$ for which $T^{N(\omega)}(\omega) = T(\omega)$, then T is "almost surely predictable". [Hint: for the second assertion let $\{T_n^{(k)}\}$ announce $T^{(k)}$ such that $P(T_n^{(k)} + 2^{-n} < T) < 2^{-n-k}$ for all k and n. Let $S_n = \inf_k T_n^{(k)}$; then $\{S_n\}$ announces T almost surely.]

9. If S is optional and T is arbitrary positive, then
$$\mathscr{F}_{S+} \cap \{S < T\} \in \mathscr{F}_{T-};$$
namely; for each $\Lambda \in \mathscr{F}_{S+}$, $\Lambda \cap \{S < T\} \in \mathscr{F}_{T-}$.

10. If S and T are both optional, then exactly one of the following four relations is in general false:
$$\mathscr{F}_{(S \wedge T)+} = \mathscr{F}_{S+} \wedge \mathscr{F}_{T+};$$
$$\mathscr{F}_{(S \wedge T)-} = \mathscr{F}_{S-} \wedge \mathscr{F}_{T-};$$
$$\mathscr{F}_{(S \vee T)+} = \mathscr{F}_{S+} \vee \mathscr{F}_{T+};$$
$$\mathscr{F}_{(S \vee T)-} = \mathscr{F}_{S-} \vee \mathscr{F}_{T-}.$$

11. Let $\{\mathscr{F}_t\}$ be an increasing family of σ-fields, and $\mathscr{G}_t$ be the σ-field generated by $\mathscr{F}_t$ and all P-null sets. Suppose S is optional relative to $\{\mathscr{G}_t\}$. Prove that there exists T which is optional relative to $\{\mathscr{F}_t\}$ and $P(S = T) = 1$. Furthermore if $\Lambda \in \mathscr{G}_S$ then there exists $M \in \mathscr{F}_{T+}$ such that $P(\Lambda \triangle M) = 0$. [Hint: approximate S as in (18); for the second assertion, consider $\{S_\Lambda = T < \infty\}$ where S_Λ is defined as in (4).]

12. If T is predictable, there exists an "almost surely" announcing sequence $\{T_n\}$ such that each T_n is countably-valued with values in the dyadic set. Namely, we have for each n, $T_n \leq T_{n+1} \leq T$, $T_n < T$ on $\{T > 0\}$, and $P\{\lim_n T_n = T\} = 1$. Furthermore each T_n is predictable. [Hint: let $T_n^{(k)} = ([2^k T_n] + 1)/2^k$. Show that there exists $\{k_n\}$ such that $P\{T_n^{(k_n)} \geq T \text{ i.o.}\} = 0$. Put $S_m = \inf_{n \geq m} T_n^{(k_n)}$ and show that S_m is dyadic-valued. For the last assertion use Problems 1 and 8.]

1.4. Martingale Theorems

Martingale theory is an essential tool in the analysis of Markov processes. In this section we give a brief account of some of the basic results in continuous time, with a view to applications in later sections. It is assumed that the reader is acquainted with the discrete time analogues of these results, such as discussed in Chapter 9 of *Course*. A theorem stated there for a submartingale will be cited for a supermartingale without comment, since it amounts only to a change of signs for the random variables involved.

Let $(\Omega, \mathscr{F}, P)$ be a probability space; $\{\mathscr{F}_t\}$ an increasing family of σ-fields (of sets in $\mathscr{F}$) to which a stochastic process $\{X_t\}$ is adapted, as in §1.1. The index set for t is $\mathbf{T} = [0, \infty)$ when this is not specified. We shall use the symbol $\mathbf{N}$ to denote the set of positive (≥ 0) integers, and $\mathbf{N}_m$ the subset of $\mathbf{N}$ not exceeding m, where $m \in \mathbf{N}$. As usual, indices such as m, n and k denote members of $\mathbf{N}$ without explicit mention.

Definition. $\{X_t, \mathscr{F}_t, t \in \mathbf{T}\}$ is a martingale iff

(i) $\forall t: X_t$ is an integrable real-valued random variable;
(ii) if $s < t$, then

$$X_s = E\{X_t \mid \mathscr{F}_s\}. \tag{1}$$

X is a supermartingale iff the "$=$" in (1) is replaced by "$\geq$"; and a submartingale iff it is replaced by "$\leq$".

We write $s \downarrow\downarrow t$ to mean "$s > t$, $s \to t$", and $s \uparrow\uparrow t$ to mean "$s < t$, $s \to t$". Let S be a countable dense subset of $\mathbf{T}$; for convenience we suppose $S \supset \mathbf{N}$.

Theorem 1. *Let $\{X_t, \mathscr{F}_t\}$ be a supermartingale. For P a.e. ω, the sample function $X(\cdot, \omega)$ restricted to the set S has a right limit at every $t \in [0, \infty)$, and* *a left limit at every $t \in (0, \infty)$. Thus we have*:

$$\begin{aligned} X(t+, \omega) &= \lim_{\substack{s \in S \\ s \downarrow\downarrow t}} X(s, \omega) \quad \textit{exists for } t \in [0, \infty); \\ X(t-, \omega) &= \lim_{\substack{s \in S \\ s \uparrow\uparrow t}} X(s, \omega) \quad \textit{exists for } t \in (0, \infty). \end{aligned} \tag{2}$$

Furthermore, for each finite interval $I \subset \mathbf{T}$, the set of numbers

$$\{X(s, \omega), s \in S \cap I\} \tag{3}$$

is bounded. Hence the limits in (2) *are finite numbers.*

Proof. Let $\{X_n, \mathscr{F}_n, n \in \mathbf{N}_m\}$ be a discrete parameter supermartingale, and let

$$U(\omega; \mathbf{N}_m; [a, b]), \quad \text{where } a < b,$$

denote the number of upcrossings from strictly below a to strictly above b by the sample sequence $\{X_n(\omega), n \in \mathbf{N}_m\}$. We recall that the upcrossing inequality states that

$$E\{U(\mathbf{N}_m; [a, b])\} \leq \frac{E\{(a - X_m)^+\}}{b - a}. \tag{4}$$

See Notes on Chapter 1. Observe that this inequality really applies to a supermartingale indexed by any finite linearly ordered set with m as the last index. In other words, the bound given on the right side of (4) depends on the last random variable of the supermartingale sequence, not on the number of terms in it (as long as the number is finite). Hence if we consider the supermartingale $\{X_t\}$ with the index t restricted to $[0, m] \cap S'$, where S' is any finite subset of S containing m, and denote the corresponding upcrossing number by $U([0, m] \cap S'; [a, b])$, then exactly the same bound as in (4) applies to it. Now let S' increase to S, then the upcrossing number increases to $U([0, m] \cap S; [a, b])$. Hence by the monotone convergence theorem, we have

$$E\{U([0, m] \cap S; [a, b])\} \leq \frac{E\{(a - X_m)^+\}}{b - a} < \infty. \tag{5}$$

It follows that for P-a.e. ω.

$$U(\omega; [0, m] \cap S; [a, b]) < \infty. \tag{6}$$

This is true for each $m \in \mathbf{N}$ and each pair of rational numbers (a, b) such that $a < b$. Therefore (6) holds simultaneously for all $m \in \mathbf{N}$ and all such pairs in a set Ω_0 with $P(\Omega_0) = 1$. We claim that for each $\omega \in \Omega_0$, the limit in (2) must exist. It is sufficient to consider the right limit. Suppose that for some t, $\lim_{s \in S, s \downarrow\downarrow t} X(s, \omega)$ does not exist. Then we have firstly

$$\varliminf_{\substack{s \in S \\ s \downarrow\downarrow t}} X(s, \omega) < \varlimsup_{\substack{s \in S \\ s \downarrow\downarrow t}} X(s, \omega),$$

and hence secondly two rational numbers a and b such that

$$\varliminf_{\substack{s \in S \\ s \downarrow\downarrow t}} X(s, \omega) < a < b < \varlimsup_{\substack{s \in S \\ s \downarrow\downarrow t}} X(s, \omega).$$

But this means that the sample function $X(\cdot, \omega)$ crosses from a to b infinitely many times on the set S in any right neighborhood of t. Since $\omega \in \Omega_0$ this behavior is ruled out by definition of Ω_0.

To prove the second assertion of the theorem, we recall the following inequality for a discrete parameter supermartingale (Theorem 9.4.1 of *Course*). For any $\lambda > 0$:

$$\lambda P\left\{\max_{n \in \mathbf{N}_m} |X_n| \geq \lambda\right\} \leq E(X_0) + 2E(X_m^-).$$

This leads by the same arguments as before to

$$\lambda P\left\{\sup_{t \in [0,m] \cap S} |X_t| \geq \lambda\right\} \leq E(X_0) + 2E(X_m^-). \tag{7}$$

Dividing by λ and letting $\lambda \to \infty$, we conclude that

$$P\left\{\sup_{t \in [0,m] \cap S} |X_t| = \infty\right\} = 0, \quad \text{or} \quad P\left\{\sup_{t \in [0,m] \cap S} |X_t| < \infty\right\} = 1.$$

This means that the sample function $X(\cdot, \omega)$ restricted to $[0, m] \cap S$ is bounded P-a.e., say in a set Λ_m with $P(\Lambda_m) = 1$. Then $P(\bigcap_{m=1}^{\infty} \Lambda_m) = 1$ and if $\omega \in \bigcap_{m=1}^{\infty} \Lambda_m$, $X(\cdot, \omega)$ is bounded in $I \cap S$ for each finite interval I. □

We shall say "almost surely" as paraphrase for "for P-a.e. ω".

Corollary 1. *If X is right continuous, then it has left limits everywhere in $(0, \infty)$ and is bounded on each finite interval, almost surely.*

Proof. The boundedness assertion in Theorem 1 and the hypothesis of right continuity imply that X is bounded on each finite interval. Next, if a sample function $X(\cdot,\ )$ is right continuous but does not have a left limit at some t, then it must oscillate between two numbers a and b infinitely many times in a left neighborhood of t. Because the function is right continuous at the points where it takes the values a and b, these oscillations also occur on the countable dense set S. This is almost surely impossible by the theorem. □

Corollary 2. *If X is right continuous and either* (a) $X_t \geq 0$ *for each t; or* (b) $\sup_t E(|X_t|) < \infty$, *then* $\lim_{t\to\infty} X_t$ *exists almost surely and is an integrable random variable.*

Proof. Under either condition we can let $m \to \infty$ in (5) to obtain

$$E\{U(S; [a,b])\} \leq \frac{A + a^+}{b - a}, \tag{8}$$

where $A = 0$ under (a), A denotes the sup under (b). Now let $U(\omega; [a,b])$ denote the total number of upcrossings of $[a,b]$ by the unrestricted sample function $X(\cdot, \omega)$. We will show that if $X(\cdot,\omega)$ is right continuous, then

$$U(\omega; S; [a,b]) = U(\omega; [a,b]). \tag{9}$$

For if the right member above be $\geq k$, then there exist $s_1 < t_1 < \cdots < s_k < t_k$ such that $X(s_j,\omega) < a$ and $X(t_j,\omega) > b$ for $1 \leq j \leq k$. Hence by right continuity at these $2k$ points, there exist $s_1' < t_1' < \cdots < s_k' < t_k'$, all members of S such that $s_1 < s_1' < t_1 < t_1' < \cdots < s_k < s_k' < t_k < t_k'$ such that $X(s_j',\omega) < a$ and $X(t_j',\omega) > b$ for $1 \leq j \leq k$. Thus the left member of (9) must also be $\geq k$. Of course it cannot exceed the right member. Hence (9) is proved. We have therefore

$$E\{U([a,b])\} \leq \frac{A + a^+}{b - a}. \tag{10}$$

It then follows as before that almost surely no sample function oscillates between any two distinct values infinitely many times in $\mathbf{T}$, and so it must tend to a limit as $t \to \infty$, possibly $\pm\infty$. Let the limit be X_∞. Under condition (a), we have by Fatou's lemma and supermartingale property:

$$E(X_\infty) = E\left(\lim_{n\to\infty} X_n\right) \leq \varliminf_n E(X_n) \leq E(X_0) < \infty.$$

Under condition (b) we have similarly:

$$E(|X_\infty|) \leq \varliminf_n E(|X_n|) \leq \sup_t E(|X_t|).$$

Hence X_∞ is integrable in either case. □

Theorem 2. *Let $(X_t, \mathscr{F}_t)$ be a supermartingale, and X_{t+} and X_{t-} be as in* (2). *We have*

$$\forall t \in [0, \infty): X_t \geq E\{X_{t+} | \mathscr{F}_t\}, \tag{11}$$

$$\forall t \in (0, \infty): X_{t-} \geq E\{X_t | \mathscr{F}_{t-}\}. \tag{12}$$

Furthermore $\{X_{t+}, \mathscr{F}_{t+}\}$ *is a supermartingale; it is a martingale if* $\{X_t, \mathscr{F}_t\}$ *is.*

Proof. Let $t_n \in S$, $t_n \downarrow\downarrow t$, then

$$X_t \geq E\{X_{t_n} | \mathscr{F}_t\}. \tag{13}$$

Now the supermartingale $\{X_s\}$ with s in the index set $\{t, \ldots, t_n, \ldots, t_1\}$ is uniformly integrable; see Theorem 9.4.7(c) of *Course*. Hence we obtain (11) by letting $n \to \infty$ in (13). Next, let $s_n \in S$, $s_n \uparrow\uparrow t$; then

$$X_{s_n} \geq E\{X_t | \mathscr{F}_{s_n}\}.$$

Letting $n \to \infty$ we obtain (12) since $\mathscr{F}_{s_n} \uparrow \mathscr{F}_{t-}$; here we use Theorem 9.4.8, (15a) of *Course*. Finally, let $u_n > u > t_n > t$, $u_n \in S$, $t_n \in S$, $u_n \downarrow\downarrow u$, $t_n \downarrow\downarrow t$, and $\Lambda \in \mathscr{F}_{t+}$. Then we have

$$\int_\Lambda X_{u_n}\, dP \leq \int_\Lambda X_{t_n}\, dP.$$

Letting $n \to \infty$ and using uniform integrability as before, we obtain

$$\int_\Lambda X_{u+}\, dP \leq \int_\Lambda X_{t+}\, dP.$$

This proves that $\{X_{t+}, \mathscr{F}_{t+}\}$ is a supermartingale. The case of a martingale is similar. □

Two stochastic processes $X = \{X_t\}$ and $Y = \{Y_t\}$ are said to be *versions* of each other iff we have

$$\forall t: P\{X_t = Y_t\} = 1. \tag{14}$$

It then follows that for any countable subset S of $\mathbf{T}$, we have also

$$P\{X_t = Y_t \text{ for all } t \in S\} = 1.$$

In particular, for any $(t_1, \ldots, t_n)$, the distributions of $(X_{t_1}, \ldots, X_{t_n})$ and $(Y_{t_1}, \ldots, Y_{t_n})$ are the same. Thus the two processes X and Y have identical finite-dimensional joint distributions. In spite of this they may have quite different sample functions, because the properties of a sample function $X(\cdot, \omega)$ or $Y(\cdot, \omega)$ are far from being determined by its values on a countable set. We shall not delve into this question but proceed to find a good version for a supermartingale, under certain conditions.

From now on we shall suppose that $(\Omega, \mathscr{F}, P)$ is a *complete* probability space. A sub-σ-field of $\mathscr{F}$ is called *augmented* iff it contains all P-null sets in $\mathscr{F}$. We shall assume for the increasing family $\{\mathscr{F}_t\}$ that $\mathscr{F}_0$ is augmented, then so is $\mathscr{F}_t$ for each t. Now if $\{Y_t\}$ is a version of $\{X_t\}$, and $X_t \in \mathscr{F}_t$, then also $Y_t \in \mathscr{F}_t$. Moreover if $\{X_t, \mathscr{F}_t\}$ is a (super) martingale, then so is $\{Y_t, \mathscr{F}_t\}$, as can be verified trivially.

Theorem 3. *Suppose that $\{X_t, \mathscr{F}_t\}$ is a supermartingale and $\{\mathscr{F}_t\}$ is right continuous, namely $\mathscr{F}_t = \mathscr{F}_{t+}$ for each $t \in \mathbf{T}$. Then the process $\{X_t\}$ has a right continuous version if and only if the function*

$$t \to E(X_t) \quad \text{is right continuous in } \mathbf{T}. \tag{15}$$

Proof. Suppose $\{Y_t\}$ is a right continuous version of $\{X_t\}$. Then if $t_n \downarrow\downarrow t$ we have

$$\lim_n E(X_{t_n}) = \lim_n E(Y_{t_n}) = E(Y_t) = E(X_t),$$

where the second equation follows from the uniform integrability of $\{Y_{t_n}\}$ and the right continuity of Y. Hence (15) is true.

Conversely suppose (15) is true; let S be a countable dense subset of $\mathbf{T}$ and define X_{t+} as in (2). Since $X_{t+} \in \mathscr{F}_{t+} = \mathscr{F}_t$, we have by (11)

$$X_t \geq E(X_{t+} \mid \mathscr{F}_{t+}) = X_{t+} \qquad P\text{-a.e.} \tag{16}$$

On the other hand, let $t_n \in S$ and $t_n \downarrow\downarrow t$; then

$$E(X_{t+}) = \lim_n E(X_{t_n}) = E(X_t) \tag{17}$$

where the first equation follows from uniform integrability and the second from (15). Combining (16) and (17) we obtain

$$\forall t \colon P(X_t = X_{t+}) = 1;$$

namely $\{X_{t+}\}$ is a version of $\{X_t\}$. Now for each ω, the function $t \to X(t+, \omega)$ as defined in (2) is right continuous in $\mathbf{T}$, by elementary analysis. Hence $\{X_{t+}\}$ is a right continuous version of $\{X_t\}$, and $\{X_{t+}, \mathscr{F}_t\}$ is a supermartingale by a previous remark. This is also given by Theorem 2, but the point is that there we did not know whether $\{X_{t+}\}$ is a version of $\{X_t\}$. □

We shall say that the supermartingale $\{X_t, \mathscr{F}_t\}$ is right continuous iff both $\{X_t\}$ and $\{\mathscr{F}_t\}$ are right continuous. An immediate consequence of Theorem 3 is the following.

Corollary. *Suppose* $\{\mathscr{F}_t\}$ *is right continuous and* $(X_t, \mathscr{F}_t)$ *is a martingale. Then* $\{X_t\}$ *has a right continuous version.*

A particularly important class of martingales is given by

$$\{E(Y \mid \mathscr{F}_t), \mathscr{F}_t\} \tag{18}$$

where Y is an integrable random variable. According to the Corollary above, the process $\{E(Y \mid \mathscr{F}_t)\}$ has a right continuous version provided that $\{\mathscr{F}_t\}$ is right continuous. In this case we shall always use this version of the process.

The next theorem is a basic tool in martingale theory, called Doob's Stopping Theorem.

Theorem 4. *Let* $\{X_t, \mathscr{F}_t\}$ *be a right continuous supermartingale satisfying the following condition. There exists an integrable random variable* Y *such that*

$$\forall t \colon X_t \geq E(Y \mid \mathscr{F}_t). \tag{19}$$

Let S *and* T *be both optional and* $S \leq T$. *Then we have*

(a) $\lim_{t \to \infty} X_t = X_\infty$ *exists almost surely;* X_S *and* X_T *are integrable, where* $X_S = X_\infty$ *on* $\{S = \infty\}$, $X_T = X_\infty$ *on* $\{T = \infty\}$;

(b) $X_S \geq E(X_T \mid \mathscr{F}_{S+})$.

In case $\{X_t, \mathscr{F}_t\}$ *is a uniformly integrable martingale, there is equality in* (b).

Proof. Let us first observe that condition (19) is satisfied with $Y \equiv 0$ if $X_t \geq 0$, $\forall t$. Next, it is also satisfied when there is equality in (19), namely for the class of martingales exhibited in (18). The general case of the theorem amounts to a combination of these two cases, as will be apparent below. We put

$$Z_t = X_t - E(Y \mid \mathscr{F}_t).$$

Then $\{Z_t, \mathscr{F}_t\}$ is a right continuous positive supermartingale by (19) and the fact that the martingale in (18) is right continuous by choice. Hence we have the decomposition

$$X_t = Z_t + E(Y \mid \mathscr{F}_t) \tag{20}$$

into a positive supermartingale and a martingale of the special kind, both right continuous. Corollary 2 to Theorem 1 applies to both terms on the right side of (20), since $Z_t \geq 0$ and $E\{|E(Y \mid \mathscr{F}_t)|\} \leq E(|Y|) < \infty$. Hence $\lim_{t \to \infty} X_t = X_\infty$ exists and is a finite random variable almost surely. Moreover it is easily verified that

$$\forall t \colon X_t \geq E(X_\infty \mid \mathscr{F}_t)$$

so that $\{X_t, \mathscr{F}_t, t \in [0, \infty]\}$ is a supermartingale. Hence for each n,

$$\left\{X\left(\frac{k}{2^n}\right), \mathscr{F}\left(\frac{k}{2^n}\right), k \in \mathbf{N}_\infty\right\}$$

is a discrete parameter supermartingale, where $\mathbf{N}_\infty = \{0, 1, 2, \ldots, \infty\}$. Let

$$S_n = \frac{[2^n S] + 1}{2^n}, \qquad T_n = \frac{[2^n T] + 1}{2^n}.$$

Then S_n and T_n are optional relative to the family $\{\mathscr{F}(k/2^n), k \in \mathbf{N}\}$, indeed strictly so, and $S_n \leq T_n$. Note that $X(S_n) = X(S) = X_\infty$ on $\{S = \infty\}$; $X(T_n) = X(T) = X_\infty$ on $\{T = \infty\}$. We now invoke Theorem 9.3.5 of *Course* to obtain

$$X(S_n) \geq E\{X(T_n) \mid \mathscr{F}(S_n)\}. \tag{21}$$

Since $\mathscr{F}_{S+} = \bigwedge_{n=1}^{\infty} \mathscr{F}_{S_n+}$ by Theorem 5 of §1.3, this implies: for any $\Lambda \in \mathscr{F}_{S+}$ we have

$$\int_\Lambda X(S_n)\, dP \geq \int_\Lambda X(T_n)\, dP. \tag{22}$$

Letting $n \to \infty$, then $X(S_n) \to X(S)$ and $X(T_n) \to X(T)$ by right continuity. Furthermore $\{X(S_n), \mathscr{F}(S_n)\}$ and $\{X(T_n), \mathscr{F}(T_n)\}$ are both supermartingales on the index set $\tilde{\mathbf{N}} = \{\ldots, n, \ldots, 1, 0\}$ which are uniformly integrable. To see this, e.g., for the second sequence, note that (21) holds when S_n is replaced by T_{n+1} since $T_{n+1} \leq T_n$; next if we put $S_n \equiv 0$ in (21) we deduce

$$\infty > E\{X(0)\} \geq E\{X(T_n)\},$$

and consequently

$$\infty > \lim_n \uparrow E\{X(T_n)\}.$$

This condition is equivalent to the asserted uniform integrability; see Theorem 9.4.7 of *Course*. The latter of course implies the integrability of the limits X_S and X_T, proving assertion (a). Furthermore we can now take the limits under the integrals in (22) to obtain

$$\int_\Lambda X(S)\, dP \geq \int_\Lambda X(T)\, dP. \tag{23}$$

The truth of this relation for each $\Lambda \in \mathscr{F}_{S+}$ is equivalent to the assertion (b). The final sentence of the theorem is proved by considering $\{-X_t\}$ as well as $\{X_t\}$. Theorem 4 is completely proved.

Let us consider the special case where

$$X_t = E(Y|\mathscr{F}_t). \tag{24}$$

We know $X_\infty = \lim_{t\to\infty} X_t$ exists and it can be easily identified as

$$X_\infty = E(Y|\mathscr{F}_\infty).$$

Now apply Theorem 4(b) with equality and with T for S, ∞ for T there. The result is, for any optional T:

$$X_T = E\{X_\infty|\mathscr{F}_{T+}\} = E\{E(Y|\mathscr{F}_\infty)|\mathscr{F}_{T+}\} = E\{Y|\mathscr{F}_{T+}\} \tag{25}$$

since $\mathscr{F}_{T+} \subset \mathscr{F}_\infty$. We may replace $\mathscr{F}_{T+}$ by $\mathscr{F}_T$ above since $\{\mathscr{F}_t\}$ is right continuous by hypothesis.

Next, let T be predictable, and $\{T_n\}$ announce T. Recalling that $\{X_t\}$ has left limits by Corollary 1 to Theorem 1, we have

$$X_{T-} = \lim_n X_{T_n} = \lim_n E\{Y|\mathscr{F}_{T_n+}\}, \tag{26}$$

where $X_{0-} = X_0$. As n increases, $\mathscr{F}_{T_n+}$ increases and by Theorem 9.4.8 of *Course*, the limits is equal to $E(Y|\bigvee_{n=1}^\infty \mathscr{F}_{T_n+}\}$. Using Theorem 8 of §1.3, we conclude that

$$X_{T-} = E\{Y|\mathscr{F}_{T-}\}. \tag{27}$$

Since $\mathscr{F}_{T-} \subset \mathscr{F}_{T+}$, we have furthermore by (25)

$$X_{T-} = E\{E(Y|\mathscr{F}_{T+})|\mathscr{F}_{T-}\} = E\{X_{T+}|\mathscr{F}_{T-}\}. \tag{28}$$

It is easy to show that these last relations do not hold for an arbitrary optional T (see Example 3 of §1.3). In fact, Dellacherie [2] has given an example in which X_{T-} is not integrable for an optional T.

The next theorem, due to P. A. Meyer [1], may be used in the study of excessive functions. Its proof shows the need for deeper notions of measurability and serves as an excellent introduction to the more advanced theory.

Theorem 5. *Let $\{X_t^{(n)}, \mathscr{F}_t\}$ be a right continuous positive supermartingale for each n, and suppose that almost surely we have*

$$\forall t\colon X_t^{(n)} \uparrow X_t. \tag{29}$$

If X_t is integrable for each t, then $\{X_t, \mathscr{F}_t\}$ is a right continuous supermartingale.

Proof. The fact that $\{X_t, \mathscr{F}_t\}$ is a positive supermartingale is trivial; only the right continuity of $t \to X_t$ is in question. Let D be the dyadic set, and put for each $t \geq 0$:

$$Y_t(\omega) = \lim_{\substack{s \downarrow\downarrow t \\ s \in D}} X_s(\omega).$$

[Note that Y_t is denoted by X_{t+} in (2) above when S is D.] By Theorem 1, the limit above exists for all $t \geq 0$ for $\omega \in \Omega_0$ where $P(\Omega_0) = 1$. From here on we shall confine ourselves to Ω_0.

Let $t_k = ([2^k t] + 1)/2^k$ for each $t \geq 0$ and $k \geq 0$. Then for each ω and t we have by (29) and the right continuity of $t \to X^{(n)}(t, \omega)$:

$$\begin{aligned} X(t, \omega) &= \sup_n X^{(n)}(t, \omega) = \sup_n \lim_{k \to \infty} X^{(n)}(t_k, \omega) \\ &\leq \lim_{k \to \infty} \sup_n X^{(n)}(t_k, \omega) = \lim_{k \to \infty} X(t_k, \omega) = Y(t, \omega). \end{aligned} \tag{30}$$

Next let T be any optional time and $T_k = ([2^k T] + 1)/2^k$. Then we have by Theorem 4,

$$E\{X^{(n)}(T)\} \geq E\{X^{(n)}(T_k)\}$$

for each $k \geq 1$. Letting $n \to \infty$ we obtain by monotone convergence:

$$E\{X(T)\} \geq E\{X(T_k)\}. \tag{31}$$

Since $X(T_k)$ converges to $Y(T)$ by the definition of the latter, it follows from (31) and Fatou's lemma that

$$E\{X(T)\} \geq \lim_k E\{X(T_k)\} \geq E\{Y(T)\}. \tag{32}$$

Now for any $\varepsilon > 0$ define

$$T(\omega) = \inf\{t \geq 0 \mid Y_t(\omega) - X_t(\omega) \geq \varepsilon\} \tag{33}$$

where $\inf \varnothing = +\infty$ as usual. The verification that T is an optional time is a delicate matter and will be postponed until the next section. Suppose this is so; then if $T(\omega) < \infty$ we have $\delta_j \geq 0$, $\delta_j \to 0$ such that

$$Y(T(\omega) + \delta_j, \omega) - X(T(\omega) + \delta_j, \omega) \geq \varepsilon.$$

Hence we have for all $n \geq 1$:

$$Y(T(\omega) + \delta_j, \omega) - X^{(n)}(T(\omega) + \delta_j, \omega) \geq \varepsilon$$

because $X \geq X^{(n)}$. Since both $t \to Y(t, \omega)$ and $t \to X^{(n)}(t, \omega)$ are right continuous the inequality above implies that

$$Y(T(\omega), \omega) - X^{(n)}(T(\omega), \omega) \geq \varepsilon$$

as $\delta_j \to 0$. Note that the case where all $\delta_j = 0$ is included in this argument. Letting $n \to \infty$ we obtain

$$Y(T(\omega), \omega) - X(T(\omega), \omega) \geq \varepsilon. \tag{34}$$

Therefore (34) holds on the set $\{T < \infty\}$, and consequently (32) is possible only if $P\{T < \infty\} = 0$. Since ε is arbitrary, this means we have almost surely:

$$\forall t \geq 0 \colon Y(t, \omega) \leq X(t, \omega). \tag{35}$$

Together with (30) we conclude that equality must hold in (35), namely that $P\{\forall t \geq 0 \colon Y_t = X_t\} = 1$. The theorem is proved since $t \to Y(t, \omega)$ is right continuous for $\omega \in \Omega_0$. □

Remark. If we do not assume in the theorem that X_t is integrable for each t, of course $\{X_t\}$ cannot be a supermartingale. But it is still true that $t \to X(t, \omega)$ is right continuous as an extended-valued function, for P-a.e. ω. The assumption that each $X^{(n)}$ is positive may also be dropped. Both these extensions are achieved by the following reduction. Fix two positive integers u and k and put

$$Y_t^{(n)} = (X_t^{(n)} - E\{X_u^{(1)} | \mathscr{F}_t\}) \wedge k, \qquad 0 \leq t \leq u$$

where the conditional expectation is taken in its right continuous version. Then $\{Y_t^{(n)}\}$ satisfies the conditions of the theorem, hence

$$Y_t = \sup_n Y_t^{(n)}$$

is right continuous almost surely in $[0, u]$. Now let first $k \uparrow \infty$, then $u \uparrow \infty$.

An important complement to Theorem 5 under a stronger hypothesis will next be given.

Theorem 6. *Under the hypotheses of Theorem 5, suppose further that the index set is $[0, \infty]$ and (29) holds for $t \in [0, \infty]$. Suppose also that for any sequence of optional times $\{T_n\}$ increasing to T almost surely, we have*

$$\lim E\{X(T_n)\} = E\{X(T)\}. \tag{36}$$

Then almost surely the convergence of $X^{(n)}$ to X_t is uniform in $t \in [0, \infty]$.

Proof. For any $\varepsilon > 0$, define

$$T_n^\varepsilon(\omega) = \inf\{t \geq 0 \mid X(t, \omega) - X^{(n)}(t, \omega) \geq \varepsilon\}.$$

Since X as well as X_n is right continuous by Theorem 5, this time it is easy to verify that T_n^ε is optional directly without recourse to the next section. Since $X_t^{(n)}$ increases with n for each t, T_n^ε increases with n, say to the limit $T_\infty^\varepsilon(\leq \infty)$. Now observe that:

$$\begin{aligned} X^{(n)} \uparrow X &\text{ uniformly in } [0, \infty],\ P\text{-a.e.} \\ &\Leftrightarrow \forall \varepsilon > 0\text{: for } P\text{-a.e. } \omega,\ \exists n(\omega) \text{ such that } T_{n(\omega)}^\varepsilon(\omega) = \infty \\ &\Leftrightarrow \forall \varepsilon > 0\text{: } P\{\exists n\text{: } T_n^\varepsilon = \infty\} = 1 \\ &\Leftrightarrow \forall \varepsilon > 0\text{: } P\{\forall n\text{: } T_n^\varepsilon < \infty\} = 0 \\ &\Leftrightarrow \forall \varepsilon > 0\text{: } \lim_n P\{T_n^\varepsilon < \infty\} = 0 \end{aligned}$$

By right continuity, we have:

$$X(T_n^\varepsilon) - X_n(T_n^\varepsilon) \geq \varepsilon \quad \text{on } \{T_n^\varepsilon < \infty\}.$$

Hence for $m \leq n$:

$$\begin{aligned} \varepsilon P\{T_n^\varepsilon < \infty\} &\leq E\{X(T_n^\varepsilon) - X_n(T_n^\varepsilon);\ T_n^\varepsilon < \infty\} \\ &\leq E\{X(T_n^\varepsilon) - X_m(T_n^\varepsilon)\}. \end{aligned}$$

Note that for any optional T, $E\{X(T)\} \leq E\{X(0)\} < \infty$ by Theorem 4. By the same theorem,

$$E\{X_m(T_n^\varepsilon)\} \geq E\{X_m(T_\infty^\varepsilon)\}.$$

Thus

$$\varepsilon P\{T_n^\varepsilon < \infty\} \leq E\{X(T_n^\varepsilon) - X_m(T_\infty^\varepsilon)\}.$$

Letting $n \to \infty$ and using (36), we obtain:

$$\varepsilon \lim_n P\{T_n^\varepsilon < \infty\} \leq E\{X(T_\infty^\varepsilon) - X_m(T_\infty^\varepsilon)\}. \tag{37}$$

Under the hypothesis (29) with $t = \infty$ included, $X_m(T_\infty^\varepsilon) \uparrow X(T_\infty^\varepsilon)$, P-a.e. Since $X(T_\infty^\varepsilon)$ is integrable, the right member of (37) converges to zero by dominated convergence. Hence so does the left member, and this is equivalent to the assertion of the theorem, as analyzed above. □

The condition (36) is satisfied for instance when all sample functions of X are left continuous, therefore continuous since they are assumed to be right continuous in the theorem. In general, the condition supplies a kind of left continuity, not for general approach but for "optional approaches". A similar condition will be an essential feature of the Markov processes we are going to study, to be called "quasi left continuity" of Hunt processes. We will now study the case where $T_n \to \infty$ in (36), and for this discussion we introduce a new term. The reader should be alerted to the abuse of the word "potential", used by different authors to mean different things. For this reason we give it a prefix to indicate that it is a special case of supermartingales. [We do not define a subpotential!]

Definition. A *superpotential* is a right continuous positive supermartingale $\{X_t\}$ satisfying the condition

$$\lim_{t \to \infty} E(X_t) = 0. \tag{38}$$

Since $X_\infty = \lim_{t \to \infty} X_t$ exists almost surely by Corollary 2 to Theorem 1, the condition (38) implies that $X_\infty = 0$ almost surely. Furthermore, it follows that $\{X_t\}$ is uniformly integrable (Theorem 4.5.4 of *Course*). Nonetheless this does not imply that for any sequence of optional times $T_n \to \infty$ we have

$$\lim_{n} E(X_{T_n}) = 0. \tag{39}$$

When this is the case, the superpotential is said to be "of class D"—a nomenclature whose origin is obscure. This class plays a fundamental role in the advanced theory, so let us prove one little result about it just to make the acquaintance.

Theorem 7. *A superpotential is of class D if and only if the class of random variables $\{X_T : T \in \mathcal{O}\}$ is uniformly integrable, where $\mathcal{O}$ is the class of all optional times.*

Proof. If $T_n \to \infty$, then $X_{T_n} \to 0$ almost surely; hence uniform integrability of $\{X_{T_n}\}$ implies (39). To prove the converse, let $n \in \mathbf{N}$ and

$$T_n = \inf\{t \geq 0 \mid X_t \geq n\}.$$

Since almost surely $X(t)$ is bounded in each finite t-interval by Corollary 1 to Theorem 1, and converges to zero as $t \to \infty$, we must have $T_n \uparrow \infty$. Hence (39) holds for this particular sequence by hypothesis. Now for any optional T, define

$$S = \begin{cases} T & \text{on } \{T \geq T_n\}, \\ \infty & \text{on } \{T < T_n\}. \end{cases}$$

Then $S \geq T_n$ and we have

$$\int_{\{X_T \geq n\}} X_T \, dP \leq \int_{\{T \geq T_n\}} X_T \, dP = \int_\Omega X_S \, dP \leq \int_\Omega X_{T_n} \, dP$$

where the middle equation is due to $X_\infty \equiv 0$ and the last inequality due to Doob's stopping theorem. Since the last term above converges to zero as $n \to \infty$, $\{X_T: T \in \mathcal{O}\}$ is uniformly integrable by definition. □

1.5. Progressive Measurability and the Section Theorem

For $t \geq 0$ let $\mathbf{T}_t = [0, t]$ and $\mathscr{B}_t$ be the Euclidean (classical) Borel field on $\mathbf{T}_t$. Let $\{\mathscr{F}_t\}$ be an increasing family of σ-fields.

Definition. $X = \{X_t, t \in \mathbf{T}\}$ is said to be *progressively measurable* relative to $\{\mathscr{F}_t, t \in \mathbf{T}\}$ iff the mapping

$$(s, \omega) \to X(s, \omega)$$

restricted to $\mathbf{T}_t \times \Omega$ is measurable $\mathscr{B}_t \times \mathscr{F}_t$ for each $t \geq 0$. This implies that $X_t \in \mathscr{F}_t$ for each t, namely $\{X_t\}$ is adapted to $\{\mathscr{F}_t\}$ (exercise).

This new concept of measurability brings to the fore the fundamental nature of $X(\cdot, \cdot)$ as a function of the pair (s, ω). For each s, $X(s, \cdot)$ is a random variable; for each ω, $X(\cdot, \omega)$ is a sample function. But the deeper properties of the process X really concern it as function of two variables on the domain $\mathbf{T} \times \Omega$. The structure of this function is complicated by the fact that Ω is not generally supposed to be a topological space. But particular cases of Ω such that $(-\infty, +\infty)$ or $[0, 1]$ should make the comparison with a topological product space meaningful. For instance, from this point of view a random time such as an optional time represents a "curve" as illustrated below:

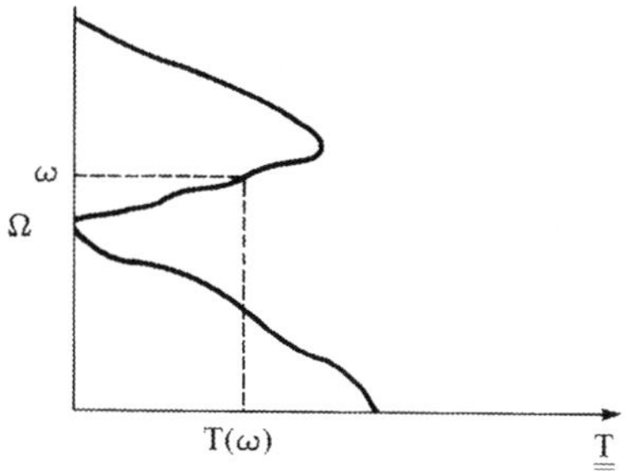

Thus the introduction of such times is analogous to the study of plane geometry by means of curves—a venerable tradition. This apparent analogy has been made into a pervasive concept in the "general theory of stochastic processes;" see Dellacherie [2] and Dellacherie-Meyer [1].

According to the new point of view, a function $X(\cdot, \cdot)$ on $\mathbf{T} \times \Omega$ which is in $\mathscr{B} \times \mathscr{F}$ is precisely a Borel measurable process as defined in §1.3. In particular, for any subset H of $\mathbf{T} \times \Omega$, its indicator function 1_H may be identified as a Borel measurable process $\{1_H(t), t \in \mathbf{T}\}$ in the usual notation, where $1_H(t)$ is just the abbreviation of $\omega \to 1_H(t, \omega)$. Turning this idea around, we call the set H progressively measurable iff the associated process $\{1_H(t)\}$ is so. This amounts to the condition that $H \cap ([0,t] \times \Omega) \in \mathscr{B}_t \times \mathscr{F}_t$ for each $t \geq 0$. Now it is easy to verify that the class of progressively measurable sets forms a σ-field $\mathscr{G}$, and that X is progressively measurable if and only if $X \in \mathscr{G}/\mathscr{E}$ in the notation of §1.1. It follows that progressive measurability is preserved by certain operations such as sequential pointwise convergence. The same is not true for, e.g., right continuous process.

Recall that $\mathscr{F}_t^0 = \sigma(X_s, 0 \leq s \leq t)$. If X is progressively measurable relative to $\{\mathscr{F}_t^0\}$, then it is so relative to any $\{\mathscr{F}_t\}$ to which it is adapted.

Theorem 1. *If $\{X_t\}$ is right [or left] continuous, then it is progressively measurable relative to $\{\mathscr{F}_t^0\}$.*

Proof. We will prove the result for right continuity, the left case being similar. Fix t and define for $n \geq 1$:

$$X^{(n)}(s, \omega) = X\left(\frac{k+1}{2^n} t, \omega\right) \quad \text{if } s \in \left[\frac{k}{2^n} t, \frac{k+1}{2^n} t\right), 0 \leq k \leq 2^n - 1;$$

$$X^{(n)}(t, \omega) = X(t, \omega).$$

Then $X^{(n)}$ is defined on $\mathbf{T}_t \times \Omega$ and converges pointwise by right continuity. For each $B \in \mathscr{E}$, we have

$$\{(s, \omega) | X^{(n)}(s, \omega) \in B\} = \bigcup_{k=0}^{2^n - 1} \left(\left[\frac{k}{2^n} t, \frac{k+1}{2^n} t\right) \times \left\{\omega \middle| X\left(\frac{k+1}{2^n} t, \omega\right) \in B\right\}\right)$$
$$\cup (\{t\} \times \{\omega | X(t, \omega) \in B\}) \in \mathscr{B}_t \times \mathscr{F}_t^0.$$

Hence the limit, which is X restricted to $\mathbf{T}_t \times \Omega$, is likewise measurable by the following lemma which is spelled out here since it requires a little care in a general state space. The latter is assumed to be separable metric and locally compact, so that each open set is the union of a countable family of closed sets.

Lemma. *Let $\mathscr{A}$ be any σ-field of subsets of $\mathbf{T} \times \Omega$. Suppose that $X^{(n)} \in \mathscr{A}$ for each n and $X^{(n)} \to X$ everywhere. Then $X \in \mathscr{A}$.*

Proof. Let $\mathbb{C}$ be the class of sets B in $\mathscr{E}$ such that $X^{-1}(B) \in \mathscr{A}$. Let G be open and $G = \bigcup_{k=1}^{\infty} F_k$ where $F_k = \{x | d(x, G^c) \geq 1/k\}$ an d is a metric. Then we have by pointwise convergence of $X^{(n)}$ to X:

$$X^{-1}(G) = \bigcup_{k=1}^{\infty} \bigcup_{n=1}^{\infty} \bigcap_{m=n}^{\infty} (X^{(m)})^{-1}(F_k).$$

This belongs to $\mathscr{A}$, and so the class $\mathbb{C}$ contains all open sets. Since it is a σ-field by properties of X^{-1}, it contains the minimal σ-field containing all open sets, namely $\mathscr{E}$. Thus $X \in \mathscr{A}/\mathscr{E}$. □

The next result is an extension of Theorem 10 of §1.3.

Theorem 2. *If X is progressively measurable and T is optional, then $X_T 1_{\{T<\infty\}} \in \mathscr{F}_{T+}$.*

Proof. Consider the two mappings:

$$\varphi : (s, \omega) \to X(s, \omega)$$
$$\psi : \omega \to (T(\omega), \omega).$$

Their composition is

$$\varphi \circ \psi : \omega \to X(T(\omega), \omega) = X_T(\omega).$$

It follows from the definition of optionality that for each $s < t$:

$$\psi^{-1}(\mathscr{B}_s \times \mathscr{F}_s) \subset \mathscr{F}_t. \tag{1}$$

For if $A \in \mathscr{B}_s$ and $\Lambda \in \mathscr{F}_s$, then

$$\{\omega \mid T(\omega) \in A;\, \omega \in \Lambda\} = \{T \in A\} \cap \Lambda \in \mathscr{F}_{s+} \subset \mathscr{F}_t.$$

Next if we write for a fixed s the restriction of φ to $\mathbf{T}_s \times \Omega$ as $\tilde{\varphi}$, we have by progressive measurability

$$\tilde{\varphi}^{-1}(\mathscr{E}) \subset \mathscr{B}_s \times \mathscr{F}_s. \tag{2}$$

Combining (1) and (2), we obtain

$$(\tilde{\varphi} \circ \psi)^{-1}(\mathscr{E}) = \psi^{-1} \circ \tilde{\varphi}^{-1}(\mathscr{E}) \subset \mathscr{F}_t.$$

In particular if $B \in \mathscr{E}$, then for each $s < t$:

$$\{\omega \mid T(\omega) \le s;\, X_T(\omega) \in B\} \in \mathscr{F}_t.$$

Taking a sequence of s increasing strictly to t, we deduce that for each t:

$$\{X_T \in B\} \cap \{T < t\} \in \mathscr{F}_t.$$

This is equivalent to the assertion of the theorem. □

Remark. A similar argument shows that if T is strictly optional, then $X_T 1_{\{T<\infty\}} \in \mathscr{F}_T$.

Let H be a subset of $\mathbf{T} \times \Omega$. We define the *debut* of H as follows:

$$D_H(\omega) = \inf\{t \geq 0 \,|\, (t, \omega) \in H\} \tag{3}$$

where $\inf \emptyset = \infty$ as usual.

Theorem 3 (The Section Theorem). *If for each t, $(\Omega, \mathscr{F}_t, P)$ is a complete measure space, then D_H as defined in (3) is an optional time for each progressively measurable set H.*

Proof. Let $\mathbf{T}_{t-} = [0, t)$ and consider the set

$$H_t = H \cap (\mathbf{T}_{t-} \times \Omega).$$

Then $H_t \in \mathscr{B}_t \times \mathscr{F}_t$ since H is progressively measurable. Let π_Ω denote the projection mapping of $\mathbf{T} \times \Omega$ onto Ω, namely for any $H \subset \mathbf{T} \times \Omega$:

$$\pi_\Omega(H) = \{\omega \in \Omega \,|\, \exists t \in \mathbf{T} \text{ such that } (t, \omega) \in H\}.$$

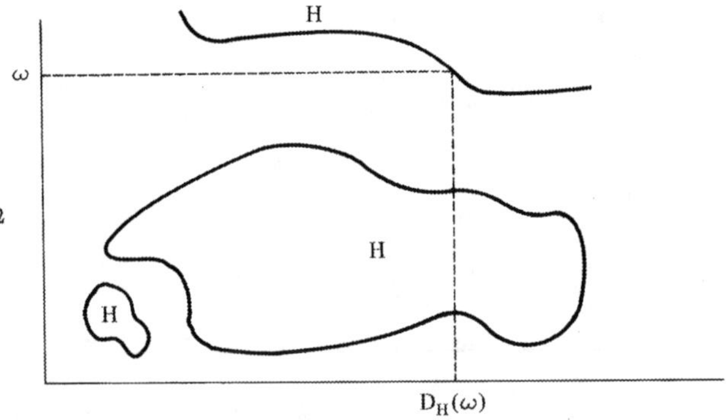

It is clear that

$$\begin{aligned} \pi_\Omega(H_t) &= \{\omega \,|\, \exists s \in [0, t) \text{ such that } (s, \omega) \in H\} \\ &= \{\omega \,|\, D_H(\omega) < t\}. \end{aligned} \tag{4}$$

[Observe that if we replace $[0, t)$ by $[0, t]$ in the second term above we cannot conclude that it is equal to the third term with "$<$" replaced by "$\leq$".] By the theory of analytical sets (see Dellacherie-Meyer [1], Chapter 3) the projection of each set in $\mathscr{B}_t \times \mathscr{F}_t$ on Ω is an "$\mathscr{F}_t$-analytic" set, hence $\mathscr{F}_t$-measurable when $(\Omega, \mathscr{F}_t, P)$ is complete. Thus for each t we have $\{D_H < t\} \in \mathscr{F}_t$, namely D_H is optional. □

The following "converse" to Theorem 3 is instructive and is a simple illustration of the general methodology. Let T be optional and consider the set

$$H = \{(t, \omega) | T(\omega) < t\}.$$

Then

$$1_H(t, \omega) = \begin{cases} 1 & \text{on } \{(t, \omega) | T(\omega) < t\}, \\ 0 & \text{on } \{(t, \omega) | T(\omega) \geq t\}. \end{cases}$$

Since T is optional, we have $1_H(t) \in \mathscr{F}_t$ for each t. Furthermore, it is *trivial* that for each ω, the sample function $1_H(\cdot, \omega)$ is left continuous. Hence 1_H is progressively measurable by Theorem 1 above. Now a picture shows that

$$D_H(\omega) = T(\omega).$$

Thus each optional time is the debut of a progressively measurable set.

We now return to the T in (33) of §1.4 and show that it is optional. Since each $X^{(n)}$ is right continuous, it is progressively measurable relative to $\{\mathscr{F}_t\}$ by Theorem 1. Therefore X is progressively measurable as the limit (or supremum) of $X^{(n)}$ by the Lemma above in which we take $\mathscr{A}$ to be the progressively measurable σ-field. On the other hand, the process $(t, \omega) \to Y(t, \omega)$, being right continuous, is progressively measurable by Theorem 1. Hence the set

$$\{(t, \omega) | Y(t, \omega) - X(t, \omega) \geq \varepsilon\}$$

is progressively measurable, and T is optional by Theorem 3.

As an illustration of the fine analysis of sample function behavior, we prove the following theorem which is needed in Chapter 2. Recall that for a discrete time positive supermartingale, almost surely every sample sequence remains at the value zero if it is ever taken. In continuous time the result has a somewhat delicate ramification.

Theorem 4. *Let $\{X_t, \mathscr{F}_t\}$ be a positive supermartingale having right continuous paths. Let*

$$T_1(\omega) = \inf\{t \geq 0 | X_t(\omega) = 0\},$$
$$T_2(\omega) = \inf\{t \geq 0 | X_{t-}(\omega) = 0\},$$
$$T = T_1 \wedge T_2.$$

Then we have almost surely $X(T + t) = 0$ for all $t \geq 0$ on the set $\{T < \infty\}$.

Proof. Since $\{X_t\}$ is right continuous, it has left limits by Corollary 1 to Theorem 1 of §1.4. By Theorem 1, both $\{X_t\}$ and $\{X_{t-}\}$ are progressively measurable relative to $\{\mathscr{F}_t\}$. If we put

$$H_1 = \{(t,\omega) | X(t,\omega) = 0\}, \qquad H_2 = \{(t,\omega) | X(t-,\omega) = 0\},$$

then $T_1 = D_{H_1}$, $T_2 = D_{H_2}$; hence both are optional. It follows from Doob's stopping theorem that for each $t \geq 0$:

$$E\{X(T_1); T_1 < \infty\} \geq E\{X(T_1 + t); T_1 < \infty\}. \tag{5}$$

But $X(T_1) = 0$ on $\{T_1 < \infty\}$ by the definition of T_1 and right continuity of paths. Hence the right member of (5) is also equal to zero. Since $X \geq 0$ this implies that

$$P\{X(T_1 + t) = 0 \text{ for all } t \in Q \cap [0, \infty); T_1 < \infty\} = P\{T_1 < \infty\}.$$

We may omit Q in the above by right continuity, and the result is equivalent to the assertion of the theorem when T is replaced by T_1.

The situation is different for T_2 because we can no longer say at once that $X(T_2) = 0$, although this is part of the desired conclusion. It is conceivable that $X(t) \to 0$ as $t \uparrow\uparrow T_2$ but jumps to a value different from 0 at T_2. To see that this does not happen, we must make a more detailed analysis of sample functions of the kind frequently needed in the study of Markov processes later. We introduce the approximating optional times as follows:

$$S_n(\omega) = \inf\left\{t \geq 0 \,|\, X_t(\omega) < \frac{1}{n}\right\}.$$

It is easy to show S_n is optional because $(-\infty, 1/n)$ is an open set without using progressive measurability. Clearly S_n increases with n, and $S_n \uparrow S \leq T$ where S is optional. Now we must consider two cases.

Case 1. $\forall n: S_n < S$. In this case $X(S-) = \lim_n X(S_n) = 0$. Since $X(t) \geq 1/n$ for $t \in [0, S_n)$, so also $X(t-) \geq 1/n$ for $t \in [0, S_n)$, it is clear that $S = T_2 = T$.

Case 2. $\exists n_0: S_{n_0} = S_{n_0 + 1} = \cdots = S$. In this case $X(S) = 0$ because $X(S_n) \leq 1/n$ for all n, and we have $S = T_1 = T$. Unless $S_{n_0} = 0$, X jumps to 0 at S_{n_0}.

We have therefore proved that $T = S$ is optional (without proving that T_1 and T_2 are). Now applying Doob's stopping theorem as follows:

$$E\{X(S_n); S_n < \infty\} \geq E\{X(S); S_n < \infty\}. \tag{6}$$

Since the left member does not exceed $1/n$, letting $n \to \infty$ we obtain

$$E\{X(S), S < \infty\} = 0.$$

The rest is as before. □

We used the fact that $\{S < \infty\} \subset \bigcap_n \{S_n < \infty\}$. The reverse inclusion is not true!

Exercises

1. Let $\{X_t, \mathscr{F}_t\}$ be a martingale with right continuous paths. Show that there exists an integrable Y such that $X_t = E(Y \mid \mathscr{F}_t)$ for each t, if and only if $\{X_t\}$ is uniformly integrable.

2. Show that if T is the first jump in a Poisson process (see Example 3 of §1.3), then $E\{X_T \mid \mathscr{F}_{T-}\} \neq X_{T-}$.

3. In the notation of the proof of Theorem 6 of §1.4, show that $X_t^{(n)}$ converges to X_t uniformly in each finite t-interval, P-a.e., if and only if for each $\varepsilon > 0$ we have $P\{\lim_n T_n^\varepsilon = \infty\} = 1$.

4. If $(t, \omega) \to X(t, \omega)$ is in $\mathscr{B} \times \mathscr{F}^0$ then for each $A \in \mathscr{E}$, the function $(t, x) \to P_t(x, A)$ is in $\mathscr{B} \times \mathscr{E}$. Hence this is the case if $\{X_t\}$ is right [or left] continuous. [Hint: consider the class of functions φ on $\mathbf{T} \times \Omega$ such that $(t, x) \to E^x\{\varphi\}$ belongs to $\mathscr{B} \times \mathscr{E}$. It contains functions of the form $1_B(t)1_\Lambda(\omega)$ where $B \in \mathscr{B}$ and $\Lambda \in \mathscr{F}^0$.]

5. If $\{X_t\}$ is adapted to $\{\mathscr{F}_t\}$, and progressively measurable relative to $\{\mathscr{F}_{t+\varepsilon}\}$ for each $\varepsilon > 0$, then $\{X_t\}$ is progressively measurable relative to $\{\mathscr{F}_t\}$.

6. Give an example of a process $\{X_t\}$ which is adapted to $\{\mathscr{F}_t\}$ but not progressively measurable relative to $\{\mathscr{F}_t\}$.

7. Suppose that $\{\mathscr{F}_t\}$ is right continuous and S and T are optional relative to $\{\mathscr{F}_t\}$, with $S \leq T$. Put

$$[[S, T)) = \{(t, \omega) \in \mathbf{T} \times \Omega \mid S(\omega) \leq t < T(\omega)\};$$

similarly for $[[S, T]]$, $((S, T))$, $((S, T]]$. Show that all these four sets are progressively measurable relative to $\{\mathscr{F}_t\}$.

The σ-field generated by $[[S, T))$ when S and T range over all optional times such that $S \leq T$ is called the *optional field*; the σ-field generated by $[[S, T))$ when S and T range over all predictable times such that $S \leq T$ is called the *predictable field*. Thus we have

$$\begin{aligned}\text{predictable field} &\subset \text{optional field}\\ &\subset \text{progressively measurable field.}\end{aligned}$$

These are the fundamental σ-fields for the general theory of stochastic processes.

Notes on Chapter 1

§1.1. The basic notions of the Markov property, as well as optionality, in the discrete parameter case are treated in Chapters 8 and 9 of the *Course*. A number of proofs carry over to the continuous parameter case, without change.

§1.2. Among the examples of Markov processes given here, only the case of Brownian motion will be developed in Chapter 4. But for dimension $d = 1$ the theory is somewhat special and will not be treated on its own merits. The case of Markov chains is historically the oldest, but its modern development is not covered by the general theory. It will only be mentioned here occasionally for peripheral illustrations. The class of spatially homogeneous Markov processes, sometimes referred to as Lévy or additive processes, will be briefly described in §4.1.

§1.3. Most of the material on optionality may be found in Chung and Doob [1] in a more general form. For deeper properties, which are sparingly used in this book, see Meyer [1]. The latter is somewhat dated but in certain respects more readable than the comprehensive new edition which is Dellacherie and Meyer [1].

§1.4. It is no longer necessary to attribute the foundation of martingale theory to Doob, but his book [1] is obsolete especially for the treatment of the continuous parameter case. The review here borrows much from Meyer [1] and is confined to later needs, except for Theorems 6 and 7 which are given for the sake of illustration and general knowledge. Meyer's proof of Theorem 5 initiated the method of projection in order to establish the optionality of the random time T defined in (33). This idea was later developed into a powerful methodology based on the two σ-fields mentioned at the end of §1.5. A very clear exposition of this "general theory" is given in Dellacherie [2].

A curious incident happened in the airplane from Zürich to Beijing in May of 1979. At the prodding of the author, Doob produced a new proof of Theorem 5 without using projection. Unfortunately it is not quite simple enough to be included here, so the interested reader must await its appearance in Doob's forthcoming book. See Doob [4; p. 477ff, p. 809].

Formula (4) in §1.4 is due to Letta in its final form; see Dellacherie–Meyer [2; p. 21ff] for its history. Instead of upcrossing we can consider downcrossing, and then we can use Doob's original Theorem 9.4.2 in *Course* after changing signs. Indeed, the number of upcrossing differs at most by one from the number of downcrossing.

Chapter 2

Basic Properties

2.1. Martingale Connection

Let a homogeneous Markov process $\{X_t, \mathscr{F}_t, t \in \mathbf{T}\}$ with transition semigroup (P_t) be given. We seek a class of functions f on $\mathbf{E}$ such that $\{f(X_t), \mathscr{F}_t\}$ is a supermartingale.

Definition. Let $f \in \mathscr{E}, 0 \leq f \leq \infty; \alpha \geq 0$ f is *α-superaveraging* relative to (P_t) iff

$$\forall t \geq 0: f \geq e^{-\alpha t} P_t f; \tag{1}$$

f is *α-excessive* iff in addition we have

$$f = \lim_{t \downarrow 0} e^{-\alpha t} P_t f. \tag{2}$$

It follows from (1) that its right member is a decreasing function of t, hence limit in (2) always exists.

Excessive functions will play a major role in Chapter 3. The connection with martingale theory is given by the proposition below.

Proposition 1. *If f is α-superaveraging and $f(X_t)$ is integrable for each t, then $\{e^{-\alpha t} f(X_t), \mathscr{F}_t\}$ is a supermartingale.*

Proof. Let $s > 0$, $t \geq 0$. We have by the Markov property:

$$f(X_s) \geq e^{-\alpha t} P_t f(X_s) = e^{-\alpha t} E^{X_s}\{f(X_t)\} = e^{-\alpha t} E\{f(X_{s+t}) | \mathscr{F}_s\}.$$

Hence

$$e^{-\alpha s} f(X_s) \geq E\{e^{-\alpha(s+t)} f(X_{s+t}) | \mathscr{F}_s\}$$

and this establishes the proposition. □

Next we present a subclass of α-superaveraging functions which plays a dominant role in the theory. We begin by stating a further condition on the transition function.

Definition. (P_t) is said to be *Borelian* iff for each $A \in \mathscr{E}$:

$$(t, x) \to P_t(x, A)$$

is measurable $\mathscr{B} \times \mathscr{E}$. This is equivalent to: for each $f \in b\mathscr{E}$:

$$(t, x) \to P_t f(x)$$

is measurable $\mathscr{B} \times \mathscr{E}$. According to Exercise 4 of §1.5, if e.g. $\{X_t\}$ is right continuous, then (P_t) is Borelian. The next definition applies to a Borelian (P_t) in general.

Definition. Let $f \in b\mathscr{E}$, $\alpha > 0$, then the α-potential of f is the function $U^\alpha f$ given by

$$U^\alpha f(x) = \int_0^\infty e^{-\alpha t} P_t f(x)\, dt$$
$$= E^x\left\{\int_0^\infty e^{-\alpha t} f(X_t)\, dt\right\}. \tag{3}$$

The integration with respect to t is possible because of the Borelian assumption above and the equality of the two expressions in (3) is due to Fubini's theorem. In particular $U^\alpha f \in \mathscr{E}$. If we denote the sup-norm of f by $\|f\|$, then we have

$$\|U^\alpha f\| \le \frac{1}{\alpha}\|f\|. \tag{4}$$

Taking $f \equiv 1$ we see that the "operator norm" satisfies

$$\|U^\alpha\| = \frac{1}{\alpha}. \tag{5}$$

U^α is also defined for $f \in \mathscr{E}_+$ but may take the value $+\infty$. The family of operators $\{U^\alpha, \alpha > 0\}$ is also known as the "resolvent" of the semigroup (P_t), or by abuse of language, of the process (X_t). We postpone a discussion of the relevant facts. For the present the next two propositions are important.

Proposition 2. *If $f \in b\mathscr{E}_+$, then for each $\alpha > 0$, $U^\alpha f$ is α-excessive ; and $\{e^{-\alpha t} U^\alpha f(X_t), \mathscr{F}_t\}$ is a supermartingale.*

Proof. We have

$$e^{-\alpha t} P_t(U^\alpha f) = e^{-\alpha t}\int_0^\infty e^{-\alpha s} P_{t+s} f\, ds = \int_t^\infty e^{-\alpha s} P_s f\, ds$$
$$\le \int_0^\infty e^{-\alpha s} P_s f\, ds = U^\alpha f.$$

This shows that $U^\alpha f$ is α-superaveraging. As $t \downarrow 0$, the third term above converges to the fourth term, hence $U^\alpha f$ is α-excessive—a fact we store for later use. Since $U^\alpha f$ is bounded the last assertion of the proposition follows from Proposition 1. □

Proposition 3. *Suppose $\{X_t\}$ is progressively measurable relative to $\{\mathscr{F}_t\}$. For $f \in b\mathscr{E}_+$ and $\alpha > 0$, define*

$$Y_t = \int_0^t e^{-\alpha s} f(X_s)\,ds + e^{-\alpha t} U^\alpha f(X_t). \tag{6}$$

Then $\{Y_t, \mathscr{F}_t\}$ is a uniformly integrable martingale which is progressively measurable relative to $\{\mathscr{F}_t\}$.

Proof. Let

$$Y_\infty = \int_0^\infty e^{-\alpha s} f(X_s)\,ds$$

which is a bounded random variable; then we have for each x:

$$E^x\{Y_\infty\} = U^\alpha f(x).$$

We have

$$E\{Y_\infty \mid \mathscr{F}_t\} = \int_0^t e^{-\alpha s} f(X_s)\,ds + E\left\{\int_t^\infty e^{-\alpha s} f(X_s)\,ds \,\middle|\, \mathscr{F}_t\right\}. \tag{7}$$

The first term on the right belongs to $\mathscr{F}_t$, because $\{X_t\}$ is progressively measurative relative to $\{\mathscr{F}_t\}$ (exercise). The second term on the right side of (7) is equal to

$$\begin{aligned} &E\left\{\int_0^\infty e^{-\alpha(t+u)} f(X_{t+u})\,du \,\middle|\, \mathscr{F}_t\right\} \\ &\quad = E\left\{e^{-\alpha t} \int_0^\infty e^{-\alpha u} f(X_u \circ \theta_t)\,du \,\middle|\, \mathscr{F}_t\right\} \\ &\quad = E\{e^{-\alpha t} Y_\infty \circ \theta_t \mid \mathscr{F}_t\} = e^{-\alpha t} E^{X_t}\{Y_\infty\} = e^{-\alpha t} U^\alpha f(X_t). \end{aligned}$$

Hence we have

$$E\{Y_\infty \mid \mathscr{F}_t\} = Y_t. \tag{8}$$

Since Y_∞ is bounded, $\{Y_t, \mathscr{F}_t\}$ is a martingale. It is an easy proposition that for any integrable Y, the family $E\{Y \mid \mathscr{G}\}$ where $\mathscr{G}$ ranges over all σ-fields of

$(\Omega, \mathscr{F}, P)$, is uniformly integrable. Here is the proof:

$$\int_{\{|E(Y|\mathscr{G})| \geq n\}} |E(Y|\mathscr{G})| \, dP \leq \int_{\{|E(Y|\mathscr{G})| \geq n\}} |Y| \, dP;$$

$$P\{|E(Y|\mathscr{G})| \geq n\} \leq \frac{1}{n} E(|Y|) \to 0.$$

Hence $\{Y_t\}$ is uniformly integrable by (8). Finally, the first term on the right side of (7) is continuous in t, hence it is progressively measurable as a process. On the other hand since $U^\alpha f \in \mathscr{E}$, it is clear that $\{U^\alpha f(X_t)\}$ as a process is progressively measurable if $\{X_t\}$ is, relative to $\{\mathscr{F}_t\}$. □

Remark. Let us change notation and put

$$A_t = \int_0^t e^{-\alpha s} f(X_s) \, ds.$$

Then $\{A_t\}$ is an increasing process, namely for each ω, $A_t(\omega)$ increases with t; and $A_\infty = Y_\infty$ by definition of the latter. The decomposition given in (6), together with (8), may be rewritten as:

$$e^{-\alpha t} U^\alpha f(X_t) = E\{A_\infty | \mathscr{F}_t\} - A_t. \tag{9}$$

This is a simple case of the Doob-Meyer decomposition of a supermartingale into a uniformly integrable martingale minus an increasing process. If we assume that the family $\{\mathscr{F}_t\}$ is right continuous, then by the corollary to Theorem 3 of §1.4, the supermartingale in (9) has a right continuous version. It is in fact a superpotential of class D (exercise).

2.2. Feller Process

Before further development of the machinery in §2.1, we turn our attention to a class of transition semigroups having nice analytic properties. We show that the sample functions of the associated Markov process have desirable regularity properties, if a proper version is taken. This will serve as a model for a more general setting in Chapter 3.

Let $\mathbf{E}$ be the state space as in §1.1, and let $\mathbf{E}_\partial = \mathbf{E} \cup \{\partial\}$ (where $\partial \notin \mathbf{E}$) be the Alexandroff one-point compactification of $\mathbf{E}$. Namely, if $\mathbf{E}$ is not compact, then a neighborhood system for the "point at infinity ∂" is the class of the complements of all compact subsets of $\mathbf{E}$; if $\mathbf{E}$ is compact, then ∂ is an isolated point in $\mathbf{E}_\partial$. $\mathbf{E}_\partial$ is a compact separable metric space. Let $\mathbb{C}$ denote the class of all continuous functions on $\mathbf{E}_\partial$. Since $\mathbf{E}_\partial$ is compact, each f in $\mathbb{C}$ is bounded. We define the usual sup-norm of f as follows:

$$\|f\| = \sup_{x \in \mathbf{E}_\partial} |f(x)|.$$

The constant function $x \to f(\partial)$ on $\mathbf{E}_\partial$ is of course in $\mathbb{C}$, and if we put

$$\forall x \in \mathbf{E}_\partial \colon f_0(x) = f(x) - f(\partial),$$

then $\lim_{x \to \partial} f_0(x) = 0$. Let $\mathbb{C}_0$ denote the subclass of $\mathbb{C}$ vanishing at ∂ ("vanishing at infinity"); $\mathbb{C}_c$ denote the subclass of $\mathbb{C}_0$ having compact supports. Recall that $\mathbb{C}$ and $\mathbb{C}_0$ are both Banach spaces with the norm $\| \ \|$; and that $\mathbb{C}_0$ is the uniform closure (or completion) of $\mathbb{C}_c$.

Let (P_t) be a submarkovian transition semigroup on $(\mathbf{E}, \mathscr{E})$. It is extended to be Markovian on $(\mathbf{E}_\partial, \mathscr{E}_\partial)$, as shown in §1.2. The following definition applies to this extension.

Definition. (P_t) is said to have the "Feller property" or "Fellerian" iff $P_0 =$ identity mapping, and

(i) $\forall f \in \mathbb{C}, t \geq 0, \qquad P_t f \in \mathbb{C};$
(ii) $\forall f \in \mathbb{C},$

$$\lim_{t \to 0} \|P_t f - f\| = 0. \tag{1}$$

It turns out that under (i) the condition in (1), which requires convergence in norm, is equivalent to the apparently weaker condition below, which requires only pointwise convergence:

(ii′) $\forall f \in \mathbb{C}, x \in E_\partial$:

$$\lim_{t \to 0} P_t f(x) = f(x). \tag{2}$$

The proof is sketched in Exercise 4 below.

Since each member of $\mathbb{C}$ is the sum of a member of $\mathbb{C}_0$ plus a constant, it is easy to see that in the conditions (i), (ii) and (ii′) above we may replace $\mathbb{C}$ by $\mathbb{C}_0$ without affecting their strength.

Theorem 1. *The function*

$$(t, x, f) \to P_t f(x)$$

on $\mathbf{T} \times \mathbf{E}_\partial \times \mathbb{C}$, *is continuous.*

Proof. Consider (t, x, f) fixed and (s, y, g) variable in the inequality below:

$$|P_t f(x) - P_s g(y)| \leq |P_t f(x) - P_t f(y)| + |P_t f(y) - P_s f(y)| + |P_s f(y) - P_s g(y)|.$$

Since $P_t f \in \mathbb{C}$, the first term converges to zero as $y \to x$; since (P_u) is Markovian, $\|P_u\| = 1$ for each u, the second term is bounded by

$$\|P_{|t-s|} f - f\|$$

which converges to zero as $|t-s| \to 0$ by (1); the third term is bounded by

$$\|P_s f - P_s g\| \le \|P_s\| \, \|f - g\| = \|f - g\|$$

which converges to zero as $g \to f$ in $\mathbb{C}$. Theorem 1 follows. □

A (homogeneous) Markov process $(X_t, \mathscr{F}_t, t \in \mathbf{T})$ on $(\mathbf{E}_\partial, \mathscr{E}_\partial)$ whose semi-group (P_t) has the Feller property is called a *Feller process*. We now begin the study of its sample function properties.

Proposition 2. *$\{X_t, t \in \mathbf{T}\}$ is stochastically continuous, namely for each $t \in \mathbf{T}$, $X_s \to X_t$ in probability as $s \to t$, $s \in \mathbf{T}$.*

Proof. Let $f \in \mathbb{C}$, $g \in \mathbb{C}$, then if $t > 0$ and $\delta > 0$ we have

$$E\{f(X_t)g(X_{t+\delta})\} = E\{f(X_t)E^{X_t}[g(X_\delta)]\} = E\{f(X_t)P_\delta g(X_t)\}$$

by the Markov property. Since $P_\delta g \in \mathbb{C}$ and $P_\delta g \to g$, we have by bounded convergence, as $\delta \downarrow 0$:

$$E\{f(X_t)g(X_{t+\delta})\} \to E\{f(X_t)g(X_t)\}. \tag{3}$$

Now if h is a continuous function on $\mathbf{E}_\partial \times \mathbf{E}_\partial$, then there exists a sequence of functions $\{h_n\}$ each of the form $\sum_{j=1}^{l_n} f_{n_j}(x)g_{n_j}(y)$ where $f_{n_j} \in \mathbb{C}$, $g_{n_j} \in \mathbb{C}$ such that $h_n \to h$ uniformly on $\mathbf{E}_\partial \times \mathbf{E}_\partial$. This is a consequence of the Stone-Weierstrass theorem, see e.g., Royden [1]. It follows from this and (3) that

$$E\{h(X_t, X_{t+\delta})\} \to E\{h(X_t, X_t)\}.$$

Take h to be a metric of the space $\mathbf{E}_\partial$. Then the limit above is equal to zero, and the result asserts that $X_{t+\delta}$ converges to X_t in probability. Next if $0 < \delta < t$, then we have for each x:

$$\begin{aligned} E^x\{f(X_{t-\delta})g(X_t)\} &= E^x\{f(X_{t-\delta})E^{X_{t-\delta}}[g(X_\delta)]\} \\ &= E^x\{f(X_{t-\delta})P_\delta g(X_{t-\delta})\} = P_{t-\delta}(fP_\delta g)(x). \end{aligned} \tag{4}$$

By Theorem 1, the last term converges as $\delta \downarrow 0$ to

$$P_t(f \cdot g)(x) = E^x\{f(X_t)g(X_t)\}.$$

It follows as before that $\forall x$:

$$E^x\{h(X_{t-\delta}, X_t)\} \to 0,$$

hence also

$$E\{h(X_{t-\delta}, X_t)\} = E\{E[h(X_{t-\delta}, X_t)|\mathscr{F}_0]\} = E\{E^{X_0}[h(X_{t-\delta}, X_t)]\} \to 0. \quad \square$$

The preceding proof affords an occasion to clarify certain obscurities. When the symbol P or E is used without a superscript, the distribution of of X_0 is unspecified. Let it be denoted by μ; thus

$$\mu(A) = P(X_0 \in A), \qquad A \in \mathscr{E}.$$

For any $\Lambda \in \mathscr{F}^0$ (not $\mathscr{F}$!), we have

$$P(\Lambda) = E\{P(\Lambda|\mathscr{F}_0)\} = E\{P^{X_0}(\Lambda)\} = \int_E \mu(dx)P^x(\Lambda) = P^\mu(\Lambda);$$

see (10) of §1.2. More generally if $Y \in L^1(\mathscr{F}^0, P)$, then

$$E(Y) = E^\mu(Y).$$

Given any probability measure μ on $\mathscr{E}$, and the transition function $\{P_t(\cdot,\cdot), t \in \mathbf{T}\}$, we can construct a Markov process with μ as its initial distribution, as reviewed in §1.2. It is assumed that such a process exists in the probability space $(\Omega, \mathscr{F}, P)$. Any statement concerning the process given in terms of P and E is therefore *ipso facto* true for any initial μ, without an explicit claimer to this effect. One may regard this usage as an editorial license to save print! For instance, with this understanding, formula (3) above contains its duplicate when E there is replaced by E^x, for each x. The resulting relation may then be written as

$$P_t(fP_\delta g)(x) \to P_t(fg)(x),$$

which may be easier to recognize than (3) itself. This is what is done in (4), since the corresponding argument for convergence as $\delta \downarrow 0$, under E instead of E^x, may be less obvious.

On the other hand, suppose that we have proved a result under P^x and E^x, for each x. Then integrating with respect to μ we obtain the result under P^μ and E^μ, for any μ. Now the identification of $P(\Lambda)$ with $P^\mu(\Lambda)$ above shows that the result is true under P and E. This is exactly what we did at the end of the proof of Proposition 2, but we have artfully concealed μ from the view there.

To sum up, for any $\Lambda \in \mathscr{F}^0$, $P(\Lambda) = 1$ is just a cryptic way of writing

$$\forall x \in \mathbf{E}\colon\ P^x(\Lambda) = 1.$$

We shall say in this case that Λ is *almost sure*. The apparently stronger statement also follows: for any probability measure μ on $\mathscr{E}$, we have $P^\mu(\Lambda) = 1$. *After Theorem* 5 *of* §2.3, *we can extend this to any* $\Lambda \in \mathscr{F}^\sim$.

The next proposition concerns the α-potential $U^\alpha f$. Let us record the following basic facts. If (P_t) is Fellerian, then

$$\forall f \in \mathbb{C}: U^\alpha f \in \mathbb{C}; \tag{5}$$

$$\forall f \in \mathbb{C}: \lim_{\alpha \to \infty} \|\alpha U^\alpha f - f\| = 0. \tag{6}$$

Since $P_t f \in \mathbb{C}$ for each t, (5) follows by dominated convergence as x tends to a limit in the first expression for $U^\alpha f$ in (3) of §2.1. To show (6), we have by a change of variables:

$$\begin{aligned}\sup_x |\alpha U^\alpha f(x) - f(x)| &\le \sup_x \int_0^\infty \alpha e^{-\alpha t} |P_t f(x) - f(x)|\, dt \\ &\le \int_0^\infty e^{-u} \|P_{u/\alpha} f - f\|\, du,\end{aligned}$$

which converges to zero as $\alpha \to \infty$ by (1) and dominated convergence. In particular, we have for each x

$$\lim_{\alpha \to \infty} \alpha U^\alpha f(x) = f(x); \tag{7}$$

this should be compared with (2). As a general guide, the properties of αU^α as $\alpha \to \infty$ and as $\alpha \to 0$ are respectively reflected in those of P_t as $t \to 0$ and $t \to \infty$. This is a known folklore in the theory of Laplace transforms, and U^α is nothing but the Laplace transform of P_t in the operator sense.

A class of functions defined in $\mathbf{E}_\partial$ is said to "separate points" iff for any two distinct points x and y in $\mathbf{E}_\partial$ there exists a member f of the class such that $f(x) \neq f(y)$. This concept is used in the Stone-Weierstrass theorem. Let $\{O_n, n \in \mathbf{N}\}$ be a countable base of the open sets of $\mathbf{E}_\partial$, and put

$$\forall x \in \mathbf{E}_\partial: \varphi_n(x) = d(x, \bar{O}_n),$$

where $d(\cdot\,, \cdot)$ denotes the distance. We know $\varphi_n \in \mathbb{C}$.

Proposition 3. *The countable subset D of $\mathbb{C}$ below separates points:*

$$D = \{U^\alpha \varphi_n : \alpha \in \mathbf{N}, n \in \mathbf{N}\}. \tag{8}$$

Proof. For any $x \neq y$, there exists O_n such that $x \in \bar{O}_n$ and $y \notin \bar{O}_n$. Hence $\varphi_n(x) = 0 < \varphi_n(y)$, namely φ_n separates x and y. We have by (7), for sufficiently large $\alpha \in \mathbf{N}$:

$$\begin{aligned}|\alpha U^\alpha \varphi_n(x) - \varphi_n(x)| &< \tfrac{1}{2}\varphi_n(y), \\ |\alpha U^\alpha \varphi_n(y) - \varphi_n(y)| &< \tfrac{1}{2}\varphi_n(y).\end{aligned}$$

Hence $\alpha U^\alpha \varphi_n(x) \neq \alpha U^\alpha \varphi_n(y)$, namely $U^\alpha \varphi_n$ separates x and y. [Of course we can use (6) to get $\|\alpha U^\alpha \varphi_n - \varphi_n\| < \frac{1}{2}\varphi_n(y)$.] □

The next is an analytical lemma separated out for the sake of clarity. We use $g \circ h$ to denote the composition of g and h; and $(g \circ h)|_S$ to denote its restriction to S.

Proposition 4. *Let $\mathbb{D}$ be a class of continuous functions from $\mathbf{E}_\partial$ to $R = (-\infty, +\infty)$ which separates points. Let h be any function on R to $\mathbf{E}_\partial$. Suppose that S is a dense subset of R such that for each $g \in \mathbb{D}$,*

$$(g \circ h)|_S \quad \textit{has right and left limits in } R. \tag{9}$$

Then $h|_S$ has right and left limits in R.

Proof. Suppose for some t, $h|_S$ does not have a right limit at t. This means that there exist $t_n \in S$, $t_n \downarrow\downarrow t$, $t'_n \in S$, $t''_n \downarrow\downarrow t$ such that

$$h(t_n) \to x \neq y \leftarrow h(t'_n).$$

There exists $g \in \mathbb{D}$ such that $g(x) \neq g(y)$. Since g is continuous, we have

$$(g \circ h)(t_n) \to g(x) \neq g(y) \leftarrow (g \circ h)(t'_n).$$

This contradicts (9) and proves the proposition since the case of the left limit is similar. □

We are ready to regulate the sample functions of a Feller process. The trick, going back to Doob, is to relate the latter with those of supermartingales obtained by composition with potentials of functions.

Proposition 5. *Let $\{X_t, \mathscr{F}_t, t \in \mathbf{T}\}$ be a Feller process, and S be any countable dense subset of $\mathbf{T}$. Then for almost every ω, the sample function $X(\cdot, \omega)$ restricted to S has right limits in $[0, \infty)$ and left limits in $(0, \infty)$.*

Proof. Let g be a member of the class $\mathbb{D}$ in (8), thus $g = U^k\varphi$ where $k \in \mathbf{N}$, $\varphi \in \mathbb{C}$. By Proposition 2 of §2.1,

$$\{e^{-kt}g(X_t), \mathscr{F}_t\}$$

is a supermartingale. Hence by Theorem 1 of §1.4, there exists Ω_g with $P(\Omega_g) = 1$, such that if $\omega \in \Omega_g$, then the sample function

$$t \to e^{-kt}g(X_t(\omega))$$

restricted to S has right and left limits as asserted. Clearly the factor e^{-kt} may be omitted in the above. Let $\Omega_* = \bigcap_{g \in \mathbb{D}} \Omega_g$, then $P(\Omega_*) = 1$. If $\omega \in \Omega_*$, then the preceding statement is true for each $g \in \mathbb{D}$. Since $\mathbb{D}$ separates points by Proposition 3, it now follows from Proposition 4 that $t \to X_t(\omega)$ restricted to S has the same property.

Now we define, for each $\omega \in \Omega_*$:

$$\forall t \geq 0: \tilde{X}_t(\omega) = \lim_{\substack{s \in S \\ s \downarrow\downarrow t}} X_s(\omega); \qquad \forall t > 0: \hat{X}_t(\omega) = \lim_{\substack{s \in S \\ s \uparrow\uparrow t}} X_s(\omega). \tag{10}$$

Elementary analysis shows that for each ω, $\tilde{X}_t(\omega)$ is right continuous in t and has left limit $\hat{X}_t$ at each $t > 0$. Similarly $\hat{X}_t$ is left continuous in t and has right limit $\tilde{X}_t$ at each $t \geq 0$. □

Theorem 6. *Suppose that each $\mathscr{F}_t$ is augmented. Then each of the processes $\{\tilde{X}_t\}$ and $\{\hat{X}_t\}$ is a version of $\{X_t\}$; hence it is a Feller process with the same transition semigroup (P_t) as $\{X_t\}$.*

Proof. By Proposition 2, for each fixed t, there exists $s_n \in S$, $s_n \downarrow\downarrow t$ such that

$$P\left\{\lim_n X_{s_n} = X_t\right\} = 1. \tag{11}$$

This is because convergence in probability implies almost sure convergence along a subsequence. But by Proposition 5, the limit in (11) is equal to $\tilde{X}_t$. Hence $P\{\tilde{X}_t = X_t\} = 1$ for each t, namely $\{\tilde{X}_t\}$ is a version of $\{X_t\}$. It follows that for each $f \in \mathbb{C}$:

$$E\{f(\tilde{X}_{s+t}) | \mathscr{F}_t\} = E\{f(X_{s+t}) | \mathscr{F}_t\} = P_s f(X_t) = P_s f(\tilde{X}_t)$$

almost surely. Augmentation of $\mathscr{F}_t$ is needed to ensure that $\tilde{X}_t \in \mathscr{F}_t$, for each t. This yields the assertion concerning $\{\tilde{X}_t\}$; the case of $\{\hat{X}_t\}$ is similar. □

Let us concentrate on the right continuous version $\tilde{X}$, and write it simply as X. Its sample functions may take ∂, namely infinity, either as a value or as a left limiting value. Even if (P_t) is strictly Markovian so that $P^x\{X_t \in E\} = P_t(x, E) = 1$ for each x and t, it does not follow that we have $X(t, \omega) \in E$ *for all* $t \in \mathbf{T}$ for any ω. In other words, it may not be bounded in each finite interval of t. The next proposition settles this question for a general submarkovian (P_t).

Theorem 7. *Let $\{X_t, \mathscr{F}_t\}$ be a Feller process with right continuous paths having left limits; and let*

$$\zeta(\omega) = \inf\{t > 0 | X_{t-}(\omega) = \partial \text{ or } X_t(\omega) = \partial\}. \tag{12}$$

Then we have almost surely $X(\zeta + t) = \partial$ for all $t \geq 0$, on the set $\{\zeta < \infty\}$.

Proof. As before let d be a metric for $\mathbf{E}_\partial$ and put

$$\varphi(x) = d(x, \partial), \qquad x \in \mathbf{E}_\partial.$$

Then φ vanishes only at ∂. Since $\varphi \in \mathbb{C}$, we have $P_t\varphi \to \varphi$ as $t \downarrow 0$, and consequently

$$U^1\varphi(x) = \int_0^\infty e^{-t}P_t\varphi(x)\,dt$$

vanishes only at ∂, like φ. Write

$$Z_t = e^{-t}\,U^1\varphi(X_t);$$

then $\{Z_t, \mathscr{F}_t\}$ is a positive supermartingale by Proposition 2 of §2.1, and $\{Z_t\}$ has right continuous paths with left limits, because $\{X_t\}$ does and $U^1\varphi$ is continuous. Furthermore $Z_{t-} = 0$ if and only if $X_{t-} = \partial$; $Z_t = 0$ if and only if $X_t = \partial$. Therefore we have

$$\zeta(\omega) = \inf\{t > 0 \,|\, Z_{t-}(\omega) = 0 \text{ or } Z_t(\omega) = 0\},$$

and the theorem follows from Theorem 4 of §1.5. □

Corollary. *If (P_t) is strictly Markovian, then almost surely the sample function is bounded in each finite t-interval. Namely not only $X(t, \omega) \in \mathbf{E}$ for all $t \geq 0$, but $X(t-, \omega) \in \mathbf{E}$ for all $t > 0$ as well.*

Proof. Under the hypothesis we have

$$P\{X_t \in \mathbf{E} \text{ for all } t \in \mathbf{T} \cap Q\} = 1,$$

It follows from the theorem that for each $t \in Q$:

$$P\{\zeta < t\} = P\{\zeta < t; X_t \in \mathbf{E}\} = 0.$$

Hence $P\{\zeta < \infty\} = 0$, which is equivalent to the assertion of the Corollary. □

Theorem 7 is a case of our discussion is §1.2 regarding ∂ as an absorbing state: cf. (13) there.

Exercises

1. Consider the Markov chain in Example 1 of §1.2, with $\mathbf{E}$ the set of positive integers, no ∂, and $\mathscr{E}$ the discrete topology on $\mathbf{E}$. Show that if $\lim_{t\downarrow 0} p_{ii}(t) = 1$ uniformly in all i, then property (1) is true for every f. Take a version which is right continuous and has left limits. Show that for each sample function

$X(\cdot,\omega)$, there is a discrete set of t at which $X(t-,\omega) \neq X(t,\omega)$, and $X(\cdot,\omega)$ is constant between two consecutive values of t in this set.

2. Show that the semigroups of Examples 2, 3 and 4 in §1.2 are all Fellerian.

3. For a Markov chain suppose that there exists an i such that $\lim_{t \downarrow 0}[1 - p_{ii}(t)]/t = +\infty$. (Such a state is called *instantaneous*. It is not particularly easy to construct such an example; see Chung [2], p. 285.) Show that for any $\delta > 0$ we have $P^i\{X(t) = i \text{ for all } t \in [0,\delta]\} = 0$. Thus under P^i, no version of the process can be right continuous at $t = 0$. Hence (P_t) cannot be Fellerian. Prove the last assertion analytically.

4. Prove that in the definition of Feller property, condition (2) implies condition (1). [Hint: by the Riesz representation theorem the dual space of $\mathbb{C}$ is the space of finite measures on $\mathbf{E}_\partial$. Use this and the Hahn-Banach thorem to show that the set of functions of the form $\{U^\alpha\varphi\}$, where $\alpha > 0$ and $\varphi \in \mathbb{C}$, is dense in $\mathbb{C}$. If f belongs to this set, (1) is true.]

2.3. Strong Markov Property and Right Continuity of Fields

We proceed to derive several fundamental properties of a Feller process. It is assumed in the following theorems that the sample functions are right continuous. The existence of left limits will not be needed until the next section. We continue to adopt the convention in §1.3 that $X_\infty = \partial$, which makes it possible to write certain formulas without restriction. But such conventions should be watched carefully.

Theorem 1. *For each optional T, we have for each $f \in \mathbb{C}$, and $u \geq 0$:*

$$E\{f(X_{T+u}) \mid \mathscr{F}_{T+}\} = P_u f(X_T). \tag{1}$$

Proof. Observe first that on $\{T = \infty\}$, both sides of (1) reduce to $f(\partial)$. Let

$$T_n = \frac{[2^n T] + 1}{2^n},$$

so that $T_n \downarrow\downarrow T$, each T_n is strictly optional, and takes values in the dyadic set D (see (19) of §1.3). Recall from Theorem 5 of §1.3 that

$$\mathscr{F}_{T+} = \bigwedge_{n=1}^{\infty} \mathscr{F}_{T_n}.$$

Hence for any $\Lambda \in \mathscr{F}_{T+}$, if we write $\Lambda_d = \Lambda \cap \{T_n = d\}$ for $d \in \mathrm{D}$, then $\Lambda_d \in \mathscr{F}_d$. Now the Markov property applied at $t = d$ yields

$$\int_{\Lambda_d} f(X_{d+u})\,dP = \int_{\Lambda_d} P_u f(X_d)\,dP,$$

see (2) of §1.2. Since $T_n = d$ on Λ_d, we have by enumerating the possible values of T_n:

$$\int_\Lambda f(X_{T_n+u})\,dP = \sum_{d\in D} \int_{\Lambda_d} f(X_{d+u})\,dP$$
$$= \sum_{d\in D} \int_{\Lambda_d} P_u f(X_d)\,dP = \int_\Lambda P_u f(X_{T_n})\,dP. \tag{2}$$

Since both f and $P_t f$ are bounded continuous, and $X(\cdot)$ is right continuous, we obtain by letting $n \to \infty$ in the first and last members of (2):

$$\int_\Lambda f(X_{T+u})\,dP = \int_\Lambda P_u f(X_T)\,dP.$$

The truth of this for each $\Lambda \in \mathscr{F}_{T+}$ is equivalent to the assertion in (1). □

When T is a constant t, the equation (1) reduces to the form of Markov property given as (iic) in §1.1, with the small but important improvement that the $\mathscr{F}_t$ there has now become $\mathscr{F}_{t+}$. The same arguments used in §1.1 to establish the equivalence of (iic) with (iib) now show that (1) remains true for $f \in b\mathscr{E}$. More generally, let us define the "post-T field" $\mathscr{F}'_T$ as follows:

$$\mathscr{F}'_T = \sigma(X_{T+u}, u \geq 0);$$

which reduces to $\mathscr{F}'_t$ when $T = t$. Then we have for each integrable Y in $\mathscr{F}'_T$:

$$E\{Y \mid \mathscr{F}_{T+}\} = E\{Y \mid X_T\}.$$

This is the extension of (iia) in §1.1. Alternatively, we have for each integrable Y in $\mathscr{F}^0$:

$$E\{Y \circ \theta_T \mid \mathscr{F}_{T+}\} = E^{X_T}\{Y\}. \tag{3}$$

Definition. The Markov process $\{X_t, \mathscr{F}_t\}$ is said to have the strong Markov property iff (3) is true for each optional time T.

Thus Theorem 1 is equivalent to the assertion that:

A Feller process with right continuous paths has the strong Markov property.

An immediate consequence will be stated as a separate theorem, which contains the preceding statement when $T \equiv 0$.

Theorem 2. *For each optional T, the process $\{X_{T+t}, \mathscr{F}_{T+t+}, t \in \mathbf{T}\}$ is a Markov process with (P_t) as transition semigroup. Furthermore, it has the strong Markov property.*

Proof. Recall that $T + t$ is optional, in fact strictly so for $t > 0$; and $X_{T+t} \in \mathscr{F}_{T+t+}$ by Theorem 10 of §1.3. Now apply (1) with T replaced by $T + t$:

$$E\{f(X_{T+t+u}) | \mathscr{F}_{T+t+}\} = P_u f(X_{T+t}). \tag{4}$$

This proves the first assertion of the theorem. As for the second, it is a consequence of the re-statement of Theorem 1 because $\{X_{T+t}, \mathscr{F}_{T+t+}\}$ is a Feller process with right continuous paths. □

There is a further extension of Theorem 1 which is useful in many applications. Indeed, the oldest case of the strong Markov property, known as Désiré André's reflection principle, requires such an extension (see Exercise 12 of §4.2).

Theorem 3. *Let $S \geq T$ and $S \in \mathscr{F}_{T+}$. Then we have for each $f \in \mathbb{C}$:*

$$E\{f(X_S) | \mathscr{F}_{T+}\} = E^{X_T}\{f(X_{S-T})\} \tag{5}$$

where $S - T$ is defined to be ∞ if $S = \infty$ (even if $T = \infty$); and $X_\infty = \partial, f(\partial) = 0$.

The quantity in the right member of (5) is the value of the function $(t, x) \to E^x\{f(X_t)\} = P_t f(x)$ at $(S(\omega) - T(\omega), X_T(\omega))$, with ω omitted. It may be written out more explicitly as

$$\int_\Omega f(X_{S(\omega)-T(\omega)}(\omega'))P^{X_{T(\omega)}}(d\omega'). \tag{6}$$

It is tempting to denote the right member of (5) by $P_{S-T}f(X_T)$ in analogy with (1), but such a notation would conflict with our later, and established, usage of the balayage operator P_T in §3.4.

Proof of Theorem 3. Observe first that (5) reduces to $f(\partial) = f(\partial)$ on $\{S = \infty\}$ by the various conventions, so we may suppose $S < \infty$ in what follows. Let

$$U_n = \frac{[2^n(S - T)] + 1}{2^n};$$

then $U_n \downarrow (S - T)$. Since S and T belong to $\mathscr{F}_{T+}$, so does U_n for each n. Consequently, for each $d \in D$ (the dyadic set) we have $\{U_n = d\} \in \mathscr{F}_{T+}$. Now let Λ be any set in $\mathscr{F}_{T+}$, and put $\Lambda_d = \Lambda \cap \{U_n = d\} \in \mathscr{F}_{T+}$. We have for each $f \in \mathbb{C}$:

$$\begin{aligned}\int_\Lambda f(X_{T+U_n})\,dP &= \sum_{d \in D} \int_{\Lambda_d} f(X_{T+d})\,dP \\ &= \sum_{d \in D} \int_{\Lambda_d} P_d f(X_T)\,dP = \int_\Lambda P_{U_n} f(X_T)\,dP\end{aligned} \tag{7}$$

where the second equation is by (1). Now $t \to P_t f(x)$ is continuous for each x and $f \in \mathbb{C}$ by Theorem 1 of §2.2. Letting $n \to \infty$ in the first and last members of (7), we obtain by right continuity:

$$\int_A f(X_S)\,dP = \int_A P_{S-T} f(X_T)\,dP.$$

This is equivalent to (5). □

The next theorem demands a thorough understanding of the concepts of "completion" and "augmentation". For the first notion see Theorem 2.2.5 of the *Course*. The second notion will now be discussed in detail.

Let $(\Omega, \mathscr{F}, P)$ be a complete probability space. The class of sets C in $\mathscr{F}$ with $P(C) = 0$ will be denoted by $\mathscr{N}$ and called the P-null sets. Completeness means: a subset of a P-null set is a P-null set. Let $\mathscr{G}$ be a sub-σ-field of $\mathscr{F}$. Put

$$\mathscr{G}^* = \sigma(\mathscr{G} \vee \mathscr{N}),$$

namely $\mathscr{G}^*$ is the minimal σ-field including $\mathscr{G}$ and $\mathscr{N}$. $\mathscr{G}^*$ is called the *augmentation* of $\mathscr{G}$ with respect to $(\Omega, \mathscr{F}, P)$. It is characterized in the following proposition, in which a "function" is from Ω into $[-\infty, +\infty]$. Recall that for a function Y on $\mathbf{E}$, $Y \in \mathscr{G}^*$ is an abbreviation of $Y \in \mathscr{G}^*/\mathscr{E}$ where $\mathscr{E}$ is the Borel field of $\mathbf{E}$.

Lemma. *A subset A of Ω belongs to $\mathscr{G}^*$ if and only if there exists $B \in \mathscr{G}$ such that $A \triangle B \in \mathscr{N}$. A function Y belongs to $\mathscr{G}^*$ if and only if there exists a function $Z \in \mathscr{G}$ such that $\{Y \neq Z\} \in \mathscr{N}$.*

Proof. Let $\mathscr{A}$ denote the class of sets A described in the lemma. Observe that $A \triangle B = C$ is equivalent to $A = B \triangle C$ for arbitrary sets A, B and C. Since $\mathscr{G}^*$ contains all sets of the form $B \triangle C$ where $B \in \mathscr{G}$ and $C \in \mathscr{N}$, it is clear that $\mathscr{G}^* \supset \mathscr{A}$. To show that $\mathscr{A} \supset \mathscr{G}^*$ it is sufficient to verify that $\mathscr{A}$ is a σ-field since it obviously includes both $\mathscr{G}$ and $\mathscr{N}$. Since $A^c \Delta B^c = A \Delta B$, $\mathscr{A}$ is closed under complementation. It is also closed under countable union by the inclusion relation

$$\left(\bigcup_n A_n\right) \triangle \left(\bigcup_n B_n\right) \subset \bigcup_n (A_n \triangle B_n),$$

and the remark that if the right member above is in $\mathscr{N}$, then so is the left member. Thus $\mathscr{A}$ is a σ-field and we have proved the first sentence of the Lemma.

Next, let $Z \in \mathscr{G}$ and $\{Y \neq Z\} \in \mathscr{N}$. Then for any subset S of $[-\infty, +\infty]$, we have the trivial inclusion

$$\{Y \in S\} \triangle \{Z \in S\} \subset \{Y \neq Z\}.$$

For each Borelian S, we have $\{Z \in S\} \in \mathscr{G}$; hence $\{Y \in S\} \in \mathscr{G}^*$ by what we have proved, and so $Y \in \mathscr{G}^*$. Conversely, let $Y \in \mathscr{G}^*$, and Y be countably-valued. Namely Y is of the form $\sum_{j=1}^{\infty} c_j 1_{A_j}$, where the c_j's are points in $[-\infty, +\infty]$, and the A_j's are disjoint sets in $\mathscr{G}^*$. Then for each j there exists $B_j \in \mathscr{G}$ such that $A_j \triangle B_j \in \mathscr{N}$. Since the A_j's are disjoint, $B_i \cap B_j \in \mathscr{N}$ if $i \neq j$. Put $B'_1 = B_1$, and $B'_j = B_1^c \cdots B_{j-1}^c B_j$ for $j \geq 2$, then $B_j \triangle B'_j \in \mathscr{N}$ and so $A_j \triangle B'_j \in \mathscr{N}$. The B'_j's are disjoint. Put $Z = \sum_{j=1}^{\infty} c_j 1_{B'_j}$. Then $Z \in \mathscr{G}$ and

$$\{Y \neq Z\} \in \bigcup_{j=1}^{\infty} \{A_j \triangle B'_j\} \in \mathscr{N}.$$

Thus Y is of the form described in the second sentence of the lemma. If $Y \in \mathscr{G}^*$ there exists countably valued $Y^{(k)} \in \mathscr{G}^*$ such that $Y^{(k)} \to Y$. Let $Z^{(k)} \in \mathscr{G}$ such that $\{Y^{(k)} \neq Z^{(k)}\} \in \mathscr{N}$; and $Z = \underline{\lim}_{k \to \infty} Z^{(k)}$. Then $Z \in \mathscr{G}$ and

$$\{Y \neq Z\} \subset \bigcup_{k=1}^{\infty} \{Y^{(k)} \neq Z^{(k)}\} \in \mathscr{N}. \qquad \square$$

Remark. If $(\Omega, \mathscr{F}, P)$ is not complete, the Lemma is true *provided* that we assume $Y \in \mathscr{F}$ in the second assertion. We leave this to the interested reader.

Corollary. $(\mathscr{G}^*)^* = \mathscr{G}^*$. *$(\Omega, \mathscr{G}^*, P)$ is complete.*

We now review the definition of a conditional expectation, although this has been used hundreds of times above. Let $Y \in \mathscr{F}$ and $E(|Y|) < \infty$. Then the conditional expectations $E(Y|\mathscr{G})$ is the class of functions Z having the following properties:

(a) $Z \in \mathscr{G}$;
(b) for each $W \in b\mathscr{G}$, we have

$$E(WZ) = E(WY). \tag{8}$$

Note that (8) is then also true for $W \in b\mathscr{G}^*$. It follows that if Z is a member of the class $E(Y|\mathscr{G})$, and Z^* is a member of the class $E(Y|\mathscr{G}^*)$, then $\{Z \neq Z^*\} \in \mathscr{N}$. We state this as follows:

$$E(Y|\mathscr{G}) \equiv E(Y|\mathscr{G}^*), \qquad \text{mod } \mathscr{N}. \tag{9}$$

We are ready to return to (1) above. First put $T \equiv t$, then shrink $\mathscr{F}_{t+}$ to $\mathscr{F}_{t+}^0$ as we may:

$$E\{f(X_{t+u})|\mathscr{F}_{t+}^0\} = P_u f(X_t). \tag{10}$$

We have seen that there is advantage in augmenting the σ-fields $\mathscr{F}_{t+}^0$ appearing in (10). Since $\mathscr{F}_{t+}^0 \supset \mathscr{F}_t^0 \supset \mathscr{F}_0^0$, it is sufficient to augment $\mathscr{F}_0^0$ to achieve this, for then $\mathscr{N}$ will be included in it, and *a fortiori* in the others.

Theorem 4. *Suppose that* $\mathscr{F}_0^0$ *is augmented, then the family* $\{\mathscr{F}_t^0, t \in \mathbf{T}\}$ *is right continuous.*

Proof. According to (10), its right member, say Z_1, is a representative of its left member. If Z_2 is any other representative of the latter, then $\{Z_1 \neq Z_2\} \in \mathscr{N}$. This is a consequence of the definition of the conditional expectation which implies its uniqueness mod $\mathscr{N}$. Hence $Z_2 \in \mathscr{F}_t^0$ by the Lemma. Thus the entire class in the left member of (10) belongs to $\mathscr{F}_t^0$. Now the same argument as used in §1.1 to show that (iic) implies (iia) there yields, for any $Y \in b\mathscr{F}_t'$:

$$E\{Y \mid \mathscr{F}_{t+}^0\} \in \mathscr{F}_t^0. \tag{11}$$

Let $Y_0 \in b\mathscr{F}_t^0$, then a representative of $E\{Y_0 \mid \mathscr{F}_{t+}^0\}$ is Y_0 itself because $\mathscr{F}_t^0 \subset \mathscr{F}_{t+}^0$. It follows as before that

$$E\{Y_0 \mid \mathscr{F}_{t+}^0\} \in \mathscr{F}_t^0. \tag{12}$$

Lemma 2 of §1.1 may be used to show that the class of sets Y satisfying (11) constitutes a σ-field. We have just seen that this class includes $\mathscr{F}_t^0$ and $\mathscr{F}_t'$. Since $\mathscr{F}^0 = \sigma(\mathscr{F}_t^0, \mathscr{F}_t')$ for each t, the class must also include $\mathscr{F}^0$. Hence (11) is true for each $Y \in \mathscr{F}^0$. For any $Y \in \mathscr{F}_{t+}^0$ $(\subset \mathscr{F}^0)$, a representative of $E\{Y \mid \mathscr{F}_{t+}^0\}$ is Y itself. Hence $Y \in \mathscr{F}_t^0$ by (11), and this means $\mathscr{F}_{t+}^0 \subset \mathscr{F}_t^0$. Therefore $\mathscr{F}_{t+}^0 = \mathscr{F}_t^0$ as asserted. □

According to the discussion in §2.2, Theorem 4 may be read as follows. For each probability measure μ on $\mathscr{E}$, let $(\Omega, \mathscr{F}^\mu, P^\mu)$ be the completion of $(\Omega, \mathscr{F}^0, P^\mu)$; and let $\mathscr{F}_t^\mu$ denote the augmentation of $\mathscr{F}_t^0$ with respect to $(\Omega, \mathscr{F}^\mu, P^\mu)$. [Note that if $Y \in \mathscr{F}^\mu$, then $E^\mu(Y)$ is defined as $E^\mu(Z)$ where $Z \in \mathscr{F}^0$ and $P^\mu\{Y \neq Z\} = 0$.] Then $\{\mathscr{F}_t^\mu, t \in \mathbf{T}\}$ is right continuous. Furthermore, we have as a duplicate of (10):

$$E^\mu\{f(X_{t+u}) \mid \mathscr{F}_t^\mu\} = P_u f(X_t),$$

and more generally for $Y \in b\mathscr{F}^0$, and each T optional relative to $\{\mathscr{F}_t^\mu\}$:

$$E^\mu\{Y \circ \theta_T \mid \mathscr{F}_{T+}^\mu\} = E^{X_T}\{Y\}; \tag{13}$$

cf. (3) above. The notation E^μ indicates that the conditional expectation is with respect to $(\Omega, \mathscr{F}^\mu, P^\mu)$; and any two representatives of the left number in

(13) differs by a P^μ-null function. It will be shown below (in the proof of Theorem 5) that (13) is also true for $Y \in b\mathscr{F}^\mu$. Now if $Y \in \mathscr{F}^0$, then $Y \circ \theta_t \in \mathscr{F}^0$. But if $Y \in \mathscr{F}^\mu$, we cannot conclude that $Y \circ \theta_t \in \mathscr{F}^\mu$, even if $Y = f(X_u)$. This difficulty will be resolved by the next step.

Namely, we introduce a smaller σ-field

$$\mathscr{F}^\sim = \bigwedge_\mu \mathscr{F}^\mu \tag{14}$$

and correspondingly for each $t \geq 0$:

$$\mathscr{F}_t^\sim = \bigwedge_\mu \mathscr{F}_t^\mu. \tag{15}$$

In both (14) and (15), μ ranges over all finite measures on $\mathscr{E}$. It is easy to see that the result is the same if μ ranges over all σ-finite measures on $\mathscr{E}$. Leaving aside a string of questions regarding these fields (see Exercises), we state the useful result as follows.

Corollary to Theorem 4. *The family* $\{\mathscr{F}_t^\sim, t \in \mathbf{T}\}$ *is right continuous, as well as the family* $\{\mathscr{F}_t^\mu, t \in \mathbf{T}\}$ *for each* μ.

Proof. We have already proved the right continuity of $\{\mathscr{F}_t^\mu\}$ and will use this to prove that of $\{\mathscr{F}_t^\sim\}$. Let $S(\mu, s)$ be a class of sets in any space, for each μ and s in any index sets. Then the following is a trivial set-theoretic identity:

$$\bigwedge_\mu \bigwedge_s S(\mu, s) = \bigwedge_s \bigwedge_\mu S(\mu, s). \tag{16}$$

Applying this to $S(\mu, s) = \mathscr{F}_s^\mu$, for all finite measures μ on $\mathscr{E}$, and all $s \in (t, \infty)$, we obtain

$$\bigwedge_\mu (\mathscr{F}_t^\mu)_+ = (\mathscr{F}_t^\sim)_+ \tag{17}$$

where

$$(\mathscr{F}_t^\mu)_+ = \bigwedge_{s>t} \mathscr{F}_s^\mu, \qquad (\mathscr{F}_t^\sim)_+ = \bigwedge_{s>t} \mathscr{F}_s^\sim.$$

These are to be distinguished from $(\mathscr{F}_{t+})^\mu$ and $(\mathscr{F}_{t+})^\sim = \bigwedge_\mu (\mathscr{F}_{t+})^\mu$, but see Exercise 6 below. Since $(\mathscr{F}_t^\mu)_+ = \mathscr{F}_t^\mu$, (17) reduces to $\mathscr{F}_t^\sim = (\mathscr{F}_t^\sim)_+$ as asserted. □

After this travail, we can now extend (13) to Y in $b\mathscr{F}^\sim$. But first we must consider the function $x \to E^x\{Y\}$. We know this is a Borel function if $Y \in b\mathscr{F}^0$; what can we say if $Y \in b\mathscr{F}^\sim$? The reader will do well to hark back to the theory of Lebesgue measure in Euclidean spaces for a clue. There we may begin with the σ-field $\mathscr{B}$ of classical Borel sets, and then complete $\mathscr{B}$ with respect to the Lebesgue measure. But we can also complete $\mathscr{B}$ with respect to other measures defined on $\mathscr{B}$. In an analogous manner, beginning with $\mathscr{F}^0$, we have completed it with respect to P^μ and called the result $\mathscr{F}^\mu$. Then we have taken the intersection of all these completions and called it $\mathscr{F}^\sim$. Exactly the same procedure can be used to complete $\mathscr{E}_\partial$ (the Borel field of $\mathbf{E}_\partial$) with respect to any finite measure μ defined on $\mathscr{E}$. The result will be denoted by $\mathscr{E}^\mu$; we then put

$$\mathscr{E}^\sim = \bigwedge_\mu \mathscr{E}^\mu. \tag{18}$$

As before, we may regard $\mathscr{E}^\mu$ and $\mathscr{E}$ as classes of functions defined on $\mathbf{E}$, as well as subsets of $\mathbf{E}$ which will be identified with their indicators. A function in $\mathscr{E}^\mu$ is called *μ-measurable*; a function in $\mathscr{E}^\sim$ is called *universally measurable.* A function f is μ-null iff $\mu(\{f \neq 0\}) = 0$. See Exercise 3 below for an instructive example.

The next theorem puts together the various extensions we had in mind. Note that $\mathscr{F}_T^\mu = (\mathscr{F}_T^\mu)_+$ as a consequence of the Corollary to Theorem 4.

Theorem 5. *If $Y \in b\mathscr{F}^\sim$, then the function φ defined on $\mathbf{E}$ by*

$$\varphi(x) = E^x(Y) \tag{19}$$

is in $\mathscr{E}^\sim$. For each μ and each T optional relative to $\{\mathscr{F}_t^\mu\}$, we have

$$Y \circ \theta_T \in \mathscr{F}^\mu, \tag{20}$$

$$E^\mu\{Y \circ \theta_T \mid \mathscr{F}_T^\mu\} = \varphi(X_T). \tag{21}$$

Proof. Observe first that for each x, since $Y \in \mathscr{F}^{\varepsilon_x}$, $E^x(Y)$ is defined. Next, it follows easily from the definition of the completion $\mathscr{F}^\mu$ that $Y \in b\mathscr{F}^\mu$ if and only if there exist Y_1 and Y_2 in $b\mathscr{F}^0$ such that $P^\mu\{Y_1 \neq Y_2\} = 0$ and $Y_1 \leq Y \leq Y_2$. Hence if we put

$$\varphi_i(x) = E^x(Y_i), \qquad i = 1, 2;$$

φ_1 and φ_2 are in $\mathscr{E}$, $\varphi_1 \leq \varphi \leq \varphi_2$ and $\mu(\{\varphi_1 \neq \varphi_2\}) = 0$. Thus $\varphi \in \mathscr{E}^\mu$ by definition. This being true for each μ, we have $\varphi \in \mathscr{E}^\sim$. Next, define a measure ν on $\mathscr{E}_\partial$ as follows, for each $f \in b\mathscr{E}$:

$$\nu(f) = \int \mu(dx) E^x\{f(X_T)\} = E^\mu\{f(X_T)\}.$$

This is the distribution of X_T when μ is the distribution of X_0. Thus for any $Z \in b\mathscr{F}^0$, we have

$$E^\mu\{Z \circ \theta_T\} = E^\mu\{E^{X_T}(Z)\} = E^\nu(Z). \tag{22}$$

If $Y \in \mathscr{F}^{\sim}$, then there exists Z_1 and Z_2 in $\mathscr{F}^0$ such that $P^{\nu}\{Z_1 \neq Z_2\} = 0$ and $Z_1 \leq Y \leq Z_2$. We have $Z_i \circ \theta_T \in \mathscr{F}^{\mu}$ by (22) of §1.3, with $\{\mathscr{F}_t^{\mu}\}$ for $\{\mathscr{F}_t\}$ (recall $\mathscr{F}^{\mu} = \mathscr{F}_{\infty}^{\mu}$), and

$$P^{\mu}\{Z_1 \circ \theta_T \neq Z_2 \circ \theta_T\} = P^{\nu}\{Z_1 \neq Z_2\} = 0 \tag{23}$$

by (22) above. This and

$$Z_1 \circ \theta_T \leq Y \circ \theta_T \leq Z_2 \circ \theta_T \tag{24}$$

imply (20). Furthermore, we have by (24):

$$E^{\mu}\{Z_1 \circ \theta_T | \mathscr{F}_T^{\mu}\} \leq E^{\mu}\{Y \circ \theta_T | \mathscr{F}_T^{\mu}\} \leq E^{\mu}\{Z_2 \circ \theta_T | \mathscr{F}_T^{\mu}\}. \tag{25}$$

Now put $\psi_i(x) = E^x(Z_i)$, $i = 1, 2$. Since $Z_i \in \mathscr{F}^0$, $\psi_i(X_T)$ is a representative of $E^{\mu}\{Z_i \circ \theta_T | \mathscr{F}_T^{\mu}\}$ by (13). On the other hand, we have $\psi_1 \leq \varphi \leq \psi_2$, hence

$$\psi_1(X_T) \leq \varphi(X_T) \leq \psi_2(X_T); \tag{26}$$

and

$$P^{\mu}\{\psi_2(X_T) \neq \psi_1(X_T)\} = P^{\mu}\{Z_2 \circ \theta_T \neq Z_1 \circ \theta_T\} = 0. \tag{27}$$

Comparing (25) and (26) we conclude that $\varphi(X_T)$ is a representative of $E\{Y \circ \theta_T | \mathscr{F}_T^{\mu}\}$. This is the assertion (21). □

We end this section with a simple but important result which is often called Blumenthal's zero-or-one law.

Theorem 6. *Let $\Lambda \in \mathscr{F}_0^{\sim}$; then for each x we have $P^x(\Lambda) = 0$ or $P^x(\Lambda) = 1$.*

Proof. Suppose first that $\Lambda \in \mathscr{F}_0^0$. Then $\Lambda = X_0^{-1}(A)$ for some $A \in \mathscr{E}$. Since $P^x\{X_0 = x\} = 1$, we have

$$P^x(\Lambda) = P^x(X_0^{-1}(A)) = 1_A(x),$$

which can only take the value 0 or 1. If $\Lambda \in \mathscr{F}_0^{\sim}$, then for each x, there exists Λ^x in $\mathscr{F}_0^0$ such that $P^x(\Lambda \triangle \Lambda^x) = 0$, so that $P^x(\Lambda) = P^x(\Lambda^x)$ which is either 0 or 1 as just proved. □

If we think that $\mathscr{F}_0^0$ is a small σ-field, and $\mathscr{F}_0^{\sim}$ is only "trivially" larger, the result appears to be innocuous. Actually it receives its strength from the fact that $\mathscr{F}_0^{\sim} = (\mathscr{F}^{\sim})_{0+}$, as part of the Corollary to Theorem 4. Whether an event will occur "instantly" may be portentous. For instance, let T be optional relative to $\{\mathscr{F}_t^{\sim}, t \in \mathbf{T}\}$, then $\{T = 0\} \in \mathscr{F}_0^{\sim}$ and consequently the event $\{T = 0\}$ has probability 0 or 1 under each P^x. When T is the hitting time of a set, this dichotomy leads to a basic notion in potential theory (see §3.4)—the regularity of a point for a set or the thinness of a set at a point. For a Brownian motion on the line, it yields an instant proof that almost every sample function starting at 0 must change sign infinitely many times in any small time interval.

Exercises

1. Prove that if $\forall t \geq 0$: $\mathscr{F}_t = \mathscr{F}_{t+}$, then for any T which is optional relative to $\{\mathscr{F}_t\}$, we have $\forall t \geq 0$: $\mathscr{F}_{T+t} = \mathscr{F}_{T+t+}$.

2. Let N^* denote the class of all sets in $\mathscr{F}^0$ such that $P^x(\Lambda) = 0$ for every $x \in \mathbf{E}$. Let $\mathscr{F}^* = \sigma(\mathscr{F}^0 \vee N^*)$. Show that $\mathscr{F}^* \subset \mathscr{F}^\sim$. The σ-field $\mathscr{F}^*$ is not to be confused with $\mathscr{F}^\sim$; see Problem 3.

3. Let $\Omega = R^1$ and $\mathbf{E} = R^1$. Define $X_t(\omega) = \omega + t$; then $\{X_t, t \geq 0\}$ is the uniform motion (Example 2 of §1.2). Show that $\mathscr{F}_t^0 = \mathscr{B}^1$, the usual Borel field on R^1, and $P^x = \varepsilon_x$, the point mass at x. Show that

$$\bigwedge_{x \in R^1} (\mathscr{F}^0)^{\varepsilon_x} \quad \text{is the class of all subsets of } R^1;$$

$$\bigwedge_{\mu} (\mathscr{F}^0)^{\mu} \quad \text{is the class of universally measurable sets of } R^1$$

where μ ranges over all probability measures on $\mathscr{B}^1$. The only member of class N^* defined in Problem 2 is the empty set so that $\mathscr{F}^* = \mathscr{B}^1$. (This example is due to Getoor.)

4. Show that

$$\bigvee_{t \in \mathbf{T}} \mathscr{F}_t^\mu = \mathscr{F}^\mu.$$

Is it true that

$$\bigvee_{t \in \mathbf{T}} \mathscr{F}_t^\sim = \mathscr{F}^\sim ?$$

5. For each n let $\mathscr{G}_n$ be a class of positive (≥ 0) functions closed under the operation of taking the lim sup of a sequence; let $\mathscr{H}$ be a class of positive functions closed under the operation of taking the lim inf of a sequence. Denote by $\mathscr{G}_n + \mathscr{H}$ the class of functions of the form $g_n + h$ where $g_n \in \mathscr{G}_n$ and $h \in \mathscr{H}$. Suppose that $\mathscr{G}_n \supset \mathscr{G}_{n+1}$ for every n and $\mathscr{G} = \bigcap_{n=1}^\infty \mathscr{G}_n$. Prove that

$$\bigcap_n (\mathscr{G}_n + \mathscr{H}) = \mathscr{G} + \mathscr{H}.$$

[Hint: if f belongs to the class in the left member, show that $f = \limsup g_n + \liminf h_n$ where $g_n \in \mathscr{G}_n$, $h_n \in \mathscr{H}$; show that $\limsup g_n \in \mathscr{G}$.]

6. Apply Problem 5 to prove that

$$\bigwedge_{s>t} (\mathscr{F}_s^0)^\mu = \left(\bigwedge_{s>t} \mathscr{F}_s^0\right)^\mu ;$$

namely that the two operations of "taking the intersection over (t, ∞)" and "augmentation with respect to P^μ" performed on $\{\mathscr{F}_s^0, s \in \mathbf{T}\}$ are commutative.

7. If both S and T are optional relative to $\{\mathscr{F}_t^\sim\}$, then so is $T + S \circ \theta_T$. [Cf. Theorem 11 of §1.3.]

2.4. Moderate Markov Property and Quasi Left Continuity

From now on we write $\mathscr{F}_t$ for $\mathscr{F}_t^\sim$ and $\mathscr{F}$ for $\mathscr{F}^\sim$, and consider the Feller process $(X_t, \mathscr{F}_t, P^\mu)$ in the probability space $(\Omega, \mathscr{F}, P^\mu)$ for an arbitrary fixed probability measure μ on $\mathscr{E}$. We shall omit the superscript μ in what follows unless the occasion requires the explicit use. For each ω in Ω, the sample function $X(\cdot, \omega)$ is right continuous in $\mathbf{T} = [0, \infty)$ and has left limits in $(0, \infty)$. The family $\{\mathscr{F}_t, t \in \mathbf{T}\}$ is right continuous; hence each optional time T is strictly optional and the pre-T field is $\mathscr{F}_T (= \mathscr{F}_{T+})$. The process has the strong Marov property. Recall that the latter is a consequence of the right continuity of paths without the intervention of left limits. We now proceed to investigate the left limits. We begin with a lemma separated out for its general utility, and valid in any space $(\Omega, \mathscr{F}, P)$.

Lemma 1. *Let $\mathscr{G}$ be a sub-σ-field of $\mathscr{F}$ and $X \in \mathscr{G}$; $Y \in \mathscr{F}$. Suppose that for each $f \in \mathbb{C}_0$ we have*

$$E\{f(Y) \mid \mathscr{G}\} = f(X). \tag{1}$$

Then $P\{X = Y\} = 1$.

Proof. It follows from (1) that for each open set U:

$$P\{1_U(Y) \mid \mathscr{G}\} = 1_U(X); \tag{2}$$

see Lemma 1 of §1.1. Take $\Lambda = \{X \notin U\}$, and integrate (2) over Λ:

$$P\{Y \in U; X \notin U\} = P\{X \in U; X \notin U\} = 0. \tag{3}$$

Let $\{U_n\}$ be a countable base of the topology; then we have by (3):

$$P\{X \neq Y\} \leq \sum_n P\{Y \in U_n; X \notin U_n\} = 0. \qquad \square$$

Corollary. *Let X and Y be two random variables such that for any $f \in \mathbb{C}_c$, $g \in \mathbb{C}_c$ we have*

$$E\{f(X)g(X)\} = E\{f(Y)g(X)\}. \tag{4}$$

Then $P\{X = Y\} = 1$.

Proof. As before (4) is true if $g = 1_A$, where A is any open set. Then it is also true if A is any Borel set, by Lemma 2 of §1.1. Thus we have

$$\int_{\{X \in A\}} f(X)\,dP = \int_{\{X \in A\}} f(Y)\,dP;$$

and consequently

$$f(X) = E\{f(X)\,|\,\mathscr{G}\} = E\{f(Y)\,|\,\mathscr{G}\},$$

where $\mathscr{G} = \sigma(X)$, the σ-field generated by X. Therefore the corollary follows from Lemma 1. □

The next lemma is also general, and will acquire further significance shortly.

Lemma 2. *Let $\{X_t, t \in R\}$ be an arbitrary stochastic process which is stochastically continuous. If almost every sample function has right and left limits everywhere in R, then for each $t \in R$, X is continuous at t almost surely.*

Proof. The assertion means that for each t,

$$P\left\{\lim_{s \to t} X_s = X_t\right\} = 1.$$

Since X is stochastically continuous at t, there exist $s_n \uparrow\uparrow t$ and $t_n \downarrow\downarrow t$ such that almost surely we have

$$\lim_n X_{s_n} = X_t = \lim_n X_{t_n}.$$

But the limits above are respectively X_{t-} and X_{t+} since the latter are assumed to exist almost surely. Hence $P\{X_{t-} = X_t = X_{t+}\} = 1$ which is the assertion. □

Remark. It is sufficient to assume that X_{t-} and X_{t+} exist almost surely, for each t.

The property we have just proved is sometimes referred to as follows: "the process has no fixed time of discontinuity." This implies stochastic continuity but is not implied by it. For example, the Markov chain in Example 1 of §1.2, satisfying condition (d) and with a *finite* state space **E**, has this property. But it may not if **E** is infinite (when there are "instantaneous states"). On the other hand, the simple Poisson process (Example 3 of §1.2) has this property even though it is a Markov chain on an infinite **E**. Of course, the said property is weaker than the "almost sure continuity" of sample functions, which means that almost every sample function is a continuous function. The latter is a rather special situation in the theory of Markov processes, but is exemplified by the Brownian motion (Example 4 of §1.2) and will be discussed in §3.1.

The next result is the left-handed companion to Theorem 1 of §2.3. The reader should observe the careful handling of the values 0 and ∞ for the T below—often a nuisance which cannot be shrugged off.

Theorem 3. *For each predictable T, we have for each $f \in \mathbb{C}$, and $u \geq 0$:*

$$E\{f(X_{T+u})1_{\{T<\infty\}} | \mathscr{F}_{T-}\} = P_u f(X_{T-})1_{\{T<\infty\}} \tag{5}$$

where $X_{0-} = X_0$.

Proof. Remember that $\lim_{t\to\infty} X_t$ need not exist at all, so that X_{T-} is undefined on $\{T = \infty\}$. Since $T \in \mathscr{F}_{T-}$, we may erase the two appearances of $1_{\{T<\infty\}}$ in (5) and state the result "on the set $\{T < \infty\}$". However, since T does not necessarily belong to $\mathscr{F}_{T_n}$, we are not allowed to argue blithely below "as if T is finite".

Let $\{T_n\}$ announce T, then we have by Theorem 1 of §2.3:

$$E\{f(X_{T_n+u}) | \mathscr{F}_{T_n}\} = P_u f(X_{T_n}). \tag{6}$$

Since $T_n \in \mathscr{F}_{T_n}$, we may multiply both members of (6) by $1_{\{T_n \leq M\}}$, where M is a positive number, to obtain

$$E\{f(X_{T_n+u})1_{\{T_n\leq M\}} | \mathscr{F}_{T_n}\} = P_u f(X_{T_n})1_{\{T_n\leq M\}}. \tag{7}$$

Since $T_n = 0$ on $\{T = 0\}$, $T_n \uparrow\uparrow T$ on $\{T > 0\}$, and f is continuous, we have

$$\lim_{n\to\infty} f(X_{T_n+u}) = \begin{cases} f(X_u) & \text{on } \{T = 0\}, \\ f(X_{T+u-}) & \text{on } \{T > 0\}. \end{cases}$$

Fortunately, $X_u = X_{u-}$ almost surely in view of Lemma 2, and so we can unify the result as $f(X_{T+u-})$ on Ω. On the right side of (7), we have similarly

$$\lim_{n\to\infty} P_u f(X_{T_n}) = P_u f(X_{T-}),$$

because $X_0 = X_{0-}$ by convention on $\{T = 0\}$, and $P_u f$ is continuous. Since also $\mathscr{F}_{T_n}$ increases to $\mathscr{F}_{T-}$ by Theorem 8 of §1.3, and f is bounded, we can take the limit in n simultaneously in the integrand and in the conditioning field in the left side of (7), by a martingale convergence theorem (see Theorem 9.4.8 of *Course*). Thus in the limit (7) becomes

$$E\{f(X_{T+u-})1_{\{T\leq M\}} | \mathscr{F}_{T-}\} = P_u f(X_{T-})1_{\{T\leq M\}}. \tag{8}$$

Letting $M \to \infty$ we obtain (5) with X_{T+u-} replacing X_{T+u} in the left member.

Next, letting $u \downarrow\downarrow 0$, since $\lim_{u\downarrow\downarrow 0} X_{T+u-} = X_T$ by right continuity and $P_u f \to f$ by Feller property, we obtain

$$E\{f(X_T)1_{\{T<\infty\}} | \mathscr{F}_{T-}\} = f(X_{T-})1_{\{T<\infty\}}.$$

Now *define* $X_{\infty-}$ (as well as X_∞) to be ∂ even where $\lim_{t\to\infty} X_t$ exists and is not equal to ∂. Then the factor $1_{\{T<\infty\}}$ in the above may be cancelled. Since $X_{T-} \in \mathscr{F}_{T-}$ by Theorem 10 of §1.3, it now follows from Lemma 1 that we have almost surely:

$$X_T = X_{T-}. \tag{9}$$

This is then true on $\{T < \infty\}$ without the arbitrary fixing of the value of $X_{\infty-}$.

For each optional T and $u > 0$, $T + u$ is predictable. Hence we have just proved that $X_{T+u} = X_{T+u-}$ almost surely on $\{T < \infty\}$. This remark enables us to replace X_{T+u-} by X_{T+u} in the left member of (8), thus completing the proof of Theorem 3. □

Corollary. *For* $Y \in b\mathscr{F}^u$,

$$E\{Y \circ \theta_T | \mathscr{F}_{T-}\} = E^{X_{T-}}\{Y\} \quad on\ \{T < \infty\}. \tag{10}$$

The details of the proof are left as an exercise. The similarity between (10) above and (13) of §2.3 prompts the following definition.

Definition. The Markov process $\{X_t, \mathscr{F}_t\}$ is said to have the moderate Markov property iff almost all sample functions have left limits and (10) is true for each predictable time T.

Thus, this is the case for a Feller process whose sample functions are right continuous and have left limits. The adjective "moderate" is used here in default of a better one; it does not connote "weaker than the strong".

There is an important supplement to Theorem 3 which is very useful in applications because it does not require T to be predictable. Let $\{T_n\}$ be a sequence of optional times, increasing (loosely) to T. Then T is optional but not necessarily predictable. Now for each ω in Ω, the convergence of $T_n(\omega)$ to $T(\omega)$ may happen in two different ways as described below.

Case (i). $\forall n: T_n < T$. In this case $X_{T_n} \to X_{T-}$ if $T < \infty$.

Case (ii). $\exists n_0: T_{n_0} = T$. In this case $X_{T_n} = X_T$ for $n \geq n_0$.

Heuristically speaking, T has the predictable character on the set of ω for which case (i) is true. Hence the result (9) above seems to imply that in both cases we have $X_{T_n} \to X_T$ on $\{T < \infty\}$. The difficulty lies in that we do not know whether we can apply the character of predictability on a part of Ω without regard to the other part. This problem is resolved in the following proof by constructing a truly predictable time, which coincides with the given optional time on the part of Ω in question.

Theorem 4. *Let T_n be optional and increase to T. Then we have almost surely*

$$\lim_n X_{T_n} = X_T \quad on\ \{T < \infty\}. \tag{11}$$

Proof. Put

$$\Lambda = \{\forall n: T_n < T\};$$

and define

$$T'_n = \begin{cases} T_n & \text{on } \{T_n < T\}, \\ \infty & \text{on } \{T_n = T\}; \end{cases} \qquad T' = \begin{cases} T & \text{on } \Lambda, \\ \infty & \text{on } \Omega - \Lambda. \end{cases}$$

Since $\{T_n < T\} \in \mathscr{F}_{T_n}$ and $\Lambda \in \mathscr{F}_T$ by Proposition 6 of §1.3, T'_n and T' are optional by Proposition 4 of §1.3. We have

$$\Lambda \cap \{T < \infty\} \subset \{T' < \infty\} \subset \Lambda. \tag{12}$$

Hence we have

$$\forall n: T'_n = T_n < T = T' \quad \text{on } \Lambda. \tag{13}$$

It follows that $T'_n \wedge n < T'$ and $T'_n \wedge n$ increases to T' on Ω. Hence T' is predictable. (Note: $\{T = 0\} \subset \Omega - \Lambda$, hence $T' > 0$ on Ω.) Applying the result (9) to T', we obtain

$$X_{T'} = X_{T'-} \quad \text{on } \{T' < \infty\};$$

and consequently

$$\lim_n X_{T_n} = X_{T'-} = X_{T'} \quad \text{on } \{T' < \infty\}. \tag{14}$$

In view of (12), (13) and (14), Equation (11) is true on $\Lambda \cap \{T < \infty\}$. But (11) is trivially true on $\Omega - \Lambda$. Hence Theorem 4 is proved. □

The property expressed in Theorem 4 is called the "quasi left continuity" of the process. The preceding proof shows that it is a consequence of the moderate Markov property.

We are now ready to prove the optionality of the hitting time for a closed set as well as an open set. For any set A in $\mathscr{E}_\partial$ define

$$T_A(\omega) = \inf\{t > 0 \mid X(t, \omega) \in A\}. \tag{15}$$

This is called the (*first*) *hitting time* of A. Compare with the first entrance time D_A defined in (25) of §1.3. The difference between them can be crucial.

They are related as follows:

$$T_A = \lim_{s \downarrow\downarrow 0} \downarrow (s + D_A \circ \theta_s). \tag{16}$$

This follows from the observation that

$$\begin{aligned} s + D_A \circ \theta_s &= s + \inf\{t \geq 0 \mid X(s + t, \omega) \in A\} \\ &= \inf\{t \geq s \mid X(t, \omega) \in A\}. \end{aligned}$$

It follows from Theorem 11 of §1.3 that if D_A is optional relative to $\{\mathscr{F}_t^0\}$, then so is $s + D_A \circ \theta_s$ for each $s > 0$, and consequently so is T_A by Proposition 3 of §1.3. It is easy to see that the assertion remains true when $\{\mathscr{F}_t^0\}$ is replaced by $\{\mathscr{F}_t^\mu\}$, for each μ.

Now it is obvious that D_A is the debut of the set

$$H_A = \{(t, \omega) \mid X(t, \omega) \in A\} \tag{17}$$

which is progressively measurable relative to $\{\mathscr{F}_t^0\} \subset \{\mathscr{F}_t^\mu\}$, since X is right continuous. For each t, $(\Omega, \mathscr{F}_t^\mu, P^\mu)$ is a complete probability space as a consequence of the definition of $\mathscr{F}_t^\mu$. Hence D_A is optional relative to $\{\mathscr{F}_t^\mu\}$ by Theorem 3 of §1.5. However that theorem relies on another which is not given in this book. The following proof is more direct and also shows why augmentation of $\mathscr{F}_t^0$ is needed.

Theorem 5. *If the Markov process $\{X_t\}$ is right continuous, then for each open set A, D_A and T_A are both optional relative to $\{\mathscr{F}_t^0\}$. If the process is right continuous almost surely and is also quasi left continuous, then for each closed set as well as each open set A, D_A and T_A are both optional relative to $\{\mathscr{F}_t^\sim\}$.*

Proof. By a remark above it is sufficient to prove the results for D_A. If A is open, then for each $t > 0$, we have the identity

$$\{D_A < t\} = \bigcup_{r \in Q_t} \{X_r \in A\} \tag{18}$$

where $Q_t = Q \cap [0, t)$, provided all sample functions are right continuous. To see this suppose $D_A(\omega) < t$, then $X(s, \omega) \in A$ for some $s \in [D_A(\omega), t)$, and by right continuity $X(r, \omega) \in A$ for some rational r in $(D_A(\omega), t)$. Thus the left member of (18) is a subset of the right. The converse is trivial and so (18) is true. Since the right member clearly belongs to $\mathscr{F}_t^0$, we conclude that $\{D_A < t\} \in \mathscr{F}_t^0$.

Now suppose that only P-almost all sample functions are right continuous. Then the argument above shows that the two sets in (18) differ by a P-null set. Hence $\{D_A < t\}$ is in the augmentation of $\mathscr{F}_t^0$ with respect to $(\Omega, \mathscr{F}, P)$. Translating this to P^μ, we have $\{D_A < t\} \in \mathscr{F}_t^\mu$ for each μ, hence $\{D_A < t\} \in \mathscr{F}_t^\sim$. Since $\{\mathscr{F}_t^\sim\}$ is right continuous we have $\{D_A \leq t\} \in \mathscr{F}_t^\sim$.

Next, suppose that A is closed. Then there is a sequence of open sets A_n such that $A_n \supset \bar{A}_{n+1}$ for each n, and

$$A = \bigcap_n A_n = \bigcap_n \bar{A}_n. \tag{19}$$

We shall indicate this by $A_n \downdownarrows A$. For instance, we may take

$$A_n = \left\{ x \in E_\partial \middle| d(x, A) < \frac{1}{n} \right\}.$$

Clearly D_{A_n} increases and is not greater than D_A; let

$$S = \lim_n \uparrow D_{A_n} \leq D_A. \tag{20}$$

We now make the important observation that for any Borel set B, we have almost surely

$$X(D_B) \in \bar{B} \quad \text{on } \{D_B < \infty\}. \tag{21}$$

For if $D_B(\omega) < \infty$, then for each $\delta > 0$, there exists $t \in [D_B(\omega), D_B(\omega) + \delta)$ such that $X(t, \omega) \in B$. Hence (21) follows by right continuity. Thus we have $X(D_{A_n}) \in \bar{A}_n$ for all n and therefore by quasi left continuity and (19):

$$X(S) = \lim_n X(D_{A_n}) \in \bigcap_n \bar{A}_n = A$$

almost surely on $\{0 \leq S < \infty\}$. The case $S = 0$ is of course trivial. This implies $S \geq D_A$; together with (20) we conclude that $D_A = S$ a.s. on $\{S < \infty\}$. But $D_A \geq S = \infty$ on $\{S = \infty\}$. Hence we have proved a.s.

$$D_A = \lim_n D_{A_n}. \tag{22}$$

Since each D_{A_n} is optional relative to $\{\tilde{\mathscr{F}}_t\}$; so is D_A by (22). □

If we replace the D's by T's above, the result (22) is no longer true in general! The reader should find out why. However, since for each Borel set B, $D_B = T_B$ on $\{X_0 \notin B\}$, we have the following consequence of (22) which is so important that we state it explicitly and more generally as follows.

Corollary. *Let A_n be closed sets, $A_n \supset A_{n+1}$ for all n, and $\bigcap_n A_n = A$. If $x \notin A$, then P^x-a.s. we have*

$$T_A = \lim_n \uparrow T_{A_n} \leq +\infty. \tag{23}$$

Here A may be empty, in which case $T_A = +\infty$.

Exercises

1. Here is a shorter proof of the quasi left continuity of a Feller process (Theorem 4). For $\alpha > 0$, $f \in b\mathbb{C}_+$, $\{e^{-\alpha t}U^\alpha f(X_t)\}$ is a right continuous positive supermartingale because $U^\alpha f$ is continuous. Hence if $T_n \uparrow T$, and $Y = \lim_n X(T_n)$ (which exists on $\{T < \infty\}$), we have a.s.

$$U^\alpha f(Y) = \lim_n U^\alpha f(X(T_n)) = E\left\{U^\alpha f(X(T)) \middle| \bigvee_n \mathscr{F}_{T_n}\right\}.$$

Now multiply by α, let $\alpha \to \infty$, and use Lemma 1.

2. Prove the following "separation" property for a Feller process. Let K be compact, G be open such that $K \subset G$. Then for any $\delta > 0$ there exists $t_0 > 0$ such that

$$\inf_{x \in K} P^x\{T_{G^c} \geq t_0\} \geq 1 - \delta.$$

[Hint: let $0 \leq f \leq 1$, $f = 1$ on K, $f = 0$ in G^c. There exists $t_0 > 0$ such that $\sup_{0 \leq t \leq t_0} \|P_t f - f\| < \delta/2$. Now consider

$$E^x\{f(X(t_0))\} \leq P^x\{T_{G^c} \geq t_0\} + E^x\{f(X(t_0)); T_{G^c} < t_0\}$$

and apply Theorem 3 of §2.3.]

3. If $Y \in \mathscr{F}^\sim/\mathscr{E}$ and $f \in \mathscr{E}^\sim/\mathscr{B}$ then $f(Y) \in \mathscr{F}^\sim/\mathscr{B}$. Hence $x \to E^x\{f(Y)\}$ is in $\mathscr{E}^\sim$ if the expectation exists. [This will be needed for Lebesgue measurability for important functions associated with the Brownian motion.]

To appreciate the preceding result it is worthwhile to mention that a Lebesgue measurable function of a Borel measurable function need not be Lebesgue measurable (see Exercise 15 in §1.3 of *Course*). Thus if we take $\mathscr{F}^0 = \mathscr{B}$ and $Y \in \mathscr{F}^0$; if $f \in \mathscr{F}^\mu$ we cannot infer that $f(Y) \in \mathscr{F}^\mu$, where μ is the Lebesgue measure on $\mathscr{F}^0$.

Notes on Chapter 2

§2.1. This chapter serves as an interregnum between the more concrete Feller processes and Hunt's axiomatic theory. It is advantageous to introduce some of the basic tools at an early stage.

§2.2. Feller process is named after William Feller who wrote a series of pioneering papers in the 1950's. His approach is essentially analytic and now rarely cited. The sample function properties of his processes were proved by Kinney, Dynkin, Ray, Knight, among others. Dynkin [1] developed Feller's theory by probabilistic methods. His book is rich in content but difficult to consult owing to excessive codification. Hunt [1] and Meyer [2] both discuss Feller processes before generalizations.

§2.3. It may be difficult for the novice to appreciate the fact that twenty five years ago a formal proof of the strong Markov property was a major event. Who is now interested in an example in which it does not hold?

A full discussion of augmentation is given in Blumenthal and Getoor [1]. This is dry and semi-trivial stuff but inevitable for a rigorous treatment of the fundamental concepts. Instead of beginning the book by these questions it seems advisable to postpone them until their relevance becomes more apparent.

§2.4. There is some novelty in introducing the moderate Markov property before quasi left continuity; see Chung [5]. It serves as an illustration of the general methodology alluded to in §1.3, where both F_{T+} and F_{T-} are considered. Historically, a moderate Markov property was first observed at the "first infinity" of a simple kind of Markov chains, see Chung [3]. Indeed, it occurred as a sudden revelation at the last stage of page-proofing the manuscript. It turns out that a strong Markov process becomes moderate when the paths are reversed in time, see Chung and Walsh [1].

A more complete discussion of the measurability of hitting times will be given in §3.3. Hunt practically began his great memoir [1] with this question, fully realizing that the use of hitting times is the principal method in Markov processes.

Chapter 3

Hunt Process

3.1. Defining Properties

Let $\{X_t, \mathscr{F}_t, t \in \mathbf{T}\}$ be a (homogeneous) Markov process with state space $(\mathbf{E}_\partial, \mathscr{E}_\partial)$ and transition function (P_t), as specified in §1.1 and §1.2. Here $\mathscr{F}_t$ is the $\mathscr{F}_t^{\sim}$ defined in §2.3. Such a process is called a *Hunt process* iff

(i) it is right continuous;
(ii) it has the strong Markov property (embodied in Theorem 1 of §2.3);
(iii) it is quasi left continuous (as described in Theorem 4 of §2.4).

Among the basic consequences of these hypotheses are the following:

(iv) $\{\mathscr{F}_t\}$ is right continuous (Corollary to Theorem 4 of §2.3);
(v) $\{X_t\}$ is progressively measurable relative to $\{\mathscr{F}_t\}$ (Theorem 1 of §1.5);
(vi) (P_t) is Borelian (Exercise 4 of §1.5).

We have shown in Chapter 2 that given a Feller semigroup (P_t), a Feller process can be constructed having all the preceding properties (and others not implied by the conditions above). Roundly stated: a Feller process is a Hunt process. Whereas a Feller process is constructed from a specific kind of transition function, a Hunt process is prescribed by certain hypotheses regarding the behavior of its sample functions. Thus in the study of a Hunt process we are pursing a deductive development of several fundamental features of a Feller process.

To begin with, we can add the following result to the above list of properties of a Hunt process.

Theorem 1. *Almost surely the sample paths have left limits in* $(0, \infty)$.

Proof. For a fixed $\varepsilon > 0$, define

$$T(\omega) = \inf\{t > 0 \,|\, d(X_t(\omega), X_0(\omega)) > \varepsilon\} \tag{1}$$

where d denotes a metric of the space $\mathbf{E}_\partial$. Our first task is to show that T is optional, indeed relative to $\{\mathscr{F}_t^0\}$. For this purpose let $\{z_k\}$ be a countable dense set in $\mathbf{E}_\partial$, and B_{kn} be the closed ball with center z_k and radius n^{-1}: $B_{kn} = \{x | d(x, z_k) \le n^{-1}\}$. Then $\{B_{kn}\}$ forms a countable base of the topology. Put

$$T_{kn}(\omega) = \inf\{t > 0 | d(X_t(\omega), B_{kn}) > \varepsilon\}.$$

Then T_{kn} is optional relative to $\{\mathscr{F}_t^0\}$ by Theorem 5 of §2.4, since the set $\{x | d(x, B_{kn}) > \varepsilon\}$ is open. We claim that

$$\{T < t\} = \bigcup_{k,n} \{X_0 \in B_{kn};\ T_{kn} < t\}. \tag{2}$$

It is clear that right member of (2) is a subset of the left member. To see the converse, suppose $T(\omega) < t$. Then there is $s(\omega) < t$ such that $d(X_s(\omega), X_0(\omega)) > \varepsilon$; hence there are n and k (both depending on ω) such that $d(X_s(\omega), X_0(\omega)) > \varepsilon + 2n^{-1}$ and $X_0(\omega) \in B_{kn}$. Thus $d(X_s(\omega), B_{kn}) > \varepsilon$ and $T_{kn}(\omega) < t$; namely ω belongs to the right member of (2), establishing the identity. Since the set in the right member of (2) belongs to $\mathscr{F}_t^0$, T is optional as claimed.

Next, we define $T_0 \equiv 0$, $T_1 \equiv T$ and inductively for $n \ge 1$:

$$T_{n+1} = T_n + T \circ \theta_{T_n} = \inf\{t > T_n | d(X(t), X(T_n)) > \varepsilon\}.$$

Each T_n is optional relative to $\{\mathscr{F}_t^0\}$ by Theorem 11 of §1.3. Since T_n increases with n, the limit $S = \lim_n T_n$ exists and S is optional. On the set $\{S < \infty\}$, we have $\lim_n X(T_n) = X(S)$ by quasi left continuity. On the other hand, right continuity of paths implies that $d(X(T_{n+1}), X(T_n)) \ge \varepsilon$ almost surely for all n, which precludes the existence of $\lim_n X(T_n)$. There would be a contradiction unless $S = \infty$ almost surely. In the latter event, we have $[0, \infty) = \bigcup_{n=0}^{\infty} [T_n, T_{n+1})$. Note that if $T_n = \infty$ then $[T_n, T_{n+1}) = \varnothing$. In each interval $[T_n, T_{n+1})$ the oscillation of $X(\cdot)$ does not exceed 2ε by the definition of T_{n+1}. We have therefore proved that for each ε, there exists Ω_ε with $P(\Omega_\varepsilon) = 1$ such that $X(\cdot)$ does not oscillate by more than 2ε in $[T_n^\varepsilon, T_{n+1}^\varepsilon)$, where

$$[0, \infty) = \bigcup_{n=0}^{\infty} [T_n^\varepsilon, T_{n+1}^\varepsilon). \tag{3}$$

Let $\Omega_* = \bigcap_{m=1}^{\infty} \Omega_{1/m}$; then $P(\Omega_*) = 1$. We assert that if $\omega \in \Omega_*$, then $X(\cdot, \omega)$ must have left limits in $(0, \infty)$. For otherwise there exists $t \in (0, \infty)$ and m such that $X(\cdot, \omega)$ has oscillation $> 2/m$ in $(t - \delta, t)$ for every $\delta > 0$. Thus $t \notin [T_n^{1/m}, T_{n+1}^{1/m})$ for all $n \ge 0$, which is impossible by (3) with $\varepsilon = 1/m$. □

Remark. We prove later in §3.3 that on $\{t < \zeta\}$ we have $X_{t-} \ne \partial$, namely $X_{t-} \in \mathbf{E}$. For a Feller process this is implied by Theorem 7 of §2.2.

Let us observe that quasi left continuity implies that $X(\cdot)$ is left continuous at each fixed $t \in (0, \infty)$, almost surely. For if the t_n's are constant such that $0 \le t_n \uparrow\uparrow t$, then each t_n is optional and so $X(t_n) \to X(t)$. Coupled with right continuity, this implies the continuity of almost all paths at each fixed t. In other words, the process has no fixed time of discontinuity. For a Feller process, this was remarked in §2.4.

Much stronger conditions are needed to ensure that almost all paths are continuous. One such condition is given below which is particularly adapted to a Hunt process. Another is given in Exercise 1 below.

Theorem 2. *Let $\{X_t\}$ be a Markov process with right continuous paths having left limits in $(0, \infty)$. Suppose that the transition function satisfies the following condition: for each $\varepsilon > 0$ and each compact $K \subset \mathbf{E}$ we have*

$$\lim_{t\to 0} \frac{1}{t} \sup_{x\in K} [1 - P_t(x, B(x, \varepsilon))] = 0 \tag{4}$$

where $B(x, \varepsilon) = \{y \in \mathbf{E}_\partial \mid d(x, y) \le \varepsilon\}$. Then almost all paths are continuous.

Proof. The proof depends on the following elementary lemma. □

Lemma. *Let f be a function from $[0,1]$ to $\mathbf{E}_\partial$ which is right continuous in $[0, 1)$ and has left limits in $(0, 1]$. Then f is not continuous in $[0, 1]$ (continuity at the endpoints being defined unilaterally) if and only if there exists $\varepsilon > 0$ such that for all $n \ge n_0(\varepsilon)$ we have*

$$\max_{0\le k\le n-1} d\left(f\left(\frac{k}{n}\right), f\left(\frac{k+1}{n}\right)\right) > \varepsilon. \tag{5}$$

Proof of the Lemma. If f is not continuous in $[0, 1]$, then there exists $t \in (0, 1]$ and $\varepsilon > 0$ such that $d(f(t-), f(t)) > 2\varepsilon$. For each $n \ge 1$ define k by $kn^{-1} < t \le (k + 1)n^{-1}$. Then for $n \ge n_0(\varepsilon)$ we have $d(f(kn^{-1}), f(t-)) < \varepsilon/2$, $d(f((k + 1)n^{-1}), f(t)) < \varepsilon/2$; hence $d(f(kn^{-1}), f((k + 1)n^{-1})) > \varepsilon$ as asserted. Conversely, if f is continuous in $[0, 1]$, then f is uniformly continuous there and so for each $\varepsilon > 0$ (5) is false for all sufficiently large n. This is a stronger conclusion than necessary for the occasion.

To prove the theorem, we put for a fixed compact K:

$$M = \{\omega \mid X(\cdot, \omega) \text{ is not continuous in } [0, 1];\ X(s, \omega) \in K \text{ for all } s \in [0, 1]\},$$

$$M_n^\varepsilon = \left\{\omega \,\middle|\, \sup_{0\le k\le n-1} d\left(X\left(\frac{k}{n}, \omega\right), X\left(\frac{k+1}{n}, \omega\right)\right) > \varepsilon;\right.$$
$$\left. X(s, \omega) \in K \text{ for all } s \in [0, 1]\right\}.$$

It follows from the lemma that $M \in \mathscr{F}^0$, and

$$M \subset \bigcup_{m=1}^{\infty} \left[\liminf_n M_n^{1/m}\right].$$

If we apply the Markov property at all kn^{-1} for $0 \le k \le n-1$, we obtain

$$P(M_n^\varepsilon) \le \sum_{k=0}^{n-1} P\left\{X\left(\frac{k}{n}\right) \in K; d\left(X\left(\frac{k}{n}\right), X\left(\frac{k+1}{n}\right)\right) > \varepsilon\right\}$$
$$\le n \sup_{x \in K} P_{1/n}(x, B(x, \varepsilon)^c).$$

Using the condition (4) with $t = n^{-1}$, we see that the last quantity above tends to zero as $n \to \infty$. Hence $P(\liminf_n M_n^\varepsilon) \le \underline{\lim}_n P(M_n^\varepsilon) = 0$ and $P(M) = 0$. Now replace the interval $[0, 1]$ by $[l/2, l/2 + 1]$ for integer $l \ge 1$, and K by $K_m \bigcup \{\partial\}$ where K_m is a compact subset of $\mathbf{E}$ and $K_m \uparrow \mathbf{E}$. Denote the resulting M by $M(l, m)$ and observe that we may replace K by $K \bigcup \{\partial\}$ in (4) because ∂ is an absorbing state. Thus we obtain

$$P\left(\bigcup_{l=1}^{\infty} \bigcup_{m=1}^{\infty} M(l, m)\right) = 0$$

which is seen to be equivalent to the assertion of the theorem. □

Example. As we have seen in §2.5, the Brownian motion in R^1 is a Feller process. We have

$$P_t(x, B(x, \varepsilon)^c) = \frac{1}{\sqrt{2\pi t}} \int_{|y-x|>\varepsilon} \exp\left[-\frac{(y-x)^2}{2t}\right] dy$$
$$= \frac{2}{\sqrt{2\pi t}} \int_\varepsilon^\infty \exp\left[-\frac{u^2}{2t}\right] du$$
$$= \sqrt{\frac{2}{\pi t}} \int_\varepsilon^\infty \frac{t}{u}\left(\exp\left[-\frac{u^2}{2t}\right]\frac{u}{t}\right) du \le \sqrt{\frac{2}{\pi t}}\frac{t}{\varepsilon} \exp\left[-\frac{\varepsilon^2}{2t}\right].$$

Hence (4) is satisfied even if we replace the K there by $\mathbf{E} = R^1$. It follows that almost all paths are continuous. Recall that we are using the fact that the sample paths of a [version of] Feller process are right continuous in $[0, \infty)$ and has left limits in $(0, \infty)$, which was proved in §2.2. Thus the preceding proof is not quite as short as it appears. The result was first proved by Wiener in 1923 in a totally different setting. Indeed it pre-dated the founding of the theory of stochastic processes.

Exercises

1. A stochastic process $\{X(t), t \ge 0\}$ is said to be *separable* iff there exists a countable dense set S in $[0, \infty)$ such that for almost every ω, the sample function $X(\cdot, \omega)$ has the following property. For each $t \ge 0$, there exists $s_n \in S$ such that $s_n \to t$ and $X(s_n, \omega) \to X(t, \omega)$. [The sequence $\{s_n\}$ depends

on ω as well as t!] It is an old theorem of Doob's that every process has a version which is separable; see Doob [1]. If $\{X(t)\}$ is stochastically continuous the proof is easy and the set S may be taken to be any countable dense set in $[0, \infty)$. Now take a sample function $X(\cdot, \omega)$ which has the separability property above. Show that if $X(s, \omega)$ with s restricted S is uniformly continuous in S, then $X(t, \omega)$ is continuous for all $t \geq 0$. Next, suppose that there exist strictly positive numbers δ, α, β and C such that for $t \geq 0$ and $0 < h < \delta$ the following condition is satisfied:

$$E\{|X(t+h) - X(t)|^{\alpha}\} \leq Ch^{1+\beta}.$$

Take S to be the dyadics and prove that $X(s, \omega)$ with $s \in S$ is continuous on S for almost every ω. Finally, verify the condition above for the Brownian motion in R^1. [Hint: for the last assertion estimate $\sum_{k=0}^{2^n-1} P\{|X((k+1)2^{-n}) - X(k2^{-n})| > n^{-2}\}$ and use the Borel-Cantelli lemma. For two dyadics s and s', $X(s) - X(s')$ is a finite sum of terms of the form $X((k+1)2^{-n}) - X(k2^{-n})$. The stated criterion for continuity is due to Kolmogorov.]

2. Give an example to show that the Lemma in this section becomes false if the condition "f has left limits in $(0, 1]$" is dropped. This does not seem easy, see M. Steele [1].

3. Let X be a homogeneous Markov process. A point x in $\mathbf{E}$ is called a "holding point" iff for some $\delta > 0$ we have $P^x\{X(t) = x \text{ for all } t \in [0, \delta]\} > 0$. Prove that in this case there exists $\lambda \geq 0$ such that $P^x\{T_{\{x\}^c} > t\} = e^{-\lambda t}$ for all $t > 0$. When $\lambda = 0$, x is called an "absorbing point". Prove that if X has the strong Markov property and continuous sample functions, then each holding point must be absorbing.

4. For a homogeneous Markov chain (Example 1 of §1.2), the state i is holding (also called *stable*) if and only if

$$\lim_{t \downarrow 0} \frac{1 - p_{ii}(t)}{t} < \infty.$$

The limit above always exists but may be $+\infty$, when finite it is equal to the λ in Problem 3. In general, a Markov chain does not have a version which is right continuous (even if all states are stable), hence it is not a Hunt process. But we may suppose that it is separable as in Exercise 1 above.

5. For a Hunt process: for each x there exists a countable collection of optional times $\{T_n\}$ such that for P^x-a.e. ω, the set of discontinuities of $X(\cdot, \omega)$ is the union $\bigcup_n T_n(\omega)$. [Hint: for each $\varepsilon > 0$ define $S^{(\varepsilon)} = \inf\{t > 0 \mid d(X_{t-}, X_t) > \varepsilon\}$. Show that each $S^{(\varepsilon)}$ is optional. Let $S_1^{(\varepsilon)} = S^{(\varepsilon)}$, $S_{n+1}^{(\varepsilon)} = S_n^{(\varepsilon)} + S^{(\varepsilon)} \circ \theta(S_n^{(\varepsilon)})$ for $n \geq 1$. The collection $\{S_n^{(1/m)}\}$, $m \geq 1$, $n \geq 1$, is the desired one.]

6. Let T be one of the T_n's in Exercise 5. Suppose that R_n is an increasing sequence of optional times such that $P^x\{\lim_n R_n = T < \infty\} = 1$, then $P^x\{\bigcup_n (R_n = T)\} = 1$. This means intuitively that T cannot be predicted. A stronger property (called "total inaccessibility") holds to the effect that for each x the two probabilities above are equal for any increasing sequence of optional $\{R_n\}$ such that $R_n \le T$ for all n; see Dellacherie and Meyer [1].
7. Prove that for a Hunt process, we have $\mathscr{F}_T = \sigma(\mathscr{F}_{T-}, X_T)$. (See Chung [5].)

3.2. Analysis of Excessive Functions

A basic tool in the study of Hunt processes is a class of functions called by Hunt "excessive". This is a far-reaching extension of the class of superharmonic functions in classical potential theory. In this section simple basic properties of these functions will be studied which depend only on the semigroup and not on the associated Markov process. Deeper properties derived from the process will be given in the following sections.

Let a Borelian transition semigroup $(P_t, t \ge 0)$ be given where P_0 is the identity. For $\alpha \ge 0$ we write

$$P_t^\alpha = e^{-\alpha t} P_t; \tag{1}$$

thus $P_t^0 = P_t$. For each α, (P_t^α) is also a Borelian semigroup. It is necessarily submarkovian for $\alpha > 0$. The corresponding potential kernel is U^α, where

$$U^\alpha f = \int_0^\infty P_t^\alpha f \, dt. \tag{2}$$

Here $f \in b\mathscr{E}_+$ or $f \in \mathscr{E}_+$. If $f \in b\mathscr{E}_+$ and $\alpha > 0$, then the function $U^\alpha f$ is finite everywhere, whereas this need not be true if $f \in \mathscr{E}_+$. This is one reason why we have to deal with U^α sometimes even when we are interested in $U^0 = U$. But the analogy is so nearly complete that we can save a lot of writing by treating the case $\alpha = 0$ and claiming the result for $\alpha \ge 0$. Only the finiteness of the involved quantities should be watched, and the single forbidden operation is "$\infty - \infty$".

The α-potential has been introduced in §2.1, as are α-superaveraging and α-excessive functions. The class of α-excessive functions will be denoted by $\mathbf{S}^\alpha$, and $\mathbf{S}^0 = \mathbf{S}$.

We begin with a fundamental lemma from elementary analysis, of which the proof is left to the reader as an essential exercise.

Lemma 1. *Let $\{u_{mn}\}$ be a double array of positive numbers, where m and n are positive integer indices. If u_{mn} increases with n for each fixed m, and increases* *with m for each fixed n, then we have*

$$\lim_m \lim_n u_{mn} = \lim_n \lim_m u_{mn} = \lim_n u_{nn}.$$

Of course these limits may be ∞.

Proposition 2. *For each* $\alpha \geq 0$, $\mathbf{S}^\alpha$ *is a cone which is closed under increasing sequential limits. The class of* α*-superaveraging functions is also such a cone which is furthermore closed under the minimum operation "*$\wedge$*".*

Proof. It is sufficient to treat the case $\alpha = 0$. To say that $\mathbf{S}$ is a cone means: if $f_i \in \mathbf{S}$, and c_i are positive constants for $i = 1, 2$, then $c_1 f_1 + c_2 f_2 \in \mathbf{S}$. This is trivial. Next let $f_n \in \mathbf{S}$ and $f_n \uparrow f$. Then for each n and t

$$f \geq f_n \geq P_t f_n;$$

letting $n \to \infty$ we obtain $f \geq P_t f$ by monotone convergence. If f is superaveraging, then $P_t f$ increases as t decreases. Hence if $f_n \in \mathbf{S}$ and $f_n \uparrow f$, we have

$$f = \lim_n f_n = \lim_n \lim_t P_t f_n = \lim_t \lim_n P_t f_n = \lim_t P_t f,$$

by Lemma 1 applied to n and a sequence of t (>0) decreasing to 0. This proves $f \in \mathbf{S}$. Finally, if f_i is superaveraging for $i = 1,2$, then $f_i \geq P_t f_i \geq P_t(f_1 \wedge f_2)$; hence $f_1 \wedge f_2 \geq P_t(f_1 \wedge f_2)$; namely that $f_1 \wedge f_2$ is superaveraging. □

It is remarkable that no simple argument exists to show that $\mathbf{S}$ is closed under "$\wedge$". A deep proof will be given in §3.4. The following special situation is very useful.

Proposition 3. *Suppose* P_t *converges vaguely to* P_0 *as* $t \downarrow 0$. *If* f *is superaveraging and also lower semi-continuous, then* f *is excessive.*

Proof. By a standard result on vague convergence, we have under the stated hypotheses:

$$f = P_0 f \leq \lim_{t \downarrow 0} P_t f \leq f. \qquad \square$$

If (P_t) is Fellerian, then the condition on (P_t) in Proposition 3 is satisfied. More generally, it is satisfied if the sample paths of the associated Markov process are right continuous at $t = 0$. For then if $f \in b\mathbb{C}$, as $t \downarrow 0$:

$$P_t f(x) = E^x\{f(X_t)\} \to E^x\{f(X_0)\} = P_0 f(x) = f(x). \tag{3}$$

Here we have a glimpse of the interplay between the semigroup and the process. Incidentally, speaking logically, we should have said "*an* associated Markov process" in the above.

Proposition 4. *If* $\alpha < \beta$, *then* $\mathbf{S}^\alpha \subset \mathbf{S}^\beta$. *For each* $\alpha \geq 0$, $\mathbf{S}^\alpha = \bigcap_{\beta > \alpha} \mathbf{S}^\beta$.

Proof. It is trivial that if $\alpha < \beta$, then $P_t^\alpha f \geq P_t^\beta f$ for $f \in \mathscr{E}_+$. Now if $f = \lim_{t \downarrow 0} P_t^\beta f$ for any value of $\beta \geq 0$, then the same limit relation holds for all values of $\beta \geq 0$, simply because $\lim_{t \downarrow 0} e^{-\beta t} = 1$ for all $\beta \geq 0$. The proposition follows quickly from these remarks. □

The next result gives a basic connection between superaveraging functions and excessive functions. It is sufficient to treat the case $\alpha = 0$. If f is superaveraging, we define its *regularization* $\hat{f}$ as follows:

$$\hat{f}(x) = \lim_{t \downarrow 0} \uparrow P_t f(x). \tag{4}$$

We have already observed that the limit above is monotone. Hence we may define $P_{t+} f(x) = \lim_{s \downarrow\downarrow t} P_s f$.

Proposition 5. $\hat{f} \in \mathbf{S}$ *and* $\hat{f}$ *is the largest excessive function not exceeding* f. *We have*

$$\forall t \geq 0\colon P_t \hat{f} = P_{t+} f. \tag{5}$$

Proof. We prove (5) first, as follows:

$$P_t \hat{f} = P_t\left(\lim_{s \downarrow\downarrow 0} \uparrow P_s f\right) = \lim_{s \downarrow\downarrow 0} P_t P_s f = \lim_{s \downarrow\downarrow 0} P_{t+s} f = P_{t+} f.$$

It is immediate from (4) that $\hat{f} \leq f$ and $\hat{f} \geq P_t f \geq P_t \hat{f}$. Furthermore by (5),

$$\lim_{t \downarrow\downarrow 0} P_t \hat{f} = \lim_{t \downarrow\downarrow 0} P_{t+} f = \lim_{t \downarrow\downarrow 0} P_t f = \hat{f},$$

where the second equation is by elementary analysis; hence $\hat{f} \in \mathbf{S}$. If $g \in \mathbf{S}$ and $g \leq f$, then $P_t g \leq P_t f$; letting $t \downarrow\downarrow 0$ we obtain $g \leq \hat{f}$. □

The next result contains an essential calculation pertaining to a potential.

Theorem 6. *Suppose* $f \in \mathbf{S}$, $P_t f < \infty$ *for each* $t > 0$, *and*

$$\lim_{t \to \infty} P_t f = 0. \tag{6}$$

Then we have

$$f = \lim_{h \downarrow 0} \uparrow U\left(\frac{f - P_h f}{h}\right). \tag{7}$$

Proof. The hypothesis $P_t f < \infty$ allows us to subtract below. We have for $h > 0$:

$$\int_0^t P_s(f - P_h f)\,ds = \int_0^t P_s f\,ds - \int_h^{t+h} P_s f\,ds = \int_0^h P_s f\,ds - \int_t^{t+h} P_s f\,ds.$$

If we divide through by h above, then let $t \uparrow \infty$, the last term converges to zero by (6) and we obtain

$$\lim_{t\uparrow\infty} \int_0^t P_s \left(\frac{f - P_h f}{h}\right) ds = \frac{1}{h} \int_0^h P_s f\,ds. \tag{8}$$

The integrand on the left being positive because $f \geq P_h f$, this shows that the limit above is the potential shown in the right member of (7). When $h \downarrow 0$, the right member increases to the limit f, because $\lim_{s\downarrow 0} \uparrow P_s f = f$. This establishes (7). □

Let us observe that there is an obvious analogue of Proposition 6 for $\mathbf{S}^\alpha$. If $\alpha > 0$ and $f \in b\mathscr{E}_+$, then the corresponding conditions in the proposition are satisfied. Hence (7) holds when U and P_h are replaced by U^α and P_h^α for such an f. This is an important case of the theorem.

An alternative approach to excessive functions is through the use of resolvents (U^α) instead of the semigroup (P_t). This yields a somewhat more general theory but less legible formulas, which will now be discussed briefly. We begin with the celebrated *resolvent equation* (9) below.

Proposition 7. *For $\alpha > 0$ and $\beta > 0$, we have*

$$U^\alpha = U^\beta + (\beta - \alpha)U^\alpha U^\beta = U^\beta + (\beta - \alpha)U^\beta U^\alpha. \tag{9}$$

Proof. This is proved by the following calculations, for $f \in b\mathscr{E}_+$ and $\alpha \neq \beta$:

$$\begin{aligned}
U^\alpha U^\beta f &= \int_0^\infty e^{-\alpha s} P_s \left[\int_0^\infty e^{-\beta t} P_t f\,dt\right] ds \\
&= \int_0^\infty \int_0^\infty e^{-\alpha s - \beta t} P_{s+t} f\,dt\,ds \\
&= \int_0^\infty \int_s^\infty e^{(\beta - \alpha)s} e^{-\beta u} P_u f\,du\,ds, \quad \text{where } u = s + t, \\
&= \int_0^\infty \left[\int_0^u e^{(\beta-\alpha)s}\,ds\right] e^{-\beta u} P_u f\,du \\
&= \int_0^\infty \frac{e^{-\alpha u} - e^{-\beta u}}{\beta - \alpha} P_u f\,du = \frac{1}{\beta - \alpha}[U^\alpha f - U^\beta f].
\end{aligned}$$

Note that the steps above are so organized that it is immaterial where $\beta - \alpha > 0$ or < 0. This remark establishes the second as well as the first equation in (9). □

If $f \in \mathbf{S}$, then since

$$\alpha U^\alpha f = \int_0^\infty \alpha e^{-\alpha t} P_t f \, dt = \int_0^\infty e^{-u} P_{u/\alpha} f \, du$$

we have

$$\forall \alpha > 0: \alpha U^\alpha f \le f, \tag{10}$$

$$\lim_{\alpha \uparrow \infty} \uparrow \alpha U^\alpha f = f. \tag{11}$$

[The case $f(x) = \infty$ in (11) should be scrutinized.] The converse is rather tricky to prove.

Proposition 8. *If $f \in \mathscr{E}_+$ and* (10) *is true, then $f^* = \lim_{\alpha \uparrow \infty} \uparrow \alpha U^\alpha f$ exists and is the largest excessive function $\le f$. If* (11) *is also true then f is excessive.*

Proof. Let $0 < \beta < \alpha$ and $f \in b\mathscr{E}_+$. Then we have by (9):

$$U^\alpha f = U^\beta [f - (\alpha - \beta) U^\alpha f]. \tag{12}$$

Here subtraction is allowed because $U^\alpha f < \infty$. If (10) is true, then $g_{\alpha\beta} = f - (\alpha - \beta) U^\alpha f \ge 0$. Since $U^\beta g_{\alpha\beta}$ is β-excessive, $U^\alpha f$ is β-excessive for all $\beta \in (0, \alpha)$. Hence $U^\alpha f \in \mathbf{S}$ by Proposition 4. Next, we see from (12) by simple arithmetic that

$$\alpha U^\alpha f = \beta U^\beta f + (\alpha - \beta) U^\beta [f - \alpha U^\alpha f] \ge \beta U^\beta f \tag{13}$$

by (10). Hence if we put

$$f^* = \lim_{\alpha \uparrow \infty} \uparrow \alpha U^\alpha f = \lim_{\alpha \uparrow \infty} \uparrow U^\alpha (\alpha f),$$

we have $f^* \in \mathbf{S}$ by Proposition 2.

For a general $f \in \mathscr{E}_+$ satisfying (10), let $f_n = f \wedge n$. Then $\alpha U^\alpha f_n \le f_n$ by (10) and the inequality $\alpha U^\alpha n \le n$. Hence we may apply what has just been proved to each f_n to obtain $f_n^* = \lim_{\alpha \uparrow \infty} \uparrow \alpha U^\alpha f_n$. Also the inequality $\alpha U^\alpha f_n \ge \beta U^\beta f_n$ leads to $\alpha U^\alpha f \ge \beta U^\beta f$, if $\alpha \ge \beta$. Hence we have by Lemma 1:

$$f^* = \lim_{\alpha \uparrow \infty} \uparrow \alpha U^\alpha f = \lim_{\alpha \uparrow \infty} \uparrow \lim_{n \uparrow \infty} \uparrow \alpha U^\alpha f_n$$

$$= \lim_{n \uparrow \infty} \uparrow \lim_{\alpha \uparrow \infty} \uparrow \alpha U^\alpha f_n = \lim_{n \uparrow \infty} \uparrow f_n^*.$$

Since each $f_n^* \in \mathbf{S}$ as shown above, we conclude that $f^* \in \mathbf{S}$. If $g \in \mathbf{S}$ and $g \leq f$, then $\alpha U^\alpha g \leq \alpha U^\alpha f$; letting $\alpha \uparrow \infty$ we obtain $g \leq f^*$. This identifies f^* as the largest excessive function $\leq f$. The second assertion of Proposition 8 is now trivial.* □

Proposition 8 has an immediate extension as follows. For each $\beta \geq 0$, $f \in \mathbf{S}^\beta$ if and only if

$$\forall \alpha > 0: \alpha U^{\alpha+\beta} f \leq f \quad \text{and} \quad \lim_{\alpha \uparrow \infty} \uparrow \alpha U^{\alpha+\beta} f = f. \tag{14}$$

This is left as an exercise.

The next result is extremely useful. It reduces many problems concerning an excessive function to the same ones for potentials, which are often quite easy.

Theorem 9. *If $f \in \mathbf{S}$, then for each $\beta > 0$, there exists $g_n \in b\mathscr{E}_+$ such that*

$$f = \lim_n \uparrow U^\beta g_n. \tag{15}$$

Proof. Suppose first that $f \in b\mathscr{E}_+$. Then as in the preceding proof, we have $U^\alpha(\alpha f) = U^\beta(\alpha g_{\alpha\beta})$. Thus we obtain (15) from (11) with $g_n = n g_{n\beta}$. In general we apply this to obtain for each k, $f_k^* = \lim_n \uparrow U^\beta g_n^{(k)}$. It follows then by Lemma 1 that

$$f = f^* = \lim_k f_k^* = \lim_k \uparrow \lim_n \uparrow U^\beta g_n^{(k)} = \lim_n \uparrow U^\beta g_n^{(n)}. \qquad □$$

We remark that this proposition is also a consequence of the extension of Theorem 6 to β-excessive functions, because then $\lim_{t\to\infty} P_t^\beta f = 0$ for $f \in b\mathscr{E}_+$, etc. But the preceding proof is more general in the sense that it makes use only of (10) and (11), and the resolvent equation, without going back to the semigroup.

It is an important observation that Theorem 9 is false with $\beta = 0$ in (15); see Exercise 3 below. An additional assumption, which turns out to be of particular interest for Hunt processes, will now be introduced to permit $\beta = 0$ there. Since we are proceeding in an analytic setting, we must begin by assuming that

$$\forall x \in \mathbf{E}: U(x, \mathbf{E}) > 0. \tag{16}$$

This is trivially true for a process with right continuous paths; for a more general situation see Exercise 2 below. Remember however that $U(\partial, \mathbf{E}) = 0$. To avoid such exceptions we shall agree in what follows that a "point" means a point in $\mathbf{E}$, and a "set" means a subset of $\mathbf{E}$, without specific mention to the

*I am indebted to Patrick Fitzsimmons for correcting a slip in the first edition.

contrary. The case of ∂ can always be decided by an extra inspection, when it is worth the pain.

Definition. The semigroup (P_t) or the corresponding U is called *transient* iff there exists a function $h \in \mathscr{E}_+$ such that

$$0 < Uh < \infty \qquad \text{on } \mathbf{E}. \tag{17}$$

An immediate consequence of (17) is as follows. There exists a sequence of $h_n \in b\mathscr{E}_+$ (with $h_n(\partial) = 0$) such that

$$Uh_n > 0 \quad \text{and} \quad Uh_n \uparrow \infty \quad \text{both in } \mathbf{E}. \tag{18}$$

We may take $h_n = nh$ to achieve this.

The probabilistic meaning of transience as defined above, as well as some alternative conditions, will be discussed later in §3.7.

Proposition 10: *If (P_t) is transient, and $f \in \mathbf{S}$, then there exists $g_n \in b\mathscr{E}_+$ such that*

$$f = \lim_n \uparrow Ug_n. \tag{19}$$

In fact, $g_n \leq n^2$ and $Ug_n \leq n$.

Proof. Put

$$f_n = f \wedge Uh_n \wedge n$$

where h_n is as in (18). Then we have

$$P_t f_n \leq P_t Uh_n = \int_t^\infty P_s h_n \, ds \downarrow 0$$

as $t \uparrow \infty$ because $Uh_n < \infty$. By Proposition 2, f_n is superaveraging; let its regularization be $\hat{f}_n$. Since $P_t \hat{f}_n \leq P_t f_n < \infty$, and $\to 0$ as $t \uparrow \infty$, we may apply Theorem 6 to obtain

$$\hat{f}_n = \lim_k \uparrow Ug_{nk}$$

where

$$g_{nk} = k(\hat{f}_n - P_{1/k}\hat{f}_n) \leq kn.$$

From the proof of Theorem 6 we see that

$$Ug_{nk} = k \int_0^{1/k} P_s \hat{f}_n \, ds \leq n.$$

For each n, Ug_{nk} increases with k; for each k, Ug_{nk} increases with n, hence we have by Lemma 1:

$$\lim_n \uparrow \hat{f}_n = \lim_n \uparrow \lim_k \uparrow Ug_{nk} = \lim_n \uparrow Ug_{nn}.$$

On the other hand, we have again by Lemma 1:

$$\lim_n \uparrow \hat{f}_n = \lim_n \uparrow \lim_{t \downarrow 0} \uparrow P_t f_n = \lim_{t \downarrow 0} \uparrow \lim_n \uparrow P_t f_n = \lim_{t \downarrow 0} P_t f = f.$$

It follows that (19) is true with $g_n = g_{nn}$. □

Exercises

1. A measure μ on $\mathscr{E}$ is said to be c-finite ("countably finite") iff there is a countable collection of finite measures $\{\mu_n\}$ such that $\mu(A) = \sum_{n=1}^{\infty} \mu_n(A)$ for all $A \in \mathscr{E}$. Show that if μ is σ-finite, then it is c-finite, but not vice versa. For instance, Fubini's theorem holds for c-finite measures. Show that for each x, $U(x, \cdot)$ is c-finite.

2. Show that for each $x \in \mathbf{E}$, $P_t(x, \mathbf{E})$ is a decreasing function of t. The following properties are equivalent:

   (a) $U(x, \mathbf{E}) = 0$; (b) $\lim_{t \downarrow\downarrow 0} P_t(x, \mathbf{E}) = 0$; (c) $P^x\{\zeta = 0\} = 1$.

3. For the Brownian motion in R^1, show that there cannot exist positive Borel (or Lebesgue) measurable functions g_n such that $\lim_{n \to \infty} Ug_n = 1$. Thus Theorem 9 is false for $\beta = 0$. [See Theorem 1 of §3.7 below for general elucidation.]

4. If $f \in \mathscr{E}_+$ and $\alpha U^\alpha f \leq f$ for all $\alpha > 0$, then $P_t f \leq f$ for (Lebesgue) almost all $t \geq 0$. [Hint: $\lim_{\alpha \uparrow \infty} \alpha U^\alpha f = f^* \leq f$ and $U^\alpha f^* = U^\alpha f$ for all $\alpha > 0$.]

3.3. Hitting Times

We have proved the optionality of T_A for an open or a closed set A, in Theorem 5 of §2.4. The proof is valid for a Hunt process. It is a major advance to extend this to an arbitrary Borel set and to establish the powerful approximation theorems "from inside" and "from outside". The key is the theory of capacity due to Choquet. We begin by identifying the probability of hitting a set before a fixed time as a Choquet capacity.

Let $\{X_t, \mathscr{F}_t, t \in \mathbf{T}\}$ be a Hunt process as specified in §3.1. For simplicity of notation, we shall write P for P^μ (for an arbitrary probability measure μ on $\mathscr{E}$); but recall that $\mathscr{F}_t \subset \mathscr{F}_t^\mu$ and $(\mathscr{F}^\mu, P^\mu)$ is complete. Fix $t \geq 0$. For any subset

A of $\mathbf{E}_\partial$, we put

$$A' = \{\omega \in \Omega \,|\, \exists s \in [0,t]: X_s(\omega) \in A\}.$$

This is the projection on Ω of the set of (s,ω) in $[0,t] \times \Omega$ such that $X(s,\omega) \in A$. The mapping $A \to A'$ from the class of all subsets of $\mathbf{E}_\partial$ to the class of all subsets of Ω has the following properties:

(i) $A_1 \subset A_2 \Rightarrow A_1' \subset A_2'$;
(ii) $(A_1 \cup A_2)' = A_1' \cup A_2'$;
(iii) $A_n \uparrow A \Rightarrow A_n' \uparrow A'$.

In view of (i) and (ii), (iii) is equivalent to

(iii′) $(\bigcup_n A_n)' = \bigcup_n A_n'$ for an arbitrary sequence $\{A_n\}$.

So far these properties are trivial and remain valid when the fundamental interval $[0,t]$ is replaced by $[0,\infty)$. The next proposition will depend essentially on the compactness of $[0,t]$.

We shall write $\Lambda_1 \Subset \Lambda_2$ to denote $P(\Lambda_1 \backslash \Lambda_2) = 0$, and $\Lambda_1 \doteq \Lambda_2$ to denote $P(\Lambda_1 \bigtriangleup \Lambda_2) = 0$. If G is open, then $G' \in \mathscr{F}^0$, so $P(G')$ is defined.

Proposition 1. *Let G_n be open, K be compact such that*

$$\forall n\colon G_n \supset \bar{G}_{n+1}, \qquad K = \bigcap_n G_n \left(= \bigcap_n \bar{G}_n\right). \tag{1}$$

Then we have

$$K' \doteq \bigcap_n G_n'. \tag{2}$$

Proof. This is essentially a re-play of Theorem 5 of §2.4. Consider the first entrance times D_{G_n}; these are optional (relative to $\{\mathscr{F}_t^0\}$). If $\omega \in \bigcap_n G_n'$, then $D_{G_n}(\omega) \le t$ for all n. Let $D(\omega) = \lim_n \uparrow D_{G_n}(\omega)$; clearly $D(\omega) \le D_K(\omega)$ and $D(\omega) \le t$. Put

$$N = \left\{\omega \,\middle|\, \lim_n X(D_{G_n}) \ne X(D)\right\};$$

we have $P(N) = 0$ by quasi left continuity. If $\omega \in \bigcap_n G_n' \backslash N$, then $X(D) \in \bigcap_n \bar{G}_n = K$ since $X(D_{G_n}) \in \bar{G}_n$ by right continuity. Thus $D_K(\omega) \le D(\omega)$ and so $D_K(\omega) = D(\omega) \le t$; namely $\omega \in K'$. We have therefore proved $\bigcap_n G_n' \Subset K'$. But $\bigcap_n G_n' \supset K'$ by (i) above, hence (2) is proved. □

Proposition 2. *Let K_n be compact, $K_n \downarrow$ and $\bigcap_n K_n = K$. Then*

$$K' \doteq \bigcap_n K_n'. \tag{3}$$

Proof. K being given, let $\{G_n\}$ be as in Proposition 1. For each m, $G_m \supset K = \bigcap_n K_n$; hence by a basic property of compactness, there exists n_m such that $G_m \supset K_{n_m}$ and consequently $G'_m \supset K'_{n_m}$ by (i). It follows that

$$K' \doteq \bigcap_m G'_m \supset \bigcap_m K'_{n_m} \supset \bigcap_n K'_n \supset K',$$

where the first equation is just (2). This implies (3). □

Corollary. *For each compact* K, *and* $\varepsilon > 0$, *there exists open* G *such that* $K \subset G$ *and*

$$P(K') \leq P(G') \leq P(K') + \varepsilon.$$

In particular

$$P(K') = \inf_{G \supset K} P(G') \tag{4}$$

where the inf *is over all open* G *containing* K.

From here on in this section, the letters A, B, G, K are reserved for arbitrary, Borel, open, compact sets respectively.

Definition. Define a set function C for all subsets of $\mathbf{E}_\partial$ as follows:

(a) $C(G) = P(G')$;
(b) $C(A) = \inf_{G \supset A} C(G)$.

Clearly C is "monotone", namely $A_1 \subset A_2$ implies $C(A_1) \leq C(A_2)$. Note also that it follows from (4) that

$$C(K) = P(K'). \tag{5}$$

We proceed to derive properties of this function C. Part (b) of its definition is reminiscent of an outer measure which is indeed a particular case; on the other hand, it should be clear from part (a) that C is not additive on disjoint sets. What replaces additivity is a strong kind of subadditivity, as given below.

Proposition 3. C *is "strongly subadditive" over open sets, namely*;

$$C(G_1 \cup G_2) + C(G_1 \cap G_2) \leq C(G_1) + C(G_2). \tag{6}$$

Proof. We have, as a superb example of the facility (felicity) of reasoning with sample functions:

$$\begin{aligned}(G_1 \cup G_2)' - G'_1 &= \{\omega \mid \exists s \in [0,t]: X_s(\omega) \in G_1 \cup G_2; \forall s \in [0,t]: X_s(\omega) \notin G_1\} \\ &\subset \{\omega \mid \exists s \in [0,t]: X_s(\omega) \in G_2; \forall s \in [0,t]: X_s(\omega) \notin G_1 \cap G_2\} \\ &= G'_2 - (G_1 \cap G_2)'.\end{aligned}$$

Taking probabilities, we obtain

$$P((G_1 \cup G_2)') - P(G_1') \leq P(G_2') - P((G_1 \cap G_2)').$$

Up to here G_1 and G_2 may be arbitrary sets. Now we use definition (a) above to convert the preceding inequality into (6). □

The same argument shows C is also strongly additive over all compact sets, because of (5). Later we shall see that C is strongly additive over all capacitable sets by the same token.

Lemma 4. *Let* $A_n\uparrow$, $A_n \subset G_n$, $\varepsilon_n > 0$ *such that*

$$\forall n\colon C(G_n) \leq C(A_n) + \varepsilon_n. \tag{7}$$

Then we have for each finite m:

$$C\left(\bigcup_{n=1}^{m} G_n\right) \leq C(A_m) + \sum_{n=1}^{m} \varepsilon_n. \tag{8}$$

Proof. For $m = 1$, (8) reduces to (7). Assume (8) is true as shown and observe that

$$A_m \subset \left(\bigcup_{n=1}^{m} G_n\right) \cap G_{m+1}. \tag{9}$$

We now apply the strong subadditivity of C over the two open sets $\bigcup_{n=1}^{m} G_n$ and G_{m+1}, its monotonicity, and (9) to obtain

$$C\left(\bigcup_{n=1}^{m+1} G_n\right) + C(A_m) \leq C\left(\left(\bigcup_{n=1}^{m} G_n\right) \cup G_{m+1}\right) + C\left(\left(\bigcup_{n=1}^{m} G_n\right) \cap G_{m+1}\right)$$

$$\leq C\left(\bigcup_{n=1}^{m} G_n\right) + C(G_{m+1}) \leq C(A_m) + \sum_{n=1}^{m} \varepsilon_n + C(A_{m+1}) + \varepsilon_{m+1}.$$

This completes the induction on m. □

We can now summarize the properties of C in the theorem below.

Theorem 5. *We have*

(i) $A_1 \subset A_2 \Rightarrow C(A_1) \leq C(A_2)$;
(ii) $A_n \uparrow A \Rightarrow C(A_n) \uparrow C(A)$;
(iii) $K_n \downarrow K \Rightarrow C(K_n) \downarrow C(K)$.

Proof. We have already mentioned (i) above. Next, using the notation of Lemma 4, we have

$$C(A) \leq C\left(\bigcup_{n=1}^{\infty} G_n\right) = \lim_m C\left(\bigcup_{n=1}^{m} G_n\right)$$

by definition (a) of C, and the monotone property of the probability measure P. Applying Lemma 4 with $\varepsilon_n = \varepsilon 2^{-n}$, we see from (8) that

$$C(A) \leq \lim_m C(A_m) + \varepsilon.$$

Since ε is arbitrary, (ii) follows from this inequality and (i). Finally, it follows from (5), Proposition 2, and the monotone property of P that

$$C(K) = P(K') = P\left(\bigcap_n K_n'\right) = \lim_n P(K_n') = \lim_n C(K_n).$$

This proves (iii). □

Definition. A function defined on all the subsets of $\mathbf{E}_\partial$ and taking values in $[-\infty, +\infty]$ is called a *Choquet capacity* iff it has the three properties (i), (ii) and (iii) in Theorem 5. A subset A of $\mathbf{E}_\partial$ is called *capacitable* iff given any $\varepsilon > 0$, there exist an open G and a compact K such that $K \subset A \subset G$ and

$$C(G) \leq C(K) + \varepsilon. \tag{10}$$

This definition (which can be further generalized) is more general than we need here, since the C defined above takes values in $[0, 1]$ only. Here is the principal theorem of capacitability.

Choquet's Theorem. *In a locally compact Hausdorff space with countable base, each analytic set is capacitable.*

A set $A \subset \mathbf{E}_\partial$ is *analytic* iff there exists a separable complete metric space (alias "Polish space") M and a continuous mapping φ of M into $\mathbf{E}_\partial$ such that $A = \varphi(M)$.

Such a definition needs much study to make sense, but we must forego it since we shall not use analytic sets below. Suffice it to say that each Borel set, namely each A in $\mathscr{E}$ in our general notation, is analytic and therefore capacitable by Choquet's theorem. We are ready for the principal result.

Theorem 6. *For each Borel set B (i.e., $B \in \mathscr{E}_\partial$) we have*

$$C(B) = P(B'). \tag{11}$$

Proof. Since B is capacitable by Choquet's theorem, for each $n \geq 1$ there exist $K_n \subset B \subset G_n$ such that

$$C(K_n) \leq C(B) \leq C(G_n) \leq C(K_n) + \frac{1}{n}. \tag{12}$$

We have $K'_n \subset B' \subset G'_n$. Let

$$\Lambda_1 = \bigcup_n K'_n \subset B' \subset \bigcap_n G'_n = \Lambda_2. \tag{13}$$

Then for each $n \geq 1$:

$$P(\Lambda_2) - P(\Lambda_1) \leq P(G'_n) - P(K'_n) = C(G_n) - C(K_n) \leq \frac{1}{n},$$

and consequently

$$\Lambda_1 \doteq \Lambda_2 \doteq B'. \tag{14}$$

It follows that

$$P(B') = P(\Lambda_1) = \lim_n P(K'_n) = \lim_n C(K_n) = C(B)$$

where the last equation is by (12). □

Corollary. $P(B') = \lim_n P(K'_n) = \lim_n P(G'_n)$.

The optionality of D_B and T_B follows quickly.

Theorem 7. *For each $B \in \mathscr{E}_\partial$, D_B and T_B are both optional relative to $\{\mathscr{F}_t\}$.*

Proof. Recall that t is fixed in the foregoing discussion. Let us now denote the B' above by $B'(t)$. For the sake of explicitness we will replace $(\Omega, \mathscr{F}, P)$ above by $(\Omega, \mathscr{F}_t, P^\mu)$ for each probability measure μ on $\mathscr{E}$, as fully discussed in §2.3. Put also $Q^t = (Q \cap [0, t)) \cup \{t\}$. Since G is open, the right continuity of $s \to X_s$ implies that

$$G'(t) = \{\omega \mid \exists s \in Q^t : X_s(\omega) \in G\}.$$

Hence $G'(t) \in \mathscr{F}_t^0 \subset \mathscr{F}_t^\mu$. Next, we have $P^\mu\{B'(t) \triangle \bigcap_n G'_n(t)\} = 0$ by (13), hence $B'(t) \in \mathscr{F}_t^\mu$ because $\mathscr{F}_t^\mu$ contains all P^μ-null sets. Finally, a careful scrutiny shows that for each $B \in \mathscr{E}_\partial$ and $t > 0$:

$$\begin{aligned}\{\omega \mid D_B(\omega) < t\} &= \bigcup_{r \in Q \cap [0,t)} \{\omega \mid \exists s \in [0, r] : X_s(\omega) \in B\} \\ &= \bigcup_{r \in Q \cap [0,t)} B'(r) \in \mathscr{F}_t^\mu.\end{aligned}$$

Hence D_B is optional relative to $\{\mathscr{F}_t^\mu\}$. The same is true of T_B by using (16) of §2.4. □

It should be recalled that the preceding theorem is also an immediate consequence of Theorem 3 of §1.5. The more laborious proof just given "tells more" about the approximation of a Borel set by compact subsets and open supersets, but the key to the argument which is concealed in Choquet's theorem, is the projection from $\mathbf{T} \times \Omega$ to Ω, just as in the case of Theorem 3 of §1.5. This is also where analytic sets enter the picture; see Dellacherie and Meyer [1].

The approximation theorem will be given first for D_B, then for T_B, in two parts. The probability measure μ is defined on $\mathscr{E}_\partial$, though the point ∂ plays only a nuisance role for what follows and may be ignored.

Theorem 8(a). *For each μ and $B \in \mathscr{E}_\partial$, there exist $K_n \subset B$, $K_n\uparrow$ and $G_n \supset B$, $G_n\downarrow$ such that*

$$D_{K_n} \downarrow D_B, \qquad D_{G_n} \uparrow D_B, \qquad P^\mu\text{-a.s.} \tag{15}$$

Proof. The basic idea is to apply the Corollary to Theorem 6 to $B'(r)$ for all $r \in Q$. Thus for each $r \in Q$, we have

$$K_{rn} \subset B \subset G_{rn} \tag{16}$$

such that

$$\lim_n P^\mu\{G'_{rn}(r) - K'_{rn}(r)\} = 0. \tag{17}$$

Let $\{r_j\}$ be an enumeration of Q, and put

$$K_n = \bigcup_{j=1}^n K_{r_jn}, \qquad G_n = \bigcap_{j=1}^n G_{r_jn}.$$

Then $K_n \subset B$, $K_n\uparrow$; $G_n \supset B$, $G_n\downarrow$; let

$$D_{K_n} \downarrow D', \qquad D_{G_n} \uparrow D''.$$

Then $D' \geq D_B \geq D''$. We have for each j:

$$\{D' > r_j > D''\} \subset \bigcap_n \{D_{K_n} > r_j > D_{G_n}\} \subset \bigcap_n \{G'_n(r_j) - K'_n(r_j)\}. \tag{18}$$

For $n \geq j$, we have

$$K_n \supset K_{r_jn}, \qquad G_n \subset G_{r_jn}$$

hence

$$K'_n(r_j) \supset K'_{r_jn}(r_j), \qquad G'_n(r_j) \subset G'_{r_jn}(r_j).$$

Hence by (17),

$$\lim_n P^\mu\{G_n'(r_j) - K_n'(r_j)\} = 0$$

and consequently by (18), $P^\mu\{D' > r_j > D''\} = 0$. This being true for all r_j, we conclude that $P^\mu\{D' = D_B = D''\} = 1$, which is the assertion in (15). □

Theorem 8(b). *For each μ and $B \in \mathscr{E}_\partial$, there exists $K_n \subset B, K_n\uparrow$ such that*

$$T_{K_n} \downarrow T_B, \qquad P^\mu\text{-a.s.} \tag{19}$$

If μ is such that $\mu(B) = 0$, then there exist $G_n \supset B$, $G_n\downarrow$ such that

$$T_{G_n} \uparrow T_B, \qquad P^\mu\text{-a.s.} \tag{20}$$

Proof. The basic formula here is (16) of §2.4:

$$T_B = \lim_{s \downarrow\downarrow 0} \downarrow (s + D_B \circ \theta_s), \tag{21}$$

applied to a sequence $s_k \downarrow\downarrow 0$. Let $\mu_k = \mu P_{s_k}$. Then for each $k \geq 1$, we have by part (a): there exist $K_{kn} \subset B$, $K_{kn}\uparrow$ as $n\uparrow$ such that

$$D_{K_{kn}} \downarrow D_B, \qquad P^{\mu_k}\text{-a.s.}$$

which means

$$D_{K_{kn}} \circ \theta_{s_k} \downarrow D_B \circ \theta_{s_k}, \qquad P^\mu\text{-a.s.} \tag{22}$$

Let $K_n = \bigcup_{k=1}^n K_{kn}$: then $K_n \subset B$, $K_n\uparrow$, and it is clear from (22) that

$$\forall k\colon D_{K_n} \circ \theta_{s_k} \downarrow D_B \circ \theta_{s_k}, \qquad P^\mu\text{-a.s.}$$

Therefore, we have by (an analogue of) Lemma 1 in §3.2:

$$\begin{aligned} T_B &= \lim_k \downarrow (s_k + D_B \circ \theta_{s_k}) = \lim_k \downarrow \lim_n \downarrow (s_k + D_{K_n} \circ \theta_{s_k}) \\ &= \lim_n \downarrow \lim_k \downarrow (s_k + D_{K_n} \circ \theta_{s_k}) = \lim_n \downarrow T_{K_n}, \qquad P^\mu\text{-a.s.} \end{aligned}$$

This proves (19) and it is important to see that the additional condition $\mu(B) = 0$ is needed for (20). Observe that $T_B = D_B$ unless $X_0 \in B$; hence under the said condition we have $P^\mu\{T_B = D_B\} = 1$. It follows from this remark that $T_G = D_G$ for an open G, because if $X_0 \in G$ we have $T_G = 0$ (which is not the case for an arbitrary B). Thus we have by part (a):

$$T_{G_n} = D_{G_n} \uparrow D_B = T_B, \qquad P^\mu\text{-a.s.}$$ □

Remark. The most trivial counterexample to (20) when $\mu(B) \neq 0$ is the case of uniform motion with $B = \{0\}$, $\mu = \varepsilon_0$. Under P^μ, it is obvious that $T_B = \infty$ but $T_G = 0$ for each open $G \supset B$. Another example is the case of Brownian motion in the plane, if $B = \{x\}$ and $\mu = \varepsilon_x$; see the Example in §3.6.

As a first application of Theorem 8, we consider the "left entrance time" and "left hitting time" of a Borel set B, defined as follows:

$$\begin{aligned} D_B^-(\omega) &= \inf\{t \geq 0 \mid X_{t-}(\omega) \in B\}, \\ T_B^-(\omega) &= \inf\{t > 0 \mid X_{t-}(\omega) \in B\}. \end{aligned} \tag{23}$$

We must make the convention $X_{0-} = X_0$ to give a meaning to D_B.

Theorem 9. *We have almost surely:*

$$\begin{aligned} D_G^- &= D_G; \qquad & T_G^- &= T_G; \\ D_B^- &\geq D_B; \qquad & T_B^- &\geq T_B. \end{aligned}$$

Proof. Since the relation (21) holds for T_B^- and D_B^- as well, it is sufficient to consider the D's.

If $X_t \in G$ and G is open, then there is an open set G_1 such that $\bar{G}_1 \subset G$ and $X_t \in G_1$. The right continuity of paths then implies that for each ω there exists $\delta_0(\omega) > 0$ such that $X_{t+\delta} \in \bar{G}_1$ for $0 < \delta \leq \delta_0$ and so $X_{t+\delta-} \in \bar{G}_1$. This observation shows $D_G^- \leq D_G$. Conversely if $t > 0$ and $X_{t-} \in G$, then there exist $t_n \uparrow\uparrow t$ such that $X_{t_n} \in G$; whereas if $X_{0-} \in G$ then $X_0 \in G$. This shows $D_G \leq D_G^-$ and consequently $D_G = D_G^-$.

For a general B, we apply the second part of (15) to obtain $G_n \supset B$ such that $D_{G_n} \uparrow D_B$, P^μ-a.s. Since $D_B^- \geq D_{G_n}^- = D_{G_n}$, it follows that $D_B^- \geq D_B$, P^μ-a.s. □

We can now settle a point remarked in §3.1, by applying Theorem 9 with $B = \{\partial\}$.

Corollary. *For a Hunt process, we have almost surely:*

$$\forall t < \zeta : X_{t-} \in \mathbf{E}. \tag{24}$$

This implies: on the set $\{t < \zeta\}$, the closure of the set of values $\bigcup_{s \in [0,t]} X(s,\omega)$ is a compact subset of $\mathbf{E}$.

We close this section by making a "facile generalization" which turns out to be useful, in view of later developments in §3.4.

Definition. A set $A \subset \mathbf{E}_\partial$ is called *nearly Borel* iff for each finite measure μ on $\mathscr{E}_\partial$, there exist two Borel sets B_1 and B_2, depending on μ, such that $B_1 \subset A \subset B_2$ and

$$P^\mu\{D_{B_2 - B_1} < \infty\} = 0. \tag{25}$$

This is equivalent to:

$$P^{\mu}\{\exists t \in [0, \infty): X_t \in B_2 - B_1\} = 0. \tag{26}$$

Of course, this definition depends on the process $\{X_t\}$.

The class of nearly Borel sets will be denoted by $\mathscr{E}^{\bullet}$. It is easy to verify that this is a σ-field. It is included in $\mathscr{E}^{\sim}$, the universally measurable σ-field, because (26) implies

$$\mu(B_2 - B_1) = P^{\mu}\{X_0 \in B_2 - B_1\} = 0.$$

One can describe $\mathscr{E}^{\bullet}$ in a folksy way by saying that the "poor" Hunt process $\{X_t\}$ cannot distinguish a set in $\mathscr{E}^{\bullet}$ from a set in $\mathscr{E}_{\partial}$. It should be obvious that the hitting time of a nearly Borel set is optional, and the approximation theorems above hold for it as well. [We do not need "nearly compact" or "nearly open" sets!] Indeed, if $A \in \mathscr{E}^{\bullet}$ then for each μ there exists $B \in \mathscr{E}_{\partial}$ such that $D_B = D_A$, P^{μ}-a.s. Needless to say, this B depends on μ.

A function f on $\mathbf{E}_{\partial}$ to $[-\infty, +\infty]$ is nearly Borel when $f \in \mathscr{E}^{\bullet}$. This is the case if and only if for each μ, there exist two Borel functions f_1 and f_2, depending on μ, such that $f_1 \leq f \leq f_2$ and

$$P^{\mu}\{\exists t \in [0, \infty): f_1(X_t) \neq f_2(X_t)\} = 0. \tag{27}$$

It follows that we may replace f_1 or f_2 in the relation above by f. We shall see in the next section that all excessive functions are nearly Borel.

3.4. Balayage and Fundamental Structure

Let $\{X_t, \mathscr{F}_t\}$ be a Hunt process, where $\{\mathscr{F}_t\}$ is as specified in §3.1.

For $f \in \mathscr{E}_+$, $\alpha \geq 0$, and optional T, we define the operator P_T^{α} as follows:

$$P_T^{\alpha} f(x) = E^x\{e^{-\alpha T} f(X_T); T < \infty\}. \tag{1}$$

We write P_T for P_T^0. If we use the convention that $X_{\infty} = \partial$ and $f(\partial) = 0$ for each f, then we may omit "$T < \infty$" in the expression above. We shall frequently do so without repeating this remark.

The following composition property is fundamental. Recall that if S and T are optional relative to $\{\mathscr{F}_t\}$, so is $S + T \circ \theta_S$: see Exercise 7 of §2.3.

Proposition 1. *We have*

$$P_S^{\alpha} P_T^{\alpha} = P_{S+T\circ\theta_S}^{\alpha}. \tag{2}$$

Proof. Since $X_T(\omega) = X(T(\omega), \omega)$, we have

$$X_T \circ \theta_S(\omega) = X(T(\theta_S\omega), \theta_S(\omega)) = X(S(\omega) + T \circ \theta_S(\omega), \omega) = X_{S+T\circ\theta_S}(\omega). \quad (3)$$

It follows that for $f \in b\mathscr{E}_+$:

$$\begin{aligned} P_S^\alpha P_T^\alpha f &= E^\cdot\{e^{-\alpha S} P_T^\alpha f(X_S)\} = E^\cdot\{e^{-\alpha S} E^{X_S}[e^{-\alpha T} f(X_T)]\} \\ &= E^\cdot\{e^{-\alpha S}[e^{-\alpha T} f(X_T)] \circ \theta_S\} \\ &= E^\cdot\{e^{-\alpha(S+T\circ\theta_S)} f(X_{S+T\circ\theta_S})\}. \end{aligned}$$

□

Here and henceforth we will adopt the notation $E^\cdot(\cdots)$ to indicate the function $x \to E^x(\cdots)$.

Two important instances of (2) are: when $S = t$, a constant; and when S and T are hitting times of a nearly Borel set A. In the latter case we write P_A^α for $P_{T_A}^\alpha$. We shall reserve the letter A for a nearly Borel set below.

Definition. A point x is said to be *regular* for the set A iff

$$P^x\{T_A = 0\} = 1. \quad (4)$$

The set of all points in $\mathbf{E}_\partial$ which are regular for A will be denoted by A^r; and the union $A \cup A^r$ is called the *fine closure* of A and denoted by A^*. The nomenclature will be justified in §3.5.

According to the zero-or-one law (Theorem 6 of §2.3), x is not regular for A iff $P^x\{T_A = 0\} = 0$ or $P^x\{T_A > 0\} = 1$. In this case we say also that the set A is *thin* at the point x. Let $\bar{A}$ denote the topological closure of A, then if $x \notin \bar{A}$, a path starting at x must remain for some time in an open neighborhood of x which is disjoint from A, hence x cannot be regular for A. Thus $A^* \subset \bar{A}$. Since in general the set $\{T_A = 0\}$ belongs to $\mathscr{F}_0 = \mathscr{F}_0^\sim$ and not to $\mathscr{F}_0^0$, the function $x \to P^x\{T_A = 0\}$ belongs to $\mathscr{E}^\sim$ rather than $\mathscr{E}$. This is a nuisance which will be ameliorated.* Observe that (4) is equivalent to

$$P_A^\alpha 1(x) \equiv E^x\{e^{-\alpha T_A}\} = 1 \quad (5)$$

for each $\alpha > 0$; indeed $P^x\{T_A = 0\}$ may be regarded as $\lim_{\alpha\uparrow\infty} P_A^\alpha 1(x)$. Finally if $x \in A^r$, then for any f we have

$$P_A f(x) = E^x\{T_A < \infty; f(X_{T_A})\} = E^x\{T_A = 0; f(X_0)\} = f(x).$$

Theorem 2. *For each x and A, the measure $P_A(x,\cdot)$ is concentrated on A^*. In other words,*

$$\forall x: P^x\{X(T_A) \in A^*; T_A < \infty\} = P^x\{T_A < \infty\}.$$

*See Exercise 3 after §3.5.

Proof. We have by definition $(A^*)^c = A^c \cap (A^r)^c$. If $y \in (A^r)^c$, then $P^y\{T_A > 0\} = 1$. It follows that

$$P^x\{T_A < \infty; X(T_A) \in (A^*)^c\} \\ \leq E^x\{T_A < \infty; X(T_A) \in A^c; P^{X(T_A)}[T_A > 0]\}. \tag{6}$$

Applying the strong Markov property at T_A, we see that the right member of (6) is equal to the P^x-probability of the set

$$\{T_A < \infty; X(T_A) \in A^c; T_A \circ \theta_{T_A} > 0\}$$

where $T_A \circ \theta_{T_A}$ means the "time lapse between the first hitting time of A and the first hitting time of A thereafter." If $X_{T_A}(\omega) \notin A$ and $T_A \circ \theta_{T_A}(\omega) > 0$ then the sample function $X(\cdot, \omega)$ is not in A for $T \in [T_A(\omega), T_A(\omega) + T_A \circ \theta_{T_A}(\omega))$, a nonempty interval. This is impossible by the definition of T_A. Hence the right member of (6) must be equal to zero, proving the assertion of the theorem. □

The following corollary for $\alpha = 0$ is just (21) of §2.4, which is true for T_B as well as D_B.

Corollary. *If A is a closed set, then $P_A^\alpha(x, \cdot)$ is concentrated in A for each $\alpha \geq 0$ and each x.*

The operator $P_A(x, \cdot)$ corresponds to what is known as "balayage" or "sweeping out" in potential theory. A unit charge placed at the point x is supposed to be swept onto the set A. The general notion is due to Poincaré, for a modern analytical definition see e.g. Helms [1]. Hunt was able to identify it with the definition given above, and that apparently convinced the potential theorists that "he's got something there"!

We interrupt the thrust of our preceding discussion by an obligatory extension of measurability. According to Theorem 5 of §2.3, for $Y \in b\mathscr{F}^\sim$ hence also for $Y \in \mathscr{F}^\sim_+$, the function $x \to E^x(Y)$ is universally measurable, namely in $\mathscr{E}^\sim$ defined in (18) of §2.3. In particular $P_A^\alpha 1$ is so measurable for $A \in \mathscr{E}$. We have therefore no choice but to enlarge the class of functions considered from $\mathscr{E}$ to $\mathscr{E}^\sim$. This could of course have been done from the outset but it is more convincing to introduce the extension when the need has arisen.

From here on, the transition probability measure $P_t(x, \cdot)$ is extended to $\mathscr{E}^\sim$. It is easy to see that for each $A \in \mathscr{E}^\sim$,

$$(t, x) \to P_t(x, A)$$

belongs to $\mathscr{B} \times \mathscr{E}^\sim$; alternately, $(t, x) \to P_t f(x)$ is in $\mathscr{B} \times \mathscr{E}^\sim$ for each $f \in b\mathscr{E}^\sim$ or $\mathscr{E}^\sim_+$. It follows that $U^\alpha f$ is in $\mathscr{E}^\sim$ for such an f and $\alpha \geq 0$. Finally, an α-excessive function is defined as before except that $f \in \mathscr{E}^\sim$, rather than $f \in \mathscr{E}$

as previously supposed. A universally measurable function is sandwiched between two Borel measurable functions (for each finite measure μ on $\mathscr{E}_\partial$) and this furnishes the key to its handling. See the proof of Theorem 5 of §2.3 for a typical example of this remark.

In what follows we fix the notation as follows: f and g are functions in $\mathscr{E}_+^\sim$; $x \in \mathbf{E}_\partial$; $\alpha \geq 0$; T is optional relative to $\{\mathscr{F}_t^\sim\}$; A is a nearly Borel set; K is a compact set; G is an open set. These symbols may appear with subscripts.

We begin with the formula:

$$P_T^\alpha U^\alpha f = E^\cdot\left\{\int_T^\infty e^{-\alpha t} f(X_t)\,dt\right\}, \tag{7}$$

which is derived as follows. Making the substitution $t = T + u$, we transform the right member of (7) by the strong Markov property into

$$E^\cdot\left\{e^{-\alpha T}\left[\int_0^\infty e^{-\alpha u} f(X_u)\,du\right] \circ \theta_T\right\} = E^\cdot\left\{e^{-\alpha T} E^{X_T}\left[\int_0^\infty e^{-\alpha u} f(X_u)\,du\right]\right\}$$
$$= E^\cdot\{e^{-\alpha T} U^\alpha f(X_T)\},$$

which is the left member of (7). It follows at once that

$$U^\alpha f = E^\cdot\left\{\int_0^T e^{-\alpha t} f(X_t)\,dt\right\} + P_T^\alpha U^\alpha f. \tag{8}$$

On $\{T = \infty\}$ we have $P_T^\alpha U^\alpha f = 0$ by our convention. Note also that when T is a constant (8) has already been given in §2.1.

We use the notation $\underline{|f}$ for the support of f, namely the set

$$\{x \in \mathbf{E}_\partial \mid f(x) > 0\}.$$

Theorem 3.

(a) $P_A^\alpha U^\alpha f \leq U^\alpha f$;
(b) $P_A^\alpha U^\alpha f = U^\alpha f$, *if* $\underline{|f} \subset A$;
(c) *If we have*

$$U^\alpha f \leq U^\alpha g \tag{9}$$

on $\underline{|f}$, *then* (9) *is true everywhere.*

Proof. Assertion (a) is obvious from (8) with $T = T_A$; so is (b) if we observe that $f(X_t) = 0$ for $t < T_A$ so that the first term on the right side of (8) vanishes. To prove (c) we need the approximation in Theorem 8(b) of §3.3. Let $A = \underline{|f}$; given x, let $K_n \subset A$ such that $T_{K_n} \downarrow T_A$, P^x-a.s. We have then by (7), as $n \to \infty$:

$$P_{K_n}^\alpha U^\alpha f(x) = E^x\left\{\int_{T_{K_n}}^\infty e^{-\alpha t} f(X_t)\,dt\right\} \uparrow E^x\left\{\int_{T_A}^\infty e^{-\alpha t} f(X_t)\,dt\right\}$$
$$= P_A^\alpha U^\alpha f(x). \tag{10}$$

This is a fundamental limit relation which makes potentials easy to handle. By the Corollary to Theorem 2, $P_{K_n}(x, \cdot)$ is concentrated on K_n, hence $U^\alpha f \leq U^\alpha g$ on K_n by the hypothesis of (c), yielding

$$P^\alpha_{K_n} U^\alpha f(x) \leq P^\alpha_{K_n} U^\alpha g(x).$$

Letting $n \to \infty$, using (a) and (b), and (10) for both f and g, we obtain

$$U^\alpha f(x) = P^\alpha_A U^\alpha f(x) \leq P^\alpha_A U^\alpha g(x) \leq U^\alpha g(x). \qquad \square$$

Assertion (c) is known as the "domination principle" in potential theory, which will be amplified after Theorem 4. Note however it deals with the potentials of functions rather than potentials of measures, which are much more difficult to deal with.

We state the next few results for $\mathbf{S}$, but they are true for $\mathbf{S}^\alpha$, *mutatis mutandis.*

Theorem 4. *Let $f \in \mathbf{S}$. Then $P_A f \in \mathbf{S}$ and $P_A f \leq f$. If $A_1 \subset A_2$, then*

$$P_{A_1} f \leq P_{A_2} f. \tag{11}$$

For each x, there exist compacts $K_n \subset A$ such that

$$P_{K_n} f(x) \uparrow P_A f(x). \tag{12}$$

Proof. Let us begin by recording an essential property of a hitting time which is not shared by a general optional time:

$$\forall t \geq 0: T_A \leq t + T_A \circ \theta_t; \quad \text{and} \quad T_A = \lim_{t \downarrow 0} \downarrow (t + T_A \circ \theta_t). \tag{13}$$

This is verified exactly like (16) of §2.4, and the intuitive meaning is equally obvious. We have from (6) and (7):

$$P^\alpha_t P^\alpha_A U^\alpha g = E^\cdot \left\{ \int_{t + T_A \circ \theta_t}^\infty e^{-\alpha s} g(X_s)\, ds \right\}.$$

It follows from this and (13) that $P^\alpha_A U^\alpha g \in \mathbf{S}$. For $f \in \mathbf{S}$, we have by Theorem 9 of §3.2, $f = \lim_k \uparrow U^\alpha g_k$ where $\alpha > 0$. Hence $P^\alpha_A f = \lim_k \uparrow P^\alpha_A U^\alpha g_k$ and so $P^\alpha_A f \in \mathbf{S}^\alpha$ by Proposition 2 of §3.2. Finally $P_A f = \lim_{\alpha \downarrow 0} \uparrow P^\alpha_A f$; hence $P_A f \in \mathbf{S}$ by Proposition 4 of §3.2. Next if $A_1 \subset A_2$, then it follows at once from (7) that $P^\alpha_{A_1} U^\alpha g \leq P^\alpha_{A_2} U^\alpha g$. Now the same approximations just shown establish (11), which includes $P_A f \leq f$ as a particular case.

Next, given x and A, let $\{K_n\}$ be as in the preceding proof. As in (10), we have

$$P^\alpha_{K_n} U^\alpha g_k(x) \uparrow P^\alpha_A U^\alpha g_k(x).$$

Hence by Lemma 1 of §3.2,

$$P_A^\alpha f = \lim_k \uparrow P_A^\alpha U^\alpha g_k = \lim_k \uparrow \lim_n \uparrow P_{K_n}^\alpha U^\alpha g_k$$
$$= \lim_n \uparrow \lim_k \uparrow P_{K_n}^\alpha U^\alpha g_k(x) = \lim_n \uparrow P_{K_n}^\alpha f(x).$$

Letting $\alpha \downarrow 0$ and using Lemma 1 once again we obtain (12). □

Corollary. *Part (c) of Theorem 3 is true if $U^\alpha g$ is replaced by any excessive function in (9).*

The next result is crucial, though it will soon be absorbed into the bigger Theorem 6 below.

Theorem 5. *Let $f \in \mathbf{S}$, and $A \in \mathscr{E}^\bullet$. If $x \in A^r$, then*

$$\inf_A f \leq f(x) \leq \sup_A f.$$

Proof. Since $x \in A^r$, $f(x) = P_A f(x)$ as noted above. It follows from Theorem 8(b) of §3.3 that for each $\varepsilon > 0$, there exists $K \subset A$ such that

$$P^x\{T_K > \varepsilon\} = P^x\{T_K > T_A + \varepsilon\} \leq \varepsilon \tag{14}$$

since $P^x\{T_A = 0\} = 1$. At the same time, we may choose K so that

$$P_K f(x) \leq P_A f(x) \leq P_K f(x) + \varepsilon$$

by (12). Hence by the Corollary to Theorem 2, we have

$$P_K(x, K) \inf_{y \in K} f(y) \leq P_K f(x) \leq \sup_{y \in K} f(y).$$

Since $P_K(x, K) \geq P^x\{T_K \leq \varepsilon\} \geq 1 - \varepsilon$ by (14), it follows from the inequalities above that

$$(1 - \varepsilon) \inf_{y \in A} f(y) \leq f(x) \leq \sup_{y \in A} f(y) + \varepsilon.$$

Since ε is arbitrary this is the assertion of the theorem. □

We have arrived at the following key result. In its proof we shall be more circumspect than elsewhere about pertinent measurability questions.

Theorem 6. *Let $f \in \mathbf{S}$. Then almost surely $t \to f(X_t)$ is right continuous on $[0, \infty)$ and has left limits (possibly $+\infty$) in $(0, \infty]$. Moreover, $f \in \mathscr{E}^\bullet$.*

Proof. We begin by assuming that $f \in \mathbf{S} \cap \mathscr{E}^{\cdot}$. Introduce the metric ρ on $[0, \infty]$ which is compatible with the extended Euclidean topology: $\rho(a, b) = |(a/1 + a) - (b/1 + b)|$, where $\infty/(1 + \infty) = 1$. For each $\varepsilon > 0$ define

$$S_\varepsilon(\omega) = \inf\{t > 0 \mid \rho(f(X_0(\omega)), f(X_t(\omega))) > \varepsilon\}.$$

An argument similar to that for the T in (2) of §3.1 shows that S_ε is optional (relative to $(\mathscr{F}_t)$). It is clear that we have

$$\bigcap_{k=1}^{\infty} \{S_{1/k} > 0\} \subset \left\{\omega \,\middle|\, \lim_{t \downarrow 0} f(X_t(\omega)) = f(X_0(\omega))\right\} = \Lambda, \tag{15}$$

say. For each x put $A = \{y \in \mathbf{E}_\partial \mid \rho(f(x), f(y)) > \varepsilon\}$. Then $A \in \mathscr{E}^{\cdot}$; and S_ε reduces to the hitting time T_A under P^x. It is a consequence of Theorem 5 that $x \notin A^r$; for obviously the value $f(x)$ does not lie between the bounds of f on A. Hence for each x, $P^x\{S_\varepsilon > 0\} = 1$ by Blumenthal's zero-one law. Now $\{S_\varepsilon > 0\} \in \mathscr{F}$, hence $P^{\cdot}\{S_\varepsilon > 0\} \in \mathscr{E}^{\sim}$ by Theorem 5 of §2.3, and this implies $P^\mu\{S_\varepsilon > 0\} = 1$ for each probability measure μ on $\mathscr{E}_\partial$. It now follows from (15) that $\Lambda \in \mathscr{F}^\mu$ since $(\mathscr{F}^\mu, P^\mu)$ is complete. This being true for each μ, we have proved that $\Lambda \in \mathscr{F}$ and Λ is an almost sure set. For later need let us remark that for each optional S, $\Lambda \circ \theta_S \in \mathscr{F}$ by Theorem 5 of §2.3.

Our next step is to define inductively a family of optional times T_α such that the sample function $t \to f(X_t)$ oscillates no more than 2ε in each interval $[T_\alpha, T_{\alpha+1})$. This is analogous to our procedure in the proof of Theorem 1 of §3.1, but it will be necessary to use transfinite induction here. The reader will find it instructive to compare the two proofs and scrutinize the difference. Let us denote by $\mathbb{O}$ the well-ordered set of ordinal numbers before the first uncountable ordinal (see e.g., Kelley [1; p. 29]). Fix ε, put $T_0 \equiv 0$, $T_1 \equiv S_\varepsilon$, and for each $\alpha \in \mathbb{O}$, define T_α as follows:

(i) if α has the immediate predecessor $\alpha - 1$, put

$$T_\alpha = T_{\alpha-1} + T_1 \circ \theta_{T_{\alpha-1}};$$

(ii) if α is a limit ordinal without an immediate predecessor, put

$$T_\alpha = \sup_{\beta < \alpha} T_\beta \quad \text{where } \beta \text{ ranges over all ordinals} < \alpha.$$

Then each T_α is optional by induction. [Note that the sup in case (ii) may be replaced by the sup of a countable sequence.] Now for each $\alpha \in \mathbb{O}$, we have for each μ:

$$P^\mu\left\{T_\alpha < \infty; \lim_{t \downarrow 0} f(X_{T_\alpha + t}) = f(X_{T_\alpha})\right\}$$

$$= E^\mu\left\{T_\alpha < \infty; P^{X(T_\alpha)}\left[\lim_{t \downarrow 0} f(X_t) = f(X_0)\right]\right\} = P^\mu\{T_\alpha < \infty\} \tag{16}$$

by the strong Markov property and the fact that $P^x(\Lambda) = 1$ for each x, proved above. It follows that P^μ-a.e. on the set $\{T_\alpha < \infty\}$, the limit relation in the first member of (16) holds; hence $T_1 \circ \theta_{T_\alpha} > 0$ by the definition of T_1, hence $T_{\alpha+1} > T_\alpha$. Consider now the numbers

$$b_\alpha = P^\mu\{T_\alpha < \infty\}, \qquad c_\alpha = E^\mu\{e^{-T_\alpha}\}, \qquad \alpha \in \mathbb{O}.$$

We have just shown that if $b_\alpha > 0$, then

$$c_{\alpha+1} = E^\mu\{e^{-T_{\alpha+1}}; T_{\alpha+1} < \infty\} < E^\mu\{e^{-T_\alpha}; T_\alpha < \infty\} = c_\alpha.$$

Now the set $\mathbb{O}$ has uncountable cardinality. If $b_\alpha > 0$ for all $\alpha \in \mathbb{O}$, then the uncountably many intervals $(c_{\alpha+1}, c_\alpha)$, $\alpha \in \mathbb{O}$, would be all nonempty and disjoint, which is impossible because each must contain a rational number. Therefore there exists $\alpha^* \in \mathbb{O}$ for which $b_{\alpha^*} = 0$; namely that

$$P^\mu\{T_{\alpha^*} = \infty\} = 1.$$

For each $\alpha < \alpha^*$, if s and t both belong to $[T_\alpha, T_{\alpha+1})$, then $\rho(f(X_s), f(X_t)) \leq 2\varepsilon$. This means: there exists Ω_ε^μ with $P^\mu(\Omega_\varepsilon) = 1$ such that if $\omega \in \Omega_\varepsilon^\mu$, then the sample function $t \to f(X(t,\omega))$ does not oscillate more than 2ε *to the right of any* $t \in [0, \infty)$. It follows that the set Ω_ε^* of ω satisfying the latter requirement belongs to $\mathscr{F}$ and is an almost sure set. Since $\bigcap_{k=1}^\infty \Omega_{1/k}^*$ is contained in the set of ω for which $t \to f(X_t(\omega))$ is right continuous in $[0, \infty)$, we have proved that the last-mentioned set belongs to $\mathscr{F}$ and is an almost sure set. This is the main assertion of the theorem.

In order to fully appreciate the preceding proof, it is necessary to reflect that no conclusion can be drawn as to the existence of left limits of the sample function in $(0, \infty)$. The latter result will now follows from an earlier theorem on supermartingales. In fact, if we write $f_n = f \wedge n$ then $\{f_n(X_t), \mathscr{F}_t\}$ is a bounded supermartingale under each P^x, by Proposition 1 of §2.1, since f_n is superaveraging. Almost every sample functions $t \to f_n(X_t)$ is right continuous by what has just been proved, hence it has left limits in $(0, \infty]$ by Corollaries 1 and 2 to Theorem 1 of §1.4. Since n is arbitrary, it follows (why?) that $t \to f(X_t)$ has left limits (possibly $+\infty$) as asserted.

It remains to prove that $f \in \mathbf{S}$ implies $f \in \mathscr{E}^\bullet$. In view of Theorem 9 of §3.2, it is sufficient to prove this for $U^\alpha g$ where $\alpha > 0$ and $g \in b\mathscr{E}_+^\sim$. For each μ, there exist g_1 and g_2 in $\mathscr{E}_\partial$ such that $g_1 \leq g \leq g_2$ and $\mu U^\alpha(g_2 - g_1) = 0$. For each $t \geq 0$, we have

$$P_t U^\alpha(g_2 - g_1) = e^{\alpha t} P_t^\alpha U^\alpha(g_2 - g_1) \leq e^{\alpha t} U^\alpha(g_2 - g_1).$$

It follows that

$$E^\mu\{U^\alpha(g_2 - g_1)(X_t)\} = \mu P_t U^\alpha(g_2 - g_1) \leq e^{\alpha t} \mu U^\alpha(g_2 - g_1) = 0.$$

Thus we have for each $t \geq 0$:

$$U^\alpha(g_2 - g_1)(X_t) = 0, \qquad P^\mu\text{-a.s.} \tag{17}$$

Therefore (17) is also true simultaneously for all rational $t \geq 0$. But $U^\alpha(g_2 - g_1) \in \mathbf{S}^\alpha \cap \mathscr{E}$, hence the first part of our proof extended to $\mathbf{S}^\alpha$ shows that the function of t appearing in (17) is right continuous, hence must vanish identically in t, P^μ-a.s. This means $U^\alpha g \in \mathscr{E}^\bullet$ as asserted. Theorem 6 is completely proved. □

In the proof of Theorem 6, we have spelled out the $\mathscr{F}$-measurability of "the set of all ω such that $f(X(\cdot, \omega))$ is right continuous". Such questions can be handled by a general theory based on $\mathbf{T} \times \Omega$-analytic sets and projection, as described in §1.5; see Dellacherie and Meyer [1], p. 165.

If $f \in \mathbf{S}$, when can $f(X_t) = \infty$? The answer is given below. We introduce the notation

$$\forall t \in (0, \infty]: f(X_t)_- = \lim_{s \uparrow\uparrow t} f(X_s). \tag{18}$$

This limit exists by Theorem 6, and is to be carefully distinguished from $f(X_{t-})$. Note also that in a similar notation $f(X_t)_+ = f(X_t) = f(X_{t+})$.

Theorem 7. *Let $F = \{f < \infty\}$. Then we have almost surely*:

$$f(X_t)_- < \infty \quad \textit{and} \quad f(X_t) < \infty \qquad \textit{for all } t > D_F, \tag{19}$$

where D_F is the first entrance time into F.

Proof. If $x \in F$, then under P^x, $\{f(X_t), \mathscr{F}_t\}$ is a right continuous positive supermartingale by Theorem 6. Hence by Corollary 1 to Theorem 1 of §1.4, $f(X_t)$ is bounded in each finite t-interval, and by Corollary 2 to the same theorem, converges to a finite limit as $t \to \infty$. Therefore (19) is true with $D_F = 0$. In general, let $F_n = \{f \leq n\}$, then $F_n \in \mathscr{E}^\bullet$ by Theorem 6 and so D_{F_n} is optional; also $f(X(D_{F_n})) \leq n$ by right continuity. It follows by the strong Markov property that (19) is true when the D_F there is replaced by D_{F_n}. Now an easy inspection shows that $\lim_n \downarrow D_{F_n} = D_F$, hence (19) holds as written. □

It is obvious that on the set $\{D_F > 0\}$, $f(X_t) = \infty$ for $t < D_F$, and $f(X_t)_- = \infty$ for $t \leq D_F$; but $f(X(D_F))$ may be finite or infinite.

There are numerous consequences of Theorem 6. One of the most important is the following basic property of excessive functions which can now be easily established.

Theorem 8. *The class of excessive functions is closed under the minimum operation.*

Proof. Let $f_1 \in \mathbf{S}$, $f_2 \in \mathbf{S}$, then we know $f_1 \wedge f_2$ is superaveraging. On the other hand,

$$\begin{aligned}\lim_{t \downarrow 0} P_t(f_1 \wedge f_2)(x) &= \lim_{t \downarrow 0} E^x\{f_1(X_t) \wedge f_2(X_t)\} \\ &\geq E^x\left\{\lim_{t \downarrow 0} [f_1(X_t) \wedge f_2(X_t)]\right\} \\ &= E^x\{f_1(X_0) \wedge f_2(X_0)\} = (f_1 \wedge f_2)(x);\end{aligned}$$

by Fatou's lemma followed by the right continuity of $f_i(X_t)$ at $t = 0$, $i = 1, 2$. Hence there is equality above and the theorem is proved. □

In particular, if $f \in \mathbf{S}$ then $f \wedge n \in \mathbf{S}$ for any n. This truncation allows us to treat bounded excessive functions before a passage to the limit for the general case, which often simplifies the argument.

As another application of Theorem 6, let us note that $P_A^\alpha 1$ is α-excessive for each $\alpha \geq 0$ and each $A \in \mathscr{E}^\bullet$, hence $P_A^\alpha 1 \in \mathscr{E}^\bullet$. Since $A^r = \{x \mid P_A^\alpha 1 = 1\}$ as remarked in (5), we have $A^r \in \mathscr{E}^\bullet$, $A^* \in \mathscr{E}^\bullet$. The significance of these sets will be studied in the next section.

Exercises

1. Let $f \in b\mathscr{E}_+$, $T_n \uparrow T$. Prove that for a Hunt process

$$\lim_{n \to \infty} U^\alpha f(X(T_n)) = E\left\{U^\alpha f(X(T)) \,\middle|\, \bigvee_n \mathscr{F}_{T_n}\right\}.$$

Hence if we assume that

$$X(T) \in \bigvee_n \mathscr{F}_{T_n}, \tag{20}$$

then we have

$$\lim_n U^\alpha f(X(T_n)) = U^\alpha f(X(T)).$$

Remark. Under the hypothesis that (20) holds for any optional $T_n \uparrow T$, the function $t \to U^\alpha f(X(t-))$ is left continuous. This is a complement to Theorem 6 due to P. A. Meyer: the proof requires projection onto the predictable field, see Chung [5].

2. Let $A \in \mathscr{E}^\bullet$ and suppose $T_n \uparrow T_A$. Then we have for each $\alpha > 0$, a.s. on the set $\bigcap_n \{T_n < T_A < \infty\}$:

$$\lim_{n \to \infty} E^{X(T_n)}\{e^{-\alpha T_A}\} = 1.$$

[Hint: consider $E^x\{e^{-\alpha T_A}; T_n < T_A \mid \mathscr{F}_{T_n}\}$ and observe that $T_A \in \bigvee_n \mathscr{F}_{T_n}$.]

3. Let x_0 be a holding but not absorbing point (see Problem 3 of §3.1). Define

$$T = \inf\{t > 0 : X(t) \neq x_0\}.$$

For an increasing sequence of optional T_n such that $T_n \le T$ and $T_n \uparrow T$ we have

$$P^{x_0}\left\{\bigcap_{n=1}^{\infty} [T_n < T]\right\} = 0.$$

Indeed, it follows that for any optional S, $P^{x_0}\{0 < S < T\} = 0$. [Hint: the second assertion follows from the first, by contraposition and transfinite induction, as suggested by J. B. Walsh.]

4. Let $f \in \mathscr{E}_+$ and $f \ge P_K f$ for every compact K. Let $g \in \mathscr{E}$ such that $U(g^+) \wedge U(g^-) < \infty$ (namely $U(g)$ is defined). Then

$$f \ge Ug \quad \text{on } \underline{|g^+}$$

implies $f \ge Ug$ (everywhere). [Hint: $f + U(g^-) \ge U(g^+)$; for each $K \subset \underline{|g^+}$ we have $f + U(g^-) \ge P_K(f + U(g^-)) \ge P_K U(g^+)$, now approximate $\underline{|g^+}$ by compacts.]

5. Suppose U is transient (see (17) of §3.2). For the f in Problem 4 we have $f \ge \alpha U^\alpha f$ for every $\alpha > 0$. [Hint: we may suppose f bounded with compact support; put $g = f - \alpha U^\alpha f$, then $f \ge U(\alpha g) = U^\alpha(\alpha f)$ on $\{g \ge 0\}$. This is a true *leger-demain* due to E. B. Dynkin [1].]

6. Prove Theorem 6 for a Feller process by using Theorem 5 of §1.4. [Hint: show first $U^\alpha f(X_t)$ is right continuous for $f \in b\mathscr{E}_+$, which is trivially true for $f \in C_0$.]

7. Let f be excessive and $A = \{f = \infty\}$. If $f(x_0) < \infty$, then $P_A 1(x_0) = 0$.

8. The optional random variable T is called *terminal* iff for every $t \ge 0$ we have

$$T = t + T \circ \theta_t \quad \text{on } \{T > t\}$$

and

$$T = \lim_{t \downarrow 0} (t + T \circ \theta_t).$$

Thus the notion generalizes that of a hitting time. Prove that if T is terminal, then $x \to P^x\{T < \infty\}$ is excessive; and $x \to E^x\{e^{-\alpha T}\}$ is α-excessive for each $\alpha > 0$.

3.5. Fine Properties

Theorem 5 of §3.4 may be stated as follows: if $f \in \mathbf{S}$, then the bounds of f on the fine closure of A are the same as those on A itself. Now if f is continuous in any topology, then this is the case when the said closure is taken in that topology. We proceed to define such a topology.

Definition. An arbitrary set A is called *finely open* iff for each $x \in A$, there exists $B \in \mathscr{E}^{\bullet}$ such that $x \in B \subset A$, and

$$P^x\{T_{B^c} > 0\} = 1. \tag{1}$$

Namely, almost every path starting at a point in A will remain in a nearly Borel subset of A for a nonempty initial interval of time. Clearly an open set is finely open. It is immediately verified that the class of finely open sets meets the two requirements below to generate a topology:

(a) the union of any collection of finely open sets is finely open;
(b) the intersection of any finite collection of finely open sets is finely open.

We shall denote this class by **O**. Observe that if we had required a finely open set to belong to $\mathscr{E}^{\bullet}$, condition (a) would fail. On the other hand, there is a fine base consisting of sets in $\mathscr{E}^{\bullet}$; see Exercise 5 below.

Since ∂ is absorbing, the singleton $\{\partial\}$ forms a finely open set so that ∂ is an isolated point in the fine topology. If $\varphi(x) = E^x\{e^{-\alpha\zeta}\}$, where $\zeta = T_{\{\partial\}}$, then $\varphi(\partial) = 1$. We must therefore *abandon the convention made earlier that a function defined on* $\mathbf{E}_\partial$ *vanishes at* ∂. There may well be other absorbing points besides ∂. A point x is called holding iff x is not regular for the set $\mathbf{E}_\partial - \{x\}$; in other words iff P^x-almost every path starting at x will remain at x for a nonempty initial period of time. This is the case if and only if x is isolated in the fine topology, but it need not be absorbing. A *stable* state in a Markov chain is such an example; indeed under a mild condition all states are stable (see Chung [2]).

A set is finely closed iff its complement in $\mathbf{E}_\partial$ is finely open. It is immediate that if $B \in \mathscr{E}^{\bullet}$ then B is finely closed iff $B^r \subset B$, which is in turn equivalent to $B^* = B$. Thus the operation "*" is indeed the closure operation in the fine topology, justifying the term "fine closure".

The next result yields a vital link between two kinds of continuity.

Theorem 1. *Let $f \in \mathscr{E}^{\bullet}$. Then f is finely continuous if and only if $t \to f(X_t)$ is right continuous on $[0, \infty)$, almost surely.*

Proof. Assume the right continuity. For any real constant c, put

$$A = \{f < c\}, \qquad A' = \{f > c\}.$$

For each $x \in A$, under P^x we have $f(X_0) = f(x) < c$, hence $P^x\{\overline{\lim}_{t\downarrow 0} f(X_t) < c\} = 1$ by right continuity at $t = 0$. This implies (1) with $B = A$. Hence A is finely open; similarly A' is finely open. Consequently $f^{-1}(U)$ is finely open, first for $U = (c_1, c_2)$ where $c_1 < c_2$, and then for each open set U of $(-\infty, \infty)$. Therefore f is finely continuous, namely continuous in the fine topology, by a general characterization of continuity.

Conversely, let $f \in \mathscr{E}^{\bullet}$ and f be finely continuous. For each $q \in Q$, put

$$A = \{f > q\}, \qquad \varphi_q(x) = E^x\{e^{-T_A}\}.$$

Then A is finely open; also $A \in \mathscr{E}^{\bullet}$ implies that T_A is optional, and φ_q is 1-excessive. It follows from Theorem 6 of §3.4 that there exists Ω_* with $P^x(\Omega_*) = 1$ such that if $\omega \in \Omega_*$ then

$$\forall q \in Q : t \to \varphi_q(X_t) \quad \text{is right continuous.} \tag{2}$$

We claim that for such an ω, $\overline{\lim}_{s \downarrow\downarrow t} f(X(s, \omega)) \leq f(X(t, \omega))$ for all $t \geq 0$. Otherwise there exist $t \geq 0$, $t_n \downarrow\downarrow t$, and a rational q such that

$$f(X(t, \omega)) < q, \qquad f(X(t_n, \omega)) > q.$$

Since $X(t_n, \omega) \in A$ and A is finely open, the point $X(t_n, \omega)$ is certainly regular for A, hence $\varphi_q(X(t_n, \omega)) = 1$ for all n. Therefore we have $\varphi_q(X(t, \omega)) = 1$ by (2). But $B = \{f < q\}$ is also finely open and $X(t, \omega) \in B$, hence by definition the point $X(t, \omega)$ is not regular for B^c, a fortiori not regular for A since $B^c \supset A$. Thus $\varphi_q(X(t, \omega)) < 1$. This contradiction proves the claim. A similar argument shows that $\underline{\lim}_{s \downarrow\downarrow t} f(X(s, \omega)) \geq f(X(t, \omega))$ for all $t \geq 0$, P^x-a.s. Hence $t \to f(X(t, \omega))$ is right continuous in $[0, \infty)$, a.s. □

Corollary. *If $f \in \mathbf{S}^{\alpha}$, then f is finely continuous.*

Theorem 2. *If A_1 and A_2 are in $\mathscr{E}^{\bullet}$, then*

$$(A_1 \cup A_2)^r = A_1^r \cup A_2^r. \tag{3}$$

For each $A \in \mathscr{E}^{\bullet}$, we have

$$(A^r)^r \subset A^r; \tag{4}$$

$$T_{A^r} \geq T_A \quad \textit{almost surely.} \tag{5}$$

Proof. (3) is easy. To prove (4) let $f = \mathscr{E}^{\bullet}\{e^{-T_A}\}$. Then $f(x) = 1$ if and only if $x \in A^r$. If $x \in (A^r)^r$ then $f(x) \geq \inf_{A^r} f$ by Theorem 5 of §3.4. Hence $f(x) = 1$. To prove (5), note first that $A^r \in \mathscr{E}^{\bullet}$ so that T_{A^r} is optional. We have by Theorem 2 of §3.4, almost surely:

$$X(T_{A^r}) \in (A^r)^* = A^r \cup (A^r)^r = A^r$$

where the last equation is by (4). Hence $T_A \circ \theta_{T_{A^r}} = 0$ by the strong Markov property applied at T_{A^r}, and this says that T_A cannot be strictly greater than T_{A^r}. □

We shall consider the amount of time that the sample paths of a Hunt process spend in a set. Define for $A \subset \mathbf{E}_\partial$:

$$J_A(\omega) = \{t \geq 0 \mid X(t, \omega) \in A\}. \tag{6}$$

Thus for each $t \geq 0$, "$t \in J_A(\omega)$" is equivalent to "$X(t, \omega) \in A$". We leave it as an exercise to show that if $A \in \mathscr{E}^\bullet$, then $J_A(\omega) \in \mathscr{B}$, a.s. Thus if m denotes the Borel-Lebesgue measure on $\mathbf{T}$, then $m(J_A(\omega))$ is the total amount of time that the sample function $X(\cdot, \omega)$ spends in A, called sometimes the "occupation time" of A. It follows that

$$m(J_A(\omega)) = \int_0^\infty 1_A(X(t, \omega))\, dt; \tag{7}$$

and consequently by Fubini's theorem for each x:

$$E^x\{m(J_A)\} = \int_0^\infty P_t(x, A)\, dt = U(x, A). \tag{8}$$

Thus the potential of A is the expected occupation time.

Definition. A set A in $\mathscr{E}^\bullet$ is said to be of zero potential iff $U(\cdot, A) \equiv 0$.

Proposition 3. *If $U^\alpha(\cdot, A) \equiv 0$ for some $\alpha \geq 0$, then $U^\alpha(\cdot, A) \equiv 0$ for all $\alpha \geq 0$.*

Proof. We have by the resolvent equation, for any β:

$$U^\beta(x, A) = U^\alpha(x, A) + (\alpha - \beta) U^\beta U^\alpha 1_A(x).$$

If $U^\alpha(\cdot, A) \equiv U^\alpha 1_A(\cdot) \equiv 0$, then $U^\beta U^\alpha 1_A(\cdot) \equiv 0$, hence $U^\beta(\cdot, A) \equiv 0$. □

It follows from (8) that if A is of zero potential, then $m(J_A) = 0$ almost surely; namely almost every path spends "almost no time" (in the sense of Borel-Lebesgue measure) in A.

Proposition 4. *If A is of zero potential, then $A^c = \mathbf{E}_\partial - A$ is finely dense.*

Proof. A set is dense in any topology iff its closure in that topology is the whole space; equivalently iff each nonempty open set in that topology contains at least one point of the set. Let O be a finely open set and $x \in O$. Then almost every path starting at x spends a nonempty initial interval of time in a nearly Borel subset B of O, hence we have $E^x\{m(J_B)\} > 0$. Since $E^x\{m(J_A)\} = 0$ we have $E^x\{m(J_{B \cap A^c})\} > 0$. Thus $O \cap A^c$ is not empty for each finely open O, and so A^c is finely dense. □

The following corollary is an illustration of the utility of fine concepts. It can be deduced from (11) of §3.2, but the language is more pleasant here.

Corollary. *If two excessive functions agree except on a set of zero potential, then they are identical.*

For they are both finely continuous and agree on a finely dense set. The assertion is therefore reduced to a familiar topological one.

From certain points of view a set of zero potential is not so "small" and scarcely "negligible". Of course, what is negligible depends on the context. For example a set of Lebesgue measure zero can be ignored in integration but not in differentiation or in questions of continuity. We are going to define certain small sets which play great roles in potential theory.

Definition. A set A in $\mathscr{E}^{\cdot}$ is called *thin* iff $A^r = \varnothing$; namely iff it is thin at every point of $\mathbf{E}_\partial$. A set is *semipolar* iff it is the union of countably many thin sets. A set A in $\mathscr{E}^{\cdot}$ is called *polar* iff

$$\forall x \in \mathbf{E}_\partial \colon P^x\{T_A < \infty\} = 0. \tag{9}$$

The last condition is equivalent to: $E^x\{e^{-\alpha T_A}\} \equiv P_A^\alpha 1(x) \equiv 0$ for each $\alpha > 0$. For comparison, A is thin if and only if

$$\forall x \in E_\partial \colon P^x\{T_A = 0\} = 0, \quad \text{or} \quad E^x\{e^{-\alpha T_A}\} < 1. \tag{10}$$

It is possible to extend the preceding definitions to all subsets of $\mathbf{E}_\partial$, without requiring them to be nearly Borel. For instance, a set is polar iff it is contained in a nearly Borel set which satisfies (9). We shall not consider such an extension.

The case of uniform motion on R^1 yields facile examples. Each singleton $\{x_0\}$ is thin but not polar. Next, let $x_n \downarrow\downarrow x_\infty > -\infty$, then the set $A = \bigcup_{n=1}^\infty \{x_n\}$ is semipolar but not thin. For a later discussion we note also that each compact subset of A is thin.

From the point of view of the process, the smallness of a set A should be reflected in the rarity of the incidence set J_A. We proceed to study the relations between the space sets and time sets. For simplicity we write φ_A for $P_A^\alpha 1$ below for a fixed α which may be taken as 1. A set A will be called *very thin* iff $A \in \mathscr{E}^{\cdot}$ and

$$\sup_{x \in A} \varphi_A(x) < 1. \tag{11}$$

Note that (11) does not imply that $\sup_{x \in \mathbf{E}} \varphi_A(x) < 1$. An example is furnished by any singleton $\{x_0\}$ in the uniform motion, since $\varphi_{\{x_0\}}(x_0) = 0$ and $\lim_{x \uparrow\uparrow x_0} \varphi_{\{x_0\}}(x) = 1$.

Proposition 5. *If A is very thin, then it is thin. Furthermore almost surely J_A is a discrete set (namely, finite in each finite time interval).*

Proof. Recall that φ_A is α-excessive; hence by Theorem 5 of §3.4, the sup in (11) is the same if it is taken over A^* instead of A. But if $x \in A^r$, then $\varphi_A(x) = 1$. Hence under (11) $A^r = \varnothing$ and A is thin.

Next, denote the sup in (11) by θ so that $\theta < 1$. Define $T_1 \equiv T_A$ and for $n \geq 1$: $T_{n+1} = T_n + T_A \circ \theta_{T_n}$. Thus $\{T_n, n \geq 1\}$ are the successive hitting times of A. These are actually the successive entrance times into A, because $X(T_n) \in A^* = A$ here. The adjective "successive" is justified as follows. On $\{T_n < \infty\}$, $X(T_n)$ is a point which is not regular for A; hence by the strong Markov property we must have $T_A \circ \theta_{T_n} > 0$, namely $T_{n+1} > T_n$. Now we have, taking $\alpha = 1$ in φ_A:

$$E\{e^{-T_{n+1}}\} = E\{e^{-T_n} E^{X(T_n)}[e^{-T_A}]; T_n < \infty\} \leq \theta E\{e^{-T_n}\}.$$

It follows that $E\{e^{-T_n}\} \leq \theta^n \downarrow 0$ as $n \to \infty$. Let $T_n \uparrow T_\infty$. Then $E\{e^{-T_\infty}\} = 0$; consequently $P\{T_\infty = \infty\} = 1$. Therefore, we have almost surely

$$J_A = \bigcup_{n=1}^{\infty} \{T_n\} 1_{\{T_n < \infty\}}; \tag{12}$$

Since $T_n \uparrow \infty$, this is a discrete set in $[0, \infty)$. □

Theorem 6. *Each semipolar set is the union of countably many very thin sets. For each $A \in \mathscr{E}^\bullet$, $A \backslash A^r$ is semipolar.*

Proof. For any $A \in \mathscr{E}^\bullet$, we define for $n \geq 1$:

$$A^{(n)} = \left\{x \in \mathbf{E}_\partial \,\middle|\, \varphi_A(x) \leq 1 - \frac{1}{n}\right\}. \tag{13}$$

Then we have

$$A = (A \cap A^r) \cup \left(\bigcup_{n=1}^{\infty} A \cap A^{(n)}\right). \tag{14}$$

For each n, it is obvious that

$$\sup_{x \in A \cap A^{(n)}} \varphi_{A \cap A^{(n)}}(x) \leq \sup_{x \in A^{(n)}} \varphi_A(x) \leq 1 - \frac{1}{n}.$$

Hence each $A \cap A^{(n)}$ is a very thin set. If A is thin, then (14) exhibits A as a countable union of very thin sets. Therefore each semipolar set by its definition is also such a union. Finally, (14) also shows that $A - (A \cap A^r)$ is such a union, hence semipolar. □

Corollary. *If A is semipolar then there is a countable collection of optional times $\{T_n\}$ such that* (12) *holds, where each $T_n > 0$, a.s.*

This follows at once from the first assertion of the theorem and (12). Note however that the collection $\{T_n\}$ is not necessarily well-ordered (by the increasing order) as in the case of a very thin set A.

We summarize the hierarchy of small sets as follows.

Proposition 7. *A polar set is very thin; a very thin set is thin; a thin set is semipolar; a semipolar set is of zero potential.*

Proof. The first three implications are immediate from the definitions. If A is semipolar, then J_A is countable and so $m(J_A) = 0$ almost surely by Theorem 6. Hence $U(\cdot, A) \equiv 0$ by (8). This proves the last implication. □

For Brownian motion in R^1, each Borel set of measure zero is of zero potential, and vice versa. Each point x is regular for $\{x\}$, hence the only thin or semipolar set is the empty set. In particular, it is trivial that semipolar and polar are equivalent notions in this case. The last statement is also true for Brownian motion in any dimension, but it is a deep result tantamount to Kellogg-Evans's theorem in classical potential theory, see §4.5 and §5.2.

Let us ponder a little more on a thin set. Such a set is finely separated in the sense that each of its points has a fine neighborhood containing no other point of the set. If the fine topology has a countable base (namely, satifies the second axiom of countability), such a set is necessarily countable. In general the fine topology does not even satisfy the first axiom of countability, so a finely separated set may not be so sparse. Nevertheless Theorem 6 asserts that almost every path meets the set only countably often, and so only on a countable subset (depending on the path). The question arises if the converse is also true. Namely, if A is a nearly Borel set such that almost every path meets it only countably often, or more generally only on a countable subset, must A be semipolar? Another related question is: if every compact subset of A is semipolar, must A be semipolar? [Observe that if every compact subset of A is polar, then A is polar by Theorem 8 of §3.3.]

These questions are deep but have been answered in the affirmative by Dellacherie under an additional hypothesis which is widely used, as follows.

Hypothesis (L). *There exists a measure ξ_0 on $\mathscr{E}^\bullet$, which is the sum of countably many finite measures, such that if $\xi_0(A) = 0$ then A is of zero potential. In other words, $U(x, \cdot) \ll \xi_0$ for all x.*

It follows easily that if we put $\xi = \xi_0 U^\alpha$ for any $\alpha > 0$, then A is of potential zero *if and only if* $\xi(A) = 0$. Hence under Hypothesis (L) such a measure exists. It will be called a *reference measure* and denoted by ξ below. For example, for Brownian motion in any dimension, the corresponding Borel-Lebesgue measure is a reference measure. It is trivial that there exists a probability measure which is equivalent to ξ. We can then use P^ξ with ξ as the initial distribution.

We now state without proof one of Dellacherie's results; see Dellacherie [1].

Theorem 8. *Assume Hypothesis (L). Let $A \in \mathscr{E}^{\bullet}$ and suppose that almost surely the set J_A in (6) is countable. Then A is semipolar.*

The expediency of Hypothesis (*L*) will now be illustrated.

Proposition 9. *If f_1 and f_2 are two excessive functions such that $f_1 \le f_2$ ξ-a.e., then $f_1 \le f_2$ everywhere. In particular the result is true with equalities replacing the inequalities. If f is excessive and $\int_{\mathbf{E}_\partial} f\, d\xi = 0$, then $f \equiv 0$.*

This is just the Corollary to Proposition 4 with a facile extension.

Proposition 10. *Let $A \in \mathscr{E}^{\bullet}$. Under (L) there exists a sequence of compact subsets K_n of A such that $K_n \uparrow$ and for each x:*

$$P^x\left\{\lim_n T_{K_n} = T_A\right\} = 1. \tag{15}$$

Proof. This is an improvement of Theorem 8 of §3.3 in that the sequence $\{K_n\}$ does not depend on x. Write $\xi(f)$ for $\int_{\mathbf{E}_\partial} f\, d\xi$ below; and put for a fixed $\alpha > 0$:

$$c = \sup_{K \subset A} \xi(P_K^\alpha 1).$$

There exists $K_n \subset A$, $K_n \uparrow$ such that

$$c = \lim_n \xi(P_{K_n}^\alpha 1).$$

Let $f = \lim_n \uparrow P_{K_n}^\alpha 1$; then $c = \xi(f)$ by monotone convergence. For any compact $K \subset A$, let $g = \lim_n \uparrow P_{K_n \cup K}^\alpha 1$. Both f and g are α-excessive and $g \ge f$ by Theorem 4 of §3.4 extended to $\mathbf{S}^\alpha$. But $\xi(g) \le c = \xi(f)$, hence $g = f$ by Proposition 9 (extended to $\mathbf{S}^\alpha$). Thus $P_K^\alpha 1 \le f$ for each compact $K \subset A$, and together with the definition of f we conclude that

$$\forall x: f(x) = \sup_{K \subset A} P_K^\alpha 1(x) \le P_A^\alpha 1(x). \tag{16}$$

On the other hand, for each x it follows from Theorem 4 of §3.4 that there exists a sequence of compact subsets $L_n(x)$ such that

$$\lim_n P_{L_n(x)}^\alpha 1(x) = P_A^\alpha 1(x).$$

Therefore (16) also holds with the inequality reversed, hence it holds with the inequality strengthened to an equality. Recalling the definition of f we have proved that $\lim_n P^\alpha_{K_n}1 = P^\alpha_A 1$. Let $T_{K_n} \downarrow S \geq T_A$. We have $E^x\{e^{-\alpha S}\} = E^x\{e^{-\alpha T_A}\}$, hence $P^x\{S = T_A\} = 1$ which is (15). □

The next curious result is a lemma in the proof of Dellacherie's Theorem 8.

Proposition 11. *Let A be a semipolar set. There exists a finite measure ν on $\mathscr{E}^\bullet$ such that if $B \in \mathscr{E}^\bullet$ and $B \subset A$, then B is polar if and only if $\nu(B) = 0$.*

Proof. Using the Corollary to Theorem 6, we define ν as follows:

$$\nu(B) = \sum_{n=1}^{\infty} \frac{1}{2^n} P^\xi\{X(T_n) \in B;\, T_n < \infty\}. \tag{17}$$

If $B \in \mathscr{E}^\bullet$ and $\nu(B) = 0$, then

$$P^\xi\left\{\bigcup_{n=1}^{\infty} \{X(T_n) \in B;\, T_n < \infty\}\right\} = 0.$$

If $B \subset A$, then $J_B \subset J_A = \bigcup_n \{T_n\} 1_{\{T_n < \infty\}}$ by the definition of $\{T_n\}$, hence the above implies that

$$P^\xi\{T_B < \infty\} = P^\xi\{J_B \neq \varnothing\} = 0.$$

Hence B is polar by Proposition 9 applied to $f = P_B 1$. The converse is trivial. □

The following analogous result, without Hypothesis (L), is of interest.

Proposition 12. *Let A be a semipolar set. Then for each x, there exists $F = \bigcup_{n=1}^{\infty} K_n$ where the K_n's are compact subsets of A, such that P^x-a.s. we have $T_{A-F} = \infty$.*

Proof. Replace ξ by x in (17) and call the resulting measure ν^x. Since this is a finite measure on a locally compact separable metric space, it is regular. Hence

$$\nu^x(A) = \sup_{K \subset A} \nu^x(K)$$

where K ranges over compact sets. Let K_n be a sequence of compact subsets of A such that $\nu^x(K_n)$ increases to $\nu^x(A)$. Then its union F has the property that $\nu^x(A - F) = 0$. The argument in the proof of Proposition 11 then shows that J_{A-F} is empty P^x-almost surely. □

Exercises

1. Prove that $\mathscr{E}^{\bullet}$ is a σ-field and that $\mathscr{E}^{\bullet} \subset \mathscr{E}^{\sim}$.

2. Prove that if A is finely dense, then $A^r = \mathbf{E}_\partial$. Furthermore for almost every ω, the set $J_A(\omega)$ is dense in $\mathbf{T}$. [Hint: consider $E^x\{e^{-T_A}\}$.]

3. Suppose that Hypothesis (L) holds. Then for $\alpha \geq 0$, each α-excessive function is Borelian. If $A \in \mathscr{E}$, then $A^r \in \mathscr{E}$. For each $t \geq 0$, $x \to P^x\{T_B \leq t\}$ is Borelian. [Hint: the last assertion follows from the first through the Stone-Weierstrass theorem; this is due to Getoor.]

4. Suppose that for some $\alpha > 0$, all α-excessive functions are lower semi-continuous. Then Hypothesis (L) holds. [Hint: consider λU^α where $\lambda = \sum_n 2^{-n}\varepsilon_{x_n}$ and $\{x_n\}$ is a dense set in $\mathbf{E}$.]

5. Let $A \subset \mathbf{E}$ and A be a fine neighborhood of x. Then for $\alpha > 0$ there exists an α-excessive function φ and a compact set K such that $\{\varphi < 1\} \subset K \subset A$, and $\{\varphi < 1\}$ is a fine neighborhood of x. [Hint: Let V be open with compact closure, $B = A^c \cup V^c$. By Theorem 8(a) of §3.3, there exists open $G \supset B$ such that $\varphi(x) = E^x\{e^{-\alpha T_G}\} < 1$. Take $K = G^c$.]

6. Fix $\alpha > 0$. If $\mathscr{T}$ is a topology on $\mathbf{E}_\partial$ which renders all α-excessive functions continuous, then $\mathscr{T}$ is a finer topology than the fine topology, namely each finely open set is open in $\mathscr{T}$. Thus, the fine topology is the least fine topology rendering all α-excessive functions continuous. [Hint: consider the φ in Problem 5.]

7. If f is bounded and finely continuous, then for each $\beta \geq 0$, $\lim_{\alpha\to\infty} \alpha U^{\alpha+\beta} f = f$. As a consequence, if the function f in Problem 4 of §3.2 is finely continuous, then it is excessive.

8. Let $A \in \mathscr{E}^{\bullet}$ and $x \in A^r$ and consider the following statement: "for almost every ω, $X(t,\omega) = x$ implies that there exists $t_n \downarrow\downarrow t$ such that $X(t_n,\omega) \in A$." This statement is true for a *fixed* t (namely when t is a constant independent of ω), or when $t = T(\omega)$ where T is optional; but it is false when t is *generic* (namely, when t is regarded as the running variable in $t \to X(t,\omega)$). [Hint: take $x = 0$, $A = \{0\}$ in the Brownian motion in R^1. See §4.2 for a proof that $0 \in \{0\}^r$.]

9. Let A be a finely open set. Then for almost every ω, $X(t,\omega) \in A$ implies that there exists $\delta(\omega) > 0$ such that $X(u,\omega) \in A$ for all $u \in [t, t + \delta(\omega))$. Here t is generic. The contrast with Problem 8 is interesting. [Hint: suppose $A \in \mathscr{E}^{\bullet}$ and let $\varphi(x) = E^x\{e^{-T_{A^c}}\}$. There exists $\delta'(\omega) \leq 1$ such that $\varphi(X(u,\omega)) \leq \lambda < 1$ for $u \in [t, t + \delta'(\omega))$. The set $B = A^c \cap \{\varphi \leq \lambda\}$ is very thin. By Proposition 5 of §3.5, $J_B(\omega) \cap [t, t + \delta'(\omega))$ is a finite set. Take $\delta(\omega)$ to be the minimum element of this set.]

10. If A is finely closed, then almost surely $J_A(\omega)$ is closed from the right, namely, $t_n \downarrow\downarrow t$ and $t_n \in J_A(\omega)$ implies $t \in J_A(\omega)$, for a generic t.

3.6. Decreasing Limits

The limit of an increasing sequence of excessive functions is excessive, as proved in Proposition 2 of §3.2. What can one say about a decreasing sequence? The following theorem is due to H. Cartan in the Newtonian case and is actually valid for any convergent sequence of excessive functions. It was proved by a martingale method by Doob, see Meyer [2]. The proof given below is simpler, see Chung [4].

Theorem 1. *Let $f_n \in \mathbf{S}$ and* $\lim f_n = f$. *Then f is superaveraging and the set $\{f > \hat{f}\}$ is semipolar. In case $\hat{f} < \infty$ everywhere, then*

$$\{f > \hat{f} + \varepsilon\} \quad \textit{is thin} \tag{1}$$

for every $\varepsilon > 0$.

Proof. For each $t \geq 0$, $f_n \geq P_t f_n$; letting $n \to \infty$ we obtain $f \geq P_t f$ by Fatou's lemma. Hence f is superaveraging. For each compact K, $f_n \geq P_K f_n$ by Theorem 4 of §3.4; letting $n \to \infty$ we obtain $f \geq P_K f$ as before. Now let A denote the set in (1) and let K be a compact subset of A; then by the Corollary to Theorem 2 of §3.4 we have $f > \hat{f} + \varepsilon$ on the support of $P_K(x, \cdot)$ for every x. Therefore, since $\hat{f} < \infty$ on A,

$$f \geq P_K f \geq P_K(\hat{f} + \varepsilon) = P_K \hat{f} + \varepsilon P_K 1$$

and consequently

$$P_t f \geq P_t P_K \hat{f} + \varepsilon P_t P_K 1. \tag{2}$$

Both $P_K \hat{f}$ and $P_K 1$ are excessive by Theorem 4 of §3.4. Letting $t \downarrow 0$ in (2), we obtain

$$\hat{f} \geq P_K \hat{f} + \varepsilon P_K 1. \tag{3}$$

For a fixed x there exists a sequence of compact subsets K'_n of A such that $P_{K'_n} \hat{f}(x) \uparrow P_A \hat{f}$ and another sequence K''_n such that $P_{K''_n} 1(x) \uparrow P_A 1(x)$; by (12) of §3.4. Taking $K_n = K'_n \cup K''_n$ we see that

$$P_{K_n} \hat{f}(x) \uparrow P_A \hat{f}(x), \quad \text{and} \quad P_{K_n} 1(x) \uparrow P_A 1(x).$$

Using this sequence of K_n in (3) we obtain

$$\hat{f}(x) \geq P_A \hat{f}(x) + \varepsilon P_A 1(x). \tag{4}$$

If $\hat{f} < \infty$ everywhere this relation implies that $A^r = \varnothing$; for otherwise if $x \in A^r$ it would read $\hat{f}(x) \geq \hat{f}(x) + \varepsilon$ which is impossible. Hence in this case

A is thin. Letting $\varepsilon \downarrow 0$ through a sequence we conclude that $\{f > \hat{f}\}$ is semipolar.

In the general case, let $f_n^{(m)} = f_n \wedge m$, $f^{(m)} = f \wedge m$. We have just shown that $\{f^{(m)} > \hat{f}^{(m)} + \varepsilon\}$ is thin for each $\varepsilon > 0$. Notice that $f^{(m)}$ is superaveraging and $\hat{f} \wedge m \geq \hat{f}^{(m)}$. It follows that

$$\begin{aligned}\{f > \hat{f} + \varepsilon\} &= \bigcup_{m=1}^{\infty} \{f \wedge m > (\hat{f} \wedge m) + \varepsilon\} \\ &\subset \bigcup_{m=1}^{\infty} \{f^{(m)} > \hat{f}^{(m)} + \varepsilon\}\end{aligned}$$

and therefore $\{f > \hat{f} + \varepsilon\}$ is semipolar. Hence so is $\{f > \hat{f}\}$. □

Corollary. *If the limit function f above vanishes except on a set of zero potential, then it vanishes except on a polar set.*

Proof. Let $\varepsilon > 0$, $A = \{f \geq \varepsilon\}$ and K be a compact subset of A. Then we have by the Corollary to Theorem 2 of §3.4:

$$f \geq P_K f \geq \varepsilon P_K 1. \tag{5}$$

Hence $P_K 1$ vanishes except on a set of zero potential, and so it vanishes on a finely dense set by Proposition 4 of §3.5. But $P_K 1$ being excessive is finely continuous, hence it vanishes identically by Corollary to Proposition 4 of §3.5. Thus K is polar. This being true for each compact $K \subset A$, A is polar by a previous remark. □

The corollary has a facile generalization which will be stated below together with an analogue. Observe that the condition (6) below holds for an excessive f.

Proposition 2. *Let $f \in \mathscr{E}_+^{\sim}$ and suppose that we have for each compact K:*

$$f \geq P_K f. \tag{6}$$

If $\{f \neq 0\}$ is of zero potential, then it is polar. If $\{f = \infty\}$ is of zero potential, then it is polar.

Proof. The first assertion was proved above. Let $K \subset \{f = \infty\}$, then we have as in (5):

$$f \geq P_K f \geq \infty\, P_K 1.$$

Hence $P_K 1$ vanishes on $\{f < \infty\}$. The rest goes as before. □

An alternative proof of the second assertion in Proposition 2, for an excessive function, may be gleaned from Theorem 7 of §3.4. For according to that theorem, almost surely the set $\{t: f(X(t)) = \infty\}$ is either empty or of the form $[0, D_F)$ or $[0, D_F]$, where D_F is the first entrance time of $\{f < \infty\}$. If $\{f = \infty\}$ is of zero potential then $\{f < \infty\}$ is finely dense by Proposition 4 of §3.5, which implies that $D_F = 0$ because $\{t \mid f(X(t)) = \infty\}$ cannot contain any nonempty interval (see Exercise 2 of §3.5).

A cute illustration of the power of these general theorems is given below.

EXAMPLE. Consider Brownian motion in R^2. We have $P_t(x, dy) = p_t(x - y)\,dy$ where

$$p_t(x) = \frac{1}{2\pi t}\exp\left(-\frac{\|x\|^2}{2t}\right), \qquad \|x\|^2 = x_1^2 + x_2^2.$$

Well known convolution property yields

$$P_s p_t(x) = \int_{R^2} p_s(x - y) p_t(y)\,dy = p_{s+t}(x).$$

Define for $\alpha > 0$:

$$f(x) = \int_0^\infty e^{-\alpha t} p_t(x)\,dt;$$

this is just $u^\alpha(x)$ in (10) of §3.7 below. For each $s \geq 0$, we have

$$e^{-\alpha s} P_s f(x) = \int_0^\infty e^{-\alpha(s+t)} P_s p_t(x)\,dt = \int_s^\infty e^{-\alpha t} p_t(x)\,dt,$$

from which it follows that f is α-excessive. It is obvious from the form of $p_t(x)$ that $f(x)$ is finite for $x \neq o$, and $f(o) = +\infty$. Thus $\{f = \infty\} = \{o\}$. Since $P_t(\cdot, \{o\}) = 0$ for each t, $\{o\}$ is of zero potential. By the second assertion of Proposition 2 applied to the semigroup $(e^{-\alpha t} P_t)$, we conclude that $\{o\}$ is a polar set. Therefore every singleton is polar for Brownian motion in R^2, and consequently also in R^d, $d \geq 2$, by considering the projection on R^2. We shall mention this striking fact later on more than one occasion.

For $\alpha = 0$ we have $f \equiv \infty$ in the above. It may be asked whether we can find an excessive function to serve in the preceding example. The answer will be given in Theorem 1 and Example 2 of the next section.

The next theorem deals with a more special situation than Theorem 1. It is an important part of the Riesz representation theorem for an excessive function (see Blumenthal and Getoor [1], p. 84).

Theorem 3. *Let $B_n \in \mathscr{E}^\bullet$ and $B_n \downarrow$; let $T_n = T_{B_n}$. Let $f \in \mathbf{S}$ and $g = \lim_n P_{T_n} f$. Then g is superaveraging and $g = \hat{g}$ on $\{g < \infty\} \cap (\bigcap_n B_n^r)^c$.*

Proof. We have by Fatou's lemma,

$$P_t g = P_t\left(\lim_n P_{T_n} f\right) \le \lim_n P_t P_{T_n} f \le \lim_n P_{T_n} f = g$$

since $P_{T_n} f$ is excessive by Theorem 4 of §3.4. Thus g is superaveraging. If $g(x) < \infty$, then there exists n_0 such that $P_{T_n} f(x) < \infty$ for $n \ge n_0$. We have by Proposition 1 of §3.4,

$$\begin{aligned} P_t P_{T_n} f(x) &= E^x\{f(X(t + T_n \circ \theta_t)); T_n < \infty\} \\ &\ge E^x\{f(X(t + T_n \circ \theta_t)); t < T_n < \infty\}. \end{aligned} \tag{7}$$

But for any $A \in \mathscr{E}^\bullet$

$$t + T_A \circ \theta_t = T_A \quad \text{on } \{t < T_A\}. \tag{8}$$

This relation says that the first hitting time on A after t *is* the first hitting time of A after 0, provided that A has not yet been hit at time t. It follows from (7) that

$$P_t P_{T_n} f(x) \ge E^x\{f(X(T_n)); t < T_n\},$$

and consequently by subtraction:

$$P_{T_n} f(x) - P_t P_{T_n} f(x) \le E^x\{f(X(T_n)); T_n \le t\}. \tag{9}$$

Now under P^x, $\{f(X(T_n)), \mathscr{F}_{T_n}, n \ge 1\}$ is a supermartingale, and $\{T_n \le t\} \in \mathscr{F}_{T_n}$. Therefore, by the supermartingale inequality we have

$$\begin{aligned} E^x\{f(X(T_n)); T_n \le t\} &\ge E^x\{f(X(T_{n+1})); T_n \le t\} \\ &\ge E^x\{f(X(T_{n+1})); T_{n+1} \le t\}. \end{aligned} \tag{10}$$

Thus the right member of (9) decreases as n increases. Hence for each $k \ge n_0$ we have for all $n \ge k$:

$$P_{T_n} f(x) - P_t P_{T_n} f(x) \le E^x\{f(X(T_k)); T_k \le t\}. \tag{11}$$

If $n \to \infty$, then $P_t P_{T_n} f(x) \to P_t g(x)$ by dominated convergence, because $P_{T_n} f(x) \le P_{T_k} f(x)$, and $P_t P_{T_k} f(x) \le P_{T_k} f(x) < \infty$. It follows from (11) that

$$g(x) - P_t g(x) \le E^x\{f(X(T_k)); T_k \le t\}. \tag{12}$$

If $x \notin B_k^r$, then $P^x\{T_k > 0\} = 1$. Letting $t \downarrow 0$ in (12) we obtain $g(x) - \hat{g}(x) = 0$. This conclusion is therefore true for each x such that $g(x) < \infty$ and $x \notin \bigcap_{k=1}^\infty B_k^r$, as asserted. □

Two cases of Theorem 3 are of particular interest.

Case 1. Let G_n be open, $\bar{G}_{n-1} \subset G_n$ and $\bigcup_n G_n = \mathbf{E}$. For each $x \in \mathbf{E}$, there exists n such that $x \in G_{n-1}$ so that $x \notin (G_n^c)^r$. Hence $\bigcap_n (G_n^c)^r = \{\partial\}$. If we put $f(\partial) = 0$ as by usual convention, then

$$\lim_n P_{T_{G_n^c}} f = g = \hat{g} \quad \text{on } \{f < \infty\}. \tag{13}$$

We can also take K_n compact, $K_{n-1} \subset K_n^0$ and $\bigcup_n K_n = \mathbf{E}$. Then $\bigcap_n (K_n^c)^r = \{\partial\}$ and (13) is true when G_n is replaced by K_n.

Case 2. For a fixed $x_0 \neq \partial$ let G_n be open, $G_n \supset \bar{G}_{n+1}$ and $\bigcap_n G_n = \{x_0\}$. For each $x \neq x_0$, there is an n such that $x \notin G_n^r$. Hence for any such x for which $f(x) < \infty$, we have

$$\lim_n \mathrm{P}_{T_{G_n}} f(x) = g(x) = \hat{g}(x).$$

On closer examination the two cases are really the same!

In a sense the next result is an analogue of Theorem 3 when the optional times T_n are replaced by the constant times n.

Theorem 4. *Let $f \in \mathbf{S}$ and suppose that*

$$h(x) = \lim_{t \to \infty} \downarrow P_t f(x) < \infty \tag{14}$$

for all x. Then $h = P_t h$ for each $t \geq 0$; and

$$f = h + f_0 \tag{15}$$

where f_0 is excessive with the property that

$$\lim_{t \to \infty} P_t f_0 \equiv 0. \tag{16}$$

Proof. For each x, the assumption (14) implies that there exists $t_0 = t_0(x)$ such that $P_{t_0} f(x) < \infty$. Since $P_t f \leq P_{t_0} f$ for $t \geq t_0$, and $P_s(P_{t_0} f)(x) \leq P_{t_0} f(x) < \infty$ for $s \geq 0$, the following limit relation holds by dominated convergence, for each $s \geq 0$:

$$P_s h = P_s\left(\lim_{t \to \infty} P_t f\right) = \lim_{t \to \infty} P_s P_t f = \lim_{t \to \infty} P_{s+t} f = h. \tag{17}$$

Put $f_0 = f - h$, so that $f_0(x) = \infty$ if $f(x) = \infty$. Then we have for $t \geq t_0$:

$$P_t f_0 = P_t f - P_t h = P_t f - h, \tag{18}$$

and it follows that f_0 is excessive. Letting $t \to \infty$ in (18) we obtain (16). □

This theorem gives a decomposition of an excessive function into an "invariant" part h and a "purely excessive" part f_0 which satisfies the important condition (6) in Theorem 6 of §3.2. It is an easy and rather emasculated version of the Riesz decomposition in potential theory. The reader is welcome to investigate what happens when the condition (14) is dropped.

A noteworthy instance of Theorem 4 is when $f = P_A 1$ where $A \in \mathscr{E}^\bullet$. In this case we have

$$h(x) = \lim_{t \to \infty} P_t P_A 1(x) = \lim_{t \to \infty} P^x\{t + T_A \circ \theta_t < \infty\}. \tag{19}$$

Since $t + T_A \circ \theta_t$ increases with t, we may write (19) as

$$h(x) = P^x\left\{\bigcap_{n=1}^{\infty} (T_A \circ \theta_n < \infty)\right\}. \tag{20}$$

A little reflection shows that the set within the braces in (20) is exactly the set of ω for which $J_A(\omega)$ is an unbounded subset of $\mathbf{T}$, where J_A is defined in (6) of §3.5.

Definition. The set A is *recurrent* iff the probability in (19) is equal to one for every $x \in \mathbf{E}$; and *transient* iff it is equal to zero for every $x \in \mathbf{E}$. Of course this is not a dichotomy in general, but for an important class of processes it indeed is (see Exercise 6 of §4.1).

Clearly if A is recurrent then $P_A 1 \equiv 1$ (on $\mathbf{E}$). The converse is true if and only if (P_t) is strictly Markovian. The phenomenon of recurrence may be described in a more vital way, as follows. For each $A \in \mathscr{E}^\bullet$ we introduce a new function:

$$L_A(\omega) = \sup\{t \geq 0 \mid X(t, \omega) \in A\} \tag{21}$$

where $\sup \varnothing = 0$ as by standard and meaningful convention. We call L_A the "quitting time of A", or "the last exit time from A". Its contrast to the hitting time T_A should be obvious. To see that $L_A \in \mathscr{F}^\sim$, note that for each $\omega \in \Omega$:

$$\{L_A(\omega) \leq t\} \equiv \{X(s, \omega) \notin A \text{ for } s > t\} \equiv \{t + T_A \circ \theta_t(\omega) = \infty\}. \tag{22}$$

It follows that

$$\{L_A < \infty\} = \bigcup_{n=1}^{\infty} \{T_A \circ \theta_n = \infty\}; \tag{23}$$

$$\{L_A = \infty\} = \bigcap_{n=1}^{\infty} \{T_A \circ \theta_n < \infty\}. \tag{24}$$

Thus A is recurrent [transient] if and only if $L_A = \infty$ [$< \infty$] almost surely. This may be used as an alternative definition.

It is clear from the first equivalence in (22) that L_A is not optional; indeed $\{L_A \le t\} \in \sigma(X_s, s \in (t, \infty))$, augmented. Such a random time has special properties and is a kind of dual to a hitting time. We shall see much of it in §5.1.

Exercises

1. Let $f \in \mathscr{E}_+$ and $T = T_A$ where $A \in \mathscr{E}$. Suppose that for each x and $t > 0$, we have
$$f(x) \ge E^x\{f(X_t)1_{\{t<T\}}\}.$$
Show that if $f(x) < \infty$, then $\{f(X_t)1_{\{t<T\}}, \mathscr{F}_t, P^x\}$ is a supermartingale.

2. In the notation of Case 1 of Theorem 3, show that for each compact $K \subset \mathbf{E}$, we have $g = P_{K^c}g$ on $\{g < \infty\}$.

3. Under Hypothesis (L), let F be a class of excessive functions. Then there exists a decreasing sequence $\{f_n\}$ in F such that if $u = \lim_n \downarrow f_n$ we have $u \ge \inf_{f \in F} f \ge \hat{u}$. [Hint: choose f_n decreasing and $\lim_n \xi(f_n/(1+f_n)) = \inf_{f\in F} \xi(f/(1+f))$.]

3.7. Recurrence and Transience

A Hunt process will be called *recurrent* iff any, hence every, one of the equivalent conditions in the following theorem is satisfied. We use the convention below that an unspecified point, set, function is of, in, on $\mathbf{E}$, not $\mathbf{E}_\partial$. Thus a function is said to be a constant if it is a constant on $\mathbf{E}$; its value at ∂, when defined, is not at issue. Neverthless, the proof of Theorem 1 is complicated by the fact that it is not obvious that assertion (1) holds *prima facie* under each of the conditions (i) to (iv) there. The additional care needed to avoid being trapped by ∂ is well worth the pain from the methodological point of view. Recall that $\zeta = T_{\{\partial\}}$.

Theorem 1. *The following four propositions are equivalent for a Hunt process, provided that* $\mathbf{E}$ *contains more than one point.*

(i) *Each excessive function is a constant.*
(ii) *For each* $f \in \mathscr{E}^{\cdot}_+$, *either* $Uf \equiv 0$ *or* $Uf \equiv \infty$.
(iii) *For each* B *in* $\mathscr{E}^{\cdot}$ *which is not thin, we have* $P_B 1 \equiv 1$.
(iv) *For each* B *in* $\mathscr{E}^{\cdot}$, *either* $P_B 1 \equiv 0$ *or* $P_B 1 \equiv 1$. *Namely, each nearly Borel set is either polar or recurrent as defined at the end of* §3.6.

Moreover, any of these conditions implies the following:
$$\forall x\colon P^x\{\zeta = \infty\} = 1. \tag{1}$$

In other words, (P_t) *is a strictly Markovian semigroup on E.*

Proof (i) ⇒ (ii). Since Uf is excessive, it must be a constant, say c, by (i). Suppose $c \neq \infty$. We have for $t \geq 0$:

$$P_t Uf(x) = E^x\left\{\int_t^\infty f(X_s)\,ds\right\}. \tag{2}$$

Thus $\int_0^\infty f(X_s)\,ds$ is integrable under E^x, hence finite P^x-almost surely. We now prove that (i) implies (1), as follows.

Let x_1 and x_2 be two distinct points. Then there exist open sets O_1 and O_2 with disjoint closures such that $x_1 \in O_1$, $x_2 \in O_2$. Both $P_{O_1}1$ and $P_{O_2}1$ are excessive functions which take the value one, hence both must be identically equal to 1 by (i). Put

$$T_1 = T_{O_1}, \qquad T_2 = T_1 + T_{O_2} \circ \theta_{T_1},$$

and for $n \geq 1$:

$$T_{2n+1} = T_{2n} + T_{O_1} \circ \theta_{T_{2n}}, \qquad T_{2n+2} = T_{2n+1} + T_{O_2} \circ \theta_{T_{2n+1}}.$$

Since $X(T_{O_1}) \in \bar{O}_1 \subset (\bar{O}_2)^c$, we have $T_1 < T_2$ on $\{T_1 < \infty\}$. The same argument then shows that the sequence $\{T_n\}$ is strictly increasing. Let $\lim_n T_n = T$; then on $\{T < \infty\}$, $X(T-)$ does not exist because $X(T_n)$ oscillates between $\bar{O}_1$ and $\bar{O}_2$ at a strictly positive distance apart. This is impossible for a Hunt process. It follows that almost surely we have $T = \infty$, and this implies (1) because ∂ is absorbing. Hence $P_t c = c$ for all t.

We now have by (2) and dominated convergence:

$$c = \lim_{t \to \infty} P_t c = \lim_{t \to \infty} P_t Uf = 0.$$

Remark. It is instructive to see why (i) does not imply (ii) when $\mathbf{E}$ reduces to one point.

(ii) ⇒ (iii). Fix $B \in \mathscr{E}^\bullet$ and write φ for $P_B 1$. By Theorem 4 of §3.6, we have $\varphi = h + \varphi_0$ where h is invariant and φ_0 is purely excessive. Hence by Theorem 6 of §3.2, there exists $f_n \in \mathscr{E}_+^\bullet$ such that $\varphi_0 = \lim_n \uparrow Uf_n$. Since $Uf_n \leq \varphi_0 \leq 1$, condition (ii) implies that $Uf_n \equiv 0$ and so $\varphi_0 \equiv 0$. Therefore $\varphi = P_t\varphi$ for every $t \geq 0$. If B is not thin, let $x_0 \in B^r$ so that $\varphi(x_0) = 1$. Put $A = \{x \in \mathbf{E} \mid \varphi(x) < 1\}$. We have for each $t \geq 0$:

$$1 = \varphi(x_0) = \int_{\mathbf{E}_\partial} P_t(x_0, dy)\varphi(y).$$

Since $P_t(x_0, \mathbf{E}_\partial) = 1$, the equation above entails that $P_t(x_0, A) = 0$. This being true for each t, we obtain $U1_A(x_0) = 0$. Hence condition (ii) implies that $U1_A \equiv 0$. But A is finely open since φ is finely continuous. If $x \in A$, then $U1_A(x) > 0$ by fine openness. Hence A must be empty and so $\varphi \equiv 1$.

(iii) ⇒ (iv). Under (iii), $P_B 1 \equiv 1$ for any nonempty open set B. Hence (1) holds as under (i). If B is not a polar set, there exists x_0 such that $P_B 1(x_0) = \delta > 0$. Put

$$A = \left\{ x \,\middle|\, P_B 1(x) > \frac{\delta}{2} \right\}; \tag{3}$$

then A is finely open. Since $x_0 \in A$, $x_0 \in A^r$ by fine openness. Hence $P_A 1 \equiv 1$ by (iii). Now we have

$$P_B 1 \geq P_A P_B 1 \geq \inf_{x \in A^*} P_B 1(x) \geq \frac{\delta}{2}, \tag{4}$$

where the first inequality is by Theorem 4 of §3.4, the second by Theorem 2 of §3.4, and the third by (3) and the fine continuity of $P_B 1$. Next, we have for $t \geq 0$, P^x-almost surely:

$$\begin{aligned} P^x\{T_B < \infty \mid \mathscr{F}_t\} &= P^x\{T_B \leq t \mid \mathscr{F}_t\} + P^x\{t < T_B < \infty \mid \mathscr{F}_t\} \\ &= 1_{\{T_B \leq t\}} + 1_{\{t < T_B\}} P^{X(t)}[T_B < \infty]. \end{aligned} \tag{5}$$

This is because $\{T_B \leq t\} \in \mathscr{F}_{t+} = \mathscr{F}_t$; and $T_B = t + T_B \circ \theta_t$ on $\{t < T_B\}$ (see (8) of §3.6), so that we have by the Markov property:

$$\begin{aligned} P^x\{t < T_B < \infty \mid \mathscr{F}_t\} &= P^x\{t < T_B; t + T_B \circ \theta_t < \infty \mid \mathscr{F}_t\} \\ &= 1_{\{t < T_B\}} P^{X(t)}[T_B < \infty]. \end{aligned}$$

As $t \to \infty$, the first member in (5) as well as the first term in the third member converges to $1_{\{T_B < \infty\}}$. Hence the second term in the third member must converge to zero. Now we have $X(t) \in \mathbf{E}$ for $t < \infty$ and therefore $P^{X(t)}[T_B < \infty] \geq \delta/2$ by (4). It follows that $\lim_{t \to \infty} 1_{\{t < T_B\}} = 0$, P^x-almost surely. This is just $P_B 1(x) = 1$, and x is arbitrary.

(iv) ⇒ (i). Let f be excessive. If f is not a constant, then there are real numbers a and b such that $a < b$, and the two sets $A = \{f < a\}$, $B = \{f > b\}$ are both nonempty. Let $x \in A$, then we have by Theorems 4 and 2 of §3.4:

$$a > f(x) \geq P_B f(x) = \int_{B^*} P_B(x, dy) f(y) \geq b P_B(x, B^*) = b P_B 1(x), \tag{6}$$

since $f \geq b$ on B^* by fine continuity. Thus $P_B 1(x) < 1$. But B being finely open and nonempty is not polar, hence $P_B 1 \equiv 1$ by (iv). This contradiction proves that f is a constant. □

Finally, we have proved above that (i) implies (1), hence each of the other three conditions also implies (1), which is equivalent to $P_t(x, \mathbf{E}) = 1$ for every $t \geq 0$ and $x \in \mathbf{E}$.

Remark. For the argument using (5) above, cf. *Course*, p. 344.

EXAMPLE 1. It is easy to construct Markov chains (Example 1 of §1.2) which are recurrent Hunt processes. In particular this is the case if there are only a finite number of states which communicate with each other, namely for any two states i and j we have $p_{ij} \not\equiv 0$ and $p_{ji} \not\equiv 0$. Condition (ii) is reduced to the following: for some (hence every) state i we have

$$\int_0^\infty p_{ii}(t)\,dt = \infty. \tag{7}$$

Since each i is regular for $\{i\}$, the only thin or polar set is the empty set. Hence condition (iv) reduces to: $P^i\{T_{\{j\}} < \infty\} = 1$ for every i and j. Indeed,

$$P^i\{L_{\{j\}} = \infty\} = 1 \tag{8}$$

in the notation of (21) of §3.6. For a general Markov chain (7) and (8) are equivalent provided that all states communicate, even when the process is not a Hunt process; see Chung [2].

EXAMPLE 2. For the Brownian motion in R^d, the transition probability kernel $P_t(x, dy) = p_t(x - y)\,dy$ where

$$p_t(z) = \frac{1}{(2\pi t)^{d/2}} \exp\left[-\frac{\|z\|^2}{2t}\right], \quad \text{where } \|z\|^2 = \sum_{j=1}^{d} z_j^2. \tag{9}$$

For $\alpha \geq 0$, define

$$u^\alpha(z) = \int_0^\infty e^{-\alpha t} p_t(z)\,dt; \qquad u(z) = u^0(z). \tag{10}$$

Then the resolvent kernel $U^\alpha(x, dy) = u^\alpha(x - y)\,dy$.

For $d = 1$ or $d = 2$, we have $u(z) = \infty$. Hence by the Fubini-Tonelli theorem,

$$U(x, B) = \int_B u(x - y)\,dy = 0 \quad \text{or} \quad \infty$$

according as $m(B) = 0$ or $m(B) > 0$. It follows that condition (ii) of Theorem 1 holds. Therefore, Brownian motion in R^1 or R^2 is a recurrent Hunt process. Since each nonempty ball is not thin, the sample path will almost surely enter into it and returns to it after any given time, by condition (iv). Hence it will almost surely enter into a sequence of concentric balls shrinking to a given point, and yet it will not go through the point because a singleton is a polar set. Such a phenomenon can happen only if the entrance times into the successive balls increase to infinity almost surely, for otherwise at a finite limit time the path must hit the center by continuity. These statements can be made quantitative by a more detailed study of the process.

Note: Whereas under recurrence $U1_B \not\equiv 0$ implies $P_B 1 \equiv 1$, the converse is not true in general. For Brownian motion in R^2, the boundary circle C of a nenempty disk is recurrent by the discussion above, but $U1_C \equiv 0$ because $m(C) = 0$.

Turning in the opposite direction, we call a Hunt process "transient" iff its semigroup is transient as defined at the end of §3.2. However, there are several variants of the condition (17) used there. It is equivalent to the condition that **E** is "σ-transient", namely it is the union of a sequence of transient sets. Hence the condition is satisfied if each compact subset of **E** is transient, because **E** is σ-compact. However, to relate the topological notion of compactness to the stochastic notion of transience some analytic hypothesis is needed. Consider the following two conditions on the potentials (resolvents) of a Hunt process:

$$\forall f \in \mathscr{E}^{\bullet}_{+}: Uf \text{ is lower semi-continuous;} \tag{11}$$

$$\exists \alpha > 0 \text{ such that } \forall f \in \mathscr{E}^{\bullet}_{+}: U^{\alpha} f \text{ is lower semi-continuous.} \tag{12}$$

Observe that there is no loss of generality if we use only bounded f, or bounded f with compact supports in (11) and (12), because lower semi-continuity is preserved by increasing limits. It is easy to see that (12) implies (11) by the resolvent equation.

Recall that condition (16) of §3.2 holds for a Hunt process and will be used below tacitly. Also, the letter K will be reserved for a compact set.

Theorem 2. *In general, either* (ii) *or* (iv) *below implies* (iii). *Under the condition* (11), (iii) *implies* (iv). *Under the condition* (12), (iv) *implies* (i).

(i) $\forall K: U1_K$ *is bounded.*
(ii) $\forall K: U1_K$ *is finite everywhere.*
(iii) $\exists h \in \mathscr{E}_+$ such that $0 < Uh < \infty$ on **E**.
(iv) $\forall K; \lim_{t \to \infty} P_t P_K 1 = 0.$ (13)

Proof. (ii) $\Rightarrow$ (iii). Let K_n be compact and increase to **E**, and put

$$A_{nk} = \{x \in K_n \mid U(x, K_n) \leq k\}.$$

The function $U1_{A_{nk}}$ is bounded above by k on the support of $1_{A_{nk}}$, hence by the domination principle (Corollary to Theorem 4 of §3.4), we have

$$U1_{A_{nk}} \leq k \tag{14}$$

everywhere in **E**. Define h as follows:

$$h = \sum_{n=1}^{\infty} \sum_{k=1}^{\infty} \frac{1_{A_{nk}}}{k2^{k+n}}. \tag{15}$$

Clearly $h \leq 1$. For each x, there exist n and k such that $x \in A_{nk}$ by assumption (ii). Hence $h > 0$ everywhere and so also $Uh > 0$. It follows from (14) that $Uh \leq 1$. Thus h has all the properties (and more!) asserted in (ii).

(iii) $\Rightarrow$ (iv) under (11). Since Uh is finite, we have

$$\lim_{t \to \infty} P_t Uh = 0 \tag{16}$$

as in (2). Since $Uh > 0$ and Uh is lower semi-continuous by (11), for each K we have $\inf_{x \in K} Uh(x) = C(K) > 0$. Hence by the corollary to Theorem 2 of §3.4 we have

$$P_K 1 \leq P_K\left(\frac{Uh}{C(K)}\right) = \frac{1}{C(K)} P_K Uh \leq \frac{1}{C(K)} Uh. \tag{17}$$

The assertion (13) follows from (16) and (17).

(iv) $\Rightarrow$ (iii) and under (12) (iv) $\Rightarrow$ (i). Let $K_n \subset K_{n+1}^0$ and $\bigcup_n K_n = \mathbf{E}$. Applying Theorem 6 of §3.2 to each $P_{K_n}1$ we have

$$P_{K_n}1 = \lim_k \uparrow Ug_{nk} \tag{18}$$

where $g_{nk} \leq k$ by the proof of Theorem 6 of §3.2. It follows by Lemma 1 of §3.2 that

$$1 = \lim_n \uparrow P_{K_n}1 = \lim_n \uparrow Ug_{nn}. \tag{19}$$

Define g as follows:

$$g = \sum_{n=1}^{\infty} \frac{g_{nn}}{n2^n}.$$

Then $g \leq 1$ and $0 < Ug \leq 1$ by (19). Next put $h = U^\alpha g$; then $h \leq Ug \leq 1$. It follows from the resolvent equation that

$$Uh = UU^\alpha g \leq \frac{1}{\alpha} Ug \leq \frac{1}{\alpha}. \tag{20}$$

The fact that for each x we have

$$Ug(x) = \int_0^\infty P_t g(x)\, dt > 0$$

implies that

$$h(x) = U^\alpha g(x) = \int_0^\infty e^{-\alpha t} P_t g(x)\, dt > 0.$$

Hence $Uh > 0$ and h satisfies the conditions in (iii). Finally, under (12) h is lower semi-continuous. Since $h > 0$, for each K we have $\inf_{x \in K} h(x) = C(K) > 0$; hence by (20):

$$U1_K \leq \frac{Uh}{C(K)} \leq \frac{1}{\alpha C(K)},$$

namely (i) is true. Theorem 2 is completely proved. □

Let us recall that (iv) is equivalent to "every compact subset of **E** is transient", i.e.,

$$\forall K\colon L_K < \infty \quad \text{almost surely.} \tag{21}$$

In this form the notion of transience is probabilistically most vivid.

EXAMPLE 3. For a Markov chain as discussed in Example 1 above, in which all states communicate, transience is equivalent to the condition $\int_0^\infty p_{ii}(t)\, dt < \infty$ for some (hence every) state. A simple case of this is the Poisson process (Example 3 of §1.2).

EXAMPLE 4. Brownian motion R^d, when $d \geq 3$, is transient. A simple computation using the gamma function yields

$$u(z) = \frac{\Gamma\left(\frac{d}{2} - 1\right)}{2\pi^{d/2}|z|^{d-2}}; \tag{22}$$

in particular for $d = 3$:

$$U(x, dy) = \frac{dy}{2\pi|x - y|}. \tag{23}$$

If B is a ball with center at the origin and radius r, we have by calculus:

$$U(o, B) = \frac{1}{2\pi} \int_B \frac{dy}{|y|} = \frac{\pi r^2}{2}.$$

Let f be bounded (Lebesgue) measurable with compact support K, then

$$Uf(x) = \frac{1}{2\pi} \int_K \frac{f(y)}{|x - y|}\, dy = \frac{1}{2\pi} \int_{K \cap B} \frac{f(y)}{|x - y|}\, dy + \frac{1}{2\pi} \int_{K \setminus B} \frac{f(y)}{|x - y|}\, dy \tag{24}$$

where B is the ball with center x and radius δ. The first term in the last member of (24) is bounded uniformly in x by $(2\pi)^{-1}\|f\|U(o,\delta) = 4^{-1}\|f\|\delta^2$. The second term there is continuous at x by bounded convergence. It is also bounded by $\delta^{-1}\|f\|m(K)$. Thus Uf is bounded continuous. In particular if $f = 1_K$, this implies condition (i) of Theorem 2. The result also implies condition (11) as remarked before. Hence all the conditions in Theorem 2 are satisfied. The case $d > 3$ is similar.

We can also compute for $d = 3$ and $\alpha \geq 0$:

$$u^\alpha(z) = \frac{e^{-|z|\sqrt{2\alpha}}}{2\pi|z|} \tag{25}$$

by using the following formula, valid for any real c:

$$\int_0^\infty \exp\left[-\alpha\left(t - \frac{c}{t}\right)^2\right] dt = \frac{1}{2}\sqrt{\frac{\pi}{\alpha}}.$$

Hence for any bounded measurable f, we have

$$U^\alpha f(x) = \int_{R^3} \frac{\exp[-|x-y|\sqrt{2\alpha}]}{2\pi|x-y|} f(y)\,dy. \tag{26}$$

For $\alpha > 0$, an argument similar to that indicated above shows that $U^\alpha f$ belongs to $\mathbb{C}_0$. This is stronger than the condition (12). For $d \geq 3$, $\alpha \geq 0$, u^α can be expressed in terms of Bessel functions. For odd values of d, these reduce to elementary functions.

Let us take this occasion to introduce a new property for the semigroup (P_t) which is often applicable. It will be said to have the "strong Feller property" iff for each $t > 0$ and $f \in b\mathscr{E}_+$, we have $P_t f \in \mathbb{C}$, namely continuous. [The reader is warned that variants of this definition are used by other authors.] This condition is satisfied by the Brownian motion in any dimension. To see this, we write

$$P_t f(x) = \int_B p_t(x-y)f(y)\,dy + \int_{R^d - B} p_t(x-y)f(y)\,dy. \tag{27}$$

For any x_0 and a neighborhood V of x_0, we can choose B to be a large ball containing V to make the second term in (27) less than ε for all x in V. The first term is continuous in x by dominated convergence. Hence $P_t f$ is continuous at x_0.

We can now show that $U^\alpha f \in \mathbb{C}$ for $\alpha > 0$, and $f \in b\mathscr{E}_+$ without explicit computation of the kernel U^α in (26). Write

$$U^\alpha f(x) = \int_0^\infty e^{-\alpha t} P_t f(x)\,dt$$

and observe that the integrand is dominated by $\|f\|e^{-\alpha t}$, and is continuous in x for each $t > 0$. Hence $U^\alpha f(x_n) \to U^\alpha f(x)$ as $x_n \to x$ by dominated convergence. Since $Uf = \lim_{\alpha \to 0} \uparrow U^\alpha f$ it now follows that Uf is lower semi-continuous, without the explicit form of U. This is a good example where the semigroup (P_t) is used to advantage—one which classical potential theory lacks. Of course, a Hunt process is a lot more than its semigroup.

Let us remark that for any (P_t), the semi-group (P_t^α) where $P_t^\alpha = e^{-\alpha t}P_t$ is clearly transient because $U^\alpha(x, \mathbf{E}) \le 1/\alpha$. This simple device makes available in a general context the results which are consequences of the hypothesis of transience. Theorem 9 of §3.2 is such an example by comparison with Proposition 10 there.

Exercises

1. For a Hunt process, not necessarily transient, suppose $A \in \mathscr{E}$, $B \in \mathscr{E}$ such that $\bar{B} \subset A$ and $U(x_0, A) < \infty$. Then if $\inf_{x \in \bar{B}} U(x, A) > 0$, we have $P^{x_0}\{L_{\bar{B}} < \infty\} = 1$, where $L_{\bar{B}}$ is the last exist time from $\bar{B}$. [Hint: define $T_n = n + T_{\bar{B}} \circ \theta_n$, then $U(x_0, A) \ge E^{x_0}\{U(X(T_n), A); T_n < \infty\}$.]

2. Let X be a transient Hunt process. Let $\bar{A}_1$ be compact, $A_n \downarrow$ and $\bigcap_n \bar{A}_n = B$. Put $f = \lim_n P_{A_n}1$. Prove that f is superaveraging and its excessive regularization $\hat{f}$ is equal to $P_B 1$. [Hint: $f \ge \hat{f} \ge P_B 1$. If $x \notin B$, then

$$P^x\left\{\bigcap_n [T_{A_n} < \infty]\right\} = P^x\{T_B < \infty\}$$

by transience. Hence $\hat{f} = P_B 1$ on $B^c \cup B^r$ which is a finely dense set.]

3. Let $h \in \mathscr{E}_+$ and $Uh < \infty$. In the same setting as Problem 2, prove that if $f = \lim_n P_{A_n} Uh$ then $\hat{f} = P_B Uh$.

4. Let X be a transient Hunt process and suppose Hypothesis (L) holds with ξ a reference probability measure. Let B a compact polar set. Prove that there exists a function $f \in \mathscr{E}_+$ such that $\xi(f) \le 1$ and $f = \infty$ on B. [Hint: in Problem 2 take A_n to be open, and $f_n = P_{A_n}1$. On B^c, $f_n \downarrow P_B 1$, hence $\xi(f_n) \downarrow 0$; on B, $f_n = 1$. Take $\{n_k\}$ so that $\xi(\sum_k f_{n_k}) \le 1$. This is used in classical potential theory to define a polar set as the set of "poles" of a superharmonic function.]

5. Prove that (iii) in Theorem 2 is equivalent to $\mathbf{E}$ being σ-transient. [Hint: under (iii), $\{Uh > \varepsilon\}$ is transient.]

3.8. Hypothesis (B)

This section is devoted to a new assumption for a Hunt process which Hunt listed as Hypothesis (B). What is Hypothesis (A)? This is more or less the set of underlying assumptions for a Hunt process given in §3.1. Unlike some other assumptions such as the conditions of transience, or the Feller property and its generalization or strengthening mentioned in §3.7, Hypothesis (B) is not an analytic condition but restricts the discontinuity of the sample paths. It turns out to be essential for certain questions of potential theory and Hunt used it to fit his results into the classical grain.

Hypothesis (B). *For some $\alpha \geq 0$, any $A \in \mathscr{E}^{\bullet}$, and open G such that $A \subset G$, we have*

$$P_A^\alpha = P_G^\alpha P_A^\alpha. \tag{1}$$

This means of course that the two measures are identical. Recalling Proposition 1 of §3.4, for $S = T_G$ and $T = T_A$, we see that (1) is true if

$$T_A = T_G + T_A \circ \theta_{T_G} \quad \text{almost surely on } \{T_A < \infty\}. \tag{2}$$

On the other hand, if $\alpha > 0$, then the following particular case of (1)

$$P_A^\alpha 1 = P_G^\alpha P_A^\alpha 1 \tag{3}$$

already implies the truth of (2). For (3) asserts that

$$E\{e^{-\alpha T_A}\} = E\{\exp[-\alpha(T_G + T_A \circ \theta_{T_G})]\},$$

whereas the left member in (2) never exceeds the right member which is $\inf\{t > T_G: X(t) \in A\}$. We will postpone a discussion of the case $\alpha = 0$ until the end of this section.

Next, the two members of (2) are equal on the set $\{T_G < T_A\}$. This is an extension of the "terminal" property of T_A expressed in (8) of §3.6. Since $G \supset A$, we have always $T_G \leq T_A$. Can $T_G = T_A$? [If $\bar{A} \subset G$, this requires the sample path to jump from G^c into $\bar{A}$.] There are two cases to consider.

Case (i). $T_G = T_A < \infty$ and $X(T_G) = X(T_A) \in A^r$. In this case, the strong Markov property at T_G entails that $T_A \circ \theta_{T_G} = 0$. Hence the two members of (2) are manifestly equal.

Case (ii). $T_G = T_A < \infty$ and $X(T_G) = X(T_A) \notin A^r$. Then since $X(T_A) \in A \cup A^r$, we must have $X(T_G) \in A \backslash A^r$. Consequently $T_A \circ \theta_{T_G} > 0$ and the left member of (2) is strictly less than the right.

We conclude that (2) holds if and only if case (ii) does not occur, which may be expressed as follows:

$$\forall x: P_G(x, A \backslash A^r) = 0. \tag{4}$$

Now suppose that (2) holds and A is thin. Then neither case (i) nor case (ii) above can occur. Hence for any open $G_n \supset A$ we have almost surely

$$T_{G_n} < T_A, \quad \text{on } \{T_A < \infty\}. \tag{5}$$

Suppose first that $x \notin A$; then by Theorem 8(b) of §3.3 there exist open sets $G_n \supset A$, $G_n \downarrow$ such that $T_{G_n} \uparrow T_A$, P^x-a.s. Since left limits exist everywhere for a Hunt process, we have $\lim_{n\to\infty} X(T_{G_n}) = X(T_A-)$; on the other hand, the limit is equal to $X(T_A)$ by quasi left continuity. Therefore, $X(\cdot)$ is continuous at T_A, P^x-a.s. For an arbitrary x we have

$$\begin{aligned}
P^x\{T_A < \infty; X(T_A-) = X(T_A)\} &= \lim_{t\downarrow 0} P^x\{t < T_A < \infty; X(T_A-) = X(T_A)\} \\
&= \lim_{t\downarrow 0} P^x\{t < T_A; P^{X(t)}[T_A < \infty; \\
&\qquad X(T_A-) = X(T_A)]\} \\
&= \lim_{t\downarrow 0} P^x\{t < T_A; P^{X(t)}[T_A < \infty]\} \\
&= \lim_{t\downarrow 0} P^x\{t < T_A; P_A 1(X_t)\} \\
&= P_A 1(x).
\end{aligned}$$

In the above, the first equation follows from $P^x\{T_A > 0\} = 1$; the second by Markov property; the third by what has just been proved because $X(t) \notin A$ for $t < T_A$; and the fifth by the right continuity of $t \to P_A 1(X_t)$. The result is as follows, almost surely:

$$X(T_A-) = X(T_A) \quad \text{on } \{T_A < \infty\}. \tag{6}$$

Equation (6) expresses the fact that the path is almost surely continuous at the first hitting time of A. This will now be strengthened to assert continuity whenever the path is in A. To do so suppose first that A is very thin. Then by (12) of §3.5, the incidence set J_A consists of a sequence $\{T_n, n \geq 1\}$ of successive hitting times of A. By the strong Markov property and (6), we have for $n \geq 2$, almost surely on $\{T_{n-1} < \infty\}$:

$$\begin{aligned}
P^x\{T_n < \infty; X(T_n-) = X(T_n)\} &= P^{X(T_{n-1})}\{T_A < \infty; X(T_A-) = X(T_A)\} \\
&= P^{X(T_{n-1})}\{T_A < \infty\} = P^x\{T_n < \infty\}.
\end{aligned}$$

It follows that $X(T_n-) = X(T_n)$ on $\{T_n < \infty\}$, and consequently

$$\forall t \geq 0\colon X(t-) = X(t) \quad \text{almost surely on } \{X_t \in A\}. \tag{7}$$

Since (7) is true for each very thin set A, it is also true for each semi-polar set by Theorem 6 of §3.5.

Thus we have proved that (2) implies the truth of (7) for each thin set $A \subset G$. The converse will now be proved. First let A be thin and contained in a compact K which in turn is contained in an open set G. We will show that (4) is true. Otherwise there is an x such that $P^x\{T_G < \infty; X(T_G) \in A\} > 0$. Since A is at a strictly positive distance from G^c, this is ruled out by (7). Hence

$P_G(x, A) = 0$. Next, let $K \subset G$ as before. Then $K \backslash K^r = \bigcup_n C_n$ where each C_n is (very) thin by Theorem 6 of §3.5. Since $C_n \subset K$, the preceding argument yields $P_G(x, C_n) = 0$ and consequently $P_G(x, K \backslash K^r) = 0$. By the discussion leading to (4), this is equivalent to:

$$T_K = T_G + T_K \circ \theta_{T_G} \quad \text{almost surely on } \{T_K < \infty\}. \tag{8}$$

Finally let $A \in \mathscr{E}^{\bullet}$, $A \subset G$. By Theorem 8(b) of §3.5, for each x there exist compact sets K_n, $K_n \subset A$, $K_n \uparrow$ such that $T_{K_n} \downarrow T_A$, P^x-almost surely. On $\{T_A < \infty\}$ we have $T_{K_n} < \infty$ for all large n. Hence (8) with $K = K_n$ yields (2) as $n \to \infty$.

We summarize the results in the theorem below.

Theorem 1. *For a Hunt process the following four propositions are equivalent*:

(i) *Hypothesis* (*B*), *namely* (1), *is true for some* $\alpha > 0$.
(ii) *Equation* (3) *is true for some* $\alpha > 0$.
(iii) *For each semi-polar* A *and open* G *such that* $A \subset G$, *we have* $P_G(x, A) = 0$ *for each* x.
(iv) *For each semi-polar* A, (7) *is true.*

Moreover, if in (1) *or* (3) *we restrict* A *to be a compact set, the resulting condition is also equivalent to the above.*

It should also be obvious that if Hypothesis (*B*) holds for some $\alpha > 0$, it holds for every $\alpha > 0$. We now treat the case $\alpha = 0$.

Theorem 2. *If the Hunt process is transient in the sense that it satisfies condition* (iii) *of Theorem* 2 *of* §3.7, *then the following condition is also equivalent to the conditions in Theorem* 1.

(v) *Hypothesis* (*B*) *is true for* $\alpha = 0$.

Proof. Let $h > 0$ and $Uh \le 1$. Then applying (1) with $\alpha = 0$ to the function Uh, we obtain

$$0 = P_A Uh - P_G P_A Uh = E^{\bullet}\left\{\int_{T_A}^{T_G + T_A \circ \theta_{T_G}} h(X_t)\, dt\right\}.$$

On the set $\{T_A < T_G + T_A \circ \theta_{T_G}; T_A < \infty\}$, the integral above is strictly positive. Hence the preceding equation forces this set to be almost surely empty. Thus (2) is true, which implies directly (1) for every $\alpha \ge 0$. □

Needless to say, we did not use the full strength of (v) above, since it is applied only to a special kind of function. It is not apparent to what extent the transience assumption can be relaxed.

In Theorem 9 of §3.3, we proved that for a Hunt process, $T_B \le T_{\bar{B}}$ a.s. for each $B \in \mathscr{E}^{\bullet}$. It is trivial that $T_B = T_{\bar{B}}$ if all paths are continuous. The next theorem is a deep generalization of this and will play an important role in §5.1. It was proved by Meyer [3], p. 112, under Hypothesis (L), and by Azema [1] without (L). The following proof due to Walsh [1] is shorter but uses Dellacherie's theorem.

Theorem 3. *Under Hypotheses* (L) *and* (B), *we have* $T_B = T_B^-$ a.s. *for each* $B \in \mathscr{E}^\bullet$.

Proof. The key idea is the following *constrained* hitting time, for each $B \in \mathscr{E}^\bullet$:

$$S_B = \inf\{t > 0 \mid X(t-) = X(t) \in B\}. \tag{9}$$

To see that S_B is optional, let $\{J_n\}$ be the countable collection of all jump times of the process (Exercise 5 of §3.1), and $\Lambda = \bigcup_{n=1}^{\infty} \{T_B \neq J_n\}$. Then $\Lambda \in \mathscr{F}(T_B)$ by Theorem 6 of §1.3, hence $S_B = (T_B)_\Lambda$ is optional by Proposition 4 of §1.3. It is easy to verify that S_B is a terminal time. Define $\varphi(x) = E^x\{e^{-S_B}\}$. Then φ is 1-excessive by Exercise 8 of §3.4. Now suppose first that B is compact. For $0 < \varepsilon < 1$ put

$$A = B \bigcap \{x \mid \varphi(x) \leq 1 - \varepsilon\}.$$

Define a sequence of optional times as follows: $R_1 = S_A$, and for $n \geq 1$:

$$R_{n+1} = R_n + S_A \circ \theta_{R_n}.$$

On $\{R_n < \infty\}$ we have $X(R_n) \in A$ because A is finely closed. Hence $\varphi(X(R_n)) \leq 1 - \varepsilon$ and so $P^{X(R_n)}\{S_A > 0\} = 1$ by the zero-one law. It follows that $R_n < R_{n+1}$ by the strong Markov property. On the other hand, we have as in the proof of Proposition 5 of §3.5, for each x and $n \geq 1$:

$$E^x\{e^{-R_{n+1}}\} \leq (1 - \varepsilon)E^x\{e^{-R_n}\} \leq (1 - \varepsilon)^n.$$

Hence $R_n \uparrow \infty$ a.s. Put $I(\omega) = \{t > 0 \mid X(t, \omega) \in A\}$. By the definition of R_n, for a.e. ω such that $R_n(\omega) < \infty$, the set $I(\omega) \cap (R_n(\omega), R_{n+1}(\omega))$ is contained in the set of t where $X(t-, \omega) \neq X(t, \omega)$, and is therefore a countable set. Since $R_n(\omega) \uparrow \infty$, it follows that $I(\omega)$ itself is countable. Hence by Theorem 8 of §3.5, A is a semi-polar set. This being true for each ε, we conclude that the set $B' = B \cap \{x \mid \varphi(x) < 1\}$ is semi-polar.

Now we consider three cases on $\{T_B < \infty\}$, and omit the ubiquitous "a.s."

(i) $X(T_B-) = X(T_B)$. Since B is compact, in this case $X(T_B-) \in B$ and so $T_B^- \leq T_B$.
(ii) $X(T_B-) \neq X(T_B)$ and $X(T_B) \in B - B'$. In this case $\varphi(X(T_B)) = 1$ and so $S_B \circ \theta_{T_B} = 0$. Since $T_B^- \leq S_B$ we have $T_B^- \leq T_B$.
(iii) $X(T_B-) \neq X(T_B)$ and $X(T_B) \in B'$. Since B' is semi-polar this is ruled out by Theorem 1 under Hypothesis (B).

We have thus proved for each compact B that $T_B^- \le T_B$ on $\{T_B < \infty\}$; hence $T_B^- = T_B$ by Theorem 9 of §3.3. For an arbitrary $B \in \mathscr{E}^\bullet$, we have by Theorem 8(b) of §3.3, for each x a sequence of compacts $K_n \subset B, K_n \uparrow$ such that $P^x\{T_{K_n} \downarrow T_B\} = 1$. Since $T_{K_n}^- = T_{K_n}$ as just proved and $T_B^- \le T_{K_n}^-$ for $n \ge 1$, it follows that $P^x\{T_B^- \le T_B\} = 1$. As before this implies the assertion of Theorem 3. □

Exercises

1. Give an example such that $P_A 1 = P_G P_A 1$ but $P_A \ge P_G P_A$ is false. [Hint: starting at the center of a circle, after holding the path jumps to a fixed point on the circle and then execute uniform motion around the circle. This is due to Dellacherie. It is a recurrent Hunt process. Can you make it transient?]

2. Under the conditions of Theorem 1, show that (7) remains true if the set $\{X_t \in A\}$ is replaced by $\{X_{t-} \in A\}$. [Hint: use Theorem 9 of §3.3.]

3. If (7) is true for each compact semi-polar set A, then it is true for every semi-polar set. [Hint: Proposition 11 of §3.5 yields a direct proof without use of the equivalent conditions in Theorem 1.]

4. Let $\mathbf{E} = \{0\} \cup [1, \infty)$; let 0 be a holding point from which the path jumps to 1, and then moves with uniform speed to the right. Show that this is a Hunt process for which Hypothesis (*B*) does not hold.

5. Let A be a thin compact set, G_n open, and $G_n \downarrow\downarrow A$. Show that under Hypothesis (*B*), we have under P^x, $x \notin A$:
$$\forall n\colon T_{G_n} < T_A, \qquad T_{G_n} \uparrow T_A, \quad \text{a.s. on } \{T_A < \infty\}.$$
Thus T_A is predictable under $P^x, x \notin A$.

6. Assume that the set A in Problem 5 has the stronger property that
$$\sup_{x \in \mathbf{E}_\partial} E^x\{e^{-T_A}\} < 1.$$
Then A is polar. [Hint: use Problem 2 of §3.4.]

Notes on Chapter 3

§3.1. Hunt process is named after G. A. Hunt, who was the author's classmate in Princeton. The basic assumptions stated here correspond roughly with Hypothesis (*A*), see Hunt [2]. The Borelian character of the semigroup is not always assumed. A more general process, called "standard process", is treated in Dynkin [1] and Blumenthal and Getoor [1]. Another technical variety called "Ray process" is treated in Getoor [1].

§3.2. The definition of an excessive function and its basic properties are due to Hunt [2]. Another definition via the resolvents given in (14) is used in Blumenthal and Getoor [1]. Though the former definition is more restrictive than the latter, it yields more transparent results. On the other hand, there are results valid only with the resolvents.

§3.3. We follow Hunt's exposition in Hunt [3]. Although the theory of analytical sets is essential here, we follow Hunt in citing Choquet's theorem without details. A proof of the theorem may be found in Helms [1]. But for the probabilistic applications (projection and section) we need an algebraic version of Choquet's theory without endowing the sample space Ω with a topology. This is given in Meyer [1] and Dellacherie and Meyer [1].

§3.4 and 3.5. These sections form the core of Hunt's theory. Theorem 6 is due to Doob [3], who gave the proof for a "subparabolic" function. Its extension to an excessive function is carried over by Hunt using his Theorem 5. The transfinite induction used there may be concealed by some kind of maximal argument, see Hunt [3] or Blumenthal and Getoor [1]. But the quickest proof of Theorem 6 is by means of a projection onto the optional field, see Meyer [2]. This is a good place to learn the new techniques alluded to in the Notes on §1.4.

Dellacherie's deep result (Theorem 8) is one of the cornerstones of what may be called "random set theory". Its proof is based on a random version of the Cantor-Bendixson theorem in classical set theory, see Meyer [2]. Oddly enough, the theorem is not included in Dellacherie [2], but must be read in Dellacherie [1], see also Dellacherie [3].

§3.6. Theorem 1 originated with H. Cartan for the Newtonian potential (Brownian motion in R^3); see e.g., Helms [1]. This and several other important results for Hunt processes have companions which are valid for left continuous moderate Markov processes, see Chung and Glover [1]. Since a general Hunt process reversed in time has these properties as mentioned in the Notes on §2.4, such developments may be interesting in the future.

"Last exit time" is hereby rechristened "quitting time" to rhyme with "hitting time". Although it is the obvious concept to be used in defining recurrence and transience, it made a belated entry in the current vocabulary of the theory. No doubt this is partly due to the former prejudice against a random time that is not optional, despite its demonstrated power in older theories such as random walks and Markov chains. See Chung [2] where a whole section is devoted to last exits. The name has now been generalized to "co-optional" and "co-terminal"; see Meyer-Smythe-Walsh [1]. Intuitively the quitting time of a set becomes its hitting time when the sense of time is reversed ("from infinity", unfortunately). But this facile interpretation is only a heuristic guide and not easy to make rigorous. See §5.1 for a vital application.

§3.7. For a somewhat more general discussion of the concepts of transience and recurrence, see Getoor [2]. The Hunt theory is mainly concerned with the transient case, having its origin in the Newtonian potential. Technically, transience can always be engineered by considering (P_t^α) with $\alpha > 0$, instead of (P_t). This amounts to killing the original process at an exponential rate $e^{-\alpha t}$, independently of its evolution. It can be shown that the resulting killed process is also a Hunt process.

§3.8. Hunt introduced his Hypothesis (B) in order to characterize the balayage (hitting) operator P_A in a way recognizable in modern potential theory, see Hunt [1; §3.6]. He stated that he had not found "simple and general conditions" to ensure its truth. Meyer [3] showed that it is implied by the duality assumptions and noted its importance in the dual theory. It may be regarded as a subtle generalization of the continuity of the paths. Indeed Azéma [1] and Smythe and Walsh [1] showed that it is equivalent to the quasi left continuity of a suitably reversed process. We need this hypothesis in §5.1 to extend a fundamental result in potential theory from the continuous case to the general case. A new condition for its truth will also be stated there.

Chapter 4

Brownian Motion

4.1. Spatial Homogeneity

In this chapter the state space $\mathbf{E}$ is the d-dimensional Euclidean space R^d, $d \geq 1$; $\mathscr{E}$ is the classical Borel field on R^d. For $A \in \mathscr{E}$, $B \in \mathscr{E}$, the sets $A \pm B$ are the vectorial sum and difference of A and B, namely the set of $x \pm y$ where $x \in A$, $y \in B$. When $B = \{x\}$ this is written as $A \pm x$; the set $o - A$ is written as $-A$, where o is the zero point (origin) of R^d.

Let $(P_t; t \geq 0)$ be a strictly stochastic transition semigroup such that for each $t \geq 0$, $x \in \mathbf{E}$ and $x_0 \in \mathbf{E}$ we have

$$P_t(x, A) = P_t(x_0 + x, x_0 + A). \tag{1}$$

Then we say (P_t) is spatially homogeneous. Temporally homogeneous Markov processes with such a semigroup constitute a large and important class. They are called "additive processes" by Paul Lévy, and are now also known as "Lévy processes"; see Lévy [1]. We cannot treat this class in depth here, but some of its formal structures will be presented here as a preparation for a more detailed study of Brownian motion in the following sections.

We define a family of probability measures as follows.

$$\pi_t(A) = P_t(o, A). \tag{2}$$

The semigroup property then becomes the convolution relation

$$\pi_{s+t}(A) = \int \pi_s(dy)\pi_t(A - y) = (\pi_s * \pi_t)(A) \tag{3}$$

where the integral is over $\mathbf{E}$ and $*$ denotes the convolution. This is valid for $s \geq 0$, $t \geq 0$, with π_0 the unit mass at o. We add the condition:

$$\pi_t \to \pi_0 \quad \text{vaguely as } t \downarrow 0. \tag{4}$$

For $f \in \mathscr{E}_+$ and $t \geq 0$ we have

$$P_t f(x) = \int f(x + y)\pi_t(dy). \tag{5}$$

As a consequence of bounded convergence, we see that if $f \in b\mathbb{C}$, then $P_t f \in b\mathbb{C}$; if $f \in \mathbb{C}_0$, then $P_t f \in \mathbb{C}_0$. Moreover $\lim_{t \downarrow 0} P_t f = f$ for $f \in \mathbb{C}_0$, by (4). Thus (P_t) is a Feller semigroup. Hence a spatially homogeneous process may and will be supposed to be a Hunt process.

For $A \in \mathscr{E}$, we define

$$U^\alpha(A) = \int_0^\infty e^{-\alpha t}\pi_t(A)\,dt;$$

and $U^\alpha(x, A) = U^\alpha(A - x)$. Then U^α is the α-potential kernel and we have for $f \in \mathscr{E}_+$:

$$U^\alpha f(x) = \int f(x + y)U^\alpha(dy). \tag{6}$$

We shall denote the Lebesgue measure in $\mathbf{E}$ by m. If $f \in L^1(m)$, then we have by (5), Fubini's theorem and the translation-invariance of m:

$$\begin{aligned}\int m(dx)P_t f(x) &= \int \left[\int m(dx) f(x + y)\right]\pi_t(dy)\\ &= \int \left[\int m(dx) f(x)\right]\pi_t(dy) = \int m(dx) f(x).\end{aligned}$$

For a general measure μ on $\mathscr{E}$, we define the measure μP_t or μU^α by

$$\mu P_t(\cdot) = \int \mu(dx)P_t(x, \cdot); \qquad \mu U^\alpha(\cdot) = \int \mu(dx)U^\alpha(x, \cdot). \tag{7}$$

The preceding result may then be recorded as

$$mP_t = m, \qquad \forall t \geq 0. \tag{8}$$

We say m is an invariant measure for the semigroup (P_t). It follows similarly from (6) that

$$mU^\alpha = \frac{1}{\alpha} m, \qquad \forall \alpha > 0; \tag{9}$$

namely m is also an invariant measure for the resolvent operators (αU^α).

Proposition 1. *Let $\alpha > 0$. If $U^\alpha f$ is lower semi-continuous for every $f \in b\mathscr{E}_+$, then $U^\alpha \ll m$. If $U^\alpha \ll m$, then $U^\alpha f$ is continuous for every $f \in b\mathscr{E}$.*

Proof. To prove the first assertion, suppose $A \in \mathscr{E}$ and $m(A) = 0$. Then $U^\alpha(x, A) = 0$ for m-a.e. x by (9). Since $U^\alpha 1_A$ is lower semi-continuous, this implies $U^\alpha(A) = U^\alpha(o, A) = 0$. Next, if $U^\alpha \ll m$ let u^α be a Radon-Nikodym derivative of U^α with respect to m. We may suppose $u^\alpha \geq 0$ and $u^\alpha \in \mathscr{E}$. Then (6) may be written as

$$U^\alpha f(x) = \int f(x + y) u^\alpha(y) m(dy) = \int f(z) u^\alpha(z - x) m(dz). \tag{10}$$

Since $u^\alpha \in L^1(m)$, a classical result in the Lebesgue theory asserts that

$$\lim_{x' \to x} \int |u^\alpha(z - x) - u^\alpha(z - x')| m(dz) = 0 \tag{11}$$

(see e.g. Titchmarsh [1], p. 377). Since f is bounded, the last term in (10) shows that $U^\alpha f$ is continuous by (11). □

Proposition 2. *The three conditions below are equivalent*:

(a) *For all* $\alpha \geq 0$ *all* α-*excessive functions are lower semi-continuous.*
(b) *For some* $\alpha > 0$, *all* α-*excessive functions are lower semi-continuous.*
(c) *A set* A (*in* $\mathscr{E}$) *is of zero potential if* $m(A) = 0$.

Proof. If (b) is true then $U^\alpha \ll m$ by Proposition 1. Hence if $m(A) = 0$, then for every x, $m(A - x) = 0$ and $U^\alpha(x, A) = U^\alpha(A - x) = 0$. Thus A is of zero potential.

If (c) is true, then $U^\alpha \ll m$. Hence by Proposition 1, $U^\alpha f$ is continuous for every $f \in b\mathscr{E}_+$. It follows that $U^\alpha f$ is lower semicontinuous for every $f \in \mathscr{E}_+$, by an obvious approximation. Hence (a) is true by Theorem 9 of §3.2, trivially extended to all α-excessive functions. □

We proceed to discuss a fundamental property of spatially homogeneous Markov processes (X_t): the existence of a *dual* process $(\hat{X}_t)$. The latter is simply defined as follows:

$$\hat{X}_t = -X_t. \tag{12}$$

It is clear that $(\hat{X}_t)$ is a spatially homogeneous process. Let $\hat{P}_t$, $\hat{U}^\alpha$, $\hat{\pi}_t$, $\hat{u}^\alpha$ be the quantities associated with it. Thus

$$\hat{P}_t(x, A) = P_t(-x, -A)$$
$$\hat{\pi}_t(A) = \pi_t(-A), \qquad \hat{u}^\alpha(x) = u^\alpha(-x).$$

It is convenient to introduce

$$u^\alpha(x, y) = u^\alpha(y - x), \qquad \hat{u}^\alpha(x, y) = \hat{u}^\alpha(y - x).$$

Thus we have

$$\hat{u}^\alpha(x, y) = u^\alpha(y, x) \tag{13}$$

and

$$\hat{U}^\alpha(x, A) = \int_A m(dy)u^\alpha(y, x). \tag{14}$$

(In the last relation it is tempting to write $\hat{U}^\alpha(A, x)$ for $\hat{U}^\alpha(x, A)$. This is indeed a good practice in many lengthy formulas.) Now let $f \in b\mathscr{E}_+$, then

$$\hat{U}^\alpha f(y) = \int m(dx)f(x)u^\alpha(x, y). \tag{15}$$

Recall the notation

$$U^\alpha \mu(x) = \int u^\alpha(x, y)\mu(dy) \tag{16}$$

for a measure μ. If follows from (15) and (16) that if μ is a σ-finite measure as well as m, we have by Fubini's theorem:

$$\int \hat{U}^\alpha f(y)\mu(dy) = \int m(dx)f(x)U^\alpha\mu(x);$$

or perhaps more legibly:

$$\int \hat{U}^\alpha f \, d\mu = \int (f \cdot U^\alpha \mu) \, dm. \tag{17}$$

This turns out to be a key formula worthy to be put in a general context.

We abstract the situation as follows. Let (X_t) and $(\hat{X}_t)$ be two Hunt processes on the same general $(\mathbf{E}, \mathscr{E})$; and let (P_t), $(\hat{P}_t)$; (U^α), $(\hat{U}^\alpha)$ be the associated semigroups and resolvents. Assume that there is a (reference) measure m and a function $u^\alpha \geq 0$ such that for every $A \in \mathscr{E}$ we have

$$U^\alpha(x, A) = \int_A u^\alpha(x, y)m(dy), \qquad \hat{U}^\alpha(x, A) = \int_A m(dy)u^\alpha(y, x). \tag{18}$$

Under these conditions we say the processes are in duality. In this case the relation (17) holds for any $f \in \mathscr{E}_+$ and any σ-finite measure μ; clearly it also holds with U^α and $\hat{U}^\alpha$ interchanged. Further conditions may be put on the dual potential density function $u(\cdot, \cdot)$. We content ourselves with one important result in potential theory which flows from duality.

Theorem 3. *Assume duality. Let μ and ν be two σ-finite measures such that for some $\alpha \geq 0$ we have*

$$U^\alpha \mu = U^\alpha \nu < \infty. \tag{19}$$

Then $\mu \equiv \nu$.

Proof. We have by the resolvent equation for $\beta \geq \alpha$:

$$U^\alpha \mu = U^\beta \mu + (\beta - \alpha) U^\beta U^\alpha \mu.$$

Thus if (19) is true it is also true when α is replaced by any greater value. To illustrate the methodology in the simplest situation let us first suppose that μ and ν are finite measures. Take $f \in b\mathbb{C}$, then

$$\lim_{\alpha \to \infty} \alpha \hat{U}^\alpha f = f$$

boundedly because $(\hat{X}_t)$ is a Hunt process. It follows from this, the duality relation (17), and the finiteness of μ that

$$\int f\, d\mu = \lim_{\alpha \to \infty} \alpha \int \hat{U}^\alpha f\, d\mu = \lim_{\alpha \to \infty} \alpha \int (f \cdot U^\alpha \mu)\, dm.$$

This is also true when μ is replaced by ν, hence by (19) for all large values of α we obtain

$$\int f\, d\mu = \int f\, d\nu.$$

This being true for all $f \in b\mathbb{C}$, we conclude that $\mu \equiv \nu$.

In the general case the idea is to find a function h on $\mathbf{E}$ such that $h > 0$ and $\int h\, d\mu < \infty$; then apply the argument above to the finite measure $h \cdot d\mu$. If μ is a Radon measure, a bounded continuous h satisfying the conditions above can be easily constructed, but for a general σ-finite μ we use a more sophisticated method involving an α-excessive function for the dual process (see Blumenthal and Getoor [1]). Since $U^\alpha \mu < \infty$, it is easy to see that there exists $g \in b\mathscr{E}$, $g > 0$, such that by (17):

$$\infty > \int (g \cdot U^\alpha \mu)\, dm = \int \hat{U}^\alpha g\, d\mu.$$

Put $h = \hat{U}^\alpha g$; then clearly $h \in b\mathscr{E}$, $h > 0$ and $\int h\, d\mu < \infty$. Moreover, h being an α-potential for $(\hat{X}_t)$, is α-excessive for $(\hat{P}_t)$ (called "α-co-excessive"). Thus it follows from (11) of §3.2 that $\lim_{\alpha \to \infty} \alpha \hat{U}^\alpha h = h$. But we need a bit more than this. We know from the Corollary to Theorem 1 of §3.5 that h is finely continuous with respect to $(\hat{X}_t)$ (called "co-finely continuous"); hence so is fh for any $f \in b\mathbb{C}$, because continuity implies fine continuity and continuity in any topology is preserved by multiplication. It follows (see Exercise 7 of §3.5) that

$$\lim_{\alpha \to \infty} \alpha \hat{U}^\alpha (fh) = fh.$$

Now we can apply (17) with f replaced by fh to conclude as before that $\int fh\,d\mu = \int fh\,dv$; hence $h\,d\mu = h\,dv$; hence $\mu \equiv v$. □

Next, we shall illustrate the method of Fourier transforms by giving an alternative characterization of a spatially homogeneous and temporally homogeneous Markov process (without reference to sample function regularities). A stochastic process $\{X_t, t \geq 0\}$ is said *to have stationary independent increments* iff it has the following two properties:

(a) For any $0 \leq t_0 < t_1 < \cdots < t_n$, the random variables $\{X(t_0), X(t_k) - X(t_{k-1}), 1 \leq k \leq n\}$ are independent;
(b) For any $s \geq 0$, $t \geq 0$, $X(s + t) - X(t)$ has the same distribution as $X(s) - X(0)$.

Theorem 4. *A spatially homogeneous and temporally homogeneous Markov process is a process having stationary independent increments, and conversely.*

Proof. We leave the converse part to the reader (cf. Theorem 9.2.2 of *Course*). To prove the direct assertion we use Fourier transform (characteristic functions). We write for $x \in R^d$, $y \in R^d$:

$$\langle x, y \rangle = \sum_{j=1}^{d} x_j y_j \tag{20}$$

if $x = (x_1, \ldots, x_d)$, $y = (y_1, \ldots, y_d)$. Let $i = \sqrt{-1}$ and $u_k \in R^d$; and set $X(t_{-1}) = o$. We have

$$E\left\{\exp\left[i \sum_{k=0}^{n+1} \langle u_k, X(t_k) - X(t_{k-1})\rangle\right] \middle| \mathscr{F}^0_{t_n}\right\}$$
$$= \exp\left[i \sum_{k=0}^{n} \langle u_k, X(t_k) - X(t_{k-1})\rangle\right]$$
$$\cdot E\{\exp[i\langle u_{n+1}, X(t_{n+1}) - X(t_n)\rangle] \mid X(t_n)\}. \tag{21}$$

Next we have if $u \in R^d$, $s \geq 0$, $t \geq 0$:

$$E\{\exp[i\langle u, X(s+t) - X(t)\rangle] \mid X(t)\}$$
$$= \exp[-i\langle u, X(t)\rangle] \int P_s(X(t), dy) \exp[i\langle u, y\rangle]$$
$$= \exp[-i\langle u, X(t)\rangle] \int P_s(o, dz) \exp[i\langle u, z + X(t)\rangle]$$
$$= \int P_s(o, dz) \exp[i\langle u, z\rangle] = \check{\pi}_s(u) \tag{22}$$

where $\check{\pi}_s$ is the characteristic function of the probability measure π_s defined in (2). Taking expectations in (21) we obtain

$$E\left\{\exp\left[i\sum_{k=0}^{n+1}\langle u_k, X(t_k) - X(t_{k-1})\rangle\right]\right\}$$
$$= E\left\{\exp\left[i\sum_{k=0}^{n}\langle u_k, X(t_k) - X(t_{k-1})\rangle\right]\right\}\cdot \check{\pi}_{t_{n+1}-t_n}(u_{n+1}).$$

Hence by induction on n, this is equal to

$$\check{\mu}_{t_0}(u_0)\prod_{k=1}^{n+1}\check{\pi}_{t_k - t_{k-1}}(u_k),$$

where $\check{\mu}_t$ is the characteristic function of the distribution μ_t of $X(t)$. Assertion (a) follows from this by a standard result on independence (see Theorem 6.6.1 of *Course*), and assertion (b) from taking expectations in (22). Note that π_s is the distribution of $X(s) - X(0)$. □

We close this section by mentioning an important property of a process having stationary independent increments: its remote field is trivial. The precise result together with some useful consequences are given below in Exercises 4, 5, and 6.

Exercises

1. Is m the unique σ-finite measure satisfying (8), apart from a multiplicative constant? [This is a hard question in general. For the Brownian motion semigroup the answer is "yes"; for the Poisson semigroup the answer is "no".]

2. For the uniform motion semigroup (P_t) (Example 2 of §1.2), show that $U^\alpha \ll m$ and find $u^\alpha(x, y)$. In this case $P_t(x, \cdot)$ is singular with respect to m for each t and x. For the Poisson semigroup (P_t) (Example 3 of §1.2), find an invariant measure m_0 (on $\mathbb{N}$). In this case $P_t(n, \cdot) \ll m_0$ for each t and n.

3. The property of independent increments in Theorem 4 may be extended as follows. Let $T_0 = 0$; for $n \geq 1$ suppose $T_{n-1} < T_n$ and $T_n - T_{n-1}$ is optional relative to the σ-fields $(\mathscr{A}_t)$, where $\mathscr{A}_t$ is generated by $\{X(T_{n-1} + s) - X(T_{n-1}), 0 < s \leq t\}$. Then the random variables $X(T_n) - X(T_{n-1})$, $n \geq 1$, are independent.

The next three exercises are valid for a process having stationary independent increments.

4. Put

$$\mathscr{G}_t = \sigma(X_s, s \geq t)^\sim, \qquad \mathscr{G} = \bigwedge_{0 \leq t < \infty} \mathscr{G}_t$$

where $^\sim$ denotes augmentation as in §2.3. The σ-field $\mathscr{G}$ is called the "remote field of the process. Prove that if $\Lambda \in \mathscr{G}$, then $P^x\{\Lambda\}$ is zero or one for every x. Is the result true for a process having arbitrary independent increments? [Hint: for discrete time analogues see §8.1 of *Course*.]

5. Suppose there exists $t > 0$ such that π_t has a density. If $f \in b\mathscr{E}$ and $f = P_t f$, then f is a constant. [Hint: show $f \in \mathbb{C}$ and use the Choquet-Deny-Hunt theorem (*Course*, §9.5).]

6. Under the hypothesis on (P_t) in Problem 5, a set $\mathscr{E}$ is recurrent or transient (see §3.6) according as $P_A 1 \equiv 1$ or $P_A 1 \not\equiv 1$. [Hint: let $f = \lim_{n \to \infty} \downarrow P_n P_A 1$, then f is a constant by Exercises 5; now use Problem 4.]

7. Suppose that for each $t > 0$, there exists a density π_t' of π_t with respect to m such that $\pi_t' > 0$ in $\mathbf{E}$. Then if $A \in \mathscr{E}$ and $P_A 1(x_0) = 0$ for some x_0, A must be polar. [Hint: show that $P_A 1(x) = 0$ for π_t-a.e. x, hence $P_t P_A 1(x) = 0$ for all x.]

8. Suppose that for each y, $u^\alpha(\cdot, y)$ is α-excessive. (It is possible to polish up u^α to achieve this.) Prove that for any measure μ, $U^\alpha \mu$ is α-excessive. Under this condition show that the condition (19) in Theorem 3 may be required to hold only m-a.e.

4.2. Preliminary Properties of Brownian Motion

The Brownian motion process in R^d was introduced in Example 2 of §3.7. For $d = 1$ it was introduced in Example 4 of §1.2. It is a Markov process with transition function:

$$P_t(x, dy) = p_t(x, y)\, dy, \qquad p_t(x, y) = \prod_{j=1}^{d} \frac{1}{\sqrt{2\pi t}} \left(\exp\left[-\frac{(y_j - x_j)^2}{2t} \right] \right). \tag{1}$$

Here

$$dy = dy_1 \cdots dy_d$$

is the Lebesgue measure in R^d, also written as $m(dy)$. We shall write "a.e." for "m-a.e.", and "density" for "density with respect to m". The product form in (1) shows that the Brownian motion $X(t)$ in R^d may be defined as the vector $(X_1(t), \ldots, X_d(t))$ where each coordinate $X_j(t)$ is a Brownian motion in R^1, and the d coordinates are stochastically independent. It is spatially homogeneous, and so by Theorem 4 of §4.1 may be characterized as a process having stationary independent increments, such that $X(t) - X(0)$ has the d-dimensional Gaussian (normal) distribution with mean the zero vector and covariance matrix equal to t times the $d \times d$ identity matrix. For $d = 1$, another useful characterization is by way of a Gaussian process, which is defined to be a process having Gaussian distributions for all finite-dimensional (marginal) distributions. The Brownian motion is then characterized as a Gaussian process such that for every s and t:

$$E(X(t)) = 0; \qquad E(X(s)X(t)) = s \wedge t. \tag{2}$$

The proof requires Exercise 2 of §6.6 of *Course*, which states that mutually orthogonal Gaussian random variables are actually independent.

A trivial extension of the definition consists of substituting $\sigma^2 t$ for t in (1), where $\sigma^2 > 0$. We deem this an unnecessary nuisance. In this section we review and adduce a few basic results for the Brownian motion. We begin with the most important one.

(I) There is a version of the Brownian motion with continuous sample paths.

For $d = 1$ the proof is given in the Example in §3.1. The general case is an immediate consequence of the coordinate representation (even without independence among the coordinates). This fundamental property is often taken as part of the definition of the process. For simplicity we may suppose *all* sample functions to be continuous.

(II) The Brownian motion semigroup has both the Feller property and the strong Feller property.

The latter is verified in §3.7, and the former in Exercise 2 of §2.2. As defined, the strong Feller property does not imply the Feller property! Recall also from Example 4 of §3.7 that Uf is bounded continuous if $f \in b\mathscr{E}$ and f has compact support, and $U^\alpha f$ for $\alpha > 0$ is bounded continuous for $f \in b\mathscr{E}$. These properties can often be used as substitutes for the Feller properties.

(III) The Brownian motion in R^d is recurrent for $d = 1$ and $d = 2$; it is transient for $d \geq 3$.

This is verified in Examples 2 and 4 of §3.7. Let us mention that for $d = 1$ and $d = 2$, there are discrete-time versions of recurrence which supplement the conclusions of Theorem 1 of §3.7. For each $\delta > 0$, let $\{X(n\delta), n \geq 1\}$ be a "skeleton" of the Brownian motion $\{X(t), t \geq 0\}$. Then $\{X(n\delta) - X((n-1)\delta), n \geq 1\}$ is a sequence of independent and identically distributed random variables, with mean zero and finite second moment (the latter being needed only in R^2). Hence the recurrence criteria for a random walk in R^1 or R^2 (§8.3 of *Course*) yield the following result. For any nonempty open set G:

$$P^x\{X(n\delta) \in G \text{ for infinitely many values of } n\} = 1.$$

This is just one illustration of the close tie between the theories of random walk and of Brownian motion. In fact, all the classical limit theorems have their analogues for Brownian motion, which are frequently easier to prove because there are ready sharp estimates. Since our emphasis here is on the *process* we shall not delve into those questions.

Let us also recall that for any $d \geq 1$, each set is either recurrent or transient by Exercise 6 of §4.1.

(IV) A singleton is recurrent for Brownian motion in R^1, and polar in R^d for $d \geq 2$.

The statement for R^1 is trivial by recurrence and continuity of paths. The statement for R^d, $d \geq 2$, is proved in the Example of §3.6. Another proof will be given in §4.4 below.

As a consequence, we can show that the fine topology in R^d, $d \geq 2$, is strictly finer than the Euclidean topology. For example, if Q is the set of points in R^d with rational coordinates then $R^d - Q$ is a finely open set because Q is a polar set. It is remarkable that all the paths "live" on this set full of holes.

(V) Let

$$F = \{x \in R^d | x_1 = 0\}, \qquad G_1 = \{x \in R^d | x_1 > 0\}, \qquad G_2 = \{x \in R^d | x_1 < 0\}. \tag{3}$$

Then $F \subset G_1^r \cap G_2^r$.

To prove this we need consider only the case $d = 1$, but we will argue in general. Let $x \in F$. By the symmetry of the distribution of $X_1(t)$, we have

$$P^x\{X(t) \in G_1\} = P^x\{X(t) \in G_2\}.$$

Since $P^x\{X(t) \in F\} = 0$ by the continuity of the distribution of $X_1(t)$, we deduce that $P^x\{X(t) \in \bar{G}_1\} = \frac{1}{2}$. Hence for each $u > 0$:

$$P^x\{X(t) \in \bar{G}_1 \text{ for all } t \in [0, u]\} \leq \tfrac{1}{2},$$

which implies $P^x\{T_{G_2} \leq u\} \geq \frac{1}{2}$. It follows that $P^x\{T_{G_2} = 0\} \geq \frac{1}{2}$, and consequently $x \in G_2^r$ by Blumenthal's zero-or-one law (Theorem 6 of §2.3). Interchanging G_1 and G_2 we obtain $x \in G_1^r$.

Corollary. *For the Brownian motion in R^1, each x is regular for $\{x\}$.*

By (V), the path starting at o must enter G_1 and G_2 in an arbitrarily short time interval $[0, \varepsilon]$, and therefore must cross F infinitely many times by continuity. Thus o is a point of accumulation of the set $\{o\}$ in the fine topology; the same is true for any x by spatial homogeneity.

(VI) For any Borel set B, we have

$$X(T_B) \in \partial B \quad \text{on } \{0 < T_B < \infty\}. \tag{4}$$

This is a consequence of the continuity of paths, but let us give the details of the argument. By the definition of T_B, on the set $\{T_B < \infty\}$ for any $\varepsilon > 0$ there exists $t \in [T_B, T_B + \varepsilon)$ such that $X(t) \in B$. Hence either $X(T_B) \in B$, or by the right continuity of the path,

$$X(T_B) = \lim_{t \downarrow\downarrow 0} X(T_B + t) \in \bar{B}.$$

On the other hand, on the set $\{0 < T_B\}$, $X(t) \in B^c$ for $0 < t < T_B$, hence by the left continuity of the path,

$$X(T_B) = \lim_{t \uparrow\uparrow 0} X(T_B - t) \in \overline{B^c}.$$

Since $\partial B = \bar{B} \cap \overline{B^c}$, (4) is proved.

If $x \in B^r$, then $P^x\{X(T_B) = X(0) = x\} = 1$, but of course x need not be in ∂B, for instance if B is open and $x \in B$. This trivial possibility should be remembered in quick arguments.

(VII) Let C_n be decreasing closed sets and $\bigcap_n C_n = C$. Then we have for each $x \in C^c \cup C^r$:

$$P^x\left\{\lim_n T_{C_n} = T_C \leq \infty\right\} = 1. \tag{5}$$

This is contained in the Corollary to Theorem 5 of §2.4. We begin by the alert that (5) may not be true for $x \in C - C^r$! Take for example a sequence of closed balls shrinking to a single point. If $x \in C^r$, (5) is trivial. Let $x \in C^c$. Then there exists k such that $x \in C_k^c$. Since C_k^c is open x is not regular for C_n for all $n \geq k$. The rest of the argument is true P^x-a.s. We have $T_{C_n} > 0$ for all $n \geq k$, and $T_{C_n}\uparrow$. Let $S = \lim_n \uparrow T_{C_n}$, then $0 < S \leq T_C$. On $\{0 < T_{C_n} < \infty\}$, we have $X(T_{C_n}) \in C_n$ by (VI). Hence on $\{0 < S < \infty\}$, we have by continuity of paths:

$$X(S) = \lim_n X(T_{C_n}) \in \bigcap_n C_n = C.$$

Thus $S \geq T_C$ and so $S = T_C$. On $\{S = \infty\}$, $T_C = \infty$.

(VIII) Let $\mathscr{F}_t$ be either $\mathscr{F}_t^0$ or $\mathscr{F}_t^\sim$ (see §2.3). Then for any x,

$$\{X(t), \mathscr{F}_t, P^x\} \quad \text{and} \quad \{\|X(t)\|^2 - td, \mathscr{F}_t, P^x\} \tag{6}$$

are martingales. (Here $\|x\|^2 = \sum_{j=1}^d x_j^2$.)

To verify the second, note that since

$$\|X(t)\|^2 - td = \sum_{j=1}^d (X_j(t)^2 - t)$$

we need only verify it for $d = 1$, which requires a simple computation. Observe that in the definition of a martingale $\{X_t, \mathscr{F}_t^0, P^x\}$, if the σ-field $\mathscr{F}_t^0$ is completed to $\mathscr{F}_t^x$ with respect to P^x, then $\{X_t, \mathscr{F}_t^x, P^x\}$ is also a martingale. Hence we may as well use the completed σ-field.

See Exercise 14 below for a useful addition.

(IX) For any constant $c > 0$, $(1/c)X(c^2t)$, $t \geq 0\}$ is also a Brownian motion. If we define

$$\begin{aligned} \tilde{X}(t) &= tX\left(\frac{1}{t}\right) \quad \text{for } t > 0; \\ &= 0 \qquad\quad \text{for } t = 0; \end{aligned} \tag{7}$$

then $\{\tilde{X}(t), t \geq 0\}$ is also a Brownian motion.

The first case is referred to as "scaling". The second involves a kind of reversing the time since $1/t$ decreases as t increases, and is useful in transforming limit behavior of the path as $t \uparrow \infty$ to $t \downarrow 0$. The proof that $\tilde{X}$ is a Brownian motion is easiest if we first observe that each coordinate is a Gaussian process, and then check in R^1 for $s > 0$, $t > 0$:

$$E^0\{\tilde{X}(s)\tilde{X}(t)\} = stE\left\{X\left(\frac{1}{s}\right)X\left(\frac{1}{t}\right)\right\} = st\left(\frac{1}{s} \wedge \frac{1}{t}\right) = s \wedge t.$$

Thus (2) is true and the characterization mentioned there yields the conclusion.

(X) Let B be a Borel set such that $m(B) < \infty$. Then there exists $\varepsilon > 0$ such that

$$\sup_{x \in \mathbf{E}} E^x\{\exp(\varepsilon T_{B^c})\} < \infty. \tag{8}$$

Proof. If $x \in (\bar{B})^c$, then $P^x\{T_{B^c} = 0\} = 1$ so (8) is trivial. Hence it is sufficient to consider $x \in \bar{B}$. For any $t > 0$, we have

$$\begin{aligned} \sup_{x \in \bar{B}} P^x\{T_{B^c} > t\} &\leq \sup_{x \in \bar{B}} P^x\{X(t) \in B\} \\ &= \sup_{x \in \bar{B}} \int_B p_t(x, y)\,dy \leq \frac{m(B)}{(2\pi t)^{d/2}}. \end{aligned} \tag{9}$$

This number may be made $\leq \frac{1}{2}$ if t is large enough. Fix such a value of t from here on. It follows from the Markov property that for $x \in \bar{B}$ and $n \geq 1$;

$$\begin{aligned} P^x\{T_{B^c} > (n+1)t\} &= P^x\{T_{B^c} > nt;\ P^{X(nt)}[T_{B^c} > t]\} \\ &\leq \tfrac{1}{2}P^x\{T_{B^c} > nt\} \end{aligned}$$

since $X(nt) \in B$ on $\{T_{B^c} > nt\}$. Hence by induction the probability above is $\leq 1/2^{n+1}$. We have therefore by elementary estimation:

$$E^x\{\exp(\varepsilon T_{B^c})\} \leq 1 + \sum_{n=0}^{\infty} e^{\varepsilon(n+1)t} P^x\{nt < T_{B^c} \leq (n+1)t\}$$

$$\leq 1 + \sum_{n=0}^{\infty} e^{\varepsilon(n+1)t} 2^{-n}.$$

This series converges for sufficient small ε and (8) is proved. □

Corollary. *If $m(B) < \infty$ then $E^x\{T_{B^c}^n\} < \infty$ for all $n \geq 1$; in particular $P^x\{T_{B^c} < \infty\} = 1$.*

This simple method yields other useful results, see Exercise 11 below.

For each $r > 0$, let us put

$$T_r = \inf\{t > 0 \mid \|X(t) - X(0)\| \geq r\}. \tag{10}$$

Let $B(x, r)$ denote the open ball with center x and radius r, namely

$$B(x, r) = \{y \in \mathbf{E} \mid \|x - y\| < r\}; \tag{11}$$

and $S(x, r)$ the boundary sphere of $B(x, r)$:

$$S(x, r) = \partial B(x, r) = \{y \in \mathbf{E} \mid \|x - y\| = r\}. \tag{12}$$

Then under P^x, T_r is just T_{B^c} where $B = B(x, r)$; it is also equal to $T_{\partial B}$. The next result is a useful consequence of the rotational symmetry of the Brownian motion.

(XI) For each $r > 0$, the random variables T_r and $X(T_r)$ are independent under any P^x. Furthermore $X(T_r)$ is uniformly distributed on $S(x, r)$ under P^x.

A rigorous proof of this result takes longer than might be expected, but it will be given. To begin with, we identify each ω with the sample function $X(\cdot, \omega)$, the space Ω being the class of all continuous functions in $\mathbf{E} = R^d$. Let φ denote a rotation in $\mathbf{E}$, and $\varphi\omega$ the point in Ω which is the function $X(\cdot, \varphi\omega)$ obtained by rotating each t-coordinate of $X(\cdot, \omega)$, namely:

$$X(t, \varphi\omega) = \varphi X(t, \omega). \tag{13}$$

Since φ preserves distance it is clear that

$$T_r(\varphi\omega) = T_r(\omega). \tag{14}$$

It follows from (VI) that $X(T_r) \in S(X(0), r)$. Let $X(0) = x$, $t \geq 0$ and A be a Borel subset of $S(x, r)$; put

$$\Lambda = \{\omega \mid T_r(\omega) \leq t;\ X(T_r(\omega), \omega) \in A\}. \tag{15}$$

Then on general grounds we have

$$\varphi^{-1}\Lambda = \{\omega \mid T_r(\varphi\omega) \leq t;\ X(T_r(\varphi\omega), \varphi\omega) \in A\}.$$

Using (13) and (14), we see that

$$\varphi^{-1}\Lambda = \{\omega \mid T_r(\omega) \leq t;\ X(T_r(\omega), \omega) \in \varphi^{-1}A\}. \tag{16}$$

Observe the double usage of φ in $\varphi^{-1}\Lambda$ and $\varphi^{-1}A$ above. We now claim that if x is the center of the rotation φ, then

$$P^x\{\Lambda\} = P^x\{\varphi^{-1}\Lambda\}. \tag{17}$$

Granting this and substituting from (15) and (16), we obtain

$$P^x\{T_r \leq t;\ X(T_r) \in A\} = P^x\{T_r \leq t;\ X(T_r) \in \varphi^{-1}A\}. \tag{18}$$

For fixed x and t, if we regard the left member in (18) as a measure in A, then (18) asserts that it is invariant under each rotation φ. It is well known that the unique probability measure having this property is the *uniform distribution on* $S(x, r)$ given by

$$\frac{\sigma(A)}{\sigma(r)}, \qquad A \in \mathscr{E},\ A \subset S(x, r) \tag{19}$$

where σ is the Lebesgue (area) measure on $S(x, r)$; and

$$\sigma(r) = \sigma(S(x, r)) = r^{d-1}\sigma(1), \qquad \sigma(1) = \frac{2\pi^{d/2}}{\Gamma\left(\frac{d}{2}\right)}. \tag{20}$$

Since the total mass of the measure in (18) is equal to $P\{T_r \leq t\}$, it follows that the left member of (18) is equal to this number multiplied by the number in (19). This establishes the independence of T_r and $X(T_r)$ since t and A are arbitrary, as well as the asserted distribution of $X(T_r)$.

Let us introduce also the following notation for later use:

$$v(r) = m(B(x, r)) = \int_0^r \sigma(s)\, ds = \frac{r^d}{d}\sigma(1). \tag{21}$$

It remains to prove (17), which is better treated as a general proposition (suggested by R. Durrett), as follows.

Proposition. *Let $\{X_t\}$ be a Markov process with transition function (P_t). Let φ be a Borel mapping of $\mathbf{E}$ into $\mathbf{E}$ such that for all $t \geq 0$, $x \in \mathbf{E}$ and $A \in \mathscr{E}$, we have*

$$P_t(x, \varphi^{-1}A) = P_t(\varphi x, A).$$

Then the process $\{\varphi(X_t)\}$ under P^x has the same finite-dimensional distributions as the process $\{X_t\}$ under $P^{\varphi(x)}$.

Remark. Unless φ is one-to-one, $\{\varphi(X_t)\}$ is not necessarily Markovian.

Proof. For each $f \in b\mathscr{E}$, we have

$$\int P_t(x, dy) f(\varphi y) = \int P_t(\varphi x, dy) f(y)$$

because this true when $f = 1_A$ by hypothesis, hence in general by the usual approximation. We now prove by induction on l that for $0 \leq t_1 < \cdots < t_l$ and $f_j \in b\mathscr{E}$:

$$E^x\{f_1(\varphi X_{t_1}) \cdots f_l(\varphi X_{t_l})\} = E^{\varphi(x)}\{f_1(X_{t_1}) \cdots f_l(X_{t_l})\}. \tag{22}$$

The left member of (22) is, by the Markov property and the induction hypothesis, equal to

$$\begin{aligned} &E^x\{f_1(\varphi X_{t_1}) P^{X_{t_1}}[f_2(\varphi X_{t_2-t_1}) \cdots f_l(\varphi X_{t_l-t_{l-1}})]\} \\ &\quad = E^x\{f_1(\varphi X_{t_1}) P^{\varphi X_{t_1}}[f_2(X_{t_2-t_1}) \cdots f_l(X_{t_l-t_{l-1}})]\} \end{aligned}$$

which is equal to the right member of (22). □

The proposition implies (17) if $\Lambda \in \mathscr{F}^0$ or more generally if $\Lambda \in \mathscr{F}^\sim$.

We conclude this section by discussing some questions of measurability. After the pains we took in §2.4 about augmentation it is only fair to apply the results to see why they are necessary, and worthwhile.

(XII) Let B be a (nearly) Borel set, f_1 and f_2 universally measurable, bounded numerical functions on $[0, \infty)$ and $\mathbf{E}_\partial$ respectively; then the functions

$$x \to E^x\{f_1(T_B)\}, \qquad x \to E^x\{f_2(X(T_B))\} \tag{23}$$

are both universally measurable, namely in $\mathscr{E}^\sim/\mathscr{B}$ where $\mathscr{B}$ is the Borel field on R^1.

That $T_B \in \mathscr{F}^{\sim}/\mathscr{B}$ is a consequence of Theorem 7 of §3.4; next, $X(T_B) \in \mathscr{F}^{\sim}/\mathscr{E}$ follows from the general result in Theorem 10 of §1.3. The assertions of (XII) are then proved by Exercise 3 of §2.4. In particular the functions in (23) are Lebesgue measurable. This will be needed in the following sections.

It turns out that if f_1 and f_2 are Borel measurable, then the functions in (23) are Borel measurable; see Exercise 6 and 7 below.

To appreciate the problem of measurability let f be a bounded Lebesgue measurable function from R^d to R^1, and $\{X_t\}$ the Brownian motion in R^d. Can we make $f(X_t)$ measurable in some sense? There exist two bounded Borel functions f_1 and f_2 such that $f_1 \le f \le f_2$ and $m(\{f_1 \ne f_2\}) = 0$. It follows that for any finite measure μ on $\mathscr{E}$ and $t > 0$, we have

$$E^\mu\{f_2(X_t) - f_1(X_t)\} = \iint \mu(dx)p_t(x, y)[f_2(y) - f_1(y)]m(dy) = 0.$$

Thus by definition $f(X_t) \in \bigwedge_\mu \mathscr{F}^\mu = \mathscr{F}^{\sim}$. Now let T_r be as in (10), then under P^x, $X(T_r) \in S(x, r)$ by (VI). But the Lebesgue measurability of f does not guarantee its measurability with respect to the area measure σ on $S(x, r)$, when f is restricted to $S(x, r)$. In particular if $0 \le f \le 1$ we can alter the f_1 and f_2 above to make $f_1 = 0$ and $f_2 = 1$ on $S(x, r)$, so that $E^\sigma\{f_2(X_t) - f_1(X_t)\} = \sigma(S(x, r))$. It should now be clear how the universal measurability of f is needed to overcome the difficulty in dealing with $f(X(T_r))$. Let us observe also that for a general Borel set B the "surface" ∂B need not have an area. Yet $X(T_{\partial B})$ induces a measure on ∂B under each P^μ, and if f is universally measurable $E^\mu\{f(X(T_{\partial B}))\}$ may be defined.

Exercises

Unless otherwise stated, the process discussed in the problems is the Brownian motion in R^d, and (P_t) is its semigroup.

1. If $d \ge 2$, each point x has uncountably many fine neighborhoods none of which is contained in another. For $d = 1$, the fine topology coincides with the Euclidean topology.

2. For $d = 2$, each line segment is a recurrent set; and each point on it is regular for it. For $d = 3$, each line is a polar set, whereas each point of a nonempty open set on a plane is regular for the set. [Hint: we can change coordinates to make a given line a coordinate axis.]

3. Let D be an open set and $x \in D$. Then $P^x\{T_{D^c} = T_{\partial D}\} = 1$. Give an example where $P^x\{T_{D^c} < T_{(\bar{D})^c}\} > 0$. Prove that if at each point y on ∂D there exists a line segment $\overline{yy'} \in D^c$, then $P^x\{T_{D^c} = T_{(\bar{D})^c}\} = 1$ for every $x \in D$. [Hint: use (V) after a change of coordinates.]

4. Let B be a ball. Compute $E^x\{T_{B^c}\}$ for all $x \in \mathbf{E}$.

5. If $B \in \mathscr{E}$ and $t > 0$, then $P^x\{T_B = t\} = 0$. [Hint: show $\int P^x\{T_B = s\}\, dx = 0$ for all but a countable set of s.]

6. For any Hunt process with continuous paths, and any closed set C, we have $T_C \in \mathscr{F}^0$. Consequently $x \to P^x\{T_C \leq t\}$ and $x \to E^x\{f(X(T_C))$ are in $\mathscr{E}$ for each $t \geq 0$ and $f \in b\mathscr{E}$ or $\mathscr{E}_+$. Extend this to any $C \in \mathscr{E}$ by Proposition 10 of §3.5. [Hint: let C be closed and $G_n \downarrow\downarrow C$ where G_n is open; then $\bigcap_n \{\exists t \in [a,b]: X(t) \in G_n\} = \{\exists t \in [a,b]: X(t) \in C\}$.]

7. For any Hunt process satisfying Hypothesis (L), $B \in \mathscr{E}$, $f \in \mathscr{B}$ and $f \in \mathscr{E}_+$ respectively, the functions $x \to E^x\{f(T_B)\}$ and $x \to E^x\{f(X(T_B))\}$ are both in $\mathscr{E}$. [Hint: by Exercise 3 of §3.5, $x \to E^x\{e^{-\alpha T_B}\}$ is in $\mathscr{E}$ for each $\alpha > 0$; use the Stone-Weierstrass theorem to approximate any function in $\mathbb{C}_0([0, \infty))$ by polynomials of e^{-x}. For the second function (say φ) we have $\alpha U^\alpha \varphi \to \varphi$ if $f \in b\mathbb{C}_+$. This is due to Getoor.]

8. If $f \in L^1(R^d)$ and $t > 0$, then $P_t f$ is bounded continuous.

9. If $f \in \mathscr{E}$ and $P_t|f| < \infty$ for every $t > 0$, then $P_t f$ is continuous for every $t > 0$. [Hint: for $\|x\| \leq 1$ we have $|P_t f(x) - P_t f(o)| \leq A[P_t|f|(o) + P_{4t}|f|(o)]$. This is due to T. Liggett.]

10. If $f \in b\mathscr{E}$, then $\lim_{t \to \infty} [P_t f(x) - P_t f(y)] = 0$ for every x and y. [Hint: put $x - z = \sqrt{2t}\zeta$ in $\int |p_t(x, z) - p_t(y, z)|\, dz$.]

11. In (X) if ε is fixed and $m(B) \to 0$, then the quantity in (8) converges to one.

12. Let X be the Brownian motion in R^1, $a > 0$ and B be a Borel set contained in $(-\infty, a]$. Prove that for each $t > 0$:
$$P^0\{T_{\{a\}} \leq t;\, X(t) \in B\} = P^0\{X(t) \in 2a - B\}$$
and deduce that
$$P^0\{T_{\{a\}} \leq t\} = 2P^0\{X(t) > a\}.$$
This is André's reflection principle. A rigorous proof may be based on Theorem 3 of §2.3.

13. For Brownian motion in R^1, we define the "last exit from zero before time t" as follows:
$$\gamma(t) = \sup\{s \leq t: X(s) = 0\}.$$
Prove that for $s \in (0, t)$ and $x \in R^1$, we have
$$P^0\{\gamma(t) \in ds\} = \frac{ds}{\pi\sqrt{s(t-s)}},$$
$$P^0\{\gamma(t) \in ds;\, X(t) \in dx\} = \frac{x}{2\pi\sqrt{s(t-s)^3}} e^{-x^2/2(t-s)}\, ds\, dx.$$
[This gives a glimpse of the theory of *excursions* initiated by P. Lévy. See Chung [8] for many such explicit formulas. The literature is growing in this area.]

14. For each $\lambda \in R^d$, $\{\exp(\langle \lambda, X_t \rangle - \|\lambda\|^2 t/2, \mathscr{F}_t, P^x\}$ is a martingale, where $\langle \lambda, x \rangle = \sum_{j=1}^d \lambda_j x_j$.

4.3. Harmonic Function

A *domain* in $R^d(=\mathbf{E})$ is an open and connected (nonempty!) set. Any open set in R^d is the union of at most a countable number of disjoint domains, each of which called a component. It will soon be apparent that there is some advantage in considering a domain instead of an open set. Let D be a domain. We denote the hitting time of its complement by τ_D, namely:

$$\tau_D = T_{D^c} = \inf\{t > 0 \mid X(t) \in D^c\}. \tag{1}$$

This is called the "(first) exit time" from D. The usual convention that $\tau_D = +\infty$ when $X(t) \in D$ for all $t > 0$ is meaningful; but note that $X(\infty)$ is generally not defined. By (X) of §4.2, we have $P^x\{\tau_D < \infty\} = 1$ for all $x \in D$ if $m(D) < \infty$, in particular if D is bounded.

The class of Borel measurable functions from R^d to $[-\infty, +\infty]$ or $[0, \infty]$ will be denoted by $\mathscr{E}$ or $\mathscr{E}_+$; with the prefix "b" when it is bounded. We say f is locally integrable in D and write $f \in L^1_{\text{loc}}(D)$ iff $\int_K |f|\,dm < \infty$ for each compact $K \subset D$, where m is the Lebesgue measure. We say A is "strictly contained" in B and write $A \Subset B$ iff $\bar{A} \subset B$.

Let $f \in \mathscr{E}_+$, and define a function h by

$$h(x) = E^x\{f(X(\tau_D)); \tau_D < \infty\}. \tag{2}$$

This has been denoted before by $P_{\tau_D}f$ or $P_{D^c}f$. Although h is defined for all $x \in \mathbf{E}$, and is clearly equal to $f(x)$ if $x \in (D^c)^o = (\bar{D})^c$, we are mainly concerned with it in $\bar{D}$. It is important to note that h is universally measurable, hence Lebesgue measurable by (XII) of §4.2. More specifically, it is Borel measurable by Exercise 6 of §4.2 because D^c is closed and the paths are continuous. Finally if f is defined (and Borel or universally measurable) only on ∂D and we replace τ_D by $T_{\partial D}$ in (2), the resulting function agrees with h in D (Exercise 1).

Recall the notation (11), (12) and (20) from §4.2.

Theorem 1. *If $h \not\equiv \infty$ in a domain D, then $h < \infty$ in D. For any ball $B(x,r) \Subset D$, we have*

$$h(x) = \frac{1}{\sigma(r)} \int_{S(x,r)} h(y)\sigma(dy). \tag{3}$$

Furthermore, h is continuous in D.

Proof. Write B for $B(x,r)$, then almost surely $\tau_B < \infty$; under P^x, $\tau_B < \tau_D$ so that $\tau_D = \tau_B + \tau_D \circ \theta_{\tau_B}$. Hence it follows from the fundamental balayage formula (Proposition 1 of §3.4) that

$$P_{\tau_D}f(x) = P_{\tau_B}P_{\tau_D}f(x) \tag{4}$$

which is just

$$h(x) = P_{\tau_B} h(x) \tag{5}$$

This in turn becomes (3) by the second assertion in (XI) of §4.2. Let us observe that (4) and (5) are valid even when both members of the equations are equal to $+\infty$. This is seen by first replacing f by $f \wedge n$ and then letting $n \to \infty$ there. A similar remark applies below. Replacing r by s in (3), then multiplying by $\sigma(s)$ and integrating we obtain

$$h(x) \int_0^r \sigma(s)\,ds = \int_0^r \int_{S(x,s)} h(y)\sigma(dy)\,ds = \int_{B(x,r)} h(y)m(dy). \tag{6}$$

where in the last step a trivial use of polar coordinates is involved. Recalling (21) of §4.2 we see that (6) may be written as

$$h(x) = \frac{1}{v(r)} \int_{B(x,r)} h(y)m(dy), \tag{7}$$

whether $h(x)$ is finite or not.

Now suppose $h \not\equiv \infty$ in D. Let $h(x_0) < \infty$, $x_0 \in D$; and $B(x, \rho) \subset B(x_0, r) \Subset D$. It follows from (7) used twice for x and for x_0 that

$$\begin{aligned} h(x) = \frac{1}{v(\rho)} \int_{B(x,\rho)} h(y)m(dy) &\le \frac{1}{v(\rho)} \int_{B(x_0,r)} h(y)m(dy) \\ &\le \frac{v(r)}{v(\rho)} h(x_0) < \infty. \end{aligned} \tag{8}$$

Thus $h < \infty$ in $B(x_0, r)$. This shows that the set $F = D \cap \{x \mid h(x) < \infty\}$ is open. But it also shows that F is closed in D. For if $x_n \in F$ and $x_n \to x_\infty \in D$, then $x_\infty \in B(x_n, \delta) \Subset D$ for some $\delta > 0$ and some large n, so that $x_\infty \in F$ by the argument above. Since D is connected and our hypothesis is that F is not empty, we conclude that $F = D$. The relation (7) then holds for every x in D with the left member finite, showing that $h \in L^1_{\text{loc}}(D)$.

To prove that h is continuous in D, let r be so small that both $B(x, r)$ and $B(x', r)$ are strictly contained in D. Applying (7) for x and x' we obtain

$$|h(x) - h(x')| \le \frac{1}{v(r)} \int_C h(y)m(dy) \tag{9}$$

where $C = B(x, r) \triangle B(x', r)$. If x is fixed and $x' \to x$ then it is obvious that $m(C) \to 0$, and the integral in (8) converges to zero by the proven integrability of h. Thus h is continuous and Theorem 1 is proved. □

We have shown above that the "sphere-averaging" property given in (3) entails the "ball-averaging" property given in (7). The converse is also true.

For we may suppose $h \not\equiv \infty$ in D; then h is finite continuous in D by the proof above. Now write (7) as follows:

$$h(x)v(r) = \int_0^r \int_{S(x,s)} h(y)\sigma(dy)\,ds.$$

Since h is continuous its surface integral over $S(x, s)$ is also continuous in s. Differentiating the equation above with respect to r we obtain (3) since $v'(r) = \sigma(r)$. We now proceed to unravel the deeper properties of the function h.

Definition 1. Let D be an open set. A function h is called *harmonic in* D iff

(a) it is finite continuous in D;
(b) the sphere-averaging property holds for any $B(x, r) \Subset D$.

By the proof of Theorem 1, if D is a domain, and h is locally integrable in D, then (b) entails (a). Another easy consequence is as follows.

Corollary. *Let D be a domain and f be a Borel measurable function on ∂D. If $P_{D^c}|f| \not\equiv \infty$ in D, then $P_{D^c}f$ is harmonic in D. In particular this is the case if f is bounded on ∂D.*

It is remarkable that there is another quite different characterization of the class of harmonic functions. Recall that $C^{(k)}(D)$ is the class of k-times continuously differentiable functions in D;

$$\Delta = \sum_{j=1}^{d} \left(\frac{\partial}{\partial x_j}\right)^2$$

is the *Laplacian* in R^d.

Definition 2. A function h is harmonic in the open set D iff it belongs to $C^{(2)}(D)$, and satisfies in D the Laplace equation:

$$\Delta h = 0. \tag{10}$$

Theorem 2. (Gauss-Koebe) *Definitions 1 and 2 are equivalent. Moreover a harmonic function in D belongs to $C^{(\infty)}(D)$.*

Proof. Let h be harmonic in D according to Definition 1. We show first that $h \in C^{(\infty)}(D)$. For any $\delta > 0$ there exists a function φ in $C^{(\infty)}(\mathbf{E})$ with the following properties: $\varphi(x) = \varphi(\|x\|)$: $\varphi(x) > 0$ for $\|x\| < \delta$; $\varphi(x) = 0$ for $\|x\| \geq \delta$; and $\int_{\mathbf{E}} \varphi(x) m(dx) = 1$. See Exercise 6 for an example. We have then

$$\int_0^\infty \varphi(r)\sigma(r)\,dr = 1. \tag{11}$$

For each x in D, there exists $\delta > 0$ such that h is bounded in $\overline{B(x,\delta)}$ and (3) holds for $0 < r < \delta$. It follows that

$$\begin{aligned} h(x) &= \int_0^\infty \left[\frac{1}{\sigma(r)} \int_{S(x,r)} h(y)\sigma(dy)\right] \varphi(r)\sigma(r)\,dr \\ &= \int_0^\infty \int_{S(x,r)} h(y)\varphi(\|x-y\|)\sigma(dy)\,dr \\ &= \int_{\mathbf{E}} h(y)\varphi(\|x-y\|)m(dy). \end{aligned} \tag{12}$$

Since φ has support in $B(o,\delta)$, and all partial derivatives of φ are bounded continuous in $B(o,\delta)$, we may differentiate the last-written integral with respect to x under the integral. Since $\varphi(\|x-y\|)$ is infinitely differentiable in x for each y, we see that h is infinitely differentiable.

To prove (10) let us recall Gauss's "divergence formula" from calculus, in a simple situation. Let B be a ball and h be twice continuously differentiable in a neighborhood of $\bar{B}$, then the formula may be written as follows:

$$\int_B \Delta h(y)m(dy) = \int_{\partial B} \frac{\partial h}{\partial n}(y)\sigma(dy) \tag{13}$$

where $\partial h/\partial n$ is the outward normal derivative. Now let us denote the right member of (3) by $A(x,r)$. Making the change of variables $y = x + rz$ we have

$$A(x,r) = \frac{1}{\sigma(1)} \int_{S(o,1)} h(x+rz)\sigma(dz). \tag{14}$$

Straightforward differentiation with respect to r gives for fixed x:

$$h'(x+rz) = \sum_{j=1}^{d} z_j \frac{\partial h}{\partial z_j}(x+rz) = \frac{\partial h}{\partial n}(x+rz); \tag{15}$$

$$\begin{aligned} A'(x,r) &= \frac{1}{\sigma(1)} \int_{S(o,1)} \frac{\partial h}{\partial n}(x+rz)\sigma(dz) \\ &= \frac{1}{\sigma(r)} \int_{S(x,r)} \frac{\partial h}{\partial n}(y)\sigma(dy) \\ &= \frac{1}{\sigma(r)} \int_{B(x,r)} \Delta h(y)m(dy). \end{aligned} \tag{16}$$

The differentiation of (13) under the integral is correct because h has bounded continuous partial derivatives in a neighborhood of $S(o,1)$. The first member of (16) vanishes for all sufficiently small values of r by (3), hence so does the integral in the last member (rid of the factor $1/\sigma(r)$). This being true for all sufficiently small values of r, we conclude $\Delta h(x) = 0$ by continuity. Thus h is harmonic according to Definition 2. Conversely if that is the hypothesis, then for any $B(x, r) \Subset D$, the formula (16) is valid because $h \in C^{(2)}(D)$. Since $\Delta h = 0$ in D the last member of (16) vanishes, hence so does the first member. Hence $A(x, r) = \lim_{r \downarrow 0} A(x, r)$, which is equal to $h(x)$ by the continuity of h. Therefore (3) is true and Theorem 2 is proved. □

Definition 2 allows a "local" characterization of harmonicity. The function h is harmonic at x iff it belongs to $C^{(2)}$ in a neighborhood of x and $\Delta h(x) = 0$. It also yields an improvement of Definition 1, by retaining condition (a) there but weakening condition (b) as follows:

(b′) the sphere-averaging property holds for each x in D, and all sufficiently small balls (strictly) contained in D with center at x.

Actually, even weaker conditions than (b') suffice, but these are technical problems; see Rao [1]. We add two basic propositions about harmonic functions below, both deducible from Definition 1 alone. The first is sometimes referred to as a "principle of maximum." However, there are some other results in potential theory called by that name.

Proposition 3. *Let h be harmonic in the domain D, and put*

$$M = \sup_{x \in D} h(x), \qquad m = \inf_{x \in D} h(x).$$

If there exists an x in D such that $h(x) = M$ or $h(x) = m$, then h is a constant in D.

Proof. Suppose $h(x) = M$. Then it follows at once from the ball-averaging property (7) that $h(y) = M$ first for a.e. y in a neighborhood of x, then for all y there since h is continuous. Thus the set $D \cap \{x \mid h(x) = M\}$ is open in D; it is also relatively closed in D by continuity. Hence $h = M$ in D. The proof for the infimum is exactly the same. □

Corollary. *If h is harmonic in a bounded domain D, and continuous in $\bar{D}$, then the maximum and minimum values of h in $\bar{D}$ are taken on ∂D. In particular if $h \equiv 0$ on ∂D, then $h \equiv 0$ in $\bar{D}$.*

The next proposition is one of Harnack's theorems which hark back to Theorem 1. We relegate two other Harnack's theorems to Exercises 8 and 9.

Proposition 4. *Let D be a domain and $\{h_n, n \geq 1\}$ a sequence of harmonic functions in D. Suppose h_n increases to h_∞ in D. Then either $h_\infty \equiv +\infty$ in D, or h is harmonic in D.*

Proof. Even without the use of Definition 2, it is easy to see that we need only prove that h_∞ is harmonic in each bounded subdomain $D_1 \Subset D$. On D_1,

$h_1 \geq m > -\infty$ by continuity. Let $\tilde{h}_n = h_n - m$; then $\tilde{h}_n$ increases to $\tilde{h}_\infty = h_\infty - m \geq 0$. Each $\tilde{h}_n$ has the sphere-averaging property in D_1, hence so does $\tilde{h}_\infty$ by monotone convergence. Therefore by the proof of Theorem 1 from (6) on, either $\tilde{h}_\infty \equiv +\infty$ in D_1, or $\tilde{h}_\infty$ is finite continuous, hence harmonic in D_1. □

Thanks to Definition 2, it is trivial to find explicit harmonic functions. It is well known that the real part of a complex analytic function is harmonic (in R^2), see e.g. Ahlfors [1]. Indeed the two characterizations have their analogues in the Cauchy integral formula and holomorphy, respectively. It is obvious from Definition 2 that in R^1, harmonicity is tantamount to linearity.

EXAMPLE. In R^2, $\log\|x\|$ is harmonic in any open set not containing o. In R^d, $d \geq 3$, $1/\|x\|^{d-2}$ is harmonic in any open set not containing o. Let K be a compact set, μ a measure such that $\mu(K) < \infty$; then

$$\int_K \log\|x - y\|\mu(dy), \qquad \int_K \|x - y\|^{2-d}\mu(dy) \tag{17}$$

are harmonic functions of x in $R^2 - K$ and $R^d - K$ respectively.

The first two assertions are standard exercises in calculus upon using Definition 2. The last assertion then follows either by differentiation under the integrals, or perhaps better by integration using Definition 1 and Fubini's theorem. Now consider the particular case of (17) when μ is the uniform distribution on the sphere $S(o, r)$. If $\|x\| > r$, then $z \to \log\|z\|$ is harmonic in a neighborhood of $B(x, r)$; hence the sphere-averaging property yields

$$\log\|x\| = \frac{1}{\sigma(r)} \int_{S(x,r)} \log\|z\|\sigma(dz) = \frac{1}{\sigma(r)} \int_{S(o,r)} \log\|x - y\|\sigma(dy);$$

similarly for the second integral in (17). However, to evaluate the integrals when $\|x\| < r$ we need the following proposition.

Proposition 5. *If h is harmonic and is a function of $\|x\|$ alone, then it must be of the form*

$$c_1 \log\|x\| + c_2, \qquad c_1\|x\|^{2-d} + c_2 \tag{18}$$

respectively in R^2 and R^d, $d \geq 3$: where c_1 and c_2 are constants.

Proof. Writing ρ for $\|x\|$, we see that the Laplace equation for h reduces to

$$\frac{d^2}{d\rho^2} h + \frac{d-1}{\rho} \frac{d}{d\rho} h = 0. \tag{19}$$

Solving this simple differential equation we obtain the general solutions given in (18). These are called *fundamental radial solutions* of the Laplace equation. □

Proposition 6. *We have for any $r > 0$:*

$$\frac{1}{\sigma(r)} \int_{S(o,r)} \log\|x - y\|\sigma(dy) = \log(\|x\| \vee r), \qquad d = 2; \tag{20}$$

$$\frac{1}{\sigma(r)} \int_{S(o,r)} \|x - y\|^{2-d}\sigma(dy) = (\|x\| \vee r)^{2-d}, \qquad d \geq 3. \tag{21}$$

Proof. Denote the left member of (20) by $f(x)$. It is harmonic in $B(o,r)$ and is a function of $\|x\|$ alone upon a geometric inspection. Hence by Proposition 5, $f(x) = a\log\|x\| + b$. For $x = o$ we have $f(o) = \log r$ by inspection; hence $a = 0$ and $b = \log r$. It follows that $f(x) = \log r$ for $\|x\| < r$. Similarly, $f(x) = a' \log\|x\| + b'$ for $\|x\| > r$. It is easily seen that $\lim_{\|x\| \to \infty} [f(x) - \log\|x\|] = 0$; hence $a' = 1$ and $b' = 0$. Thus $f(x) = \log\|x\|$ for $\|x\| > r$. We have already observed this above. It remains to show that f is finite continuous at $\|x\| = r$. This is a good exercise in calculus, not quite trivial and left to the reader. The evaluation of (21) is similar. □

The exact calculation of formulas like (20) and (21) forms a vital part of classical potential theory, and is the source of many interesting results.

Exercises

1. For any open set D show that $P^x\{X(\tau_D) \in \partial D\} = 1$ for all $x \in \bar{D}$. However, if $x \in \partial D$ it is possible that $P^x\{\tau_D < T_{\partial D}\} = 1$. An example is known as *Lebesgue's thorn*, see Port and Stone [1], p. 68.

2. Show that if h is harmonic then so are all its partial derivatives of all orders. Verify the harmonicity of the following functions in R^3:

$$\frac{x_1}{\|x\|}, \qquad \frac{1}{\|x\|}\tan^{-1}\frac{x_2}{x_1}, \qquad x_3 \tan^{-1}\frac{x_2}{x_1}, \qquad 2x_3^2 - x_1^2 - x_2^2.$$

3. Let h be harmonic in R^d, $d \geq 1$ such that $P_t|h| < \infty$ for all $t > 0$. Then $h = P_t h$ for all $t \geq 0$, where (P_t) is the semigroup of the Brownian motion. The process $\{h(X_t), \mathscr{F}_t, t \geq 0\}$ is a continuous martingale. [Hint: use polar coordinates and the sphere-averaging property; a by-product is the value of $\sigma(1)$.]

4. Let $h \in \mathscr{E}, 0 \leq h < \infty$ and $h = P_1 h$. Then h is a constant. [Hint: if $h \not\equiv \infty$ then h is locally integrable; now compare the values of $P_n h$ at two points as $n \to \infty$, by elementary analysis. This solution is due to Hsu Pei.]

5. (Picard's Theorem.) If h is harmonic and ≥ 0 in R^d, then it is a constant. [Hint: there is a short proof by ball-averaging. A longer proof uses martingale and Exercise 4 of §4.1 to show that $\lim_{t\to\infty} h(X_t)$ is a constant a.s.]

6. Let $\varphi_1(x) = c_1 \exp((\|x\|^2 - 1)^{-1})$ for $\|x\| < 1$, $\varphi_1(x) = 0$ for $\|x\| \geq 1$, and c_1 is so chosen that $\int_{R_d} \varphi_1(x)\,dx = 1$. Put $\varphi_\delta(x) = \varphi_1(x/\delta)(1/\delta^d)$. Show that φ_δ has all the properties required of φ in the proof of Theorem 2.

7. Let D be a domain and h harmonic in D. If D is unbounded we adjoin a point at infinity ∞ to ∂D, and say that $x \to \infty$ when $\|x\| \to \infty$. Suppose now that

$$\overline{\lim_{D \ni x \to \partial D \cup \{\infty\}}} h(x) \leq M. \tag{22}$$

Then $h(x) \leq M$ for all $x \in D$. Give an example in which the conclusion becomes false if $\{\infty\}$ is omitted in (22).

8. Let $\{h_\alpha\}$ be a family of harmonic functions in an open set D. Suppose that the family is uniformly bounded on each compact subset of D. Then from any sequence from the family we can extract a subsequence which converges to a harmonic function in D, and the convergence is uniform on each compact subset of D. [Hint: prove first equi-continuity of the family on each compact subset by using (9), then apply the Ascoli-Arzelà theorem.]

9. Let D be a domain. For each compact $K \subset D$ there exists a constant $c(K)$ such that for any harmonic function $h > 0$ in D we have for any $x_1 \in K$, $x_2 \in K$:

$$\frac{h(x_1)}{h(x_2)} \geq c(K).$$

[Hint: the classical proof uses inequalities which are derived from the Poisson representation for a harmonic function in a ball (see §4.4). But a proof follows from (8), followed by the usual "chain argument."].

10. Let $a > 0$, $b > 0$. Evaluate the following integrals by means of Proposition 6:

$$\int_0^{2\pi} \log(a^2 + b^2 - 2ab\cos\theta)\,d\theta;$$

$$\int_0^{2\pi} (a^2 + b^2 - 2ab\cos\theta)^{-1/2}\,d\theta.$$

Compare the method with contour integration in complex variables.

11. Prove the continuity of the left members of (20) and (21) as functions of x, as asserted in Proposition 6.

12. Compute

$$\int_{B(o,r)} \log\|x-y\|m(dy), \qquad \int_{B(o,r)} \|x-y\|^{2-d}m(dy)$$

in R^2 and R^d, $d \geq 3$, respectively. [Answer for the second integral:

$$\frac{\sigma(1)}{d}\frac{r^d}{\|x\|^{d-2}} \qquad \text{if } \|x\| \geq r;$$

$$\frac{\sigma(1)r^2}{2} + \sigma(1)\|x\|^2\left(\frac{1}{d} - \frac{1}{2}\right), \qquad \text{if } \|x\| \leq r.]$$

13. (Gauss) In R^d, $d \geq 3$, let $\lfloor\mu \subset K \subset B(o, r)$, then

$$\mu(K) = \frac{r^{d-2}}{\sigma_d(r)} \int_{S(o,r)} U\mu(y)\sigma(dy).$$

Here $\sigma_d(r) = \sigma(r)$ in (20) of §4.2.

14. Extend the result in Exercise 13 to R^2, using U^* (given in (14) of §4.6 below) for U.

4.4. Dirichlet Problem

Let D be an open set and $f \in b\mathscr{E}$ on ∂D, and consider the function h defined in $\bar{D}$ by (2) of §4.3. Changing the notation, we shall denote h by $H_D f$, thus

$$H_D f(x) = E^x\{f(X(\tau_D)); \tau_D < \infty\}, \qquad x \in \bar{D}. \tag{1}$$

This formula defines a measure $H_D(x, \cdot)$ on the boundary ∂D, called the *harmonic measure* (of x with respect to D). For each Borel subset A of ∂D, the function $H(\cdot, A)$ is harmonic in D by the Corollary to Theorem 1 of §4.3. We may regard the measure $H(x, \cdot)$ as defined for all Borel sets, though it is supported by ∂D. The measure $H(x, \cdot)$ is also defined for $x \in \overline{D^c}$ and is harmonic in $(\bar{D})^c = (D^c)^o$. For example if ∂D is the horizontal axis in R^2, $H_D f$ is a harmonic function both in the upper and lower open half-planes, but not necessarily in the whole plane.

We now study the behavior of $h(x)$ as x approaches the boundary. Since h is defined only in $\bar{D}$, we shall not repeat the restriction that $x \in \bar{D}$ below. Recall that a point z on ∂D is *regular* or *irregular for* D^c according as $P^z\{\tau_D = 0\} = 1$ or 0 by Blumenthal's zero-or-one law. It turns out that this is the same as its being a "regular boundary point of D" in the language of the non-probablistic treatment of the Dirichlet problem. Since z may be regular for D^c but irregular for D, the difference of terminology should be borne in mind. For instance, it is easy to construct a domain with a boundary point z which is irregular for D^c, but not so easy for z to be irregular for D.

Proposition 1. *For each $t > 0$, the function*

$$x \to P^x\{\tau_D \le t\}$$

is lower semi-continuous in $\mathbf{E}$.

Proof. Since D is open and X is continuous, we have $\{\tau_D = t\} \subset \{X(t) \in D^c\}$. It follows that

$$\{\tau_D > t\} = \{\forall s \in (0, t]: X(s) \in D\}$$
$$= \bigcap_{n=1}^{\infty} \left\{\forall s \in \left(\frac{t}{n}, t\right]: X(s) \in D\right\}.$$

The set above is in $\mathscr{F}_t^0$, and

$$P^x\{\tau_D > t\} = \lim_{n\to\infty} \downarrow P^x \left\{\forall s \in \left(\frac{t}{n}, t\right]: X(s) \in D\right\}. \tag{2}$$

The probability on the right side of (2) is of the form $P_{t/n}\varphi(x)$, where $\varphi(x) \in b\mathscr{E}$. Hence by the strong Feller property ((II) of §4.2), $x \to P_{t/n}\varphi(x)$ is bounded continuous. Therefore the left member of (2) is upper semi-continuous in x, which is equivalent to the proposition. □

Note that on account of Exercise 5 of §4.2, $P^x\{\tau_D = t\} = 0$, but we do not need this.

Theorem 2. *If z on ∂D is regular for D^c, and if f is continuous at z, then we have*

$$\lim_{\bar{D} \ni x \to z} h(x) = f(z). \tag{3}$$

Proof. We have by Proposition 1, for each $\varepsilon > 0$:

$$\varliminf_{x\to z} P^x\{\tau_D \le \varepsilon\} \ge P^z\{\tau_D \le \varepsilon\} = 1. \tag{4}$$

Hence the corresponding limit exists and is equal to one. Note that here x need not be restricted to $\bar{D}$. Let T_r be as in (10) of §4.2. For $r > 0$ and $\varepsilon > 0$, $P^x\{T_r > \varepsilon\}$ does not depend on x, and for any fixed r we have

$$\lim_{\varepsilon \downarrow 0} P^o\{T_r > \varepsilon\} = 1, \tag{5}$$

because $T_r > 0$ almost surely. Now for any x, an elementary inequality gives

$$P^x\{\tau_D \le \varepsilon < T_r\} \ge P^x\{\tau_D \le \varepsilon\} + P^x\{\varepsilon < T_r\} - 1. \tag{6}$$

Consequently we have by (4)

$$\varliminf_{x \to z} P^x\{\tau_D \le \varepsilon < T_r\} \ge P^o\{\varepsilon < T_r\} \tag{7}$$

and so by (5):

$$\lim_{x \to z} P^x\{\tau_D < T_r\} = 1. \tag{8}$$

On the set $\{\tau_D < T_r\}$, we have $\|X(\tau_D) - X(0)\| < r$ and so $\|X(\tau_D) - z\| < \|X(0) - z\| + r$. Since f is continuous at z, given $\delta > 0$, there exists η such that if $y \in \partial D$ and $\|y - z\| < 2\eta$, then $|f(y) - f(z)| < \delta$. Now let $x \in \bar{D}$ and $\|x - z\| < \eta$; put $r = \eta$. Then under P^x, $X(\tau_D) \in \partial D$ (Exercise 1 of §4.3) and $\|X(\tau_D) - z\| < 2\eta$ on $\{\tau_D < T_\eta\}$; hence $|f(X(\tau_D)) - f(z)| < \delta$. Thus we have

$$E^x\{|f(X(\tau_D)) - f(z)|; \tau_D < \infty\} < \delta P^x\{\tau_D < T_\eta\} + 2\|f\| P^x\{T_\eta \le \tau_D < \infty\}. \tag{9}$$

When $x \to z$, the last term above converges to zero by (8) with η for r. Since δ is arbitrary, we have proved that the left member of (9) converges to zero. As a consequence,

$$\lim_{x \to z} h(x) = f(z) \lim_{x \to z} P^x\{\tau_D < \infty\} = f(z)$$

by (4). □

Let us supplement this result at once by showing that the condition of regularity of z cannot be omitted for the validity of (3) in R^d, $d \ge 2$. Note: in R^1 every $z \in \partial D$ is regular for $\{z\} \subset D^c$.

Theorem 3. *Suppose $d \ge 2$ and z is irregular for D^c. Then there exists $f \in b\mathbb{C}$ on ∂D such that*

$$\varliminf_{D \ni x \to z} h(x) < f(z). \tag{10}$$

Proof. Indeed (10) is true for any $f \in b\mathbb{C}(\partial D)$ such that $f(z) = 1$ and $f < 1$ on $\partial D - \{z\}$. An example of such a function is given by $(1 - \|x - z\|) \vee 0$. Since $\{z\}$ is polar by (IV) of §4.2, and z is irregular for D^c, we have $P^z\{X(\tau_D) = z\} = 0$. This implies the first inequality below:

$$\begin{aligned} 1 &> E^z\{f(X(\tau_D)); \tau_D < \infty\} \\ &= \lim_{r \downarrow 0} E^z\{T_r < \tau_D; h(X(T_r))\} \\ &\ge \lim_{r \downarrow 0} P^z\{T_r < \tau_D\} \inf_{x \in B(z,r) \cap D} h(x). \end{aligned}$$

where the equation follows from $\lim_{r \downarrow 0} P^z\{T_r < \tau_D\} = 1$, and the strong Markov property applied at T_r on $\{T_r < \tau_D\}$. The last member above is equal to the left member of (10). □

We proceed to give a sufficient condition for z to be regular for D^c. This is a sharper form of a criterion known as Zaremba's "cone condition". We shall replace his solid cone by a flat one. Let us first observe that if $z \in \partial D$ and B is any ball with z as center, then z is regular for D^c if and only if it is regular for $(D \cap B)^c$. In this sense regularity is a local property. A "flat cone" in R^d, $d \geq 2$, is a cone in a hyperplane of dimension $d - 1$. It is "truncated" if all but a portion near its vertex has been cut off by a hyperplane perpendicular to its axis. Thus in R^2, a truncated cone reduces to a (short) line segment; in R^3, to a (narrow) fan.

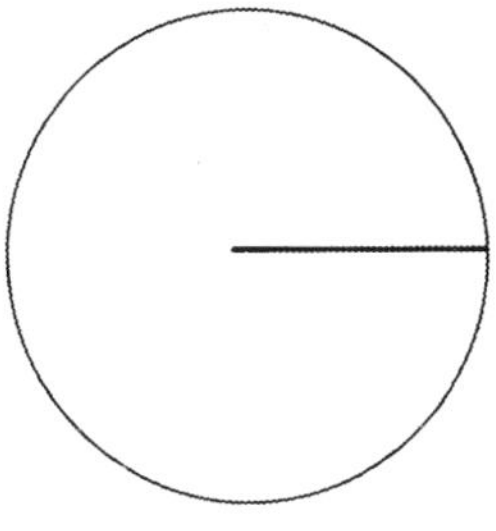

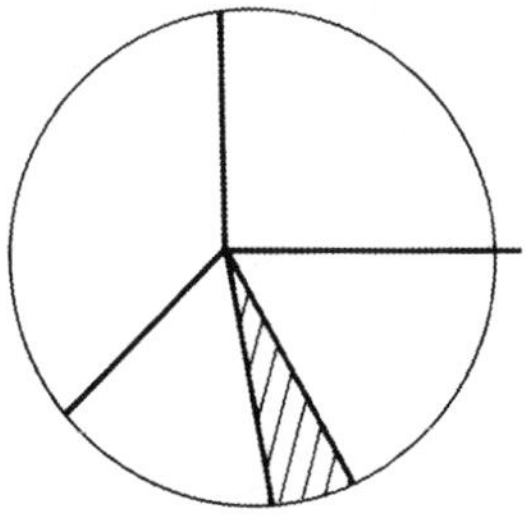

Theorem 4. *The boundary point z is regular for D^c if there is a truncated flat cone with vertex at z and lying entirely in D^c.*

Proof. We may suppose that z is the origin and the flat cone lies in the hyperplane of the first $d - 1$ coordinate variables. In the following argument all paths issue from o. Let

$$T_n = \inf\left\{t > \frac{1}{n} \,\middle|\, X_d(t) = 0\right\}.$$

Since o is regular for $\{o\}$ in R^1 by (V) of §4.2, we have $\lim_{n \to \infty} T_n = 0$, P^o-a.s. Since $X_d(\cdot)$ is independent of $Y(\cdot) = \{X_1(\cdot), \ldots, X_{d-1}(\cdot)\}$, T_n is independent of $Y(\cdot)$. It follows (proof ?) that $Y(T_n)$ has a $(d - 1)$-dimensional distribution which is seen to be rotationally symmetric. Let C be a flat cone with vertex o, and C_0 its truncation which lies in D^c. Then rotational symmetry implies that $P^o\{Y(T_n) \in C\} = \theta$, where θ is the ratio of the "angle" of the flat cone to a "full angle". Since $\lim_{n \to \infty} Y(T_n) = 0$ by continuity, it follows that $\lim_{n \to \infty} P^o\{Y(T_n) \in C_0\} = \theta$. Since $C_0 \subset D^c$, this implies

$$P^o\{\tau_D = 0\} = \lim_{n \to \infty} P^o\{\tau_D \leq T_n\} \geq \theta > 0.$$

Therefore, o is regular for D^c by the zero-or-one law. □

Let us call a domain D *regular* when every point on ∂D is regular for D^c. As examples of Theorem 4, in R^2 an open disk with a radius deleted is regular; similarly in R^3 an open ball minus a sector of its intersection with a plane is regular, see Figure (p. 165). Of course, a ball or a cube is regular; for a simpler proof, see Exercise 1.

The most classical form of the *Dirichlet boundary value problem* may be stated as follows. Given an open set D and a bounded continuous function f on ∂D, to find a function which is harmonic in D, continuous in $\bar{D}$, and equal to f on ∂D. We shall refer to this problem as (D, f). The Corollary to Theorem 1 of §4.3, and Theorem 2 above taken together establish that the function $H_D f$ is a solution to the problem, provided that D is regular as just defined. When B is a ball, for instance, this is called the "interior Dirichlet problem" if $D = B$, and the "exterior Dirichlet problem" if $D = (\bar{B})^c$. The necessity of regularity was actually discovered long after the problem was posed. If there are irregular boundary points, the problem may have no solution. The simplest example is as follows. Let $D = B(o, 1) - \{o\}$ in R^2, namely the punctured unit disk; and define $f = 0$ on $\partial B(o, 1)$, $f(o) = 1$. The function $H_D f$ reduces in this case to the constant zero (why ?), which does not satisfy the boundary condition at o. But is there a true solution to the Dirichlet problem? Call this h_1; then h_1 is harmonic in D, continuous in $\bar{D}$ and equal to 0 and 1 respectively on $\partial B(o, 1)$ and at o. Let φ be an arbitrary rotation about the origin, and put $h_2(x) = h_1(\varphi x)$ for $x \in \bar{D}$. Whether by Definition 1 or 2 of §4.3, it is easy to see that h_2 is harmonic in D and satisfies the same boundary condition as h_1. Hence h_1 is identical with h_2 (see Proposition 5 below). In other words, we have shown that h_1 is rotationally invariant. Therefore by Proposition 5 of §4.3, it must be of the form $c_1 \log\|x\| + c_2$ in D. The boundary condition implies that $c_1 = 0$, $c_2 = 1$. Thus h_1 is identically equal to one in D. It cannot converge to zero at the origin. [The preceding argument is due to Ruth Williams.] It is obvious that o is an irregular boundary point of D, but one may quibble about its being isolated so that the continuity of f on ∂D is a fiction. However, we shall soon discuss a more general kind of example by the methods of probability to show where the trouble lies. Let us state first two basic results in the simplest case, the first of which was already used in the above.

Proposition 5. *If D is bounded open and regular, then $H_D f$ is the unique solution to the original Dirichlet problem (D, f).*

Proof. We know that $h = H_D f$ is harmonic in D. Since all points of ∂D are regular, and f is continuous in ∂D, h is continuous in $\bar{D}$ by Theorem 2. Let h_1 be harmonic in D, continuous in $\bar{D}$, and equal to f on ∂D. Then $h - h_1$ is harmonic in D, continuous in $\bar{D}$, and equal to zero on ∂D. Applying the Corollary to Proposition 3 of §4.3, we obtain $h - h_1 = 0$ in $\bar{D}$. □

The following result is to be carefully distinguished from Proposition 5.

Proposition 6. *Let D be a bounded open set. Suppose that h is harmonic in D and continuous in $\bar{D}$. Then*

$$h(x) = H_D h(x), \qquad \forall x \in \bar{D}.$$

Proof. There exist regular open sets D_n such that $D_n \uparrow\uparrow D$. In fact each D_n may be taken to be the union of cells of a grid on R^d, seen to be regular by the cone condition. We have $h = H_{D_n} h$ in $\bar{D}_n$ by Proposition 5. Let first $x \in D$, then there exists m such that $x \in D_n$ for $n \geq m$; hence

$$h(x) = E^x\{h(X(\tau_{D_n}))\}.$$

Letting $n \to \infty$, we have $\tau_{D_n} \uparrow \tau_D$ P^x-a.s. by (VII) of §4.2. Hence the right member above converges to $H_D h(x)$ by bounded convergence. Next if $x \in \partial D$ and x is regular for D^c, then $H_D h(x) = h(x)$ trivially. Finally if $x \in \partial D$ and x is not regular for D^c, then $P^x\{\tau_D > 0\} = 1$. We have for each $t > 0$:

$$\begin{aligned} E^x\{t < \tau_D; h(X(\tau_D))\} &= E^x\{t < \tau_D; E^{X(t)}[h(X(\tau_D))]\} \\ &= E^x\{t < \tau_D; h(X(t))\} \end{aligned}$$

because $X(t) \in D$ on $\{t < \tau_D\}$, so the second equation follows from what we have already proved. As $t \downarrow 0$, the first term above converges to $H_D h(x)$, the third term to $E^x\{0 < \tau_D; h(X(0))\} = h(x)$ by bounded convergence. □

We are now ready to discuss a general kind of unsolvable Dirichlet problem. Let D_0 be a bounded open domain in R^d, $d \geq 2$, and S a compact polar set contained in D_0. For instance, if $d = 2$, S may be any countable compact set; if $d = 3$, S may be a countable collection of linear segments. Let $D = D_0 - S$, and define $f = 1$ on ∂D_0, $f = 0$ on S. Note that $\partial D = (\partial D_0) \cup S$ since a polar set cannot have interior points. There exists a sequence of open neighborhoods $B_n \downarrow\downarrow S$ such that $\bar{B}_n \subset D_0$. Let $D_n = D_0 - B_n$. Suppose that h is harmonic in D, continuous in $\bar{D}$, and equal to one on ∂D_0. Nothing is said about its value on S. Since h is harmonic in D_n and continuous in $\bar{D}_n$, we have by Proposition 6 for $x \in D_n$:

$$\begin{aligned} h(x) &= E^x\{h(X(\tau_{D_n}))\} \\ &= P^x\{T_{\partial D_0} < T_{\partial B_n}\} + E^x\{T_{\partial B_n} < T_{\partial D_0}; h(X(T_{\partial B_n}))\}. \end{aligned} \tag{11}$$

Since S is polar, we have

$$\lim_n P^x\{T_{\partial B_n} < T_{\partial D_0}\} = P^x\{T_S < T_{\partial D_0}\} = 0. \tag{12}$$

Since h is bounded in $\bar{D}$, it follows that the last term in (11) converges to zero as $n \to \infty$. Consequently we obtain

$$h(x) = \lim_n P^x\{T_{\partial D_0} < T_{\partial B_n}\} = P^x\{T_{\partial D_0} < \infty\} = 1. \tag{13}$$

This being true for x in D_n for every n, we have proved that $h \equiv 1$ in D. Thus the Dirichlet problem (D, f) has no solution.

More generally, let f be continuous on ∂D_0, and arbitrary on S. Then $H_D f$ is harmonic in D, and the argument above shows that $H_{D_0} f$ is the unique continuous extension of $H_D f$ to D_0 (which is harmonic in D_0 to boot). In particular, the Dirichlet problem (D, f) is solvable only if $f = H_{D_0} f$ on S.

In the example above we constructed an irregular part of ∂D by using the polar set S. We shall see in §4.5 below that for any open D, the set of irregular points of ∂D is always a polar set. This deep result suggests a reformulation of the Dirichlet problem. But let us consider some simple examples first.

EXAMPLE 1. In R^d, $d \geq 3$, let $B = \overline{B(o, b)}$. Consider the function $P_B 1(x)$ for $x \in B^c$. This function is harmonic in B^c and is obviously a function of $\|x\|$ alone. Hence by Proposition 5 of §4.3, it is of the form $c_1 \|x\|^{2-d} + c_2$. The following equation illustrates a useful technique and should be argued out directly:

$$\forall x \in B^c \colon U(x, B) = P^x\{T_B < \infty\} U(y, B) \tag{14}$$

where y is any fixed point on ∂B. But actually this is just the identity $U1_B = P_B U 1_B$ contained in Theorem 3(b) of §3.4, followed by the observation that $U(y, B)$ is a constant for $y \in \partial B$ by rotational symmetry. Now the left member of (14), apart from a constant, is equal to

$$\int_B \frac{dy}{\|x - y\|^{d-2}}$$

which converges to zero as $\|x\| \to \infty$. Hence $P_B 1(x) \to 0$ as $\|x\| \to \infty$, and so $c_2 = 0$. Since $P_B 1 = 1$ on ∂B, $c_1 = b^{d-2}$. Thus

$$P^x\{T_B < \infty\} = \left(\frac{b}{\|x\|}\right)^{d-2}, \qquad \|x\| \geq b. \tag{15}$$

Next, let $0 < a < b < \infty$ and

$$D = \{x \mid a < \|x\| < b\}. \tag{16}$$

Consider the Dirichlet problem (D, f), where f is equal to one on $S(o, a)$, zero on $S(o, b)$. The probabilistic solution given by $P_{D^c} f$ signifies the probability that the path hits $S(o, a)$ before $S(o, b)$. It is the unique solution by Proposition 5. Since it is a function of $\|x\|$ alone it is of the form $c_1 \|x\|^{2-d} + c_2$. The constants c_1 and c_2 are easily determined by its value for $\|x\| = a$ and $\|x\| = b$. The result is recorded below:

$$P^x\{T_{S(o,a)} < T_{S(o,b)}\} = \frac{\|x\|^{2-d} - b^{2-d}}{a^{2-d} - b^{2-d}}, \qquad a \leq \|x\| \leq b \tag{17}$$

If $b \uparrow +\infty$ then $T_{S(o,b)} \uparrow \infty$ (why?), and we get (15) again with b replaced by a.

Exactly the same method yields for $d = 2$;

$$P^x\{T_{S(o,a)} < T_{S(o,b)}\} = \frac{\log\|x\| - \log b}{\log a - \log b}, \qquad a \le \|x\| \le b. \tag{18}$$

If we let $b \uparrow +\infty$, this time the limit of the above probability is 1, as it should be on account of recurrence. On the other hand, if we fix b and let $a \downarrow 0$ in (18), the result is $P^x\{T_{\{0\}} < T_{S(o,b)}\} = 0$ (why?). Now let $b \uparrow \infty$ to conclude that $P_{\{o\}}1(x) = 0$ for $x \neq o$. Since $P_t(x, \{o\}) = 0$ for any x in $\mathbf{E}$ and $t > 0$, it follows that $P_t P_{\{o\}}1(x) = 0$; hence $P_{\{o\}}1(x) = \lim_{t \downarrow 0} P_t P_{\{o\}}1(x) = 0$. Thus $\{o\}$ is a polar set. This is the second proof of a fundamental result reviewed in (IV) of §4.2.

EXAMPLE 2. The harmonic measure $H_B(x, \cdot)$ for a ball $B = B(o, r)$ can be explicitly computed. It is known as *Poisson's formula* and is one of the key formulas in the analytical developments of potential theory. In R^2 it is an analogue of, and can be derived from, the Cauchy integral formula for an analytic function in B:

$$f(x) = \frac{1}{2\pi i} \int_{S(o,r)} \frac{f(z)}{x - z}\, dz \tag{19}$$

where the integral is taken counterclockwise in the sense of complex integration. We leave this as an exercise and proceed to derive the analogue in R^d, $d \ge 3$. The method below works also in R^2.

For fixed z we know that $\|x - z\|^{2-d}$ is harmonic at all x except z. Hence so are its partial derivatives, and therefore also the following linear combination:

$$\sum_{j=1}^{d} \frac{2z_j}{d-2} \frac{\partial}{\partial x_j} \frac{1}{\|x - z\|^{d-2}} - \frac{1}{\|x - z\|^{d-2}} = \frac{\|z\|^2 - \|x\|^2}{\|x - z\|^d}.$$

Now consider the integral:

$$\int_{S(o,r)} \frac{r^2 - \|x\|^2}{\|x - z\|^d}\, \sigma(dz).$$

Then this is also harmonic for $x \in B(o, r)$ (why?). Inspection shows that it is a function of $\|x\|$ alone, hence it is of the form $c_1\|x\|^{2-d} + c_2$. Since its value at $x = 0$ is equal to $r^{2-d}\sigma(r)$ we must have $c_1 = 0$ and $c_2 = r^{2-d}\sigma(r) = r\sigma(1)$. Recall the value of $\sigma(1)$ from (20) of §4.2 which depends on the dimension d and will be denoted by $\sigma_d(1)$ in this example. Now let f be any continuous function on $S(o, r)$ and put

$$I(x, f) = \frac{1}{r\sigma_d(1)} \int_{S(o,r)} \frac{r^2 - \|x\|^2}{\|x - z\|^d} f(z)\sigma(dz). \tag{20}$$

This is the *Poisson integral* of f for $B(o, r)$. We have just shown that $I(x, 1) = 1$ for $x \in B(o, r)$. We will now prove that $I(x, f) = H_B(x, f)$ for all $f \in \mathbb{C}(\partial B)$, so that the measure $H_B(x, \cdot)$ is given explicitly by $I(x, A)$ for all $A \in \mathscr{E}$, $A \subset \partial B$.

The argument below is general and will be so presented. Consider

$$h(x, z) = \frac{1}{r\sigma_d(1)} \frac{r^2 - \|x\|^2}{\|x - z\|^d}, \qquad \|x\| < r, \|z\| = r,$$

then $h > 0$ where it is defined, and $\int_{S(o,r)} h(x, z)\sigma(dz) = 1$. Furthermore if $\|y\| = r$, $y \neq z$, then $\lim_{x \to z} h(x, y) = 0$ boundedly for y outside any neighborhood of z. It follows easily that the probability measures $I(x, dz) = h(x, z)\sigma(dz)$ converge vaguely to the unit mass at z as $x \to z$. Namely for any $f \in \mathbb{C}(\partial B)$, $I(x, f)$ converges to $f(z)$ as $x \to z$. Since $I(\cdot, f)$ is harmonic in $B(o, r)$, $I(\cdot, f)$ is a solution to the Dirichlet problem (D, f). Hence it must coincide with $H_B f = H_B(\cdot, f)$ by Proposition 5.

Let us remark that $H_B(x, \cdot)$ is the distribution of the "exit place" of the Brownian motion path from the ball $B(o, r)$, if it starts at x in the ball. This is a simple-sounding problem in so-called "geometric probability". An explict formula for the joint distribution of τ_B and $X(\tau_B)$ is given in Wendel [1].

In view of the possible unsolvability of the original Dirichlet problem, we will generalize it as follows. Given D and f as before, we say that h is a solution to the *generalized Dirichlet problem* (D, f) iff h is harmonic in D and converges to f at every point of ∂D which is regular for D^c. Thus we have proved above that this generalized problem always has a solution given by $H_D f$. If D is bounded, this is the unique bounded solution. The proof will be given in Proposition 11 of §4.5. We shall turn our attention to the general question of non-uniqueness for an unbounded domain D.

Consider the function g defined by

$$g(x) = P^x\{\tau_D = \infty\} = 1 - P_{D^c}1(x). \tag{21}$$

If D is bounded, then $\tau_D < \infty$ almost surely so that $g \equiv 0$. In general, g is a solution of the generalized Dirichlet problem $(D, 0)$. The question is whether it is a trivial solution, i.e., $g \equiv 0$ *in* D. Recall from §3.6 that (for a Hunt process) a Borel set A is recurrent if and only if $P_A 1 \equiv 1$. It turns out that this identity holds if it holds in A^c. For then we have for any x,

$$\begin{aligned} P^x\{T_A < \infty\} &\geq P^x\{X(t) \in A\} + E^x\{X(t) \in A^c; P^{X(t)}[T_A < \infty]\} \\ &= P^x\{X(t) \in A\} + P^x\{X(t) \in A^c\} = 1. \end{aligned}$$

Applying this remark to $A = D^c$, we obtain the following criterion.

Proposition 7. *We have* $g \equiv 0$ *in* D *if and only if* D^c *is recurrent.*

As examples: in R^d, $d \geq 3$, if D is the complement of a compact set, then $g \not\equiv 0$ in D. In R^2, $g \not\equiv 0$ in D if and only if D^c is a polar set, in which case $g \equiv 1$.

When $g \not\equiv 0$ in D, the generalized Dirichlet problem (D, f) has the solutions $H_D f + cg$ for any constant c. The next result implies that there are no other bounded solutions. We state it in a form dictated by the proof.

Theorem 8. *Let D be an open set and f be a bounded Borel measurable function on ∂D. If h is bounded and harmonic in D, and $\lim_{D \ni x \to z} h(x) = f(z)$ for all $z \in \partial D$, then h must be of the form below:*

$$h(x) = H_D f(x) + cg(x), \qquad x \in D; \tag{22}$$

where g is defined in (21) *and c is any constant. This h in fact converges to f at each point z of ∂D which is regular for D^c and at which f is continuous. In particular if f is continuous on ∂D, h is a generalized solution to the Dirichlet problem (D, f).*

Proof. There exists a sequence of bounded regular open sets D_n such that $D_n \Subset D_{n+1}$ for all n, and $\bigcup_n D_n = D$. Such a sequence exists by a previous remark in the proof of Proposition 6. Suppose that h is bounded and harmonic in D. Consider the Dirichlet problem (D_n, h). It is plain that h is a solution to the problem, hence we have by Proposition 5:

$$h(x) = H_{D_n} h(x), \qquad x \in \bar{D}_n. \tag{23}$$

This implies the first part of the lemma below, in which we write T_n for τ_{D_n}, T for τ_D to lighten the typography; also $T_0 = 0$.

Lemma 9. *For each $x \in D$, the sequence $\{h(X(T_n)), \mathscr{F}(T_n), P^x; n \geq 0\}$ is a bounded martingale. We have P^x-a.s.:*

$$\lim_{n \to \infty} h(X(T_n)) = \begin{cases} f(X(T)) & \text{on } \{T < \infty\}; \\ c & \text{on } \{T = \infty\}; \end{cases} \tag{24}$$

where c is a constant not depending on x.

Proof. Since h is bounded in D by hypothesis, we have by the strong Markov property, for each $x \in D$ and $n \geq 1$:

$$\begin{aligned} E^x\{h(X(T_n)) \mid \mathscr{F}(T_{n-1})\} &= E^{X(T_{n-1})}\{h(X(T_n))\} \\ &= H_{D_n} h(X(T_{n-1})) = h(X(T_{n-1})) \end{aligned}$$

where the last equation comes from (23). This proves the first assertion, and consequently by martingale convergence theorem the limit in (24) exists.

We now prove that there exists a random variable Z belonging to the remote field $\mathscr{G}$ of the Brownian motion process, such that the limit in (24) is equal to Z almost surely on the set $\{T = \infty\}$. [Readers who regard this as "intuitively obvious" should do Exercise 10 first.] Recall that $\mathscr{G}$ may be defined as follows. From each integer $k \geq 1$ let $\mathscr{G}_k$ be the σ-field generated by $X(n)$ for $n \geq k$, and augmented as done in §2.3, namely $\mathscr{G}_k = \sigma(X(n), n \geq k)^{\sim}$. Then $\mathscr{G} = \bigwedge_{k=1}^{\infty} \mathscr{G}_k$. By the Corollary to Theorem 8.1.4 of *Course* (see also the discussion on p. 258 there), $\mathscr{G}$ is trivial. [This is the gist of Exercise 4 of §4.1.]

Define $\tilde{h}$ in R^d to be h in D, and zero in D^c. We claim that there is a random variable Z (defined in Ω) such that

$$\lim_{n \to \infty} \tilde{h}(X(T_n)) = Z, \qquad P^x\text{-a.s.} \tag{25}$$

for every $x \in R^d$. For $x \in D$, under P^x we have $\tilde{h}(X(T_n)) = h(X(T_n))$, so the limit in (25) exists and is the same as that in (24). For $x \in D^c$, under P^x we have $\tilde{h}(X(T_n)) = \tilde{h}(x) = 0$, hence the limit in (25) exists trivially and is equal to zero. Since (25) holds P^x-a.s. for every $x \in R^d$, we can apply the shift to obtain for each $k \geq 1$:

$$\lim_{n \to \infty} \tilde{h}(X(T_n)) \circ \theta_k = Z \circ \theta_k, \qquad P^x\text{-a.s.} \tag{26}$$

for every $x \in R^d$. Let $x \in D$; then under P^x and on the set $\{T = \infty\}$, we have $T_n > k$ and consequently $\tilde{h}(X(T_n)) \circ \theta_k = h(X(T_n))$ for all sufficiently large values of n. Hence in this case the limit in (26) coincides with that in (24). Therefore the latter is equal to $Z \circ \theta_k$ which belongs to $\mathscr{G}_k$. This being true for each $k \geq 1$, the limit in (24) is also equal to $\overline{\lim}_{k \to \infty} Z \circ \theta_k$ which belongs to $\mathscr{G}$. Since $\mathscr{G}$ is trivial, the upper limit is a constant P^x-a.s. for each $x \in R^d$. This means: for each x there is a number $c(x)$ such that

$$P^x\{T = \infty; Z = c(x)\} = 1. \tag{27}$$

By Exercise 4 below, $g > 0$ in at most one component domain D_0 of D, and $g \equiv 0$ in $D - D_0$. Choose any x_0 in D_0 and put for all x in D:

$$\varphi(x) = P^x\{T = \infty; Z \neq c(x_0)\}.$$

It is clear from the definition of Z that for any ball $B = B(x, r) \Subset D$, $\varphi(x) = P_{B^c}\varphi(x)$ as in Theorem 1 of §4.3. Since φ is bounded, it is harmonic in D. Since $\varphi \geq 0$ and $\varphi(x_0) = 0$, as shown above, it follows by Proposition 3 of §4.3 (minimum principle) that $\varphi \equiv 0$ in D_0. As $g \equiv 0$ in $D - D_0$, we have $\varphi \equiv 0$ in D. Thus we may replace $c(x)$ by $c(x_0)$ in (27), proving the second line in (24). For $x \in D$, P^x-a.s. on the set $\{T < \infty\}$, we have $T_n \uparrow T$, $X(T_n) \in D$ and $X(T_n) \to X(T)$; since $h(y) \to f(z)$ as $y \in D$, $y \to z \in \partial D$ by hypothesis, we have $h(X(T_n)) \to f(X(T))$. This proves the first line of (24). The lemma is proved.

Now it follows by (23), (25) and bounded convergence that for each $x \in D$:

$$h(x) = \lim_{n\to\infty} E^x\{h(X(T_n))\}$$

$$= E^x\left\{\lim_{n\to\infty} h(X(T_n))\right\} = E^x\{Z\} = E^x\{Z; T < \infty\} + E^x\{Z; T = \infty\}$$

which reduces to (22) by (24). The rest of Theorem 8 is contained in Theorem 2 and is stated only for a recapitulation.

Exercises

1. Without using Theorem 4, give a simple proof that a ball or a cube in R^d is regular. Generalize to any solid with a "regular" surface so that there is a normal pointing outward at each point. [Hint: (V) of §4.2 is sufficient for most occasions.]

2. Let D be the domain obtained by deleting a radius from the ball $B(o, 1)$ in R^3; and let $f = \|x\|$ on ∂D. The original Dirichlet problem (D, f) has no solution.

3. In classical potential theory, a bounded domain D is said to be regular iff the Dirichlet problem (D, f) is solvable for every continuous f on D. Show that this definition is equivalent to the definition given here.

4. For any open set D, show that the function g defined in (21) is either identically zero or strictly positive in each connected component (domain) of D. Moreover there is at most one component in which $g > 0$. Give examples to illustrate the various possibilities.

5. What is the analogue of (17) and (18) for R^1?

6. Derive the Poisson integral in R^2 from Cauchy's formula (19), or the Taylor series of an analytic function (see Titchmarsh [1]).

7. Derive the Poisson integral corresponding to the exterior Dirichlet problem for the ball $B = B(o, r)$; namely find the explicit distribution of $X(T_{\partial B})1_{\{T_{\partial B} < \infty\}}$ under P^x, where $x \in B^c$.

8. Let $d \geq 2$; A be the hyperplane $\{x \in R^d \mid x_d = 0\}$; and let $D = \{x \in R^d \mid x_d > 0\}$. Compute $H_D(x, \cdot)$, namely the distribution of $X(T_A)$ under P^x, for $x \in D$. [Hint: we need the formula for Brownian motion in $R^1: P^x\{T_{\{o\}} \in dt\} = |x|/(2\pi t)^{3/2} \exp[-x^2/2t]\, dt$; the rest is an easy computation. Answer: $H_D(x, dy) = \Gamma(d/2) x_d \pi^{-d/2} \|x - y\|^{-d} \sigma(dy)$ where σ is the area on A.]

9. Let D be open in $R^d, d \geq 2$. Then D^c is polar if and only if there does not exist any nonconstant bounded harmonic function in D. [Hint: if D^c is not polar, there exists a bounded continuous function f on ∂D which is not constant.]

10. Let $\{x_n, n \geq 1\}$ be independent, identically distributed random variables taking the values ± 1 with probability 1/2 each; and let $X_n = \sum_{j=1}^n x_j$. Define for $k \geq 1$: $T_k = T_{\{k\}}$ on $\{X_1 = +1\}$; $T_k = T_{\{-k\}}$ on $\{X_1 = -1\}$; and $\Lambda = \{\lim_{k\to\infty} X(T_k) = +\infty\}$. Show that each T_k is optional with respect to $\{\mathscr{F}_n\}$, where $\mathscr{F}_n = \sigma(X_j, 1 \leq j \leq n)$; $T_k \uparrow \infty$ almost surely; but Λ does not belong to the remote field $\bigwedge_{m=1}^\infty \sigma(X_n, n \geq m)$. Now find a similar example for the Brownian motion. (This is due to Durrett.)

4.5. Superharmonic Function and Supermartingale

We introduce a class of functions which stand in the same relation to harmonic functions as supermartingales to martingales.

Definition. Let D be an open set. A function u is called *superharmonic in D* iff

(a) $-\infty < u \leq +\infty$; $u \not\equiv +\infty$ in each component domain of D; and u is lower semi-continuous in D;
(b) for any ball $B(x, r) \Subset D$ we have

$$u(x) \geq \frac{1}{\sigma(r)} \int_{S(x,r)} u(y)\sigma(dy). \tag{1}$$

[A function u is *subharmonic in D* iff $-u$ is superharmonic in D; but we can dispense with this term here.] Thus the sphere-averaging property in (3) of §4.3 is replaced here by an inequality. The assumption of lower semi-continuity in (a) implies that u is bounded below on each compact subset of D. In particular the integral in the right member of (1) is well defined and not equal to $-\infty$. We shall see later that it is in fact finite even if $u(x) = +\infty$.

In consequence of (1), we have also for any $B(x, r) \Subset D$:

$$u(x) \geq \frac{1}{v(r)} \int_{B(x,r)} u(y)m(dy). \tag{2}$$

Since u is bounded below in $B(x, r)$ we may suppose $u \geq 0$ there, and $u(x) < \infty$. We then obtain (2) from (1) as we did in (6) of §4.3. It now follows that if u is superharmonic in a *domain D*, then it is locally integrable there. To see this we may again suppose $u \geq 0$ because u is bounded below on each compact subset of D. Let A be the set of points x in D such that u has a finite integral over some ball with center at x. It is trivial that A is open. Observe that if $x \in A$, then there is x_n arbitrarily near x for which $u(x_n) < \infty$, and consequently u has a finite integral over any $B(x_n, r) \Subset D$. Therefore this is also true for any $B(x, r) \Subset D$ by an argument of inclusion. Now let $x_n \in A$, $x_n \to x \in D$. Then for sufficiently large n and small r we have $x \in B(x_n, r) \Subset D$. Thus $x \in A$ and so A is closed in D. Since D is a domain and A is not empty we have proved $A = D$ as claimed.

We now follow an unpublished method of Doob's to analyse superharmonic functions by means of supermartingales. Suppose first that D is bounded and u is superharmonic in an open set containing $\bar{D}$. Let ρ denote the distance function in R^d. For a fixed $\varepsilon > 0$ and each x in D we put

$$r(x) = \tfrac{1}{2}\rho(x, \partial D) \wedge \varepsilon, \qquad S(x) = S(x, r(x));$$

thus $S(x)$ is the sphere with center x and radius $r(x)$ varying with x as prescribed. Define a sequence of space-dependent hitting times $\{T_n, n \geq 0\}$ as follows: $T_0 = 0$ and for $n \geq 1$:

$$T_n = \inf\{t > T_{n-1} \,|\, X(t) \in S(X(T_{n-1}))\} \tag{3}$$

where $\inf \varnothing = +\infty$ as usual. Each T_n is optional with respect to $\{\mathscr{F}_t\}$ (exercise!). The strong Markov property of X and the inequality (1) imply the key relation:

$$u(X(T_{n-1})) \geq E^x\{u(X(T_n)) \,|\, \mathscr{F}(T_{n-1})\}, \tag{4}$$

provided that $u(x) < \infty$. The integrability of $u(X(T_n))$ is proved by induction on n under this proviso. Thus $\{u(X(T_n)), \mathscr{F}(T_n), n \geq 1\}$ is a supermartingale under P^x. Since D is bounded all the terms of the supermartingale are bounded below by a fixed constant. Hence it may be treated like a positive supermartingale; in particular, it may be extended to the index set "$1 \leq n \leq \infty$" in a trivial way.

Let us first prove the crucial fact that for each $x \in D$, we have

$$P^x\left\{\lim_{n\to\infty} T_n = \tau_D\right\} = 1. \tag{5}$$

For it is clear that P^x-a.s., $T_n \uparrow$ and $T_n < \tau_D < \infty$ because D is bounded. Let $T_\infty = \lim_n \uparrow T_n$. Then $X(T_\infty) = \lim_{n\to\infty} X(T_n)$ by continuity and so $X(T_\infty) \in \bar{D}$. Suppose if possible that $X(T_\infty) \in D$. Then for all sufficiently large values of n, we have $\rho(X(T_{n-1}), \partial D) > 2^{-1}\rho(X(T_\infty), \partial D) > 0$ and $\rho(X(T_{n-1}), X(T_n)) < 4^{-1}\rho(X(T_\infty), \partial D) \wedge \varepsilon$. But by (3), $\rho(X(T_{n-1}), X(T_n)) = 2^{-1}\rho(X(T_{n-1}), \partial D) \wedge \varepsilon$. These inequalities are incompatible. Hence $X(T_\infty) \in \partial D$ and consequently $T_\infty = \tau_D$.

Now fix $t > 0$ and define a stopping random variable as follows:

$$N = \inf\{n \geq 1 \mid T_n \geq t\}. \tag{6}$$

where $\inf \varnothing = +\infty$. For each integer $k \geq 1$, $\{N = k\}$ belongs to the σ-field generated by $T_1, \ldots, T_k$; hence to $\mathscr{F}(T_k)$. Thus N is optional with respect to $\{\mathscr{F}(T_n), n \geq 1\}$. Therefore by Doob's stopping theorem for a positive (hence closable) supermartingale (Theorem 9.4.5 of *Course*):

$$\{u(X(T_{n \wedge N})), \mathscr{F}(T_{n \wedge N}), 1 \leq n \leq \infty\} \tag{7}$$

is a supermartingale under P^x provided $u(x) < \infty$. Here $X(T_\infty) = X(\tau_D)$, $\mathscr{F}(T_\infty) = \mathscr{F}(\tau_D)$ by (5). It follows that

$$u(x) \geq \varliminf_n E^x\left\{u(X(T_{n \wedge N}))\right\} \geq E^x\left\{\varliminf_n u(X(T_{n \wedge N}))\right\} \tag{8}$$

by Fatou's lemma since all the terms are bounded below. Now $\lim_n X(T_{n \wedge N}) = X(T_N)$ whether N is finite or infinite, and by the lower semi-continuity of u in $\bar{D}$ (because u is superharmonic in an open set containing $\bar{D}$) we conclude that

$$u(x) \geq E^x\{u(X(T_{N(\varepsilon)}))\}, \tag{9}$$

where we have indicated the dependence of N on ε. On the set $\{t < \tau_D\}$, there exists n such that $t < T_n$; hence $N(\varepsilon) < \infty$ and $T_{N(\varepsilon)-1} < t \leq T_{N(\varepsilon)}$ by (6). By the definition of $T_{N(\varepsilon)}$, this implies $|X(t) - X(T_{N(\varepsilon)})| \leq \varepsilon$. Thus we have proved

$$\lim_{\varepsilon \downarrow 0} X(T_{N(\varepsilon)}) = X(t \wedge \tau_D) \tag{10}$$

P^x-a.s. on $\{t < \tau_D\}$. On the set $\{t \geq \tau_D\}$, $N(\varepsilon) = +\infty$ for all $\varepsilon > 0$ and $X(T_{N(\varepsilon)}) = X(\tau_D)$ by (5). Hence (10) is also true. Using (10) in (9) and the lower semi-continuity of u again, we obtain

$$u(x) \geq E^x\{u(X(t \wedge \tau_D))\}. \tag{11}$$

The next step is standard: if $0 < s < t$, we have P^x-a.s.

$$\begin{aligned} E^x\{u(X(t \wedge \tau_D)) \mid \mathscr{F}_s\} &= E^{X(s)}\{u(X((t - s) \wedge \tau_D))\} \leq u(X(s)) \\ &= u(X(s \wedge \tau_D)) \end{aligned} \tag{12}$$

on $\{s < \tau_D\}$, since $X(s) \in D$; whereas on $\{s \geq \tau_D\}$ the two extreme members of (12) both reduce to $u(X(\tau_D))$ and there is equality. We record this important result below.

Theorem 1. *If D is bounded and u is superharmonic in an open set containing $\bar{D}$, then*

$$\{u(X(t \wedge \tau_D)), \mathscr{F}(t), 0 \leq t \leq \infty\} \tag{13}$$

is a supermartingale under P^x for each $x \in D$ for which $u(x) < \infty$. When u is harmonic in an open set containing $\bar{D}$, then (13) *is a martingale under P^x for each $x \in D$.*

We proceed to extend this result to an unbounded D and a function u which is superharmonic in D only. At the same time we will prove that the associated supermartingale has right continuous paths with left limits. To do this we use the basic machinery of a Hunt process given in §3.5. First, we must show that the relevant process is a Hunt process. For the Brownian motion in R^d this is trivial; indeed we verified that it is a Feller process. But what we need here is a Brownian motion somehow living on the open set D, as follows (see Exercise 9 for another way).

Let D be any open subset of R^d, and $D_\partial = D \cup \{\partial\}$ the one-point compactification of D. Thus ∂ plays the role of "the point at infinity": $x \to \partial$ iff x leaves all compact subsets of D, namely $x \to \partial D$; and all points of ∂D are identified with ∂. Define a process $\{\tilde{X}(t); t \geq 0\}$ living on the state space D_∂ as follows:

$$\tilde{X}(t) = \begin{cases} X(t), & \text{if } t < \tau_D; \\ \partial, & \text{if } \tau_D \leq t \leq \infty. \end{cases} \tag{14}$$

By this definition ∂ is made an absorbing state, and τ_D the lifetime of $\tilde{X}$; see (13) and (14) of §1.2. Define a transition semigroup $(Q_t, t \geq 0)$ as follows; for any bounded Borel function φ on D_∂ and $x \in D$:

$$Q_t\varphi(x) = E^x\{t < \tau_D; \varphi(X_t)\} + P^x\{t \geq \tau_D\}\varphi(\partial). \tag{15}$$

In particular, $Q_0\varphi = \varphi$. We call $\tilde{X}$ the "Brownian motion killed outside D" or "killed at the boundary ∂D". It is easy to check that (Q_t) is the transition semigroup of $\tilde{X}$, but the next theorem says much more. Recall $\{\mathscr{F}_t\}$ is the family of augmented σ-fields associated with the unrestricted Brownian motion $\{X_t, t \geq 0\}$.

Theorem 2. $\{\tilde{X}_t, \mathscr{F}_t, t \geq 0\}$ *is a Hunt process.*

Proof. First of all, almost all sample paths are continuous in $[0, \infty)$, in the topology of D_∂. To see this we need only check the continuity at τ_D when $\tau_D < \infty$. This is true because as $t \uparrow \tau_D < \infty$, $X(t) \to \partial = X(\tau_D)$. Since continuity clearly implies quasi left continuity, it remains only to check the strong Markov property. Let f be a continuous function on D_∂, then it is bounded because D_∂ is compact. Let T be optional with respect to $\{\mathscr{F}_t\}$, and $\Lambda \in \mathscr{F}_T$.

We have for $x \in D_\partial$ and $t \geq 0$:

$$E^x\{\Lambda; Q_t f(\tilde{X}_T); T < \infty\}$$
$$= E^x\{\Lambda; T < \tau_D; Q_t f(X_T)\} + P^x\{\Lambda; \tau_D \leq T < \infty\} f(\partial). \qquad (16)$$

since $Q_t f(\partial) = f(\partial)$ for $t \geq 0$. By (15), the first term on the right side above is equal to

$$E^x\{\Lambda; T < \tau_D; E^{X(T)}[t < \tau_D; f(X_t)]\} + E^x\{\Lambda; T < \tau_D; E^{X(T)}[t \geq \tau_D]\} f(\partial),$$

which is equal to

$$E^x\{\Lambda; T + t < \tau_D; f(X_{T+t})\} + P^x\{\Lambda; T < \tau_D \leq T + t\} f(\partial)$$

by the strong Markov property of the unrestricted Brownian motion X applied at T. Substituting into (16) we obtain

$$E^x\{\Lambda; T + t < \tau_D; f(X_{T+t})\} + P^x\{\Lambda; \tau_D \leq T + t < \infty\} f(\partial).$$

This is none other than $E^x\{\Lambda; f(\tilde{X}_{T+t}); T + t < \infty\}$ upon inspection. Since Λ is an arbitrary set from $\mathscr{F}_T$, we have verified that

$$E^x\{f(\tilde{X}_{T+t}) \mid \mathscr{F}_T\} = Q_t f(\tilde{X}_T)$$

for every continuous f on D_∂. Therefore, $\tilde{X}$ has the strong Markov property (cf. Theorem 1 of §2.3), indeed with respect to a larger σ-field than that generated by $\tilde{X}$ itself. □

We are ready for the *dénouement*.

Theorem 3. *Let D be an open set and u be superharmonic and ≥ 0 in D; put $u(\partial) = 0$. Then u is excessive with respect to (Q_t). Conversely if u is excessive with respect to (Q_t), and $\not\equiv \infty$ in each component domain of D, then u restricted to D is superharmonic and ≥ 0.*

Proof. Let each D_n be bounded open and $D_n \uparrow\uparrow D$. Write $\tau_n = \tau_{D_n}$, $\tau = \tau_D$ below. Theorem 1 above is applicable to each D_n. Hence if $x \in D$ and $u(x) < \infty$ we have

$$u(x) \geq E^x\{u(X(t \wedge \tau_n))\} \geq E^x\{u(X(t)) 1_{\{t < \tau_n\}}\} \qquad (17)$$

where the second inequality is trivial because $u \geq 0$. As $n \to \infty$ we have P^x-a.s. $\tau_n \uparrow \tau$ and $\{t < \tau_n\} \uparrow \{t < \tau\}$. Hence it follows by monotone convergence that

$$u(x) \geq E^x\{u(X(t)) 1_{\{t < \tau\}}\} = Q_t u(x) \qquad (18)$$

where the equation is due to $u(\partial) = 0$. Since u is lower semi-continuous at x we have P^x-a.s. $u(x) \le \underline{\lim}_{t\downarrow 0} u(X(t))$; while $\lim_{t\downarrow 0} 1_{\{t<\tau\}} = 1$. Hence by Fatou:

$$u(x) \le E^x\left\{\underline{\lim}_{t\downarrow 0} u(X(t))1_{\{t<\tau\}}\right\} \le \underline{\lim}_{t\downarrow 0} Q_t u(x). \tag{19}$$

The two relations (18) and (19), together with the banality $u(\partial) = Q_t u(\partial)$, show that u is excessive with respect to (Q_t).

Conversely, if u is excessive with respect to (Q_t), then $u \ge 0$; and the inequality (1) is a very special case of $u \ge P_A u$ in Theorem 4 of §3.4. Observe that $Q_{A^c}(x, \cdot) = P_{A^c}(x, \cdot)$ if $x \in D$ and $\bar{A} \subset D$, where Q_A is defined for $\tilde{X}$ as P_A is for X. To show that u is superharmonic in D it remains to show that u is lower semi-continuous at each x in D. Let $B(x, 2r) \Subset D$. Then for all $y \in B(x, r) = B$, we have $P^y\{\tau \le t\} \le P^y\{T_r \le t\}$ and the latter does not depend on y (T_r is defined in (10) of §4.2). It follows that

$$\lim_{t\downarrow 0} P^y\{\tau \le t\} = 0 \quad \text{uniformly for } y \in B. \tag{20}$$

Now write for $y \in B$:

$$\begin{aligned} Q_t u(y) &= E^y\{t < \tau; u(X_t)\} + P^y\{\tau \le t\} u(\partial) \\ &= P_t u(y) - E^y\{\tau \le t; u(X_t) - u(\partial)\}. \end{aligned} \tag{21}$$

Suppose first that u is bounded, then $P_t u$ is continuous by the strong Feller property; while the last term in (21) is less than ε if $0 < t < \delta(\varepsilon)$, for all $y \in B$. We have then

$$u(y) \ge Q_t u(y) \ge P_t u(y) - \varepsilon;$$

$$\underline{\lim}_{y\to x} u(y) \ge \lim_{y\to x} P_t u(y) - \varepsilon = P_t u(x) - \varepsilon \ge Q_t u(x) - \varepsilon.$$

Letting $t \downarrow 0$ we see that u is lower semi-continuous at x. For a general excessive u, we have $u = \lim_n \uparrow (u \wedge n)$, hence its lower semi-continuity follows at once from that of $u \wedge n$. □

Here is an application to a well known result in potential theory known as the *barrier theorem*. It gives a necessary and sufficient condition for the regularity at a boundary point of an open set. We may suppose the set to be bounded since regularity is a local property. A function u defined in D is called a *barrier* at z iff it is superharmonic and >0 in D, and $u(x)$ converges to zero as x in D tends to z.

Proposition 4. *Let $z \in \partial D$; then z is regular for D^c if and only if there exists a barrier at z.*

Proof. Let u be a barrier at z and put $u(\partial) = 0$. Then u is excessive with respect to (Q_t). We may suppose u bounded by replacing it with $u \wedge 1$. Suppose z is not regular for D^c. Let B_1 and B_2 be two balls with center at z, $B_1 \subset B_2$. We have then

$$E^z\{\tau_{B_2} < \tau_D; u(X(\tau_{B_2}))\} = E^z\{\tau_{B_1} < \tau_D; E^{X(\tau_{B_1})}[\tau_{B_2} < \tau_D; u(X(\tau_{B_2}))]\}. \quad (22)$$

On $\{\tau_{B_1} < \tau_D\}$, we have $X(\tau_{B_1}) \in D$ and

$$\begin{aligned} E^{X(\tau_{B_1})}[\tau_{B_2} < \tau_D; u(X(\tau_{B_2}))] &\leq E^{X(\tau_{B_1})}[u(\tilde{X}(\tau_{B_2} \wedge \tau_D)] \\ &= Q_C u(\tilde{X}(\tau_{B_1})) \leq u(\tilde{X}(\tau_{B_1})) = u(X(\tau_{B_1})), \end{aligned}$$

where $C = (B_2 \cap D)^c \cup \{\partial\}$, explicitly. Substituting into (22) we obtain

$$E^z\{\tau_{B_2} < \tau_D; u(X(\tau_{B_2}))\} \leq E^z\{\tau_{B_1} < \tau_D; u(X(\tau_{B_1}))\}. \quad (23)$$

Since P^z-a.s., $\tau_D > 0$ while $\tau_{B_2} \downarrow 0$ if B_2 shrinks to z, we may choose B_2 so that the left member of (23) has a value >0 because $u > 0$ in D. Now fix B_2 and let B_1 shrink to z. Then $X(\tau_{B_1}) \to z$ and so $u(X(\tau_{B_1})) \to 0$ by hypothesis; hence the right member converges to zero by bounded convergence. This contradiction proves that z must be regular for D^c.

Conversely if z is regular for D^c, put $f(x) = \|x - z\|$ on ∂D. Then f is bounded on ∂D since D is bounded; and $f(x) = 0$ if and only if $x = z$. By the solution of the Dirichlet problem for (D, f), $u = P_{D^c} f$ is harmonic in D and converges to $f(z) = 0$ as $x \to z$. Since $f > 0$ on ∂D except at one point (which forms a polar set), it is clear that $u > 0$ in D. Thus u is the desired barrier at z. □

Proposition 5. *Let u be superharmonic and bounded below in a domain D. Let B be open and $\bar{B} \subset D$. Then $P_{B^c}u$ is harmonic in B, and superharmonic in D.*

Proof. We may suppose $u \geq 0$ by adding a constant and put $u(\partial) = 0$. Then u is excessive with respect to (Q_t) and so by Theorem 4 of §3.4, for every $x \in D$:

$$u(x) \geq Q_{B^c}u(x) = P_{B^c}u(x). \quad (24)$$

Since u is locally integrable, there exists x in each component of B for which $u(x) < \infty$. Theorem 1 of §4.3 applied with $f = u$ then establishes the harmonicity of $P_{B^c}u$ in B. $P_{B^c}u$ is excessive with respect to (Q_t) by Theorem 4 of §3.4, hence superharmonic in D by Theorem 3 above. □

When B is a ball with $\bar{B} \subset D$, the function P_{B^c} is known as the *Poisson integral of u for B* and plays a major role in classical potential theory. Since B is regular we have $P_{B^c}u = u$ in $D - B$. Thus the superharmonic function u is transformed into another one which has the nicer property of being harmonic in B. This is the simplest illustration of the method of balayage: "sweeping all the charge in B onto its boundary". (There is no charge where $\Delta u = 0$, by a law of electromagnetism.) Poisson's formula in Example 2 of §4.4 gives the exact form of the Poisson integral. A closely related notion will now be discussed.

EXAMPLE. Recall from Example 4 of §3.7, for $d \geq 3$:

$$u(x, y) = \frac{A_d}{\|x - y\|^{d-2}}, \qquad A_d = \frac{\Gamma\left(\frac{d}{2} - 1\right)}{2\pi^{d/2}}.$$

Then $u(x, y)$ denotes the potential density at y for the Brownian motion starting from x. For a fixed y, $u(\cdot, y)$ is harmonic in $R^d - \{y\}$ by Proposition 5 of §4.3. Since it equals $+\infty$ at y it is trivial that the inequality (1) holds for $x = y$; since it is continuous in the extended sense, $u(\cdot, y)$ is superharmonic in R^d. Now let D be an open set and for a fixed $y \in D$ put

$$g_D(x, y) = u(x, y) - P_{D^c}u(x, y) \tag{25}$$

where P_{D^c} operates on the function $x \to u(x, y)$. It follows from the preceding discussion that $g_D(\cdot, y)$ is superharmonic in D, harmonic in $D - \{y\}$, and vanishes in $(\bar{D})^c$ and at points of ∂D which are regular for D^c; if D is regular then it vanishes in D^c. Since $u(\cdot, y)$ is superharmonic and ≥ 0 in R^d, it is excessive with respect to (P_t) by an application of Theorem 3 with $D = R^d$. Of course this fact can also be verified directly without the intervention of superharmonicity; cf. the Example in §3.6. Now $g_D \geq 0$ by Theorem 4 of §3.4. Hence it is excessive with respect to (Q_t) by another application of Theorem 3. For any Borel measurable function f such that $U|f|(x) < \infty$ we have

$$\int_{R^d} g_D(x, y) f(y)\, dy = Uf(x) - P_{D^c}Uf(x) = E^x\left\{\int_0^{\tau_D} f(X_t)\, dt\right\} \tag{26}$$

where X_t may be replaced by $\tilde{X}_t$ in the last member above. Thus $g_D(x, y)$ plays the same role for $\tilde{X}$ as $u(x, y)$ for X; it is the potential density at y for the Brownian motion starting from x and killed outside D.

The function g_D is known as the *Green's function* for D and the quantity in (26), which may be denoted by $G_D f$, is the associated *Green's potential* of f. It is an important result that $g_D(x, y) = g_D(y, x)$ for all x and y; indeed Hunt [1] proved that $Q_t(x, dy)$ has a density $q(t; x, y)$ which is symmetric in (x, y) and continuous in $(0, \infty) \times D \times D$. This is a basic result in the deeper study of the killed process, which cannot be undertaken here. In R^2, Green's function can also be defined but the situation is complicated by the fact that the Brownian motion is recurrent so that the corresponding

u in (25) is identically $+\infty$. [This is excessive but not superharmonic!] Finally, for a ball in any dimension Green's function can be obtained by a device known as Kelvin's transformation; see e.g. Kellogg [1]. The case $d = 1$ is of course quite elementary.

As a corollary of Proposition 5, we see now that the sphere-average in the right member of (1) is finite for every x in D. For it is none other than $P_{B^c}u(x)$ when $B = B(x, r)$. The next proposition is somewhat deeper.

Proposition 6. *If u is excessive with respect to (Q_t), and $\not\equiv \infty$ in each component of D, with $u(\partial) < \infty$, then $Q_t u < \infty$ for each $t > 0$.*

Proof. Let $x \in D$, $B = B(x, r) \Subset D$, and $\tau = \tau_D$. Then we have

$$E^x\{t < \tau; u(X_t)\} = E^x\{t \leq \tau_B; u(X_t)\} + E^x\{\tau_B < t < \tau; u(X_{\tau_B \vee t})\}. \quad (27)$$

The first term on the right side of (27) does not exceed

$$E^x\{X_t \in \bar{B}; u(X_t)\} = \int_B p(t; x, y) u(y)\, dy \leq \frac{1}{(2\pi t)^{d/2}} \int_B u(y)\, dy < \infty$$

because u is locally integrable. The second term does not exceed

$$E^x\{u(X(\tau_B))\} = P_{B^c}u(x) < \infty$$

by the supermartingale stopping theorem (Theorem 4 in §1.4) applied to $u(\tilde{X}_t)\}$. Since

$$E^x\{t \geq \tau;\ \ u(\tilde{X}_t)\} = P^x\{t \geq \tau\} u(\partial) < \infty$$

$Q_t u(x) < \infty$ *by* (15). □

We can now extend Theorem 1 as follows.

Theorem 7. *Let D be an open set and u be superharmonic and ≥ 0 in D. Then for any $x \in D$,*

$$w = \lim_{t \uparrow\uparrow \tau_D} u(X(t)) \quad \textit{exists } P^x\text{-a.s.} \quad (28)$$

Put for $0 \leq t \leq \infty$:

$$w(t) = \begin{cases} u(X(t)), & t < \tau_D; \\ w, & t \geq \tau_D; \end{cases} \quad (29)$$

thus $w(\infty) = w$. Then for any $x \in D$, we have P^x-a.s.: $t \to w(t)$ is right continuous in $[0, \infty)$ and has left limits in $(0, \infty]$; and $\{w(t), \mathscr{F}(t), 0 < t \leq \infty\}$ is a supermartingale under P^x. In case $u(x) < \infty$, the parameter value $t = 0$ may be added to the supermartingale. If u is harmonic and bounded in D, then $\{w(t), \mathscr{F}(t), 0 \leq t \leq \infty\}$ is a continuous martingale under P^x for any $x \in D$.

Proof. By Theorem 3, u is excessive with respect to (Q_t). Hence by Theorem 6 of §3.4 applied to the Hunt process $\tilde{X}$, $t \to u(\tilde{X}_t)$ is right continuous in $[0, \infty)$ and has left limits in $(0, \infty]$, P^x-a.s. for any $x \in D$. It follows that the limit in (28) exists. The asserted continuity properties of $t \to w(t)$ then follow trivially. Next, let D_n be bounded open, $\bar{D}_n \subset D_{n+1}$ and $\bigcup_n D_n = D$. Given x in D there exists n for which $x \in D_n$. It follows from Proposition 5 that $E^x\{u(X(\tau_{D_m}))\} < \infty$ for all $m \geq n$, and is decreasing since u is excessive. Hence, $E^x\{w\} < \infty$ by Fatou. We have either by Theorem 1 or by the stopping theorem for excessive functions, under P^x:

$$\begin{aligned} w(0) = u(x) &\geq \varliminf_n E^x\{u(X(t \wedge \tau_{D_n}))\} \geq E^x\left\{\varliminf_n u(X(t \wedge \tau_{D_n}))\right\} \\ &= E^x\{u(X(t)); t < \tau_D\} + E^x\{w; t \geq \tau_D\} \\ &= E^x\{w(t)\}. \end{aligned}$$

For any $x \in D$ and $t > 0$;

$$E^x\{w(t)\} = Q_t u(x) + E^x\{t \geq \tau_D; w\} < \infty.$$

These relations imply the assertions of the theorem, the details of which are left to the reader. □

There is a major improvement on Theorem 7 to the effect that almost all sample functions of $w(t)$ defined in (29) are actually *continuous*. This is Doob's theorem and the proof given below makes use of a new idea, the reversal of time. We introduce it for a spatially homogeneous Markov process discussed in §4.1, as follows. The transition probability function is said to have a *density* iff for every $t > 0$ there exists $p_t \geq 0$, $p_t \in \mathscr{E} \times \mathscr{E}$ such that

$$P_t(x, dy) = p_t(x, y)\, dy. \tag{30}$$

Recall that the Lebesgue measure m is invariant in the sense of (8) of §4.1. In the case of (30) this amounts to $\int m(dx) p_t(x, y) = 1$ for m-a.e. y. Although m is not a finite measure, it is trivial to define P^m and E^m in the usual way; thus $P^m(\Lambda) = \int m(dx) P^x(\Lambda)$ for each $\Lambda \in \mathscr{F}^{\sim}$. Now for a fixed $c > 0$ define a *reverse process* $\tilde{X}_c$ in $[0, c]$ as follows:

$$\tilde{X}_c(t) = X(c - t), \qquad t \in [0, c]. \tag{31}$$

There is no difficulty in the concept of a stochastic process on a parameter set such as $[0, c]$. The sample functions of $\tilde{X}_c$ are obtained from those of X by reversing the time from c. Since X as a Hunt process is right continuous and has left limits in $(0, c)$, $\tilde{X}_c$ is left continuous and has right limits there. This kind of sample functions has been considered earlier in Chapter 2, for instance in Theorem 6 of §2.2.

Proposition 8. *Let X be a spatially homogeneous Markov process satisfying (30), where for each $t > 0$, p_t is a symmetric function of (x, y). Then under P^m, $\tilde{X}_c$ has the same finite dimensional distributions as X in $[0, c]$.*

Proof. Let $0 \le t_1 < \cdots < t_n \le c$, and $f_j \in \mathscr{E}_+$ for $1 \le j \le n$. To show that

$$E^m\left\{\prod_{j=1}^{n} f_j(X(t_j))\right\} = E^m\left\{\prod_{j=1}^{n} f_j(\tilde{X}_c(t_j))\right\} \tag{32}$$

we rewrite the right member above as $E^m\{\prod_{j=n}^{1} f_j(X(c - t_j))\}$. Since $mP_{c-t_j} = m$, the latter can be evaluated as follows:

$$\int \cdots \int dx_n f_n(x_n) \prod_{j=n-1}^{1} p(t_{j+1} - t_j; x_{j+1}, x_j) f_j(x_j)\, dx_j$$

$$= \int \cdots \int f_1(x_1)\, dx_1 \prod_{j=1}^{n-1} p(t_{j+1} - t_j; x_j, x_{j+1}) f_{j+1}(x_{j+1})\, dx_{j+1}.$$

The last expression is equal to the left member of (32) by an analogous evaluation. □

We return to Brownian motion X below.

Theorem 9. *Let u be a positive superharmonic function in R^d. Then $t \to u(X(t))$ is* a.s. *continuous in $[0, \infty]$ (i.e., continuous in $(0, \infty)$ and has limits at 0 and at ∞).*

Proof. Consider the reverse process $\tilde{X}_c$ in (31). Although under P^m it is a copy of Brownian motion in $[0, c]$ under P^m by Proposition 8, it is not absolutely clear that we can apply Theorem 6 of §3.4 to $u(\tilde{X}_c(t))$ since that theorem is stated for a Hunt process in $[0, \infty)$. However, it is easy to extend $\tilde{X}_t$ to $[0, \infty)$ as follows. Define

$$Y_c(t) = \begin{cases} \tilde{X}_c(t), & \text{for } t \in [0, c], \\ \tilde{X}_c(c) + X(t) - X(c), & \text{for } t \in [c, \infty). \end{cases} \tag{33}$$

It follows from the independent increments property of X and Proposition 8 that $\{Y_c(t), t \ge 0\}$ under P^m is a copy of Brownian motion under P^m. Therefore $t \to u(Y_c(t))$ is right continuous in $[0, \infty)$ and has left limits in $(0, \infty]$. In particular, $t \to u(\tilde{X}_c(t)) = u(X(c - t))$ is right continuous in $(0, c)$, which means that $t \to u(X(t))$ is left continuous in $(0, c)$. But $t \to u(X(t))$ is also right continuous in $(0, c)$, hence it is in fact continuous there. This being true for each c, we conclude that under P^m, $t \to u(X(t))$ is continuous in $[0, \infty]$. By the definition of P^m, this means that for m-a.e. x we have the same result under P^x. For each $s > 0$ and x, $P_s(x, \cdot)$ is absolutely continuous with respect to m. Hence for $P_s(x,\cdot)$-a.e. y, the result holds under P^y. For $s \geq 0$ let

$$\Lambda_s = \{\omega \mid u(X(\cdot,\omega)) \text{ is continuous in } [s,\infty]\}.$$

Then $\Lambda_s \in \mathscr{F}^\sim$ (proof?). We have for each x,

$$P^x\{\Lambda_s\} = E^x\{P^{X(s)}[\Lambda_0]\} = 1$$

because $X(s)$ has $P_s(x,\cdot)$ as distribution. Letting $s \downarrow 0$, we obtain $P^x\{\Lambda_0\} = 1$. □

Theorem 9 is stated for u defined in the whole space. It will now be extended as follows. Let D be an open set, and u be positive and superharmonic in D. The function $t \to u(X(t,\omega))$ is defined in the set $I(\omega) = \{t > 0 \mid X(t,\omega) \in D\}$, which is a.s. an open set in $(0,\infty)$. We shall say that $u(X)$ is (right, left) continuous *wherever defined* iff $t \to u(X(t,\omega))$ is so on $I(\omega)$ for a.e. ω. Let r and r' be rational numbers, $r < r'$. It follows from Theorem 7 that for each x

$$P^x\{X(t) \in D \text{ for all } t \in [r,r'], \text{ and } t \to u(X(t)) \text{ is right continuous in } (r,r')\} = P^x\{X(t) \in D \text{ for all } t \in [r,r']\}. \tag{34}$$

For we can apply Theorem 7 to the Brownian motion starting at time r in D, on the set $\{r' < \tau_D\}$. Since every generic t in $I(\omega)$ is caught between a pair $[r,r']$ in the manner shown in (34), we have proved that for a Brownian motion X with any initial distribution, $u(X)$ is right continuous wherever defined. In particular, the last assertion is true for the Brownian motion Y_c defined in (33) under P^m. Thus $t \to u(Y_c(t))$ is right continuous wherever defined, hence $t \to u(X(c-t))$ is right continuous for $t \in (0,c)$ and $c - t \in I(\omega)$. Since c is arbitrary, this implies that $t \to u(X(t))$ is left continuous for $t \in I(\omega)$. Thus $u(X)$ is in fact continuous wherever defined under P^m. As in Theorem 9, we can replace P^m by P^x for every x. In the context of Theorem 7, we have proved that $t \to u(X(t))$ is continuous in $(0,\tau_D)$, hence $t \to w(t)$ is continuous in $[0,\infty]$. We state this as a corollary, although it is actually an extension of Theorem 9.

Corollary. *In Theorem 7, $t \to w(t)$ is continuous in $[0,\infty]$*, a.s.

Using the left continuity of an excessive function along Brownian paths, we can prove a fundamental result known as the Kellogg-Evans theorem in classical potential theory (see §5.1).

Theorem 10. *For the Brownian motion process, a semi-polar set is polar.*

Proof. By the results of §3.5, it is sufficient to prove that a thin compact set K is polar. Put $\varphi(x) = E^x\{e^{-T_K}\}$. Since K is thin, $\varphi(x) < 1$ for all x. Let first $x \notin K$, and G_n be open sets such that $G_n \downdownarrows K$. Then $\lim_n \uparrow T_{G_n} = T_K \leq \infty$ and $T_{G_n} < T_K$ for all n, P^x-a.s. The strict inequality above is due to the continuity of the Brownian paths and implies $T_K = T_{G_n} + T_K \circ \theta(T_{G_n})$ on $\{T_{G_n} < \infty\}$. It follows by the strong Markov property that

$$E^x\{e^{-T_K} | \mathscr{F}(T_{G_n})\} = e^{-T_{G_n}}\varphi(X(T_{G_n})). \tag{35}$$

Since $T_{G_n} \in \mathscr{F}(T_{G_n})$, we have $T_K \in \bigvee_{n=1}^{\infty} \mathscr{F}(T_{G_n})$. Letting $n \to \infty$ in (35), we obtain a.s.

$$e^{-T_K} = e^{-T_K}\varphi(X(T_K)) \tag{36}$$

where Theorem 9 is crucially used to ensure that $\varphi(X(\cdot))$ is left continuous at T_K if $T_K < \infty$, and has a limit at T_K if $T_K = \infty$. In the latter case of course $\varphi(X(T_K))$ is meant to be $\lim_{t \uparrow \infty} \varphi(X(t))$. However, we need (36) only on the set $\{T_K < \infty\}$, on which the relation (36) entails $\varphi(X(T_K)) = 1$. Since $\varphi < 1$ everywhere this forces $P^x\{T_K < \infty\} = 0$. Now let x be arbitrary and $t > 0$. We have

$$P^x\{t < T_K < \infty\} = E^x\{t < T_K;\, P^{X(t)}[T_K < \infty]\} = 0$$

because $X(t) \in K^c$ on $\{t < T_K\}$. Letting $t \downarrow 0$ we obtain $P^x\{T_K < \infty\} = 0$ since x is not regular for K. Thus K is polar. □

Corollary. *For any Borel set A, $A \backslash A^r$ is a polar set.*

This follows because $A \backslash A^r$ is semi-polar by Theorem 6 of §3.5. We can now prove the uniqueness of the solution of the generalized Dirichlet problem in §4.4.

Proposition 11. *Let D be a bounded open set; f be a bounded measurable function defined on ∂D and continuous on $\partial^* D = \partial D \cap (D^c)^r$. Then the function $H_D f$ in* (1) *of* §4.4 *is the unique bounded solution to the generalized Dirichlet problem.*

Proof. We know from §4.4 that $H_D f$ is a solution. Let h be another solution and put $u = h - H_D f$. Then u is harmonic in D and converges to zero on $\partial^* D$, the set of "regular boundary points". By the Corollary to Theorem 10, $D^c \backslash (D^c)^r$ is a polar set. Observe that it contains $\partial D - \partial^* D$. For each x in D, the following assertions are true P^x-a.s. As $t \uparrow\uparrow \tau_D$, $X(t) \to X(\tau_D) \in \partial^* D$ because $\partial D - \partial^* D$ being a subset of $D^c \backslash (D^c)^r$ cannot be hit. Hence $u(X(t))$ as $t \uparrow\uparrow \tau_D$ converges to zero, and consequently the random variable w defined in (28) is equal to zero. Applying the last assertion of Theorem 7 to u, we conclude that $u(x) = E^x\{w\} = 0$. In other words, $h \equiv H_D f$ in D. □

We close this section by discussing the analogue of Definition 2 for a harmonic function in §4.3. It is trivial that the limit of an increasing sequence of superharmonic functions in a domain D is superharmonic provided it is not identically infinite there. On the other hand, any superharmonic function in D is the limit of an increasing sequence of infinitely differentiable superharmonic functions in the sense made precise in Exercise 12 below. Finally, we have the following characterization for a twice continuously differentiable superharmonic function.

Theorem 12. *If $u \in \mathbb{C}^{(2)}(D)$ and $\Delta u \leq 0$ in D, then u is superharmonic in D. If u is superharmonic and belong to $\mathbb{C}^{(2)}$ in a neighborhood of x, then $\Delta u(x) \leq 0$.*

Proof. To prove the first assertion let $B(x,r) \Subset D$. Then u belongs to $\mathbb{C}^{(2)}$ in an open neighborhood of $B(x,r)$. Hence formula (16) of §4.3 holds with h replaced by u, and yields $(d/d\delta)A(x,\delta) \leq 0$, for $0 < \delta \leq r$. It follows that $A(x,r) \leq \lim_{\delta \downarrow 0} A(x,\delta) = u(x)$, which is the inequality (1). The other conditions are trivially satisfied and so u is superharmonic in D. Next, suppose that u is superharmonic and belongs to $\mathbb{C}^{(2)}$ in $B(x,r)$. We may suppose $u \geq 0$ there, and then by Theorem 3 u is excessive with respect to the Brownian motion killed outside $B(x,r)$. It follows by Theorem 4 of §3.4 that $P_{B(x,\delta)^c}u(x)$ for $0 < \delta < r$ decreases as δ increases. Hence in the notation above $(d/d\delta)A(x,\delta) \leq 0$ for m-a.e. δ in $(0,r)$, and so by (16) of §4.3 we have $\int_{S(x,\delta)} \Delta u(y)\sigma(dy) \leq 0$ for a sequence of $\delta \downarrow 0$. Therefore $\Delta u(x) \leq 0$ by continuity as $\delta \downarrow 0$. □

Exercises

1. If u is superharmonic in D, prove that the right members of (1) and of (2) both increase to $u(x)$ as $r \downarrow 0$.

2. If u is superharmonic in D, then for each $x \in D$ we have $u(x) = \underline{\lim}_{x \neq y \to x} u(y)$.

3. Let u be superharmonic in D, and $D_r = \{x \in D \mid \rho(x, \partial D) > r\}$, $r > 0$. Define
$$u_r(x) = \int_{B(x,r)} u(y)m(dy).$$
Prove that u_r is a continuous superharmonic function *in* D_r.

4. (Minimum principle). Let u be superharmonic in a domain D. Suppose that there exists $x_0 \in D$ such that $u(x_0) = \inf_{x \in D} u(x)$; then $u \equiv u(x_0)$ in D. Next, suppose that for every $z \in \partial D \cup \{\infty\}$ we have $\underline{\lim}_{x \to z} u(x) \geq 0$, where $x \to \infty$ means $\|x\| \to \infty$. Then $u \geq 0$ in D.

5. Let D be a domain and u have the properties in (a) of the definition of a superharmonic function. Then a necessary and sufficient condition for u to be superharmonic in D is the following. For any ball $B \Subset D$, and any function h which is harmonic in B and continuous in $\bar{B}$, $u \geq h$ on ∂B implies $u \geq h$ in $\bar{B}$. [Hint: for the sufficiency part take f_n continuous

on ∂B and $\uparrow u$ there, then $H_B f_n \leq u$ on ∂B; for the necessity part use Proposition 5 of §4.4.]

6. Extend the necessity part of Exercise 5 to any open set B with compact $\bar{B} \subset D$. [Hint: if B is not regular, Proposition 5 of §4.4 is no longer sufficient, but Proposition 6 is. Alternatively, we may apply Exercise 4 to $u - h$.]

7. If u is superharmonic in R^d and $t > 0$, then $P_t u$ is continuous [Hint: split the exponential and use $P_{t/2}u(0) < \infty$.]

8. Let D be bounded, u be harmonic in D and continuous in $\bar{D}$. Then the process in (13) is a martingale. Without using the martingale verify directly that $u(x) = E^x\{u(X(t \wedge \tau_D))\}$ for $0 \leq t \leq \infty$. Note that D need not be regular.

9. Let D be an open set and define for $t \geq 0$:

$$\check{X}(t) = X(t \wedge \tau_D).$$

The process $\check{X}$ is called the *Brownian motion stopped at the boundary of D*. Its appropriate state space is $\bar{D}$. Find the transition semigroup of X. Are all the points of ∂D absorbing states? Prove that X is a Hunt process. [Hint: we need Theorem 10; but try first to see the difficulty!]

10. If D is open bounded and z is regular for D^c. Prove that the function $u(x) = E^x\{\tau_D\}$ is a barrier at z for D.

11. Show that the first function in (17) of §4.3 is subharmonic in R^2; the second superharmonic in R^d, $d \geq 3$.

12. Let u be superharmonic in D. For $\delta > 0$ put

$$u_\delta(x) = \int_{R^d} \varphi_\delta(\|x - y\|)u(y)\,dy$$

where φ_δ is given in Exercise 6 of §4.3. Let D_δ be as in Exercise 3 above. Prove that u_δ is superharmonic and infinitely differentiable in D_δ, and increases to u as $\delta \downarrow 0$.

13. For each $t_0 > 0$, the function $x \to P^x\{\tau_D > t_0\}$ is excessive with respect to (Q_t). Hence it is continuous in D. [Hint: use Proposition 1 of §4.4.]

14. For each bounded Borel measurable f, and $t > 0$, $Q_t f$ is continuous in D. Hence if D is regular, the Brownian motion killed outside D is a Feller process. [Hint: for $0 < s < t$ we have $Q_t f = P_s Q_{t-s} f - E^x\{s \geq \tau_D; Q_{t-s} f(X_s)\}$; use (20).]

15. Under the conditions of Theorem 1, prove that the function $P_{D^c}u$ is superharmonic in an open set D_0 containing $\bar{D}$. [Hint: u being superharmonic in D_0, $P_{D^c}u$ is excessive with respect to the Brownian motion killed outside D_0.]

16. In R^2, a positive superharmonic function is a constant.

In the next two exercises, proofs of some of the main results above are sketched which do not use the general Hunt theory as we did.

17. Let u be a positive superharmonic function in R^d. We may suppose it bounded by first truncating it. Prove that almost surely:
    (a) $t \to u(X(t))$ is lower semi-continuous;
    (b) if Q is the set of positive rationals, then
    $$\lim_{Q \ni t \downarrow 0} u(X(t)) = u(X(0));$$
    (c) the preceding relation is true without restricting t to Q, on account of (a);
    (d) $t \to u(X(t))$ is right continuous by the transfinite induction used in Theorem 6 of §3.4.

18. Let $u(x) = E^x\{e^{-T_A}\}$ where $A \in \mathscr{E}$. Prove that u is lower semi-continuous (for the Brownian motion). Show that $\{e^{-t}u(X_t), \mathscr{F}_t, t \geq 0\}$ is a supermartingale, and that it is right continuous, hence also has left limits. [Hint: Use the *proof* in Problem 17; we do not need the superharmonicity.]

Theorem 9 follows from Problem 17 as before, and Theorem 10 from Problem 18.

4.6. The Role of the Laplacian

The Dirichlet problem deals with the solution of Laplace's equation $\Delta\varphi = 0$ in D with the boundary condition $\varphi = f$ on ∂D. The inhomogeneous case of this equation is called Poisson's equation: $\Delta\varphi = g$ in D where g is a given function. Formally this equation can be solved if there is an operator Δ^{-1} inverse to Δ, so that $\Delta(\Delta^{-1}g) = g$. We have indeed known such an inverse under a guise, and now proceed to unveil it.

Suppose that g is bounded continuous and $U|g| < \infty$. Then we have

$$\lim_{t \downarrow 0} \frac{1}{t}(P_t - I)Ug = \lim_{t \downarrow 0} \frac{-1}{t} \int_0^t P_s g\, ds = -g \tag{1}$$

where $I \equiv P_0$ is the identity operator. Thus the operator

$$\mathscr{A} = \lim_{t \downarrow 0} \frac{1}{t}(P_t - I) \tag{2}$$

acts as an inverse to the potential operator U. It turns out $\mathscr{A} = \frac{1}{2}\Delta$ when acting on a suitable class of functions specified below.

For $k \geq 0$ we denote by $\mathbb{C}^{(k)}$ the class of functions which have continuous partial derivatives of order $\leq k$. Note that it does not imply boundedness of the function or its partial derivatives. The subclass of $\mathbb{C}^{(k)}$ with bounded kth partial derivatives is denoted by $\mathbb{C}_b^{(k)}$; the subclass of $\mathbb{C}^{(k)}$ having compact supports is denoted by $\mathbb{C}_c^{(k)}$. Clearly $\mathbb{C}_c^{(k)} \subset \mathbb{C}_b^{(k)}$. These conditions may be restricted to an open subset D of R^d, in which case the notation $\mathbb{C}^{(k)}(D)$, etc. is used.

Theorem 1. *Let f be bounded Lebesgue measurable and belong to $\mathbb{C}_b^{(2)}$ in an open neighborhood of x. Then we have*

$$\lim_{t \downarrow 0}\left(\frac{P_t - I}{t}\right)f(x) = \frac{\Delta}{2} f(x), \tag{3}$$

boundedly in the neighborhood.

Proof. Let x be fixed and f belong to $\mathbb{C}_b^{(2)}$ in the ball $B(x,r)$. By Taylor's theorem with an integral remainder term, we have for $\|y\| < r$:

$$f(x+y) - f(x) = \sum_{i=1}^{d} y_i f_i(x) + \sum_{i,j=1}^{d} y_i y_j \int_0^1 f_{ij}(x+sy)(1-s)\,ds \tag{4}$$

where f_i and f_{ij} denote the first and second partial derivatives of f. Note that they are bounded in $B(x,r)$. Now write

$$\frac{1}{t}\{P_t f(x) - f(x)\} = \frac{1}{t}\int_{R^d} p(t;o,y)[f(x+y) - f(x)]\,dy; \tag{5}$$

put $y = z\sqrt{t}$, and split the range of the integral into two parts: $\|y\| = \|z\|\sqrt{t} < r$ and $\|y\| = \|z\|\sqrt{t} \geq r$ respectively. Observe that $p(t;o,z\sqrt{t})\,d(z\sqrt{t}) = p(1;o,z)\,dz$. The second part of the integral is bounded by

$$\frac{2\|f\|}{t}\int_{\|z\| \geq r/\sqrt{t}} p(1;o,z)\,dz$$

which converges to zero boundedly as $t \downarrow 0$. Substituting (4) into the first part of the integral, we obtain

$$\frac{1}{t}\int_{\|z\|<r/\sqrt{t}} p(1;o,z)\left\{\sqrt{t}\sum_{i=1}^{d} z_i f_i(x) + t\sum_{i,j=1}^{d} z_i z_j \int_0^1 f_{ij}(x+sz\sqrt{t})(1-s)\,ds\right\}dz$$

$$= \int_{\|z\|<r/\sqrt{t}} p(1;o,z)\sum_{i,j=1}^{d} z_i z_j \int_0^1 f_{ij}(x+sz\sqrt{t})(1-s)\,ds\,dz, \tag{6}$$

because by spherical symmetry we have for each i:

$$\int_{\|z\|<r/\sqrt{t}} p(1;o,z) z_i\,dz = 0.$$

The right member of (6) may be written as

$$\sum_{i,j=1}^{d} \int_{R^d} p(1; o, z) z_i z_j \varphi_{ij}(t, z)\, dz$$

$$\text{where } \varphi_{ij}(t, z) = \begin{cases} \int_0^1 f_{ij}(x + sz\sqrt{t})(1 - s)\, ds, & \text{if } \sqrt{t}\|z\| < r; \\ 0, & \text{if } \sqrt{t}\|z\| \geq r. \end{cases} \tag{7}$$

We have by bounded convergence,

$$\lim_{t \downarrow 0} \varphi_{ij}(t, z) = \int_0^1 f_{ij}(x)(1 - s)\, ds = \frac{1}{2} f_{ij}(x);$$

and by the properties of the normal distribution,

$$\int_{R^d} p(1; o, z) z_i z_j\, dz = \delta_{ij}.$$

It follows that as $t \downarrow 0$, the sum in (7) converges boundedly to

$$\frac{1}{2} \sum_{i,j=1}^{d} \delta_{ij} f_{ij}(x) = \frac{1}{2} \Delta f(x). \qquad \square$$

We wish to apply Theorem 1 to $f = Ug$. When does this function belong to $\mathbb{C}^{(2)}$? First of all, we need $U|g| < \infty$. When $U1 = +\infty$, it is not sufficient that g be bounded. Let $\mathbb{B}_c$ denote the class of functions on R^d which are Lebesgue measurable and bounded with compact supports. If $g \in \mathbb{B}_c$ then we have shown in §3.7, for $d = 3$, that

$$Ug(x) = \frac{1}{2\pi} \int \frac{g(y)}{\|x - y\|}\, dy$$

is bounded continuous in x. In fact $Ug \in \mathbb{C}^{(1)}$. The same argument there applies to any $d \geq 3$. Thus in this case Ug is smoother than g. In order that Ug belong to $\mathbb{C}^{(2)}$, we need a condition which is stronger than $\mathbb{C}^{(0)}$ (continuity) but weaker than $\mathbb{C}^{(1)}$. There are several forms of such a condition but we content ourselves with the following. The functions g is said to satisfy a *Hölder condition* iff for each compact K there exists $\alpha > 0$ and $M < \infty$ (α and M may depend on K) such that

$$|g(x) - g(y)| \leq M\|x - y\|^{\alpha} \tag{8}$$

for all x and y in K. If $\alpha = 1$ this is known as *Lipschitz condition*. If g also has compact support then it is clearly bounded in R^d. We shall denote this class of functions by $\mathbb{H}_c$. For a proof of the following result under slightly more general assumptions, see Port and Stone [1], pp. 115–118. An easy case of it is given in Exercise 1 below.

Theorem 2. *In* R^d, $d \geq 3$, *if* $g \in \mathbb{B}_c$ *then* $Ug \in \mathbb{C}^{(1)}$; *if* $g \in \mathbb{H}_c$ *then* $Ug \in \mathbb{C}^{(2)}$.

We are now ready to show that the two operators $-\Delta/2$ and U act as inverses to each other on certain classes of functions.

Theorem 3 (for R^d, $d \geq 3$). *If* $g \in \mathbb{H}_c$, *then* $-(\Delta/2)(Ug) = g$. *If* $g \in \mathbb{C}_c^{(2)}$, *then* $U(-(\Delta/2)g) = g$.

Proof. If $g \in \mathbb{H}_c$ or $g \in \mathbb{C}_c^{(2)}$ then $Ug \in \mathbb{C}^{(2)}$ by Theorem 2. In either case g and $U|g|$ are both bounded continuous in R^d. Applying Theorem 1 with $f = Ug$, together with (1), we obtain $(\Delta/2)(Ug) = -g$. Next if $g \in \mathbb{C}_c^{(2)}$, we have by Theorem 1 applied to g and bounded convergence:

$$P_t g - g = \lim_{h \downarrow 0} \int_0^t P_s\left(\frac{P_h g - g}{h}\right) ds = \int_0^t P_s\left(\lim_{h \downarrow 0} \frac{P_h g - g}{h}\right) ds$$
$$= \int_0^t P_s\left(\frac{\Delta}{2}\, g\right) ds. \tag{9}$$

If K is the compact support of g, then $|P_t g| \leq \|g\| P_t 1_K \to 0$ as $t \to \infty$ because K is transient in R^d, $d \geq 3$. Since $|\Delta g|$ is bounded and has compact support, the last member of (9) converges to $U((\Delta/2)g)$ as $t \to \infty$. We have thus proved that $-g = U((\Delta/2)g)$. □

Since $\mathbb{C}_c^{(2)} \subset \mathbb{H}_c$, Theorem 3 implies that $-\Delta/2$ and U are inverses to each other on $\mathbb{C}_c^{(2)}$.

Next, recalling (26) of §4.5 but writing g for the f there, we know that if $U|g| < \infty$, then

$$Ug - P_{D^c} Ug = G_D g \tag{10}$$

where

$$G_D g(x) = E^x\left\{\int_0^{\tau_D} g(X_t)\, dt\right\}, \tag{11}$$

and D is any bounded open set in R^d, $d \geq 3$. This is then true if $g \in \mathbb{H}_c$. Moreover if we apply the Laplacian to all the terms in (10), we obtain by Theorem 3

$$\Delta(G_D g) = -2g \quad \text{in } D \tag{12}$$

because $P_{D^c} Ug$ is harmonic in D, being the solution to the generalized Dirichlet problem for (D, Ug). If D is regular, then $P_{D^c} Ug$ is continuous in $\bar{D}$ as the classical solution to the Dirichlet problem. Hence $G_D g$ is continuous in $\bar{D}$ and vanishes on ∂D, by (10). We summarize the result as follows.

Theorem 4 (for R^d, $d \geq 3$). *If $g \in \mathbb{H}_c$ and D is a bounded regular open set, then the function given in (11) is the unique solution of Poisson's equation $-(\Delta/2)\varphi = g$ in D which is continuous in $\bar{D}$ and vanishes on ∂D.*

To see the uniqueness, suppose φ is such a solution. Then $\varphi - G_D g$ is harmonic in D, continuous in $\bar{D}$, and vanishes on ∂D. Hence it vanishes in $\bar{D}$ by the maximum (minimum) principle for harmonic functions.

We proceed to obtain similar results for R^2. Since the Brownian motion in R^2 is recurrent, the potential kernel U is of no use. We use a substitute culled from Port and Stone [1]. Let $\|x_0\| = 1$ and put for all $x \in R^2$:

$$p_t^*(x) = p_t(x) - p_t(x_0), \qquad u^*(x) = \int_0^\infty p_t^*(x)\,dt.$$

We have

$$\begin{aligned} u^*(x) &= \frac{1}{2\pi}\int_0^\infty (e^{-\|x\|^2/2t} - e^{-1/2t})t^{-1}\,dt \\ &= \frac{1}{2\pi}\int_0^\infty (e^{-\|x\|^2 s} - e^{-s})s^{-1}\,ds \\ &= \frac{1}{2\pi}\int_0^\infty \int_{\|x\|^2 s}^{s} e^{-r}\,dr\,s^{-1}\,ds \\ &= \frac{1}{2\pi}\int_0^\infty \int_r^{r/\|x\|^2} s^{-1}\,ds\,e^{-r}\,dr = \frac{1}{\pi}\log\frac{1}{\|x\|}. \end{aligned} \tag{13}$$

Now put

$$p_t^*(x, y) = p_t^*(y - x), \qquad u^*(x, y) = u^*(y - x),$$
$$P_t^*(x, dy) = p_t^*(x, y)\,dy;$$

$$U^*(x, dy) = u^*(x, y)\,dy = \frac{1}{\pi}\log\frac{1}{\|x - y\|}\,dy. \tag{14}$$

The kernel U^* is the *logarithmic potential*. Note that not only $u^*(x)$ is positive or negative according as $\|x\| \leq 1$ or ≥ 1; but the same is true for $p_t^*(x)$ for each $t \geq 0$. If g is bounded with compact support, then U^*g is bounded continuous in R^2. In this respect the analogy of U^* with U in R^d for $d \geq 3$ is perfect; Theorem 2 also goes over as follows.

Theorem 2′. *In R^2, if $g \in \mathbb{B}_c$, then $U^*g \in \mathbb{C}^{(1)}$; if $g \in \mathbb{H}_c$ then $U^*g \in \mathbb{C}^{(2)}$.*

The following computations should be handled with care because the quantities may be positive or negative. If g is Lebesgue integrable, we have

$$P_s^* g = P_s g - p_s(x_0)\int g(y)\,dy, \qquad P_t P_s^* g = P_{t+s} g - p_s(x_0)\int g(y)\,dy.$$

It follows that if $g \in \mathbb{H}_c$ so that $\int |u^*(x, y)g(y)|\,dy < \infty$, we have

$$U^*g - P_t U^*g = \int_0^t P_s g\,ds - \lim_{A \to \infty} \int_A^{A+t} P_s g\,ds = \int_0^t P_s g\,ds \tag{15}$$

because $|P_s g| \le 1/2\pi s \int |g(y)|\,dy \to 0$ as $s \to \infty$.

Theorem 5 (for R^2). *If* $g \in \mathbb{H}_c$, *then* $-(\Delta/2)(U^*g) = g$. *If* $g \in \mathbb{C}_c^{(2)}$, *then* $U^*(-(\Delta/2)g) = g$.

Proof. The first assertion follows from Theorems 1 and 2′ and (15), in exactly the same way as the corresponding case in Theorem 3. To prove the second assertion we can use an analogue of (9), see Exercise 3 below. But we will vary the technique by first proving the assertion when $\Delta g \in \mathbb{H}_c$. Under this stronger assumption we may apply the first assertion to Δg to obtain

$$-\frac{\Delta}{2}\,U^*(\Delta g) = \Delta g.$$

Hence $\Delta(g + \frac{1}{2}U^*(\Delta g)) = 0$. Since both g and $\frac{1}{2}U^*(\Delta g)$ are bounded it follows by Picard's theorem (Exercise 5 of §4.3) that $g + \frac{1}{2}U^*(\Delta g) = 0$ as asserted. For a general $g \in \mathbb{C}_c^{(2)}$, we rely on the following lemma which is generally useful in analytical questions of this sort. Its proof is left as an exercise.

Lemma 6. *Let g be bounded Lebesgue measurable and put for $\delta > 0$:*

$$g_\delta(x) = \int g(x - y)\varphi_\delta(y)\,dy$$

where φ_δ is defined in Exercise 6 of §4.3. Then $g_\delta \in \mathbb{C}^{(\infty)}$. If in addition g belongs to $\mathbb{C}_b^{(2)}$ in an open set D, then $g_\delta \to g$ and $\Delta g_\delta \to \Delta g$ both boundedly in D as $\delta \downarrow 0$.

We now return to the proof of the second assertion of Theorem 5. If $g \in \mathbb{C}_c^{(2)}$, then Δg and Δg_δ for all $\delta > 0$ have a compact support K. Since $|U^* 1_K| < \infty$ we have $U^*(\Delta g_\delta) \to U^*(\Delta g)$ by dominated convergence because of the bounded convergence in Lemma 6. Since $g_\delta \in \mathbb{C}^{(\infty)}$, we have already proved that $g_\delta + \frac{1}{2}U^*(\Delta g_\delta) = 0$ for all $\delta > 0$; as $\delta \downarrow 0$ this yields

$$g + \tfrac{1}{2}U^*(\Delta g) = 0. \qquad \square$$

Next we consider Poisson's equation in R^2. Since U^* is not a true potential, the analogue of (10) is in doubt. We make a detour as follows. Let $g \in \mathbb{C}_c^{(2)}$, then (9) is true. We now rewrite (9) in terms of the process

$$E^x\{g(X_t)\} - g(x) = E^x\left\{\int_0^t \frac{\Delta}{2}\,g(X_s)\,ds\right\}. \tag{16}$$

If we put

$$M_t = g(X_t) - \int_0^t \frac{\Delta}{2} g(X_s)\, ds \tag{17}$$

then a general argument (Exercise 10) shows that $\{M_t, \mathscr{F}_t, t \geq 0\}$ is a martingale under P^x for each x. Let D be a bounded open set, then $\{M_{t \wedge \tau_D}, \mathscr{F}_t, t \geq 0\}$ is a martingale by Doob's stopping theorem. Since g and Δg are bounded and $E^x\{\tau_D\} < \infty$ for $x \in D$, it follows by dominated convergence that if $x \in D$:

$$\begin{aligned} g(x) = E^x\{M_0\} &= \lim_{t \to \infty} E^x\{M_{t \wedge \tau_D}\} = E^x\{M_{\tau_D}\} \\ &= E^x\{g(X_{\tau_D})\} - E^x\left\{\int_0^{\tau_D} \frac{\Delta}{2} g(X_s)\, ds\right\}. \end{aligned} \tag{18}$$

In particular if $g = U^*f$ where $f \in \mathbb{H}_c$, then g is bounded, $g \in \mathbb{C}^{(2)}$ by Theorem 2′, and $(\Delta/2)g = -f$ by Theorem 5. Hence (18) becomes

$$U^*f(x) = P_{D^c}U^*f + G_D f. \tag{19}$$

This is the exact analogue of (10). It follows that Theorem 4 holds intact in R^2.

We have made the assumption of compact support in several places above such as Theorems 2 and 2′, mainly to lighten the exposition. This assumption can be replaced by an integrability condition, but the following lemma (valid in R^d for $d \geq 1$) shows that in certain situations there is no loss of generality in assuming compact support.

Lemma 7. *Let $f \in \mathbb{C}^{(k)}$ $(k \geq 0)$ or satisfy a Hölder condition, and D be a bounded open set. Then there exists g having compact support which satisfies the same condition and coincides with f in D.*

Proof. Let D_0 be a bounded open set such that $D_0 \supset \bar{D}$ and $\rho(\partial D, \partial D_0) = \delta_0$. Let $0 < \delta < \delta_0$ and put

$$\psi(x) = \int_{D_0} \varphi_\delta(x - y)\, dy$$

where φ_δ is as in Lemma 6. Then $\psi \in \mathbb{C}^{(\infty)}$. If $\rho(x, D_0) > \delta$, then $\psi(x) = 0$. If $x \in D$, then $\psi(x) = 1$. It is clear that the function $g = f \cdot \psi$ has the required properties. □

Example 1. Let D be a bounded regular open set in R^d, $d \geq 1$. Solve the equation

$$\Delta\varphi = 1 \quad \text{in } D; \qquad \varphi = 0 \quad \text{in } D^c.$$

By Lemma 7, there exists $g \in \mathbb{C}_c^{(\infty)}$ such that $g = 1$ in $\bar{D}$. Put

$$\varphi(x) = -\tfrac{1}{2}E^x\left\{\int_0^{\tau_D} g(X_t)\,dt\right\} = -\tfrac{1}{2}G_D g(x).$$

We do not need Theorem 2 or 2′ here, only the easier Exercise 1 to conclude that Ug or U^*g belongs to $\mathbb{C}^{(\infty)}$; hence $G_D g \in \mathbb{C}^{(\infty)}$ by (10) or (19). We have by Theorem 4,

$$\Delta\varphi = g = 1 \quad \text{in } D.$$

It is clear that $\varphi(x) = 0$ if $x \in D^c$ by the regularity of D.

Since $g = 1$ in D, we have $\varphi(x) = -\tfrac{1}{2}E^x\{\tau_D\}$. Using (VIII) of §4.2, we have

$$E^x\{\tau_D\} = \frac{1}{d}E^x\{\|X(\tau_D)\|^2\} - \frac{1}{d}\|x\|^2. \tag{20}$$

Since the first term on the right side of (20) is harmonic in D, we have

$$-\frac{\Delta}{2}E^x\{\tau_D\} = \frac{1}{2d}\Delta\|x\|^2 = 1,$$

a neat verification.

The preceding discussion does not extend to R^1. In general, Brownian motion in R^1 is a rather special case owing to the simple topological structure of the line. On the analytical side, since Δ reduces to the second derivative many results simplify. For instance it is trivial that given any continuous g there exists f such that $\Delta f = g$; think of the high dimensional analogue! For these reasons it is possible to obtain far more specific results in R^1 by methods which do not easily generalize. We leave a couple of exercises to suggest the possibilities without developing them because a creditable treatment of the Brownian motion in R^1 in its own right would take a volume or two. However, we will describe the solution of Poisson's equation by a different method attuned to one dimension, which is useful for other harder problems. Let g be bounded continuous in (a, b), $a < b$; we put $\tau = \tau_{(a,b)}$ and

$$\varphi(x) = E^x\left\{\int_0^{\tau} g(X_t)\,dt\right\}.$$

Recall the notation $T_h = \inf\{t > 0: |X(t) - X(0)| \geq h\}$. For each $x \in (a, b)$, let $[x - h, x + h] \subset (a, b)$. Then $P^x\{T_h < \tau < \infty\} = 1$. We have by the strong Markov property:

$$\begin{aligned}\varphi(x) &= E^x\left\{\left(\int_0^{T_h} + \int_{T_h}^{\tau}\right)g(X_t)\,dt\right\}\\ &= E^x\left\{\int_0^{T_h} g(X_t)\,dt\right\} + P_{T_h}\varphi(x).\end{aligned} \tag{21}$$

It is obvious by symmetry that $P^x\{X(T_h) = x + h\} = P^x\{X(T_h) = x - h\} = \frac{1}{2}$; hence

$$P_{T_h}\varphi(x) = \tfrac{1}{2}\{\varphi(x+h) + \varphi(x-h)\}.$$

Substituting into (21) we obtain

$$\frac{1}{h^2}\{\varphi(x+h) - 2\varphi(x) + \varphi(x-h)\} = \frac{-2}{h^2} E^x\left\{\int_0^{T_h} g(X_t)\,dt\right\}. \tag{22}$$

Without loss of generality we may suppose that $g \geq 0$. Then it follows from (21) that

$$\varphi(x) \geq \tfrac{1}{2}\{\varphi(x+h) + \varphi(x-h)\}.$$

Since φ is bounded this implies that φ is continuous, indeed concave (see e.g. Courant [1], Vol. 2, p. 326). Next as $h \downarrow 0$, it is easy to see by dominated convergence that the right member of (22) converges to

$$\lim_{h\downarrow 0} \frac{1}{h^2} E^x\{T_h\} g(x) = g(x)$$

because $E^x\{T_h\} = h^2$ by a well-known particular case of (20). Therefore, the left member of (22) converges as $h \downarrow 0$ to $g(x)$. Now the limit is known as the generalized second derivative. A classical result by Schwarz (a basic lemma in Fourier series, see Titchmarsh [1], p. 431) states that if φ is continuous and the limit exists as a continuous function of x in (a, b), then φ'' exists in (a, b) and is equal to the limit. Thus $\varphi'' = -2g$. □

It turns out that in higher dimensions there are also certain substitutes for the Laplacian similar to the generalized second derivative above. Some of the analytical difficulties disappear if these substitutes are used instead; on the other hand under more stringent conditions they can be shown to be the Laplacian after all, as in the case above.

Example 2. Let $0 < a < b < \infty$. Compute the expected occupation times of $B(o, a)$ before the first exit time from $B(o, b)$.

We will treat the problem in R^2 which is the most interesting case since the expected occupation time of $B(o, a)$ is infinite owing to the recurrence of the Brownian motion. In the notation of (19), the problem is to compute $G_D f$ when $D = B(o, b)$, $f = 1$ in $B(o, a)$ and $f = 0$ in $B(o, b) - B(o, a)$. This f is only "piecewise continuous" but we can apply (19) to two domains separately. We seek φ such that

$$\begin{aligned} \Delta\varphi(x) &= -2, && \text{for } \|x\| < a, \\ \Delta\varphi(x) &= 0, && \text{for } a < \|x\| < b, \\ \varphi(x) &= 0, && \text{for } \|x\| \geq b. \end{aligned}$$

Clearly, φ depends on $\|x\|$ only, nence we can solve the equations above by the polar form of the Laplacian given in (19) of §4.3. Let $\|x\| = r$; we obtain

$$\begin{aligned} \varphi(r) &= -\frac{r^2}{2} + c_1 + c_2 \log r, \qquad & 0 < r < a; \\ &= c_3 + c_4 \log r, & a < r < b. \end{aligned}$$

How do we determine these four constants? Obviously $\varphi(0) < \infty$ gives $c_2 = 0$, and $\varphi(b) = 0$ gives $c_3 = -c_4 \log b$. It is now necessary to have recourse to the first assertion in Theorem 2′ to know that both φ and $\partial\varphi/\partial r$ are continuous. Hence $\varphi(a-) = \varphi(a+)$, $(\partial\varphi/\partial r)(a-) = (\partial\varphi/\partial r)(a+)$. This yields the solution:

$$\begin{aligned} \varphi(r) &= \frac{a^2 - r^2}{2} + a^2 \log \frac{b}{a}, \qquad & 0 \le r \le a; \\ &= a^2 \log \frac{b}{r}, & a \le r \le b. \end{aligned}$$

Exercises

1. In R^d, $d \ge 3$, if $g \in \mathbb{C}_c^{(k)}$, $k \ge 1$, then $Ug \in \mathbb{C}^{(k)}$. In R^2, if $g \in \mathbb{C}_c^{(k)}$, $k \ge 1$, then $U^*g \in \mathbb{C}^{(k)}$.
2. Prove Lemma 6. [Hint: use Green's formula for $B(x, \delta_0)$, $\delta_0 > \delta$ to show $\Delta f_\delta(x) = \int \Delta f(x - y)\varphi_\delta(y)\,dy$.]
3. Prove the analogue of (9) for R^2:
$$P_t g - g = \int_0^t P_s^*\left(\frac{\Delta}{2} g\right) ds$$
and deduce the analogue of (10) from it.
4. Is the following dual of (15) true?
$$U^*g - U^*P_t g = \int_0^t P_s g\, ds.$$
5. Define $u^*(x)$ in R^1 in the same way as in R^2. Compute $u^*(x)$ and extend Theorem 5 to R^1. [Hint: $u^*(x) = 1 - |x|$.]
6. Compute $u^\alpha(x)$ in R^1. Prove that if g is bounded continuous (in R^1), then for each $\alpha > 0$, $U^\alpha g$ has bounded first and second derivatives, and $(\Delta/2)(U^\alpha g) = \alpha U^\alpha g - g$. Conversely, if g and its first and second derivatives are bounded continuous, then $U^\alpha((\Delta/2)g) = \alpha U^\alpha g - g$. [Hint: $u^\alpha(x) = e^{-|x|\sqrt{2\alpha}}/\sqrt{2\alpha}$. These results pertain to the definition of an "infinitesimal generator" which is of some use in R^1; see, e.g., Ito [1].]

7. Let $D = (a, b)$ in R^1. Derive Green's function $g_D(x, y)$ by identifying the solution of Poisson's equation for an arbitrary bounded continuous f:

$$\int g_D(x, y) f(y)\, dy = E^x\left\{\int_0^{\tau_D} f(X_t)\, dt\right\}.$$

[Hint: $g_D(x, y) = 2(x - a)(b - y)/(b - a)$ if $a < x \leq y \leq b$; $2(b - x)(y - a)/(b - a)$ if $a < y \leq x < b$.]

8. Solve the problem in Example 2 for R^3.

9. Let μ be a σ-finite measure in R^d, $d \geq 2$. Suppose $U\mu$ is harmonic in an open set D. Prove that $\mu(D) = 0$. [Hint: let $f \in \mathbb{C}^{(2)}$ in $B \Subset D$ and $f = 0$ outside B where B is a ball; $\int_B (U\mu) \Delta f\, dm = \int_{R_d} U(\Delta f)\, d\mu$; use Green's formula.]

10. The martingale in (17) is a case of a useful general proposition. Let $\{M_t, t \geq 0\}$ be associated with the Markov process $\{X_t, \mathscr{F}_t, t \geq 0\}$ as follows: (i) $M_0 = 0$; (ii) $M_t \in \mathscr{F}_t$; (iii) $M_{s+t} = M_s + M_t \circ \theta_s$ where $\{\theta_s, s \geq 0\}$ is the shift; (iv) for each x, $E^x\{M_t\} = 0$. Then $\{M_t, \mathscr{F}_t, t \geq 0\}$ is a martingale under each P^x. Examples of M_t are $g(X_t) - g(X_0)$ and $\int_0^t \varphi(X_s)\, ds$, where $g \in b\mathscr{E}$, $\varphi \in b\mathscr{E}$; and their sum. Condition (iii) is the additivity in *additive functionals.*

4.7. The Feynman-Kac Functional and the Schrödinger Equation

In this section we discuss the boundary value problem for the Schrödinger equation. This includes Dirichlet's problem in §4.4 as a particular case. The probabilistic method is based on the following functional of the Brownian motion process. Let $q \in b\mathscr{E}$, and put for $t \geq 0$:

$$e_q(t) = \exp\left[\int_0^t q(X_s)\, ds\right], \tag{1}$$

where $\{X_t\}$ is the Brownian motion in R^d, $d \geq 1$. Let D be a bounded domain, τ_D the first exit time from D defined in (1) of §4.4, and $f \in \mathscr{E}_+(\partial D)$. We put for all x in R^d:

$$u(x) = E^x\{e_q(\tau_D) f(X(\tau_D))\}. \tag{2}$$

Since the integrand above belongs to $\mathscr{F}^\sim$, u is universally measurable, (hence Lebesgue measurable) by Exercise 3 of §2.4. In fact $u \in \mathscr{E}$ because $\tau_D \in \mathscr{F}^0$ by Exercise 6 of §4.2, and $e_q(t) f(X(t))$ as a function of (t, ω) belongs to $\mathscr{B} \times \mathscr{F}^0$. The details are left as Exercise 1. Of course $u \geq 0$ everywhere, but u may be $+\infty$. Our principal result below is the dichotomy that if $f \in b\mathscr{E}_+$ then either

$u \equiv +\infty$ in D, or u is bounded in $\bar{D}$. This will be proved in several steps. We begin with a theorem usually referred to as *Harnack's inequality*.

Let $\mathcal{U}$ denote the class of functions defined in (2), for a fixed D, and all q and f as specified, subject furthermore to $\|q\|_D \le Q$, where Q is a fixed constant. For $\varphi \in \mathscr{E}$ and $A \in \mathscr{E}$, $\|\varphi\|_A = \sup_{x \in A} |\varphi(x)|$; when A is the domain of definition of φ, it may be omitted from the notation.

Theorem 1. *Each u in $\mathcal{U}$ is either identically $+\infty$ in D, or everywhere finite in D. For each compact subset K of D, there is a positive constant A depending only on D, K and Q, such that for all finite u in $\mathcal{U}$, and any two points x and x′ in K, we have*

$$u(x') \le Au(x).$$

Proof. Fix a $\delta > 0$ so small that

$$\tfrac{1}{2} \le E^{\cdot}\{e^{\pm QT_{2\delta}}\} \le 2 \tag{3}$$

where T_r is defined in (10) of §4.2. This is possible by Exercise 11 of §4.2. We now proceed to prove that if $u(x_0) < \infty$, and $0 < r < \delta \wedge (\rho_0/2)$, where ρ_0 is the distance from x_0 to ∂D, then for all $x \in B(x_0, r)$ we have

$$u(x) \le 2^{d+2} u(x_0). \tag{4}$$

For $0 < s \le 2r$, we have $T_s < \tau_D$ under P^{x_0}, since $B(x_0, 2r) \Subset D$. Hence by the strong Markov property

$$\infty > u(x_0) = E^{x_0}\{e_q(T_s)u(X(T_s))\} \ge E^{x_0}\{e^{-QT_s}u(X(T_s))\}. \tag{5}$$

The next crucial step is the stochastic independence of T_s and $X(T_s)$ under any P^x, proved in (XI) of §4.2. Using this and (3) we obtain

$$u(x_0) \ge \tfrac{1}{2}E^{x_0}\{u(X(T_s))\} = \frac{1}{2\sigma(s)} \int_{S(x_0,s)} u(y)\sigma(dy). \tag{6}$$

The step leading from (3) to (7) in §4.3 then yields

$$u(x_0) \ge \frac{1}{2v(2r)} \int_{B(x_0,2r)} u(y)\, dy. \tag{7}$$

For all $x \in B(x_0, r)$, we have $B(x, r) \subset B(x_0, 2r) \Subset D$. Hence we obtain similarly:

$$\begin{aligned} u(x) &= E^x\{e_q(T_s)u(X(T_s))\} \\ &\le E^x\{e^{QT_s}u(X(T_s))\} \le 2E^x\{u(X(T_s))\}. \end{aligned} \tag{8}$$

This leads to the first inequality below, and then it follows from (7) that

$$u(x) \le \frac{2}{v(r)} \int_{B(x,r)} u(y)\,dy \le \frac{2}{v(r)} \int_{B(x_0,2r)} u(y)\,dy$$
$$\le \frac{4v(2r)}{v(r)} u(x_0). \tag{9}$$

We have thus proved (4), in particular that $u(x) < \infty$. As a consequence, the set of x in D for which $u(x) < \infty$ is open in D. To show that it is also closed relative to D, let $x_n \to x_\infty \in D$ where $u(x_n) < \infty$ for all n. Then for a sufficiently large value of n, we have $\|x_\infty - x_n\| < \delta \wedge (\rho(x_n, \partial D)/2)$. Hence the inequality (4) is applicable with x and x_0 replaced by x_∞ and x_n, yielding $u(x_\infty) < \infty$. Since D is connected, the first assertion of the theorem is proved.

Now let D_0 be a subdomain with $\bar{D}_0 \subset D$. Let $0 < r < \delta \wedge (\rho(D_0, \partial D)/2)$. Using the connectedness of D_0 and the compactness of $\bar{D}_0$, we can prove the existence of an integer N with the following property. For any two points x and x' in D_0, there exist n points $x_1, \ldots, x_n$ in D_0 with $2 \le n \le N + 1$, such that $x = x_1$, $x' = x_n$, $\|x_{j+1} - x_j\| < r$ and $\rho(x_{j+1}, \partial D) > 2r$ for $1 \le j \le n - 1$. A detailed proof of this assertion is not quite trivial, nor easily found in books, hence is left as Exercise 3 with sketch of solution. Applying (4) successively to x_j and x_{j+1}, $1 \le j \le n - 1$ $(\le N)$, we obtain Theorem 1 with $A = 2^{(d+2)N}$. If K is any compact subset of D, then there exists D_0 as described above such that $K \subset D_0$. Therefore the result is *a fortiori* true also for K. □

Theorem 1 is stated in its precise form for comparison with Harnack's inequality in the theory of partial differential equations. We need it below only for a fixed u. The next proposition is Exercise 11 of §4.2 and its proof is contained in that of (X) of §4.2. It turns the trick of transforming Theorem 1 into Theorem 3.

Proposition 2. *If $m(D)$ tends to zero, then $E^x\{e^{Q\tau_D}\}$ converges to one, uniformly for $x \in R^d$.*

Theorem 3. *Let u be as in* (2), *but suppose in addition that f is bounded on ∂D. If $u \not\equiv \infty$ in D, then u is bounded in $\bar{D}$.*

Proof. Let K be a compact subset of D, and $E = D - K$. It follows from Proposition 2 that given any $\varepsilon > 0$, we can choose K to make $m(E)$ so small that

$$\sup_{x \in R^d} E^x\{e^{Q\tau_E}\} \le 1 + \varepsilon. \tag{10}$$

Put for $x \in \bar{E}$:

$$u_1(x) = E^x\{e_q(\tau_D) f(X(\tau_D));\ \tau_E < \tau_D\},$$
$$u_2(x) = E^x\{e_q(\tau_D) f(X(\tau_D));\ \tau_E = \tau_D\}.$$

By the strong Markov property, since $\tau_D = \tau_E + \tau_D \circ \theta_{\tau_E}$ on the set $\{\tau_E < \tau_D\}$:

$$u_1(x) = E^x\{e_q(\tau_E)u(X(\tau_E)); \tau_E < \tau_D\}.$$

On $\{\tau_E < \tau_D\}$, $X(\tau_E) \in K$. Since $u \not\equiv \infty$ in D, u is bounded on K by Theorem 1. Together with (10) this implies

$$u_1(x) \le E^x\{e^{Q\tau_E}\}\|u\|_K \le (1 + \varepsilon)\|u\|_K.$$

Since f is bounded we have by (10)

$$u_2(x) \le E^x\{e^{Q\tau_E}\}\|f\| \le (1 + \varepsilon)\|f\|.$$

Thus for all $x \in \bar{E}$ we have

$$u(x) = u_1(x) + u_2(x) \le (1 + \varepsilon)(\|u\|_K + \|f\|). \tag{11}$$

Since $\bar{D} = \bar{E} \cup K$, u is bounded in $\bar{D}$. □

We shall denote the $u(x)$ in (2) more specifically by $u(D, q, f; x)$.

Corollary. *Let $f \in b\mathscr{E}$. If $u(D, q, |f|; x) \not\equiv \infty$ in D, then $u(D, q, f; x)$ is bounded in $\bar{D}$.*

In analogy with the notation used in §4.6, let $\mathscr{B}(D)$ denote the class of functions defined in D which are bounded and Lebesgue measurable; $\mathbb{H}(D)$ the class of functions defined in D which are bounded and satisfy (8) of §4.6 for each compact $K \subset D$. Thus $b\mathbb{C}^{(1)}(D) \subset \mathbb{H}(D) \subset b\mathbb{C}^{(0)}(D)$. Let us state the following analytic lemma.

Proposition 4. *If $g \in \mathscr{B}(D)$, then $G_D g \in \mathbb{C}^{(1)}(D)$. For $d = 1$ if $g \in b\mathbb{C}^{(0)}(D)$, for $d \ge 2$ if $g \in \mathbb{H}(D)$, then $G_D g \in \mathbb{C}^{(2)}(D)$.*

For $d = 1$ this is elementary. For $d \ge 2$, the results follow from (10) and (19) of §4.6, via Theorems 2 and 2′ there, with the observation that $G_D g = G_D(1_D g)$ by (11) of §4.6. Note however that we need the versions of these theorems localized to D, which are implicit in their proofs referred to there. For a curious deduction see Exercise 4 below.

Theorem 5. *For $d = 1$ let $q \in b\mathbb{C}^{(0)}(D)$; for $d \ge 2$ let $q \in \mathbb{H}(D)$. Under the conditions of the Corollary to Theorem 3, the function $u(D, q, f; \cdot)$ is a solution of the following equation in D:*

$$\left(\frac{\Delta}{2} + q\right)\varphi = 0. \tag{12}$$

Proof. Since u is bounded in $\bar{D}$, we have for $x \in D$:

$$E^x\left\{\int_0^{\tau_D} |q(X_t)u(X_t)|\,dt\right\} \le \|qu\|_D E^x\{\tau_D\} < \infty. \tag{13}$$

[Note that in the integral above the values of q and u on ∂D are irrelevant.] This shows that the function $1_{\{t<\tau_D\}}|q(X_t)u(X_t)|$ of (t, ω) is dominated in the integration with respect to the product measure $m_1 \times P^x$ over $[0, \infty) \times \Omega$, where m_1 denotes the Lebesgue measure on $[0, \infty)$. Therefore, we can use Fubini's theorem to transform the following integral as shown below:

$$E^x\left\{\int_0^{\tau_D} q(X_t)u(X_t)\,dt\right\} = E^x\left\{\int_0^{\infty} 1_{\{t<\tau_D\}} q(X_t)E^{X_t}[e_q(\tau_D)f(X(\tau_D))]\,dt\right\}$$

$$= E^x\left\{f(X(\tau_D))\int_0^{\infty} 1_{\{t<\tau_D\}} q(X_t)\exp\left[\int_t^{\tau_D} q(X_s)\,ds\right]dt\right\}.$$

The third member above is obtained from the second by first reversing the order of the two integrations, then applying the Markov property at each t under E^x, noting that $\tau_D = t + \tau_D \circ \theta_t$ on $\{t < \tau_D\}$. We can now perform the trivial integration with respect to t to obtain

$$E^x\left\{f(X(\tau_D))\left[\exp\int_0^{\tau_D} q(X_s)\,ds - 1\right]\right\} = u(x) - E^x\{f(X(\tau_D))\}.$$

This result may be recorded in previous notation as follows:

$$G_D(qu) = u - H_D f. \tag{14}$$

Recall that $H_D f$ is harmonic in D hence belongs to $\mathbb{C}^{(\infty)}$ there. Since $qu \in \mathscr{B}(D)$, it follows from Proposition 4 that $u \in \mathbb{C}^{(1)}(D)$. Then $qu \in b\mathbb{C}^{(0)}(D)$ for $d = 1$, $qu \in \mathbb{H}(D)$ for $d \ge 2$, hence $u \in \mathbb{C}^{(2)}(D)$. Taking the Laplacian in (14) and using (12) and (19) of §4.6, we conclude that $-2qu = \Delta u$ in D as asserted. □

The equation (12) is the celebrated Schrödinger's equation in quantum physics. For $q \equiv 0$ it reduces to Laplace's equation.

The next result includes Theorem 2 of §4.4 as a particular case.

Theorem 6. *Under the hypotheses of Theorem 5, if $z \in \partial D$ and z is regular for D^c, and if f is continuous at z, then we have*

$$\lim_{\bar{D} \ni x \to z} u(x) = f(z). \tag{15}$$

Proof. We may suppose $f \geq 0$ by using $f = f^+ - f^-$. Given $\varepsilon > 0$, there exists $r > 0$ such that

$$1 - \varepsilon \leq E^{\cdot}\{e^{\pm 2QT_r}\} \leq 1 + \varepsilon; \tag{16}$$

$$\alpha(\varepsilon) = \sup_{y \in B(z,2r) \cap (\partial D)} |f(y) - f(z)| \leq \varepsilon. \tag{17}$$

Let $x \in \bar{D} \cap B(z, r)$. Put

$$u_1(x) = E^x\{T_r < \tau_D; e_q(\tau_D) f(X(\tau_D))\} = E^x\{T_r < \tau_D; e_q(T_r) u(X(T_r))\},$$
$$u_2(x) = E^x\{\tau_D \leq T_r; e_q(\tau_D) f(X(\tau_D))\}.$$

We have $X(T_r) \in D$ on $\{T_r < \tau_D\}$, hence by Theorem 3 followed by Schwarz's inequality:

$$\begin{aligned} u_1(x) &\leq E^x\{T_r < \tau_D; e^{QT_r}\} \|u\|_D \\ &\leq P^x\{T_r < \tau_D\}^{1/2} E^x\{e^{2QT_r}\}^{1/2} \|u\|_D. \end{aligned}$$

As $x \to z$, this converges to zero by (8) of §4.4, and (16) above. Next we have for $x \in B(z, r)$:

$$\begin{aligned} |u_2(x) - f(z)| \leq{} & E^x\{\tau_D \leq T_r; e_q(\tau_D)\} \alpha(\varepsilon) \\ & + E^x\{\tau_D \leq T_r; |e_q(\tau_D) - 1|\} |f(z)| \\ & + P^x\{T_r < \tau_D\} |f(z)|. \end{aligned}$$

The first term on the right is bounded by $E^x\{e^{QT_r}\}\varepsilon \leq (1 + \varepsilon)\varepsilon$ by (16) and (17); the second by $|1 - E^x\{e^{\pm QT_r}\}| |f(z)| \leq \varepsilon |f(z)|$ by (16); and the third converges to zero as $x \to z$ by (8) of §4.4. The conclusion (15) follows from these estimates. □

Putting together Theorems 3, 5 and 6, we have proved that for every $f \in \mathbb{C}(\partial D)$, a $\mathbb{C}^{(2)}(D)$-solution of the equation (12) is given by $u(D, q, f; \cdot)$ provided that $u(D, q, 1; \cdot) \not\equiv \infty$ in D. Moreover if D is regular, then this solution belongs to $C^{(0)}(\bar{D})$. Thus we have solved the Dirichlet boundary value problem for the Schrödinger equation by the explicit formula given in (2). It turns out that under the conditions stated above, this is the *unique* solution. For the proof and other closely related results we refer the reader to the very recent paper by Chung and Rao [3]. A special case is given in Exercise 7 below.

Let us remark that contrary to the Laplace case, the uniqueness of solution in the Schrödinger case is in general false. The simplest example is given in R^1 by the equation $u'' + u = 0$ in $D = (0, \pi)$. The particular solution $u(x) = \sin x$ vanishes on ∂D! In general, unicity depends on the size of the domain D as well as the function q. Such questions are related to the eigenvalue problem associated with the Schrödinger operator. Here we see that the quantity $u(D, q, 1, x)$ serves as a *gauge* in the sense that its finiteness for some x in D ensures the unique solvability of all continuous boundary value problems.

Exercises

1. Prove that the function u in (2) is Borel measurable.

2. If $f \in \mathscr{E}_+$ in (2), then either $u \equiv 0$ in D or $u > 0$ in D. [Here D need not be bounded.]

3. (a) Let D be a domain. Then there exist domains D_n strictly contained in D and increasing to D. [Hint: let U_n be the set of points in D at distance $> 1/n$ from ∂D. Fix an x_0 in D and let D_n be the connected component of U_n which contains x_0. Show that $\bigcup_n D_n$ is both open and closed relative to D.]
   (b) Let D_0 be a bounded domain strictly contained in D. Let $0 < r < \frac{1}{2}\rho(D_0, \partial D)$, and $\bar{D}_0 \subset \bigcup_{i=1}^{N} B(x_i, r/2)$ where all $x_i \in \bar{D}_0$. Define a *connection* "$\rightarrow$" on the set of centers $S = \{x_i, 1 \le i \le N\}$ as follows: $x_i \rightarrow x_j$ if $\|x_i - x_j\| < r$. Use the connectedness of D_0 to show that for any two elements x_a and x_b of S, there exist distinct elements x_{i_j}, $1 \le j \le l$, such that $x_{i_1} = x_a, x_{i_l} = x_b$, and $x_{i_j} \rightarrow x_{i_{j+1}}$ for $1 \le j \le l - 1$. In the language of graph theory, the set S with the connection $\rightarrow$ forms a *connected graph*. [This formulation is due to M. Steele.]
   (c) Show that the number N whose existence is asserted at the end of the proof of Theorem 1 may be taken to be the number N in (b) plus one.

   In the following problems D is a bounded domain in R^d, $d \ge 1$; $q \in b\mathscr{E}$.]

4. (a) Let D_1 be a subdomain $\Subset D$. If $\lfloor g \subset D - D_1$, then $G_D g$ is harmonic in D_1.
   (b) Let $D_1 \Subset D_2 \Subset D$. If $g \in \mathbb{H}(D)$ then there exists g_1 such that $g_1 \in \mathbb{H}_c(R^d)$ and $g_1 = g$ in $\bar{D}_1$, $g_1 = 0$ in $R^d - D_2$. [Hint: multiply g by a function in $C^{(\infty)}$ as in Lemma 7 of §4.6.]
   (c) Prove Proposition 4 by using Theorems 2 and 2′ of §4.6. [This may be putting the horse behind the cart as alluded to in the text, but it is a good exercise!]

5. If $u(D, q, 1; \cdot) \not\equiv \infty$ in D, then it is bounded away from zero in D. Moreover there exists a constant $C > 0$ such that

   $$u(D, q, 1; x) \ge Cu(E, q, 1; x)$$

   for all subdomains $E \Subset D$, and all $x \in \bar{D}$.

6. Prove that $u(D, q, 1; \cdot) \not\equiv \infty$ if and only if for all x in D we have

   $$\int_0^\infty E^x\{e_q(t); t < \tau_D\}\, dt < \infty.$$

[Hint: for some $t_0 > 0$ we have two constants $C_1 > 0$ and $C_2 > 0$ such that $C_1 \le E^x\{e_q(\tau_D);\ 0 < \tau_D \le 1\} \le C_2$ for all $x \in D$; now estimate $E^x\{e_q(\tau_D); n < \tau_D \le n+1\}$ and add.]

7. Suppose D is regular and $E^x\{e^{\|q\|\tau_D}\} < \infty$ for some $x \in D$. Then for any $f \in \mathbb{C}(\partial D)$, $u(D, q, f, \cdot)$ is the unique solution of (12) with boundary value f. [Hint: let φ be a solution which vanishes on ∂D. Show that $\varphi = G_D(q\varphi)$. Prove by induction on n that

$$\varphi(x) = \frac{1}{n!} E^x\left\{\int_0^{\tau_D} q(X_t)\left(\int_0^t q(X_s)\,ds\right)^n \varphi(X_t)\,dt\right\}$$

for all $n \ge 0$. Now estimate $|\varphi(x)|$.]

8. In R^1 let $D = (a, b)$, and $q \in \mathbb{C}(\bar{D})$. Put

$$u_z(x) = E^x\{e_q(\tau_D);\ X(\tau_D) = z\}$$

for $z = a$ and b. Prove that if $u_a \not\equiv \infty$ in D, then both u_a and u_b are bounded in $\bar{D}$. [Hint: to prove $u_b \not\equiv \infty$, use the following result from the elementary theory of differential equations. Either the boundary value problem for the equation $\varphi'' + q\varphi = 0$ in D has a nonzero solution with $\varphi(a) = \varphi(b) = 0$; or it has a unique solution with any given values $\varphi(a)$ and $\varphi(b)$. This is due to M. Hogan.]

An extension of Exercise 8 to higher dimensions has been proved by Ruth Williams. If ∂D is sufficiently smooth, A is an "open" subset of ∂D; and $u(D, q, 1_A; \cdot) \not\equiv \infty$ in D, then $u(D, q, 1; \cdot) \not\equiv \infty$ in D.

Notes on Chapter 4

§4.1 The theory of spatially homogeneous Markov processes is an extension of that of random walks to the continuous parameter case. This is an old theory due largely to Paul Lévy [1]. Owing to its special character classical methods of analysis such as the Fourier transform are applicable; see Gihman and Skorohod [1] for a more recent treatment.

For the theory of dual processes see Blumenthal and Getoor [1], which improved on Hunt's original formulation. Much of the classical Newtonian theory is contained in the last few pages of the book in a condensed manner, but it is a remarkable synthesis not fully appreciated by the non-probabilists.

§4.2. For lack of space we have to de-emphasize the case of dimension $d = 1$ or 2 in our treatment of Brownian motion. So far as feasible we use the general methods of Hunt processes and desist from unnecessary short-cuts. More coverage is available in the cognate books by K. M. Rao [1] and Port and Stone [1]. The former exposition takes a more general probabilistic approach while the latter has more details on several topics discussed here.

§4.3 and §4.4. The force of the probabilistic method is amply illustrated in the solution of the Dirichlet problem. The reader who learns this natural approach first may indeed wonder at the tour de force of the classical treatments, in which some of the basic definitions such as the regularity of boundary would appear to be rather contrived.

As an introduction to the classical viewpoint the old book by Kellogg [1] is still valuable, particularly for its discussion of the physical background. A simpler version may be found in Wermer [1]. Ahlfors [1] contains an elementary discussion of harmonic functions and the Dirichlet problem in R^2, and the connections with analytic functions. Brelot [1] contains many modern developments as well as an elegant (French style) exposition of the Newtonian theory.

For a neater proof of Theorem 2 in §4.4, due to G. Letta, see Chung [11].

The proof of Theorem 8 by means of Lemma 9 may be new. The slow pace adopted here serves as an example of the caution needed in certain arguments. This is probably one of the reasons why even probabilists often bypass such proofs.

§4.5. Another method of treating superharmonic functions is through approximation with smooth ones, based on results such as Theorem 12; see the books by Rao and Port-Stone. This approach leads to their deeper analysis as Schwartz distributions. We choose Doob's method to give further credance to the viability of paths. This method is longer but ties several items together. The connections between (sub)harmonic functions and (sub)martingales were first explored in Doob [2], mainly for the logarithmic potential. In regard to Theorems 2 and 3, a detailed study of Brownian motion killed outside a domain requires the use of Green's function, namely the density of the kernel Q_t defined in (15), due to Hunt [1]. Here we regard the case as a worthy illustration of the general methodology (∂ and all).

Doob proved Theorem 9 in [2] using H. Cartan's results on Newtionian capacity. A non-probabilistic proof of the Corollary to Theorem 10 can be found in Wermer [1]. The general proposition that "semipolar implies polar" is Hunt's Hypothesis (H) and is one of the deepest results in potential theory. Several equivalent propositions are discussed in Blumenthal and Getoor [1]. A proof in a more general case than the Brownian motion will be given in §5.2.

§4.6. The role of the infinitesimal generator is being played down here. For the one-dimensional case it is quite useful, see e.g., Ito [1] for some applications. In higher dimensions the domain of the operator is hard to describe and its full use is neither necessary nor sufficient for most purposes. It may be said that the substitution of integral operators (semigroup, resolvent, balayage) for differential ones constitutes an essential advance of the modern theory of Markov processes. Gauss and Koebe made the first fundamental step in identifying a harmonic function by its averaging property (Theorem 2 in §4.3). This is indeed a lucky event for probability theory.

§4.7. This section is added as an afterthought to show that "there is still sap from the old tree". For a more complete discussion see Chung and Rao [3] where D is not assumed to be bounded but $m(D) < \infty$. The one-dimensional case is treated in Chung and Varadhan [1]. The functional $e_q(t)$ was introduced by Feynman with a purely imaginary q in his "path integrals"; by Kac [1] with a nonpositive q. Its application to the Schrödinger equation is discussed in Dynkin [1] with $q \leq 0$, Khas'minskii [1] with $q \geq 0$. The general case of a bounded q requires a new approach partly due to the lack of a maximum principle.

Let us alert the reader to the necessity of a meticulous verification of domination, such as given in (13), in the sort of calculations in Theorem 5. Serious mistakes have resulted from negligence on this score. For instance, it is *not* sufficient in this case to verify that $u(x) < \infty$, as one might be misled to think after a preliminary (illicit) integration with respect t.

Comparison of the methods used here with the classical approach in elliptic partial differential equations should prove instructive. For instance, it can be shown that the finiteness of $u(D, q, 1; \cdot)$ in D is equivalent to the existence of a strictly positive solution belonging to $\mathbb{C}^{(2)}(D) \cap \mathbb{C}^{(0)}(\bar{D})$. This is also equivalent to the proposition that all eigenvalues λ of the Schrödinger operator, written in the form $(\Delta/2 + q)\varphi = \lambda\varphi$, are strictly negative; see a forthcoming paper by Chung and Li [1]. Further results are on the way.

Chapter 5

Potential Developments

5.1. Quitting Time and Equilibrium Measure

In the next two sections our aim is to establish a number of notable results in classical potential theory by the methods developed in the earlier chapters. In contrast with the preceding sections of last chapter, the horizon will be widened to reach far beyond Brownian motion. We shall deal with Hunt processes satisfying certain general hypotheses and the results will apply to classes of potential kernels including the M. Riesz potentials (see Exercises below) as well as the logarithmic and Newtonian. There are usually different sets of overlapping conditions to yield a particular result. The theory of dual processes in Blumenthal and Getoor [1] gives a framework which has a considerable range, but it is a long and sometimes technically complicated passage. Here instead we offer a more direct and relatively new approach to a number of selected topics, with a view to further development. The case of Brownian motion will be discussed toward the end.

We begin in this section with the analysis of the sample path when and where it quits a transient set. This will lead to a basic result known as the *equilibrium principle* in potential theory.

Let $\{X_t, t \geq 0\}$ be a Hunt process on $(\mathbf{E}, \mathscr{E})$, as in §3.1. [We banish ∂ from sight if not from thought.] It turns out that initially we need only a Markov process with paths which have left limits in $(0, \infty)$, and for which hitting times have the measurability properties valid for a Hunt process. The other hypotheses for a Hunt process will not be explicitly used until further notice. Let ξ be a σ-finite measure on $\mathscr{E}$, and suppose that the potential kernel U has a density u with respect to ξ, as follows:

$$U(x, A) = \int_A u(x, y)\xi(dy), \qquad x \in \mathbf{E}, A \in \mathscr{E}. \tag{1}$$

Of course $u \geq 0$ and $u \in \mathscr{E} \times \mathscr{E}$; but no further condition will be imposed until later. Let $A \in \mathscr{E}$ and

$$\Delta_A = \{\omega \mid \exists t > 0 : X_t(\omega) \in A\};$$

see §3.3 for the measurability of Δ_A. We define

$$\gamma_A(\omega) = \begin{cases} \sup\{t > 0 \mid X_t(\omega) \in A\}, & \text{if } \omega \in \Delta_A, \\ 0, & \text{if } \omega \in \Omega - \Delta_A; \end{cases}$$

contrast this with

$$T_A(\omega) = \begin{cases} \inf\{t > 0 \mid X_t \in A\}, & \text{if } \omega \in \Delta_A, \\ \infty, & \text{if } \omega \in \Omega - \Delta_A. \end{cases}$$

T_A is the hitting time, γ_A is the *quitting time* (or *last exit time*) of A (denoted by L_A in (21) of §3.6). Their dual relationship is obvious from the above. Recall that A is transient if and only if

$$\forall x \in \mathbf{E}: P^x\{\gamma_A < \infty\} = 1. \tag{3}$$

We have for each $t > 0$:

$$\{\gamma_A > t\} = \{T_A \circ \theta_t < \infty\}; \tag{4}$$

from which it follows that $\gamma_A \in \mathscr{F}_\infty$. The next result is trivial but crucial:

$$\{\gamma_A > 0\} = \{T_A < \infty\}. \tag{5}$$

For $x \in \mathbf{E}$, $B \in \mathscr{E}$, define the *quitting kernel* L_A as follows:

$$L_A(x, B) = P^x\{\gamma_A > 0; X(\gamma_A -) \in B\}. \tag{6}$$

and write as usual for $f \in \mathscr{E}_+$:

$$L_A f(x) = \int L_A(x, dy) f(y) = E^x\{\gamma_A > 0; f(X(\gamma_A -))\}.$$

From here on we fix the transient set A and omit it from the notation until further notice. Let $f \in b\mathbb{C}$ (bounded continuous), $\varepsilon > 0$, and consider the following approximation of $f(X(\gamma -))$:

$$\int_0^\infty \frac{1}{\varepsilon} E^x\{f(X_t); t < \gamma \le t + \varepsilon\}\, dt. \tag{7}$$

Put

$$\psi_\varepsilon(x) = \frac{1}{\varepsilon} P^x\{0 < \gamma \le \varepsilon\}; \tag{8}$$

ψ_ε is universally measurable. Now a basic property of γ may be expressed by means of the shift as follows:

$$\gamma \circ \theta_t = (\gamma - t)^+. \tag{9}$$

For if $\gamma_A \le t$, then $T_A \circ \theta_t = \infty$, namely $\gamma_A \circ \theta_t = 0$; if $\gamma_A > t$, then $\gamma_A = t + \gamma_A \circ \theta_t$. As a consequence of (9), we have for each $t \ge 0$:

$$\{t < \gamma \le t + \varepsilon\} = \{0 < \gamma \circ \theta_t \le \varepsilon\}. \tag{10}$$

Using this we apply the simple Markov property at time t to see that the integrand in (7) is equal to

$$E^x\left\{f(X_t)\frac{1}{\varepsilon}E^{X_t}[0 < \gamma \le \varepsilon]\right\} = E^x\{f(X_t)\psi_\varepsilon(X_t)\}.$$

Hence the quantity in (7) is equal to $U(f \cdot \psi_\varepsilon)(x)$. Let us verify that this is uniformly bounded with respect to ε. Since f is bounded it is sufficient to check the following:

$$\begin{aligned}\int_0^\infty \frac{1}{\varepsilon} P^x\{t < \gamma \le t + \varepsilon\}\,dt &= \int_0^\infty \frac{1}{\varepsilon}\int_\Omega 1_{\{t<\gamma\le t+\varepsilon\}}(\omega)P^x(d\omega)\,dt \\ &= \int_{\{\gamma>0\}}\left(\frac{1}{\varepsilon}\int_{(\gamma-\varepsilon)^+}^{\gamma} 1\,dt\right)P^x(d\omega) \le P^x\{\gamma > 0\} \le 1.\end{aligned}$$

Here we have used Fubini's theorem to reverse the order of integration of the double integral above with respect to $P^x \times m$, where m is the Lebesgue measure on $[0, \infty)$. Exactly the same evaluation of the quantity in (7) yields

$$\int_{\{\gamma>0\}}\frac{1}{\varepsilon}\int_{(\gamma-\varepsilon)^+}^{\gamma} f(X_t)\,dt\,P^x(d\omega). \tag{11}$$

Since $f \in b\mathbb{C}$, and X has a left limit at γ ($< \infty$), as $\varepsilon \downarrow 0$ the limit of the quantity in (11) is equal to

$$\int_{\{\gamma>0\}} f(X(\gamma-))P^x(d\omega) = E^x\{\gamma > 0; f(X(\gamma-))\}.$$

We have therefore proved that

$$\int L(x, dy)f(y) = \lim_{\varepsilon\downarrow 0} U(f\psi_\varepsilon)(x) = \lim_{\varepsilon\downarrow 0}\int u(x, y)f(y)M_\varepsilon(dy) \tag{12}$$

where M_ε is the measure given by

$$M_\varepsilon(dy) = \psi_\varepsilon(y)\xi(dy).$$

Now we make the following assumptions on the function u, which will be referred to as (R).

$$\begin{aligned}&\text{(i)}\quad \text{For each } x \in \mathbf{E},\ y \to u(x, y)^{-1} \text{ is finite continuous;}\\ &\text{(ii)}\quad u(x, y) = \infty \text{ if and only if } x = y; \text{ we put } u(x, x)^{-1} = 0.\end{aligned} \tag{R}$$

It is clear that condition (i) is equivalent to the following: $u(x, y) > 0$ for all x and y in $\mathbf{E}$; and for each $x \in \mathbf{E}$, $y \to u(x, y)$ is extended continuous in $\mathbf{E}$. Here are some preliminary consequences of the conditions. Since $\int u(x, y) M_\varepsilon(dy) < \infty$ by (12), it follows from (ii) that $M_\varepsilon(\{x\}) = 0$ for every x; namely M_ε is diffuse. Next since $\inf_{y \in K} u(x, y) > 0$ for each compact K by (i), we have $M_\varepsilon(K) < \infty$. Thus M_ε is a Radon measure. Now let $\varphi \in \mathbb{C}_c$; then the function $y \to \varphi(y) u(x, y)^{-1}$ belongs to $\mathbb{C}_c$ for each x, by (i). Substituting this for f in (12), we obtain

$$\int \frac{L(x, dy)}{u(x, y)} \varphi(y) = \lim_{\varepsilon \downarrow 0} \int \varphi(y) M_\varepsilon(dy) \tag{13}$$

because $u(x, y) u(x, y)^{-1} = 1$ for $y \in \mathbf{E} - \{x\}$, and the point set $\{x\}$ may be ignored since M_ε is diffuse. Since for each x, $L(x, \cdot)$ is a finite measure and $u(x, y)^{-1}$ is bounded on each compact, $L(x, dy) u(x, y)^{-1}$ is a Radon measure. It is well known that two Radon measures are identical if they agree on all φ in $\mathbb{C}_c$ (Exercise 1). Hence the relation (13) implies that there exists a single Radon measure μ on $\mathscr{E}$ such that

$$\int_B \frac{L(x, dy)}{u(x, y)} = \mu(B), \qquad \forall x \in \mathbf{E},\ B \in \mathscr{E}. \tag{14}$$

Let us pause to marvel at the fact that the integral above does not depend on x. This suggests an ergodic phenomenon which we shall discuss in §5.2.

Since M_ε is a Radon measure for each $\varepsilon > 0$, it follows also from (13) that μ is the unique vague limit of M_ε as $\varepsilon \downarrow 0$; but this observation is not needed here. We are now going to turn (14) around:

$$L(x, B) = \int_B u(x, y) \mu(dy), \qquad x \in \mathbf{E},\ B \in \mathscr{E}. \tag{15}$$

When $B = \{x\}$ in (14) the left member is equal to zero by condition (ii). Hence μ is diffuse. Next putting $B = \{y\}$ with $y \neq x$ in (14) we obtain

$$L(x, \{y\}) = u(x, y) \mu(\{y\}) = 0. \tag{16}$$

For an arbitrary $B \in \mathscr{E}$, we have

$$\begin{aligned} L(x, B) &= \int_{B \setminus \{x\}} u(x, y) \frac{L(x, dy)}{u(x, y)} + L(x, B \cap \{x\}) \\ &= \int_{B \setminus \{x\}} u(x, y) \mu(dy) + L(x, B \cap \{x\}) \\ &= \int_B u(x, y) \mu(dy) + L(x, B \cap \{x\}) \end{aligned} \tag{17}$$

since μ is diffuse. Therefore (15) is true if and only if

$$\forall x: L(x, \{x\}) = P^x\{\gamma > 0; X(\gamma-) = x\} = 0. \tag{18}$$

In the proof of (18) we consider two cases according as the point x is holding or not. Recall that x is a holding point if and only if almost every path starting at x remains at x for a strictly positive time. It follows by the zero-one law that if x is not holding point, then almost every path starting at x must be in $\mathbf{E} - \{x\}$ for some rational value of t in $(0, \delta)$ for any $\delta > 0$, by right continuity of the path (this is the first explicit use of right continuity). Define a sequence of rational-valued optional times $\{S_n, n \geq 1\}$ as follows:

$$S_n = \min\left\{\frac{m}{2^n}\,\middle|\, m \geq 1;\ X\left(\frac{m}{2^n}\right) \neq x\right\}.$$

If x is not holding, then $P^x\{\lim_n S_n = 0\} = 1$. We have

$$P^x\{\gamma > S_n; X(\gamma-) = x\} = E^x\{P^{X(S_n)}[\gamma > 0; X(\gamma-) = x]\}. \tag{19}$$

(Simple Markov property is sufficient here since S_n is countably valued.) Since $X(S_n) \neq x$, the right member of (19) equals zero by (16) with x and y interchanged. Letting $n \to \infty$ in (19) we obtain (18).

If x is a holding point, then it is clear that

$$0 < U(x, \{x\}) = u(x, x)\xi(\{x\}).$$

It follows firstly that $\xi(\{x\}) > 0$ and secondly $U(x, \{x\}) = \infty$. Together with the hypothesis that x is holding the latter condition implies that $\{x\}$ is a recurrent set under P^x. This is a familiar result in the theory of Markov chains (where the state space is a countable set). Moreover, another basic result in the latter theory asserts that it is almost impossible for the path to go from a recurrent set to a transient set. Both results can be adapted to the case in question, the details of which are left in Exercise 8 below (strong Markov property is needed). In conclusion, we have proved that if x is holding then $P^x\{T_A < \infty\} = 0$, which implies (18). We summarize the results above with an important addition as follows.

Theorem 1. *Let X be a Hunt process with the potential kernel in* (1) *satisfying conditions* (i) *and* (ii). *Then for each transient set A, there exists a Radon measure μ_A such that for any $x \in \mathbf{E}$ and $B \in \mathscr{E}$:*

$$E^x\{\gamma_A > 0; X(\gamma_A-) \in B\} = \int_B u(x, y)\mu_A(dy). \tag{20}$$

If almost all paths of the process are continuous, then μ_A has support in ∂A. In general if A is open then μ_A has support in $\bar{A}$.

Proof. If the paths are continuous, then clearly we have on $\Delta_A : X(\gamma_A -) = X(\gamma_A) \in \partial A$. In general if A is open, then on Δ_A it is impossible for $X(\gamma_A) \in A$ by right continuity of the paths. Hence there is a sequence of values of t strictly increasing to γ_A at which $X(t) \in A$; consequently the left limit $X(\gamma_A -) \in \bar{A}$. By (6), $L(x, \cdot)$ has support in ∂A in the first case, and in $\bar{A}$ in the second case. Hence so does μ_A by (14). □

It is essential to see why for a compact A the argument above does not show that μ_A has support in A. For it is possible on Δ_A that $X(\gamma_A) \in A$ while $X(\gamma_A -) \notin \bar{A}$; namely the path may jump from anywhere to ∂A and then quit A forever.

Corollary. *We have*

$$P^x\{T_A < \infty\} = \int u(x, y)\mu_A(dy). \tag{21}$$

This follows from (20) when $B = \mathbf{E}$, in view of (5).

The measure μ_A is called the *equilibrium measure* for A. Its exact determination in electrostatics is known as Robin's problem. Formula (14) above gives the stochastic solution to this problem in R^d, $d \geq 3$. In order to amend Theorem 1 and its corollary so that μ_A will have support in $\bar{A}$ for an arbitrary Borel set A, there are several possibilities. The following expedient is due to John B. Walsh. Consider the left-hitting time T_A^- defined in (23) of §3.3, and the corresponding *left quitting time* γ_A^-;

$$\gamma_A^-(\omega) = \sup\{t > 0 \mid X_{t-}(\omega) \in A\} \tag{22}$$

where $\sup \emptyset = 0$. Since for a Hunt process left limits exist in $(0, \infty)$, and $t \to X_{t-}$ is left continuous, we have $X(\gamma_A^- -) \in \bar{A}$, regardless if the sup in (22) is attained or not. This is the key to the next result.

Theorem 2. *Under the hypotheses of Theorem* 1 *there exists a Radon measure* μ_A^- *with support in* $\bar{A}$ *such that for every* $x \in \mathbf{E}$ *and* $B \in \mathscr{E}$:

$$E^x\{\gamma_A^- > 0; X(\gamma_A^- -) \in B\} = \int u(x, y)\mu_A^-(dy). \tag{23}$$

Proof. Let us beware that "left" notions are not necessarily the same as "right" ones! The definition of transience is based on $X_t = X_{t+}$, not X_{t-}, and it is not obvious that the transient set A will remain "transient" for the left limits of the process. The latter property means $P^x\{\gamma_A^- < \infty\} = 1$ for every x. That this is indeed true is seen as follows. We have by Theorem 9 of §3.3, $T_A^- \geq T_A$ a.s. for any Borel set A. On the other hand, the left analogue of (4) is true: $\{\gamma_A^- > t\} = \{T_A^- \circ \theta_t < \infty\}$. It follows that

$$\{\gamma_A^- > t\} \subset \{T_A \circ \theta_t < \infty\} = \{\gamma_A > t\}.$$

Therefore, $\{\gamma_A^- = \infty\} \subset \{\gamma_A = \infty\}$, namely A is left-transient if it is (right)-transient. [Is the converse true?]

The rest of the proof of Theorem 2 is exactly the same as that of Theorem 1, and the question of support of μ_A^- is settled by the remark preceding the theorem. □

Since left notions are in general rather different from their right analogues, Theorem 2 would require re-thinking of several basic notions such as "left-regular" and "left-polar" in the developments to follow. Fortunately under Hypotheses (B) and (L), we know by Theorem 3 of §3.8 that $T_A = T_A^-$ a.s. It follows that $\gamma_A = \gamma_A^-$ a.s. as well (Exercise 6) and so under these hypotheses Theorem 2 contains Theorem 1 as a particular case. We state this as a corollary. We shall denote the support of a measure μ by $\lfloor \mu$.

Corollary. Under Hypothesis (B), (20) and (21) hold with $\lfloor \mu_A \subset \bar{A}$.

Why is Hypothesis (L) not mentioned? Because it is implied by the conditions of the potential kernel, provided we use ξ as the reference measure (Exercise 2).

It is known that Hypothesis (B) holds under certain duality assumptions (see Meyer [3]). In order to state a set of conditions in the context of this section under which Hypothesis (B) holds, we need the next proposition.

Proposition 3. *Under the conditions of Theorem* 1, *for each* y *the function* $x \to u(x, y)$ *is superaveraging. If it is lower semi-continuous then it is excessive.*

Proof. For each $f \in b\mathscr{E}_+$, Uf is excessive by Proposition 2 of §2.1. Hence for each $t > 0$:

$$P_t Uf(x) = \int P_t u(x, y) f(y) \xi(dy) \le \int u(x, y) f(y) \xi(dy)$$
$$= Uf(x), \tag{24}$$

where

$$P_t u(x, y) = \int P_t(x, dz) u(z, y). \tag{25}$$

Since (24) is true for all $f \in b\mathscr{E}_+$, it follows that for each x there exists N_x with $\xi(N_x) = 0$ such that if $y \notin N_x$:

$$P_t u(x, y) \le u(x, y).$$

Now the measure ξ charges every nonempty open set (why?). Hence for an arbitrary y we have $y_n \notin N_x$, $y_n \to y$, so that $u(z, y_n) \to u(z, y)$ for every z by condition (i). Therefore by Fatou:

$$P_t u(x, y) \le \varliminf_n P_t u(x, y_n) \le \varliminf_n u(x, y_n) = u(x, y).$$

This proves the first assertion of the proposition. The second follows from Proposition 3 of §3.2 and the remark following it. □

Since $u(\cdot, y)$ is superaveraging, we denote its regularization by $\underline{u}(\cdot, y)$:

$$\underline{u}(x, y) = \lim_{t \downarrow 0} P_t u(x, y).$$

For each y, $x \to \underline{u}(x, y)$ is excessive by Proposition 5 of §3.2. Observe that the function $\underline{u}$ may not satisfy the conditions (i) and (ii), in particular $\underline{u}(x, x)$ may not be infinite. The following results is proved in Chung and Rao [1] but the proof is too difficult to be given here. A simpler proof would be very interesting indeed.

Theorem 4. *Under the conditions of Theorem* 1, *if we assume also that*

(a) *each compact is transient,*
(b) *for each* x, $\underline{u}(x, x) = +\infty$,

then Hypothesis (*B*) *is true.*

In view of Proposition 3, condition (b) above holds if $u(\cdot, y)$ is lower semi-continuous for each y. In this case condition (a) is satisfied if the process is transient according to the definition given in §3.2, by part of Theorem 2 of §3.7. We shall return to these conditions in §5.2.

It is clear in the course of this book that the systematic use of hitting times (balayage) constitutes a major tool in the theory of Hunt processes. By comparison, the notion of quitting times was of recent origin and its potentials remain to be explored. The next result serves as an illustration of the method. Further work along this line should prove rewarding.

Theorem 5. *Assume Hypothesis* (*B*) *as well as* (*R*). *Let* A_n *be transient sets such that* $A_n \downarrow A$ *and* $\bigcap_n \overline{A}_n = A$. *Then we have for each* $x \in A^c \cup A^r$ *and each* $f \in b\mathbb{C}$:

$$\lim_n L_{A_n}(x, f) = L_A(x, f). \tag{26}$$

In other words the sequence of measures $L_{A_n}(x, \cdot)$ *converges tightly to* $L_A(x, \cdot)$.

Proof. We begin with the following basic relation. For each $x \in A^c \cup A^r$:

$$\bigcap_n \{T_{A_n} < \infty\} = \{T_A < \infty\}, \qquad P^x\text{-a.s.} \tag{27}$$

This has been proved before in this book (where?), though perhaps not exactly in this form. The reader should ponder why the conditions on x and on the transience of A_n are needed, as well as the quasi left continuity of the process. Now we write (27) in terms of quitting times as follows:

$$\bigcap_n \{\gamma_{A_n} > 0\} = \{\gamma_A > 0\}. \tag{28}$$

We will omit below the cliche "almost surely" when it is obvious. Clearly $\gamma_{A_n}\downarrow$ and $\gamma_{A_n} \geq \gamma_A$. Let $\beta = \lim_n \gamma_{A_n}$. Then on $\{\gamma_A > 0\}$, we have $X(\gamma_{A_n}-) = X(\gamma_{\bar{A}_n}-) \in \bar{A}_n$ as shown above on account of Hypothesis (B). It follows by right continuity that if $\beta < \gamma_{A_n}$ for all n then

$$X(\beta) = \lim_n X(\gamma_{A_n}-) \in \bigcap_n \bar{A}_n = A. \tag{29}$$

Thus $\beta \leq \gamma_A$ and so $\beta = \gamma_A$. The last equation is trivial if $\beta = \gamma_{A_n}$ for some n. Next, we prove that on $\{\gamma_A > 0\}$ we have

$$\lim_n X(\gamma_{A_n}-) = X(\gamma_A-). \tag{30}$$

This is trivial if X is continuous at γ_A. The general argument below is due to John B. Walsh and is somewhat delicate. If X has a jump at γ_A, this jump time must be one of a countable collection of optional times $\{\alpha_n\}$, by Exercise 5 of §3.1. This means that for almost every ω, $\gamma_A(\omega) = \alpha_n(\omega)$ where n depends on ω. We can apply the strong Markov property at α_n for all n to "cover γ_A" whenever X is discontinuous there. (We cannot apply the property at γ_A!) Two cases will be considered for each α_n, written as α below.

Case 1. $X(\alpha) \notin A$. Applying (27) with $x = X(\alpha)$, we see that since $T_A \circ \theta_\alpha = \infty$ on $\{\alpha = \gamma_A\}$, there exists $N(\omega) < \infty$ such that $T_{A_n} \circ \theta_\alpha = \infty$, hence $\alpha = \gamma_{A_n}$ for $n \geq N(\omega)$. Thus (30) is trivially true because $\gamma_{A_n} = \gamma_A$ for all sufficiently large values of n.

Case 2. $X(\alpha) \in A$. Then on $\{\alpha = \gamma_A\}$ we must have $X(\alpha) \in A \backslash A^r$ because the path does not hit A at any time strictly after α. Since α is a jump time and $A \backslash A^r$ is semipolar by Theorem 6 of §3.5, this possibility is ruled out under Hypothesis (B), by Theorem 1 (iv) of §3.8. This ends the proof of (30).

Now let $x \in A^c \cup A^r$ and $f \in b\mathbb{C}$. Then we have by (28), $\{\gamma_{A_n} > 0\} \downarrow \{\gamma_A > 0\}$ P^x-a.s.; hence by (30) and bounded convergence:

$$\lim_n E^x\{\gamma_{A_n} > 0; f(X(\gamma_{A_n}-))\} = E^x\{\gamma_A > 0; f(X(\gamma_A-))\}. \tag{31}$$

Recalling (6) this is the assertion in (26). □

Corollary. *If* A_1 *is compact then*

$$\lim_n \mu_{A_n}(\mathbf{E}) = \mu_A(\mathbf{E}). \tag{32}$$

Proof. We may suppose $A_1 \neq \mathbf{E}$. There exists $x_0 \notin A_1$, and $u(x_0, \cdot)^{-1}$ is bounded continuous on A_1. Hence we have by (26).

$$\lim_n \int_{A_n} \frac{L_{A_n}(x_0, dy)}{u(x_0, y)} = \int_A \frac{L_A(x_0, dy)}{u(x_0, y)}$$

which reduces to (32). □

The corollary will be applied in the next section to yield important results, under further conditions on the potential kernel.

Exercises

1. Let $(\mathbf{E}, \mathscr{E})$ be as in §1.1. A measure μ on $\mathscr{E}$ is called a Radon measure iff $\mu(K) < \infty$ for each compact subset K on $\mathbf{E}$. Prove that if μ and ν are two Radon measures such that $\int f\, d\mu = \int f\, d\nu$ for all $f \in \mathbb{C}_c$, then $\mu \equiv \nu$. [Hint: use Lemma 1 of §1.1 to show that $\mu(D) = \nu(D) < \infty$ for each relatively compact open D; then use Lemma 2 of §1.1 to show that $\mu(B \cap D) = \nu(B \cap D)$ for all $B \in \mathscr{E}$. Apply this to a sequence of D_k such that $D_k \uparrow \mathbf{E}$. This exercise is given here because it does not seem easy to locate it in textbooks.]

In Exercises 2 and 5, we assume the conditions of Theorem 1.

2. Let A be transient. If $\mu_A(\mathbf{E}) = 0$, then A is polar. Conversely if $P_A 1(x) = 0$ for some x, then $\mu_A(\mathbf{E}) = 0$.
3. Assume $U(x, K) < \infty$ for each x and compact K. Let f be any excessive function. If $f(x) > 0$ for some x then $f(x) > 0$ for all x. [Hint: use Proposition 10 of §3.2.]
4. Assume each compact set is transient. Let f be an excessive function. If $f(x) < \infty$ for some x then the set $\{x \mid f(x) = \infty\}$ is a polar set. [Hint: use Theorem 7 of §3.4 and $u > 0$.]
5. Under the same conditions as Exercise 4 prove that each singleton is polar. [Hint: let D_n be open relatively compact, $D_n \downarrow\downarrow \{x_0\}$; $P_{D_n}1 = U\mu_n$ with $\underline{|\mu_n} \subset \overline{D}_n$. Show that (a subsequence of) $\{\mu_{n_j}\}$ converges vaguely to $\lambda\delta_{x_0}$, and $\lambda u(x, x_0) \leq \underline{\lim}_n\, U\mu_n(x) \leq 1$ so that $\lambda = 0$. For $x_1 \notin \overline{D}_1$, $\lim_n U\mu_n(x_1) = 0$ because $u(x_1, \cdot)$ is bounded continuous in $\overline{D}_1$. Now use Exercise 2.]
6. Prove that for a Hunt process, $\gamma_A \geq \gamma_A^-$ a.s. for any $A \in \mathscr{E}$. If $T_A = T_A^-$ a.s. then $\gamma_A = \gamma_A^-$ a.s.
7. Prove that a polar set is left-polar; a thin set is left-thin.
8. The following is true for any Hunt process (even more generally). Let x be a holding but not absorbing point. Define the *sojourn time* $S = \inf\{t > 0 \mid X_t \neq x\}$, and the re-entry time $R = \inf\{t > S \mid X_t = x\}$. Show that

$P^x\{S > t\} = e^{-\lambda t}$ for some $\lambda: 0 < \lambda < \infty$. Let $P^x\{R < \infty\} = \rho$. Show that $U(x, \{x\}) = \lambda^{-1}(1 - \rho)^{-1}$; hence $U(x, \{x\}) = \infty$ if and only if $\{x\}$ is "recurrent under P^x". Let $A \in \mathscr{E}$; prove that if $P^x\{T_A < \infty\} > 0$ then $P^x\{T_A < R\} > 0$. Consequently if $\{x\}$ is recurrent under P^x, then so is A. Note: without further assumption on the process, a holding point x may not be hit under P^y, $y \neq x$; hence the qualification "under P^x" above.

9. Derive a formula for $E^x\{e^{-\lambda\gamma}f(X_\gamma -); \gamma > 0\}$ for $\lambda > 0$, $\gamma = \gamma_A$ as in Theorem 1.

10. Find the equilibrium measure μ_A when A is a ball in the Newtonian case (namely Brownian motion in R^3).

5.2. Some Principles of Potential Theory

In this section we continue the study of a Hunt process with the potential kernel given in (1) of §5.1, under further conditions. Recall that Hypothesis (L) is in force with ξ as the reference measure. The principal new condition is that of symmetry:

$$\forall (x, y): u(x, y) = u(y, x). \tag{S}$$

Two alternative conditions of transience will be used:

$$\forall x, \forall \text{ compact } K: U(x, K) < \infty; \tag{T_1}$$

$$\text{each compact set is transient.} \tag{T_2}$$

Two alternative conditions of regularity (besides (R) in §5.1) will be used:

$$\forall y: x \to u(x, y) \text{ is lower semi-continuous;} \tag{U_1}$$

$$\forall y: x \to u(x, y) \text{ is excessive.} \tag{U_2}$$

There are various connections between the conditions. Recall that (T_1) and (U_1) imply (T_2) by Theorem 2 of §3.7; (R) and (U_1) imply (U_2) by Proposition 3 of §5.1; (R), (T_2) and either (U_1) or (U_2) imply Hypothesis (B) by Theorem 4 of §5.1; (R) and (S) imply (U_1) trivially.

Readers who do not wish to keep track of the various conditions may assume (R), (S) and (T_1), and be assured that all the results below hold true with the possible exception of those based on the energy principle. However, it is one of the fascinating features of potential theory that the basic results are interwoven in the manner to be illustrated below. This has led to the development of *axiomatic potential theory* by Brelot's school.

The *Equilibrium Principle* of potential theory may be stated as follows. For each compact K, there exists a finite measure μ_K with support in K such that

$$P_K 1 = U\mu_K. \tag{E}$$

The principle holds under (R), (T_2) and either (U_1) or (U_2), because then Hypothesis (B) holds and (E) holds by the Corollary to Theorem 2 of §5.1.

We are interested in the mutual relationship between several major principles such as (E). Thus we may assume the validity of (E) itself in some of the results below. Let us first establish a basic relation known as Hunt's "switching formula" in the duality theory. The proof is due to K. M. Rao. From here on in this section the letters K, D, and B are reserved respectively for a compact, relatively compact open, and (nearly) Borel set.

Theorem 1. *Under* (S), (T_1) *and* (U_2), *we have for each* B:

$$P_B u(x, y) = P_B u(y, x). \tag{1}$$

Proof. We begin by noting that we have by (S)

$$P_B u(x, y) = \int u(y, z) P_B(x, dz); \tag{2}$$

hence for each x, $y \to P_B u(x, y)$ is excessive by simple analysis (cf., Exercise 8 of §4.1). On the other hand for each y, $x \to P_B u(x, y)$ is excessive by (U_2) and a general property of excessive function (Theorem 4 of §3.4).

Let $K \subset D$, we have

$$U 1_K = P_D U 1_K < \infty$$

by Theorem 3 of §3.4 and (T_1); namely

$$\int_K u(x, y)\xi(dy) = \int_K P_D u(x, y)\xi(dy) < \infty. \tag{3}$$

Since $u(x, y) \geq P_D u(x, y)$ by Theorem 4 of §3.4, and K is arbitrary for a fixed D, it follows from (3) that

$$u(x, y) = P_D u(x, y) \tag{4}$$

for each x and ξ-a.e. y in D. Both members of (4) are excessive in y for each fixed x, hence (4) holds for each x in $\mathbf{E}$ and $y \in D$, by the Corollary to Proposition 4 of §3.5. [This property of u is of general importance.]

Next we have

$$P_K u(x, y) = \int P_K(x, dz) u(y, z) = \int P_K(x, dz) P_D u(y, z) \tag{5}$$

because $P_K(x,\cdot)$ is supported by K, and $u(y,z) = P_D u(y,z)$ for $z \in K \subset D$ by (4) as just proved. It follows by Fubini and (S) that the quantity in (5) is equal to

$$\begin{aligned}\int P_D(y,dw)\int P_K(x,dz)u(z,w) &= \int P_D(y,dw)P_K u(x,w)\\ &\le \int P_D(y,dw)u(x,w)\\ &= P_D u(y,x).\end{aligned}$$

Thus we have proved that for all x and y, and $K \subset D$:

$$P_K u(x,y) \le P_D u(y,x). \tag{6}$$

Integrating this with respect to $\xi(dx)$ over an arbitrary compact set C, we have

$$\int_C \xi(dx)P_K u(x,y) \le \int_C P_D u(y,x)\xi(dx) = P_D U 1_C(y). \tag{7}$$

Taking a sequence of such sets $D_n \downarrow\downarrow K$, we have $P_{D_n}U1_C(y) \downarrow P_K U 1_C(y)$, provided $y \notin K\backslash K^r$, by the Corollary to Theorem 5 of §2.4. Using this sequence in (7), we obtain

$$\int_C \xi(dx)P_K u(x,y) \le P_K U 1_C(y) = \int_C P_K u(y,x)\xi(dx), \qquad y \notin K\backslash K^r. \tag{8}$$

It is essential to notice that the last member above is finite by (T_1) because it does not exceed $U(y,C)$. Since C is arbitrary we deduce from (8) that

$$P_K u(x,y) \le P_K u(y,x) \tag{9}$$

for each $y \notin K\backslash K^r$, and ξ-a.e. x. Both members of (9) are excessive in each variable when the other variable is fixed and $\xi(K\backslash K^r) = 0$ because $K\backslash K^r$ is semipolar. Therefore (9) holds in fact for all x and y: since it is symmetric in x and y, we conclude that

$$\forall (x,y)\colon P_K u(x,y) = P_K u(y,x). \tag{10}$$

Now given B and x, there exist compacts $K_n \subset B$, such that $T_{K_n} \downarrow T_B$, P^x-a.s. by Theorem 8(b) of §3.3. Since for each y, $u(X_t, y)$ is right continuous in t under (U_2) by Theorem 6 of §3.4, it follows by Fatou and (10) that

$$P_B u(x,y) \le \underline{\lim}_n P_{K_n} u(x,y) = \underline{\lim}_n P_{K_n} u(y,x) \le P_B u(y,x). \tag{11}$$

Interchanging x and y we obtain (1). □

We are now ready to establish another major principle, known as the Maria-Frostman *Maximum Principle* as follows. For any σ-finite measure μ supported by K, we have

$$\sup_{x \in \mathbf{E}} U\mu(x) = \sup_{x \in K} U\mu(x) \tag{M}$$

This principle must be intuitively obvious to a physicist, since it says that the potential induced by a charge is greatest where the charge lies. Yet its proof seems to depend on some sort of duality assumption on the distribution of the charge. This tends to show that physical processes carry an inherent duality suggested by the reversal of time.

Theorem 2. *The maximum principle* (M) *holds under* (S), (T_1) *and* (U_2).

Proof. There is nothing to show if the right member of (M) is infinite, so we may suppose it finite and equal to M. For $\varepsilon > 0$ define the set

$$B = \{x \in \mathbf{E} \,|\, U\mu(x) \le M + \varepsilon\}. \tag{12}$$

Since $U\mu$ is excessive under (U_2), it is finely continuous (Corollary 1 to Theorem 1 of §3.5) and so B is finely closed. The fine continuity of $U\mu$ also implies that $K \subset B^r$ (why?). Therefore we have by Theorem 1 and Fubini:

$$\begin{aligned} P_B U\mu(x) &= \int P_B u(x, y)\mu(dy) = \int P_B u(y, x)\mu(dy) \\ &= \int u(y, x)\mu(dy) = U\mu(x), \end{aligned} \tag{13}$$

because K supports μ and for each $y \in K$, $P_B u(y, x) = u(y, x)$ trivially since $y \in B^r$. On the other hand, $P_B(x, \cdot)$ has support in B by Theorem 2 of §3.4, since B is finely closed. Hence we have

$$P_B U\mu(x) = \int P_B(x, dy) U\mu(y) \le \sup_{y \in B} U\mu(y) \le M + \varepsilon. \tag{14}$$

Putting (13) and (14) together we conclude that $U\mu \le M$ since ε is arbitrary. □

The argument leading to (13) is significant and recorded below.

Corollary. *For any σ-finite measure μ with support contained in B^r, we have*

$$U\mu = P_B U\mu. \tag{15}$$

In conjunction with (E), an interesting consequence of (15) is the following particular case of Hypothesis (B):

$$P_K 1 = P_G P_K 1. \tag{16}$$

In some situations this can take the place of Hypothesis (B).

The next principle will be named the *Polarity Principle.*For Brownian notion it is proved in Theorem 10 of §4.5. Its general importance was recognized by Hunt. The depth of this principle is evident from the following formulation, in which we follow a proof of K. M. Rao's with some modification.

Theorem 3. *Assume (T_2), Hypothesis (B), and both the equilibrium principle (E) and the maximum principle (M). Then the polarity principle holds as follows:*

$$\textit{Every semipolar set is polar.} \tag{P}$$

Remark. For a thin set this says that if it cannot be hit immediately from some point, then it cannot be hit at all from any point!

Proof. By Theorem 8(b) of §3.3 and Theorem 6 of §3.5, it is sufficient to prove that each compact thin set is polar. Let K be such a set, and define for each $n \geq 1$:

$$A_n = K \cap \left\{ x \in \mathbf{E} \,\middle|\, P_K 1(x) \leq 1 - \frac{1}{n} \right\}.$$

Since $P_K 1$ is finely continuous, each A_n is finely closed. Let L be a compact subset of A_n. Then $P_L 1(x) \leq 1 - 1/n$ for $x \in L$; hence for all x by (E) and (M). This implies $E^x\{e^{-T_L}\} \leq 1 - 1/n$ for all x; hence L is polar by Exercise 6 of §3.8, where Hypothesis (B) is used. Thus A_n is polar and so is $A = \bigcup_{n=1}^{\infty} A_n$. Put $C = K - A$. Then C is finely closed (why?); and we have

$$\forall x \in \mathbf{E}: P_K 1(x) = P_C 1(x); \qquad \forall x \in C: P_C 1(x) = 1. \tag{17}$$

Put $T_1 = T_K$; if α is a countable ordinal which has the predecessor $\alpha - 1$, put

$$T_\alpha = T_{\alpha-1} + T_K \circ \theta_{T_{\alpha-1}};$$

if α is a limit countable ordinal, put

$$T_\alpha = \sup_{\beta < \alpha} T_\beta$$

where β ranges over all ordinals preceding α. The following assertions are almost surely true for each α. Since A is polar, we have $X(T_\alpha) \in C$ on

$\{T_\alpha < \infty\}$. For a limit ordinal α this is a consequence of quasi left continuity. In view of (17), we have for each α:

$$\{T_\alpha < \infty\} = \bigcap_{n=1}^{\infty} \{T_{\alpha+n} < \infty\}. \tag{18}$$

On the other hand, since K is thin, strong Markov property implies that $T_\alpha < T_{\alpha+1}$ on $\{T_\alpha < \infty\}$. It follows as in the proof of Theorem 6 of §3.4 that there exists a first countable ordinal α^* (not depending on ω) for which $T_{\alpha^*} = \infty$. This α^* must be a limit ordinal by (18). Therefore, on the set of ω where K is ever hit, it is hit at a sequence of times which are all finite and increase to infinity. This contradicts the transience of K. Hence K cannot be hit at all, so it is polar. □

Corollary. *For any $B \in \mathscr{E}^\bullet$ the set $B\backslash B^r$ is polar.*

In particular for any open D, the set of points on ∂D which are not regular for D^c is a polar set, because it is a subset of $D^c\backslash(D^c)^r$. This is the form of the result known in classical potential theory as Kellogg-Evans theorem. The general form of (P) goes back to Brelot, Choquet and H. Cartan. Its relevance to the Dirichlet problem has been discussed in Proposition 11 of §4.5.

Now we return to the setting of §5.1, and adduce some important consequences about the equilibrium measure under the new assumptions (S), (T_2) and (M). For each transient set A, we define

$$C(A) = \mu_A(\mathbf{E}) \tag{19}$$

to be the *capacity* of A. Under (T_2), $C(K)$ is defined for each compact K and is a finite number since μ_K is a Radon measure. For any two σ-finite measures λ_1 and λ_2, we put

$$\langle \lambda_1, \lambda_2 \rangle = \iint \lambda_1(dx) u(x, y) \lambda_2(dy) = \int U\lambda_2 \, d\lambda_1. \tag{20}$$

The condition of σ-finiteness is made to ensure the applicability of Fubini's theorem. Under (S) it then follows that

$$\langle \lambda_1, \lambda_2 \rangle = \langle \lambda_2, \lambda_1 \rangle \tag{21}$$

whether finite or infinite. This symmetry plays a key role in what follows. The next result is a characterization of the capacity well known in classical potential theory.

Theorem 4. *Let ν be any σ-finite measure with $\lfloor \nu \subset K$, such that*

$$\forall x \in K: U\nu(x) \leq 1. \tag{22}$$

Then we have

$$\nu(K) \leq C(K). \tag{23}$$

Proof. Let $K \subset D$; under (T_2) both K and D are transient and so μ_D as well as μ_K exist. Under (S) we have as a case of (20):

$$\int U\mu_D \, d\nu = \int U\nu \, d\mu_D. \tag{24}$$

The left member above equals $\nu(K)$ because $U\mu_D = P_D 1 = 1$ on K. The right member does not exceed $\mu_D(\mathbf{E})$ because $U\nu \leq 1$ by (22) and (M). Now let $D_n \downarrow\downarrow K$; then we have proved that $\nu(K) \leq \mu_{D_n}(\mathbf{E}) = C(D_n)$ for each n, and consequently (23) follows by the Corollary to Theorem 5 of §5.1 □

Note that if the inequality in (22) holds for all $x \in \mathbf{E}$, then the intervention of (M) is not needed in the preceding proof. This is the case for Corollary 1 below.

Corollary 1. *If* $K_1 \subset K_2$, *then* $C(K_1) \leq C(K_2)$.

Corollary 2. *If* ν *is a* σ*-finite measure on* $\mathscr{E}$ *such that* $U\nu < \infty$, ν-a.e, *then* ν *does not charge any polar set.*

Proof. Let $A_n = \{x \mid U\nu(x) \leq n\}$. If ν charges a polar set it must charge a polar subset of A_n for some n, and so also a compact polar subset of A_n, because ν is necessarily inner regular (Exercise 1). Let ν_1 be the restriction of ν to such a set K, and $\nu_2 = \nu_1/n$. Then $U\nu_2 \leq 1$ on K, hence by the theorem $\nu_2(K) \leq C(K)$. But $C(K) = 0$ by Exercise 2 of §5.1. Thus $\nu_2 \equiv 0$ and ν does not charge K after all. □

Let λ and ν be two finite measures on $\mathscr{E}$; then $\lambda - \nu$ is well defined by $(\lambda - \nu)(B) = \lambda(B) - \nu(B)$; integration with respect to $\lambda - \nu$ is also well defined by $\int f \, d(\lambda - \nu) = \int f \, d\lambda - \int f \, d\nu$ when both $\int f \, d\lambda$ and $\int f \, d\nu$ are finite. We extend the definition (20) as follows. The *energy* of $\lambda - \nu$ is defined to be

$$\langle \lambda - \nu, \lambda - \nu \rangle = \langle \lambda, \lambda \rangle - \langle \lambda, \nu \rangle - \langle \nu, \lambda \rangle + \langle \nu, \nu \rangle \tag{25}$$

provided that all four terms on the right are finite. This amounts to the condition that $\langle \lambda + \nu, \lambda + \nu \rangle < \infty$. Let us remark that if $\lambda_1 - \lambda_2 = \lambda_3 - \lambda_4$ and both $\langle \lambda_1 - \lambda_2, \lambda_1 - \lambda_2 \rangle$ and $\langle \lambda_3 - \lambda_4, \lambda_3 - \lambda_4 \rangle$ are defined then they are equal. If the energies of $\lambda_1 - \lambda_2$ and of $\lambda_3 - \lambda_4$ are defined then so is the energy of $(\lambda_1 - \lambda_2) - (\lambda_3 - \lambda_4) = (\lambda_1 + \lambda_4) - (\lambda_2 + \lambda_3)$. In particular we can define the energy of a signed measure using its Hahn-Jordan decomposition, but there is no need for that. We are ready to announce another principle.

Energy Principle. For any two finite measures λ and v on $\mathscr{E}$, we have $\langle \lambda - v, \lambda - v \rangle \geq 0$ provided it is defined; if it is equal to zero then $\lambda \equiv v$.

Theorem 5. *Suppose that the Hunt process has a transition density function p_t with respect to ξ which is symmetric, namely for $t > 0$:*

$$P_t(x, dy) = p_t(x, y)\xi(dy); \qquad p_t(x, y) = p_t(y, x) \tag{26}$$

for all (x, y); where $p_t \geq 0$, $p_t \in \mathscr{E} \times \mathscr{E}$. Then the energy principle holds.

Proof. We have

$$u(x, y) = 2\int_0^\infty p_{2t}(x, y)\,dt$$

and

$$p_{2t}(x, y) = \int p_t(x, z)p_t(z, y)\xi(dz).$$

Writing μ for $\lambda - v$ purely for abbreviation (not as "signed measure"), we have if $\langle \mu, \mu \rangle < \infty$:

$$\begin{aligned}\langle \mu, \mu \rangle &= \iint \mu(dx)u(x, y)\mu(dy)\\ &= 2\int_0^\infty dt \int \left[\iint \mu(dx)p_t(x, z)p_t(z, y)\mu(dy)\right]\xi(dz)\\ &= 2\int_0^\infty dt \int \left(\int \mu(dx)p_t(x, z)\right)^2 \xi(dz),\end{aligned}$$

by the symmetry of p_t. Hence $\langle \mu, \mu \rangle \geq 0$. If $\langle \mu, \mu \rangle = 0$ then there exists a set N of t with $m(N) = 0$ and for each $t \notin N$ a set Z_t of z with $\xi(Z_t) = 0$ such that

$$\int \mu(dx)p_t(x, z) = 0 \quad \text{if } t \notin N,\ z \notin Z_t. \tag{27}$$

Let $f \in b\mathbb{C}$; it follows from (27) that

$$\int \mu(dx)P_t f(x) = 0, \quad \text{if } t \notin N. \tag{28}$$

Since X is right continuous, $\lim_{t \downarrow 0} P_t f(x) = f(x)$ for each x. Taking a sequence of $t \notin N$, $t \downarrow 0$ in (28) we obtain

$$\forall f \in b\mathbb{C}: \int f\,d\lambda - \int f\,dv = 0,$$

where both integrals are finite. Hence $\lambda \equiv v$ (see Exercise 1 of §5.1). □

From here on all the results so far proved will be used wherever appropriate. Let K be a fixed compact set and $\Phi(K)$ denote the class of finite measures ν such that $\lfloor \nu \subset K$ and $\langle \nu, \nu \rangle < \infty$. It follows from Corollary 2 to Theorem 4 that such a ν does not charge any polar set, since $\int U\nu \, d\nu < \infty$. For the equilibrium measure μ_K, we have

$$\langle \mu_K, \mu_K \rangle = \int U\mu_K \, d\mu_K = \int P_K 1 \, d\mu_K \leq \mu_K(K) = C(K) < \infty. \tag{29}$$

Hence $\mu_K \in \Phi(K)$. But then μ_K does not charge the set A below:

$$A = K \cap \{x \in \mathbf{E} \mid P_K 1(x) < 1\} \tag{30}$$

because A is polar as in the proof of Theorem 3. Consequently the inequality in (29) is an equality:

$$\langle \mu_K, \mu_K \rangle = C(K). \tag{31}$$

We know (Exercise 2 of §5.1) that $C(K) = 0$ if and only if K is polar. If $C(K) > 0$ let $\nu \in \Phi(K)$ with $\nu(K) = C(K)$. Let us compute

$$\langle \nu - \mu_K, \nu - \mu_K \rangle = \langle \nu, \nu \rangle - \langle \nu, \mu_K \rangle - \langle \mu_K, \nu \rangle + \langle \mu_K, \mu_K \rangle. \tag{32}$$

Since ν does not charge A we have

$$\langle \nu, \mu_K \rangle = \langle \mu_K, \nu \rangle = \int U\mu_K \, d\nu = \int P_K 1 \, d\nu = \nu(K). \tag{33}$$

It follows from (31) and (33) that the energy of $\nu - \mu_K$ is equal to $\langle \nu, \nu \rangle - C(K)$. We have therefore proved the following result under the energy principle.

Theorem 6. *Let $C(K) > 0$. For any $\nu \in \Phi(K)$ with $\nu(K) = C(K)$, we have*

$$C(K) \leq \langle \nu, \nu \rangle \tag{34}$$

where equality holds if and only if $\nu = \mu_K$.

In other words, the equilibrium measure for K is the unique measure among all measures on K having the same total mass which minimizes the energy. This is the way it was found by Gauss and Frostman for Newtonian and M. Riesz potentials (see Exercises 7–12 below). Here we have turned the table around by first establishing the existence of the equilibrium measure in §5.1 under fairly general conditions, then verifying its minimization property as shown above. Although more restrictive assumptions, in particular the symmetry of transition density, are imposed to ensure Theorem 6, these conditions still cover the classical cases with room to spare.

Another uniqueness theorem for the equilibrium measure based on duality (Theorem 3 of §4.1) is given in Exercise 2.

The word "equilibrium" connotes in physics a "steady state", namely a stationary distribution of some random process. Such an identification seems missing in the literature and is supplied below. Although it is a kind of "Columbus' egg" stood on its head, there may be possibilities for exploration: an extension to the asymmetric case is discussed in Chung and Rao [2].

We assume that the compact K is not a polar set and is regular: $K = K^r$. The latter condition amounts to the absence of the polar set A in (30). In this case the quitting kernel L_K in §5.1 is a strict probability kernel, namely:

$$\forall x \in K: L_K(x, K) = \int_K u(x, y)\mu_K(dy) = 1. \tag{35}$$

Hence we can construct a discrete parameter homogeneous Markov process on the state space K, with L_K as its transition probability function. Under (R) and (S), it is easy to see that the family of functions $\{u(x, \cdot), x \in K\}$, restricted to K, is uniformly integrable with respect to μ_K. For if $x_n \in K$ and $x_n \to x$, then $\lim_n u(x_n, y) = u(x, y)$ for each y; at the same time $L(x_n, K) = L(x, K)$ for all n. The asserted uniform integrability then follows from an elementary result (Theorem 4.5.4 of *Course*). This is stronger than the following condition, known as Doblin's Hypothesis (D); see Doob [1], p. 192. There exists $\varepsilon > 0$ such that if $A \subset K$, $A \in \mathscr{E}$ with $\mu_K(A) \le \varepsilon$, then $L_K(x, A) \le 1 - \varepsilon$ for all $x \in K$. It is an easy case of Doblin's general results that there exists a probability measure π on A such that for every $A \subset K$, $A \in \mathscr{E}$ and every $x \in K$:

$$\lim_n \frac{1}{n} \sum_{j=1}^n L_K^{(j)}(x, A) = \pi(A); \tag{36}$$

where the $L_K^{(j)}$'s are the iterates of $L_K = L_K^{(1)}$. The measure π is an invariant (stationary) measure for L, namely $\pi L_K = \pi$. Now under (S) there is an obvious invariant measure obtained by normalizing the equilibrium measure:

$$\mu_K^\circ = \frac{1}{\mu_K(\mathbf{E})}\mu_K = \frac{1}{C(K)}\mu_K. \tag{37}$$

It follows that $\pi = \mu_K^\circ$ by the remarks below.

The results above can be improved in several respects, for an arbitrary probability kernel on a compact space K, of the form $L(x, dy) = u(x, y)\mu(dy)$ where $\mu(K) < \infty$. If $u > 0$ and satisfies (U_1) (without (R) or (S)), then there is a unique invariant probability measure π and the Césaro mean in (36) may be replaced by a simple limit; see Chung and Rao [2]. Thus the hypotheses of symmetry imposed in the preceding discussion are merely expedient to ensure the polarity and energy principles. A theory of energy for non-symmetric kernels would be a major advance in this circle of ideas.

Let us now return briefly to the case of Brownian motion in R^3 which corresponds to the Newtonian potential. This is also referred to as the electrostatic potential induced by charges on conductors. In this case the equilibrium principle may be stated as follows. For each bounded Borel set A, there exists a finite measure μ_A supported by ∂A such that

$$P_A 1(x) = \frac{1}{2\pi} \int_{\partial A} \frac{\mu_A(dy)}{\|x - y\|}. \tag{38}$$

This representation is very fruitful. First, it yields the following formula for the capacity of A:

$$C(A) = 2\pi \lim_{x \to \infty} \|x\| P_A 1(x). \tag{39}$$

It follows at once that if $A_1 \subset A_2$, then $C(A_1) \le C(A_2)$. Next, if A_n is a sequence of compacts decreasing to A, then the convergence in (39) is uniform with respect to n by an easy estimate based on (38). Consequently we can interchange limits in x and n to obtain

$$\lim_n C(A_n) = C(A). \tag{40}$$

Finally, for any two Borel sets A and B we have the inequality:

$$P_A 1 + P_B 1 \ge P_{A \cup B} 1 + P_{A \cap B} 1, \tag{41}$$

which is proved as follows. For each x, $P_{A \cup B} 1(x) - P_B 1(x)$ is the probability that the path from x hits $A \cup B$ but not B. Such a path must hit A but not $A \cap B$, the probability of which is $P_A 1(x) - P_{A \cap B} 1(x)$, hence the inequality. Applying (39) to (41) we obtain

$$C(A) + C(B) \ge C(A \cup B) + C(A \cap B). \tag{42}$$

This is the strong subadditivity of the capacity functions discussed in Proposition 3 of §3.3. Together with the other properties verified above, the development in §3.3 shows that we can extend the function $C(\cdot)$ to all subsets of R^3 to be a Choquet capacity. Cf. Helms [1, §7.5].

As another illustration, let us determine the equilibrium measure for a ball. Take $B = B(o, r)$ and $x = o$ in (14) of §5.1. Since $u(x, y) = (2\pi\|x - y\|)^{-1}$, we have for any Borel set A:

$$\mu_B(A) = 2\pi \int_A \|y\| L(o, dy). \tag{43}$$

Rotational symmetry implies that $L(o, \cdot)$ is the uniform distribution on ∂B. A formal proof may be given along the lines of Proposition (XI) of §4.2. It follows that

$$\mu_B(A) = 2\pi r \frac{\sigma(\partial B \cap A)}{4\pi r^2} = \frac{1}{2r}\sigma(\partial B \cap A);$$

in particular

$$C(B) = C(\partial B) = \mu_B(\partial B) = 2\pi r. \tag{44}$$

We can check this from (41) and (16) of §4.4:

$$\lim_{x\to\infty} 2\pi\|x\| P_B 1(x) = \lim_{x\to\infty} 2\pi\|x\| \frac{r}{\|x\|} = 2\pi r.$$

For another application of (39), let A_1 be a Borel set with its closure contained in the interior of A. Then it is obvious that $P_A 1(x) = P_{A-A_1}1(x)$ for $x \in A^c$. Hence $C(A) = C(A - A_1)$ by (39). It is interesting to see how this result can be sharpened and extended to the class of processes considered in §5.1, with continuous paths. We have by (19)

$$C(A) = \int_{\mathbf{E}} \frac{L_A(x, dy)}{u(x, y)}$$

for any x. Choose $x \in A^c$. Then the continuity of paths implies that under P^x we have $\gamma_A = \gamma_{A-A_1}$ almost surely, because the path from outside A cannot hit A without hitting $A - A_1$, and cannot quit $A - A_1$ (forever) without quitting A. Therefore, $L_A(x,\cdot) = L_{A-A_1}(x,\cdot)$ which says that the equilibrium measures for A and $A - A_1$ are the same. The fact that $\mu_{\bar{A}} = \mu_{\partial A}$ may be regarded as a limiting case.

In the language of electrostatics, if the interior of a solid conductor is partially hollowed out, distribution of the induced charge on the outside surface remains unchanged when it is grounded (in equilibrium). An analytic proof of this physical observation may be based on the energy minimizing characterization of the equilibrium measure (Theorem 6). To some of us such a proof may seem more devious than the preceding one, but this is a matter of subjective judgment and previous conditioning. Be it as it may, here is an appropriate place to end these notes (for the moment), leaving the reader with thoughts on the empirical origin of mathematical theory, the grand old tradition of analysis, and the relatively new departure founded on the theory of probability.

Exercises

1. Let $(\mathbf{E}, \mathscr{E})$ be as specified in §3.1. Show that any σ-finite measure ν on $\mathscr{E}$ is inner regular; namely: $\nu(B) = \sup_{K \subset B} \nu(K)$. [Hint: for a finite measure this follows from Exercise 12 of §2.1 of *Course*.]

2. Show that under the hypothesis of Theorem 5 the duality relation (18) of §4.1 holds with $\hat{u}^\alpha = u^\alpha$. Assume also (P). Prove the uniqueness of the equilibrium measure μ_K in the following form: if $U\nu = U\mu_K$ on K and $\lfloor \nu \subset K^r$, then $\nu = \mu_K$. [Hint: use the Corollary to Theorem 2; and note that $\lfloor \mu_K \subset K^r$ by (P).]

3. Under (S) and (E), prove that if μ and ν are σ-finite measures such that $U\mu \leq U\nu$, then $\mu(\mathbf{E}) \leq \nu(\mathbf{E})$. [Hint: integrate with respect to μ_K.]

4. Let $\Phi^{\pm}(K)$ denote the class of $\lambda = \lambda_1 - \lambda_2$ where $\lambda_1 \in \Phi(K)$, $\lambda_2 \in \Phi(K)$. Under the energy principle prove that μ_K is the unique member of $\Phi^{\pm}(K)$ which minimizes G below:

$$G(\lambda) = \langle \lambda, \lambda \rangle - 2\lambda(K).$$

This is the quadratic actually used by Gauss.

5. Suppose that $\langle \lambda, \lambda \rangle \geq 0$ for every $\lambda \in \Phi^{\pm}(K)$ as defined in Exercise 4. This is called the *Positivity Principle* and forms part of the energy principle. Show that under (S) we have the Cauchy-Schwarz inequality: $\langle \lambda, \nu \rangle^2 \leq \langle \lambda, \lambda \rangle \langle \nu, \nu \rangle$ for any λ and ν in $\Phi^{\pm}(K)$.

6. Generalize (34) as follows. Let K be compact: ν a finite measure on K; u defined on $K \times K$ is symmetric and ≥ 0. Define $L(x, dy) = u(x, y)\nu(dy)$ and assume that for every $x \in K$ we have $L(x, K) \leq 1$. Prove that for any measure λ on K with $\lambda(K) \leq 1$ we have

$$\langle \lambda L, \lambda L \rangle \leq \langle \lambda, \lambda L \rangle.$$

Hence if the positivity principle in Exercise 5 holds, then $\langle \lambda L, \lambda L \rangle \leq \langle \lambda, \lambda \rangle$. Thus the energy of a subprobability measure decreases after each transformation by L. This is due to J. R. Baxter. [Hint: put $\varphi(y) = \int \lambda(dx) u(x, y)$. Then $\int (L\varphi)^2 \, d\nu \leq \int L(\varphi^2) \, d\nu \leq \int \varphi^2 \, d\nu$; hence $\langle \lambda L, \lambda L \rangle = \int (\varphi L \varphi) \, d\nu \leq \int \varphi^2 \, d\nu = \langle \varphi, \varphi L \rangle$.]

The next few problems are taken from Frostman [1] and meant to give some idea of the classical approach to the equilibrium problem. They are analytical propositions without reference to a process, where $(\mathbf{E}, \mathscr{E}) = (R^d, \mathscr{B}^d)$, $\xi = m$ (Lebesgue measure).

7. Let $u \geq 0$ and $(x, y) \to u(x, y)$ be lower semi-continuous. Prove that there exists a probability measure ν such that

$$\langle \nu, \nu \rangle = \inf_{\lambda} \langle \lambda, \lambda \rangle$$

where λ ranges over all probability measures on a compact K. [Hint: take a vaguely convergent subsequence of a sequence $\{\lambda_n\}$ such that $\langle \lambda_n, \lambda_n \rangle$ tends to the infinum.]

8. Let ν and λ be as in Exercise 7. Assume in addition that u is symmetric. Prove that $\langle \nu, \nu \rangle \leq \langle \nu, \lambda \rangle$ for any λ. Deduce that if $\langle \nu, \nu \rangle = \langle \lambda, \lambda \rangle$ then

$\nu \equiv \lambda$ under the energy principle. [Hint: consider the energy of $(1-\varepsilon)\nu + \varepsilon\lambda$ for $\varepsilon \downarrow 0$.]

9. Suppose $\langle \lambda, \lambda \rangle < \infty$. Then $U\nu = \langle \nu, \nu \rangle$, λ-a.e. on K. [Hint: let $\langle \nu, \nu \rangle = c > b > a > 0$ and suppose that $U\nu < a$ on A with $\lambda(A) > 0$. It follows from lower semi-continuity that there exists a ball B with $\nu(B) > 0$ such that $U\nu > b$ on B. Now transfer the mass $\nu(B)$ and re-distribute it over A proportionately with respect to λ; namely put for all S: $\nu'(S) = \nu(S \backslash B) + \nu(B)\lambda(S \cap A)\lambda(A)^{-1}$. Compute $\langle \nu, \nu' \rangle$ to get a contradiction with Exercise 8. Hence $U\nu \geq c$, λ-a.e. on K.]

Note. It follows from (31) and (37) that $U\nu_K = \langle \nu_K, \nu_K \rangle$ on $K \cap K^r$, hence except for a polar set by (P). If λ has finite energy then λ does not charge any polar set, hence the result in Exercise 9 is true by the methods used in §§5.1–5.2. The notion of "capacity" was motivated by such considerations.

10. Let $u(x, y) = \|x - y\|^{-\alpha}$ where $\alpha > 0$ in R^2, and $\alpha > 1$ in R^3. Prove that if $U\mu$ is continuous, then (M) holds. [Hint: the Laplacian of $U\mu$ at a point where maximum is attained must be <0 by calculus. This is called the "elementary maximum principle" by Frostman and is an illustration of the humble origin of a noble result.]

11. Let u be as in Exercise 10 but $1 < \alpha < 3$ in R^3. Let μ have the compact support K and suppose that $U\mu$ considered as a function on K only is continuous there. Then it is continuous everywhere. This is known as the *Continuity Principle* and is an important tool in the classical theory. [Hint: $U\mu$ is continuous in K^c. It is sufficient to show that $\varphi_\delta(x) = \int_{B(x,\delta)} u(x, y)\mu(dy) \downarrow 0$ uniformly in x as $\delta \downarrow 0$. For $x \in K$ this is true by the hypothesis and Dini's theorem. For $x \notin K$ let y be a point in K nearest x. Then $\|y - z\| \leq 2\|x - z\|$ for all $z \in K$; if $\|x - y\| \leq \delta$ then $\varphi_\delta(x) \leq 2^\alpha \varphi_{2\delta}(y)$.]

12. Establish the celebrated M. Riesz convolution formula below:

$$\int_{R^d} \|x - y\|^{\alpha - d} \|y - z\|^{\beta - d}\, dy = C(d, \alpha, \beta)\|x - y\|^{\alpha + \beta - d}$$

for $d \geq 1$ and $0 < \alpha + \beta < d$, where C is a constant. [Hint: put $x = o$, $\|z\| = 1$ to see that the existence of C is trivial. For its explicit evaluation use the formula

$$k(\alpha)\|x - y\|^{\alpha - d} = \int_0^\infty t^{(\alpha/2) - 1} p_t(x, y)\, dt$$

where p_t is given in (1) of §4.2, and $k(\alpha)$ is easily computed.]

The next three exercises concern Brownian motion in R^3.

13. Let $B = B(o, r)$. Prove that

$$P^o\{\gamma_B \in dt\} = r(2\pi t^3)^{-1/2} e^{-r^2/2t}\, dt.$$

[Hint: $P^o\{\gamma_B > t\} = E^o\{P^{X(t)}[T_B < \infty]\}$; use (16) of §4.4.]

14. As in Exercise 13, prove that

$$E^o\{e^{-\alpha T_B}\} = 2r\sqrt{2\alpha}(1 - e^{-r\sqrt{2\alpha}})^{-1}.$$

[Hint: by Exercise 13 of §4.2, with $\lambda = (\alpha, 0, 0)$, we have

$$E^o\{e^{-\alpha T_B}\}^{-1} = E^o\{\exp[\sqrt{2\alpha}X_1(t)]\} = \frac{1}{\sigma(r)}\int_{\partial B}\exp[\sqrt{2\alpha}x_1]\sigma(dy).$$

Both results above for R^d, $d \geq 3$ are in Getoor [3]].

15. For any bounded Borel set A, and any Borel set S, prove that

$$\int_0^\infty dx\, P^x\{0 < \gamma_A \leq t;\, X(\gamma_A) \in S\} = t\mu_A(S)$$

where μ_A is the equilibrium measure for A. [Hint: use

$$\int_0^\infty dx(u(x, y) - P_t u(x, y)) = t.$$

Apart from the use of γ_A the result (valid for R^d, $d \geq 3$) is implicit in Spitzer [1].]

Notes on Chapter 5

The title of this chapter is a *double entendre.* It is not the intention of this book to treat potential theory except as a concomitant of the underlying processes. Nevertheless we shall proceed far enough to show several facets of this old theory in a new light.

§5.1 The content of this section is based on Chung [6]. This was an attempt to utilize the inherent duality of a Markov process as evidenced by its hitting and quitting times. Formula (20) or its better known corollary (21) is a particular case of the representation of an excessive function by the potential of a measure plus a generalized harmonic function. Such a representation is proved in Chung and Rao [1] for the class of potentials which satisfies the conditions of Theorem 1. In classical potential theory this is known as F. Riesz's decomposition of a subharmonic function; see Brelot [1]. We stop short of this fundamental result as it would take us too far into a well-entrenched field.

Let us point out that Robin's problem of determining the equilibrium charge distribution is *solved* by the formula (14). One might employ a kind of Monte Carlo method to simulate the last exit probability there to obtain empirical results.

§5.2. The various principles of Newtonian potential theory are discussed in Rao [1]. The extension of these to M. Riesz potentials was a major advance made by Frostman [1]. Their mutual relations form the base of an axiomatic potential theory founded by Brelot. Wermer [1] gives a brief account of the physical concepts of capacity and energy leading to the equilibrium potential.

In principle, it should be possible to treat questions of duality by the method of time reversal. Although a general theory of reversing has existed for some time, it seems still too difficult for applications. A prime example is the polarity principle (Theorem 3) which is not *prima facie* such a problem. Yet all known probabilistic proofs use some kind of reversing (cf. Theorem 10 of §4.5). It would be extremely interesting to obtain this and perhaps also the maximum principle (Theorem 2) by a manifest reversing argument. Such an approach is also suggested by the concept of reversibility of physical processes from which potential theory sprang.

For some historical perspective on the material in Chapters 4 and 5, read Chung [10].

Chapter 6

Generalities

There are some results that we will need later which are short enough to be developed as and when needed, but which are useful enough to stand on their own. There are others which, on the contrary, require much more time and space than we have available for their proper development, but which can at least be stated in enough generality for our use, even if we cannot prove them in detail. We collect some of both kinds in this chapter.

6.1. Essential Limits

Let m and m^* be Lebesgue measure and outer measure, respectively, on the line. If f is an extended-real-valued function on $\boldsymbol{R}^+$ and $A \subset \boldsymbol{R}^+$ is a set of positive Lebesgue measure, then the **essential supremum** of f over A is the supremum of λ for which $m^*\{t \in A : f(t) > \lambda\} > 0$, and is denoted by $\operatorname{ess\,sup}_A f$; the **essential infimum** is the infimum of λ for which $m^*\{t \in A : f(t) < \lambda\} > 0$, denoted $\operatorname{ess\,inf}_A f$.

Definition 6.1.
$$\operatorname*{ess\,lim\,sup}_{s\to t+} f(s) = \inf_{\varepsilon>0}(\operatorname*{ess\,sup}_{s\in(t,t+\varepsilon)} f(s));$$
$$\operatorname*{ess\,lim\,inf}_{s\to t+} f(s) = \sup_{\varepsilon>0}(\operatorname*{ess\,inf}_{s\in(t,t+\varepsilon)} f(s)).$$

Then the **essential right-hand limit** of $f(s)$ as $s \to t+$ exists if $\operatorname{ess\,lim\,sup}_{s\to t+} f(s) = \operatorname{ess\,lim\,inf}_{s\to t+} f(s)$, and it equals the common value. It is denoted $\operatorname{ess\,lim}_{s\to t+} f(s)$. The **essential left-hand limit** is defined analogously, and the **essential limit** of $f(s)$ as $s \to t$ exists if the essential left- and right-hand limits are equal, and if so, it equals the common value. It is denoted by $\operatorname{ess\,lim}_{s\to t} f(s)$.

There is a topology behind this limit. A set G is **essentially open** if there is an open set O such that $G \subset O$ and $m^*(G - O) = 0$. (See the exercises.)

These have the usual properties of limits, which we will take for granted. Let us summarize some less obvious properties.

Proposition 6.2. *Let f be an extended-real-valued function on $\boldsymbol{R}^+$, such that for every $t \in \boldsymbol{R}^+$, $\varphi(t) \stackrel{\text{def}}{=} \operatorname{ess\,lim}_{s\to t+} f(s)$ exists. Then*

(i) *φ is right-continuous;*
(ii) *$\varphi(t) = f(t)$ for a.e. (Lebesgue) t;*
(iii) *if f is bounded, then*

$$\varphi(t) = \lim_{\delta\downarrow 0} \frac{1}{\delta} \int_t^{t+\delta} f(s)\,ds$$
$$= \lim_{p\to\infty} \int_0^\infty pe^{-ps} f(t+s)\,ds. \tag{6.1}$$

Proof.

(i) Let $t \ge 0$. By definition, if $\varepsilon > 0$, there is $\delta > 0$ such that $|\varphi(t) - f(s)| < \varepsilon$ for a.e. (Lebesgue) $s \in (t, t+\delta)$. If $t < t' < t+\delta$, then

$$|\varphi(t') - \varphi(t)| = \operatorname*{ess\,lim}_{s\to t'+} |f(s) - \varphi(t)|$$
$$\le \operatorname*{ess\,sup}_{s\in(t',t+\delta)} |f(s) - \varphi(t)|$$
$$\le \varepsilon.$$

(ii) We leave it to the reader to verify that φ must be Lebesgue measurable. If f is bounded, then (6.1) follows exactly as it would for ordinary limits.

(iii) Assume f is bounded – if not, replace it by arctan f. Let $g(t) = \int_0^t f(s)\,ds$. By Lebesgue's theorem, for a.e. t, g' exists and

$$f(t) = g'(t) = \lim_{\delta\downarrow 0} \frac{1}{\delta} \int_0^\delta f(s)\,ds = \varphi(t),$$

and it is easily seen that it has to equal the second limit in (6.1).

□

These ideas extend to functions with values in more general spaces. If f is a function on $\boldsymbol{R}^+$ with values in a metric space $\boldsymbol{E}$, we say $\operatorname{ess\,lim}_{s\to t+} f(s) = L$ if for each continuous g on $\boldsymbol{E}$, $\operatorname{ess\,lim} g(f(s)) = g(L)$. If $\boldsymbol{E}$ is separable, there is a countable determining set (g_n), and it is enough that this hold for each (g_n).

Doob introduced the idea of the separability of a stochastic process as a way to assure that the process is determined by its values on a countable parameter set. We will look at separability for essential limits, which turns out to be quite a bit simpler than it is in the classical case.

Let $\boldsymbol{E}$ be a separable metric space and let $\{X_t, t \in \boldsymbol{T}\}$ be a stochastic process with values in $\boldsymbol{E}$.

Definition 6.3. *(X_t) is* **essentially separable** *if there exists a countable dense subset $D \subset \boldsymbol{R}^+$ with the property that for a.e. ω,*

$$\operatorname*{ess\,sup}_{t\in(a,b)} f(X_t(\omega)) = \sup_{t\in(a,b)\cap D} f(X_t(\omega))$$

for all $(a,b) \subset \boldsymbol{R}^+$ and $f \in C(\boldsymbol{E})$. We call D an **essential separability set**.

The main consequence of this is that the existence of limits along the separability set implies the existence of essential limits.

Theorem 6.4. *Let $\{X_t, t \in \boldsymbol{T}\}$ be a Borel measurable stochastic process with values in $\boldsymbol{E}$. Then (X_t) is essentially separable.*

Proof. Let $T_1, T_2, \dots$ be a sequence of i.i.d. exponential random variables defined on another probability space $(\Omega', \mathcal{F}', P')$. We claim that for a.e. $\omega' \in \Omega'$, the set $D \equiv \{T_1(\omega'), T_2(\omega'), \dots\}$ is an essential separability set for (X_t).

Fix $f \in C(\boldsymbol{E})$, $0 \le a \le b$, and $\lambda \in \boldsymbol{R}$. We also fix (as we can) $\omega \in \Omega$. To simplify the notation, let $F_t = f(X_t(\omega))$ – we omit the ω, but we will remember that it is there – and let $A_\lambda = \{t \in (a,b) : F_t > \lambda\}$.

Since ω is fixed, F is a deterministic function and A_λ is a deterministic set, so

$$P'\{F_{T_n} > \lambda, a < t < b\} = P'\{T_n \in A_\lambda\} = \int_{A_\lambda} \mathrm{e}^{-s}\, ds.$$

As e^{-s} is decreasing in s,

$$\mathrm{e}^{-b} m\{A_\lambda\} \le P'\{F_{T_n} > \lambda, a < T_n < b\} \le \mathrm{e}^{-a} m\{A_\lambda\}, \tag{6.2}$$

where m is Lebesgue measure.

If $\lambda > \operatorname{ess\,sup}_{t\in(a,b)} F_t$, then $m(A_\lambda) = 0$, and the above probability is zero for all n. It follows that for P'-a.e. ω'

$$\lambda > \operatorname*{ess\,sup}_{t\in(a,b)} F_t \Longrightarrow \lambda \ge \sup_{n:\, T_n(\omega')\in(a,b)} F_{T_n(\omega')}. \tag{6.3}$$

On the other hand, if $\lambda < \operatorname{ess\,sup}_{t\in(a,b)} F_t$, then $m(A_\lambda) > 0$, and the probability above is strictly positive. By independence of the T_n,

$$\begin{aligned} P'\{\sup_{n:T_n\in(a,b)} F_{T_n} \le \lambda\} &= \prod_{n=1}^{\infty} P'\{F_{T_n} \le \lambda, T_n \in (a,b)\} \\ &\le \lim_{N\to\infty} (1 - \mathrm{e}^{-b} m(A_\lambda))^N \\ &= 0. \end{aligned}$$

Thus for P'-a.e. ω',

$$\lambda < \operatorname*{ess\,sup}_{t\in(a,b)} F_t \Longrightarrow \lambda < \sup_{n:\, T_n(\omega')\in(a,b)} F_{T_n(\omega')}. \tag{6.4}$$

Let λ, a and b run through rational values to see that for P'-a.e. ω', we have for all $0 \leq a < b$

$$\operatorname*{ess\,sup}_{t\in(a,b)} F_t = \sup_{n:T_n(\omega')\in(a,b)} F_{T_n(\omega')}.$$

Now let F run through a countable determining set (f_n) in $C(\boldsymbol{E})$ to see that for each ω, and for P'-a.e. ω', $D(\omega')$ is a separability set for the function $t \mapsto X_t(\omega)$.

The final step is an application of Fubini's theorem, so we need to define all processes on the same probability space. We embed them in the product probability space $(\Omega \times \Omega', \mathcal{F} \times \mathcal{F}', P \times P')$ by defining $X_t(\omega, \omega') = X_t(\omega)$, $T_n(\omega, \omega') = T_n(\omega')$. Let

$$\Lambda = \{(\omega, \omega') : (6.3) \text{ and } (6.4) \text{ hold for all rational } \lambda, a, \text{ and } b, \text{ and all } f_n\}.$$

Clearly $\Lambda \in \mathcal{F} \times \mathcal{F}'$, for the set is defined by countably many conditions. From what we have shown above, for each ω,

$$\int_{\Omega'} I_\Lambda(\omega, \omega')P'(d\omega') = 1.$$

It follows by Fubini's theorem that

$$\int_{\Omega \times \Omega'} I_\Lambda d(P \times P') = 1,$$

and then that for P'-a.e. ω',

$$\int_{\Omega} I_\Lambda(\omega, \omega')P(d\omega) = 1.$$

This last statement implies the theorem, for it says that for a.e. ω', we have with P-probability one that

$$\sup_{t\in D\cap(a,b)} f(X_t) = \operatorname*{ess\,sup}_{t\in(a,b)} f(X_t)$$

for all $0 \leq a < b$ and all $f \in C(\boldsymbol{E})$, i.e. D is an essential separability set. □

For a quick consequence, consider what it implies for supermartingales.

Corollary 6.5. *Let* $\{X_t, \mathcal{F}_t,\ t \in \boldsymbol{T}\}$ *be a measurable supermartingale. Then a.e. sample path has essential left and right limits at all* $t > 0$.

Proof. It is well-known that a supermartingale has left and right limits along any countable parameter set. In particular it has limits along a separability set, and hence has essential right and left limits at all times. □

6.2. Penetration Times

There is a type of stopping time which naturally accompanies essential limits, called a penetration time. These are analogous to first-hitting times and debuts, but are less delicate. They don't require the theory of analytic sets to handle them (see §1.5), just Fubini's theorem, and they have some pleasant measurability properties. In particular, they exist for a larger class of sets than first hitting times.

Let $(\Omega, \mathcal{F}, P)$ be a probability space and let $\{X_t, t \in \boldsymbol{T}\}$ be a stochastic process with values in a locally compact metric space $(\boldsymbol{E}, \mathcal{E})$. Let $A \subset \boldsymbol{E}$.

Definition 6.6. *The* **first penetration time of** A *is*

$$\pi_A = \inf \left\{ t \in \boldsymbol{T} : m^*\{s \le t : X_s \in A\} > 0 \right\}.$$

Let $\mathcal{F}_t^\circ = \sigma\{X_s : s \le t\}$ be the natural (uncompleted) filtration of (X_t), and let $(\mathcal{F}_t)$ be the augmented filtration (see §1.4). We assume $(\mathcal{F}_t)$ (but not $(\mathcal{F}_t^0)$) is right-continuous.

Proposition 6.7. *Suppose (X_t) is progressive relative to $(\mathcal{F}_t^\circ)$.*

(i) *If $A \in \mathcal{E}$, π_A is an $\mathcal{F}_{t+}^\circ$-stopping time;*
(ii) *if $A \in \mathcal{E}^\sim$, then π_A is an $\mathcal{F}_t$-stopping time.*

Proof. Since (X_t) is progressive, $(s, \omega) \mapsto I_A \circ X_{s \wedge t}(\omega)$ is $\mathcal{B} \times \mathcal{F}_t^\circ$-measurable, so $L_t \equiv \int_0^t I_A \circ X_s\, ds$ is $\mathcal{F}_t^\circ$-measurable by Fubini. Thus $\{\pi_A \ge t\} = \{L_t = 0\} \in \mathcal{F}_t^\circ$, which implies (i).

For (ii), let ν be the probability measure on $\mathcal{E}$ given by

$$\nu(C) = \int_0^\infty e^{-t} I_C \circ X_s ds.$$

As $A \in \mathcal{E}^\sim \subset \mathcal{E}^\nu$, there are B and C in $\mathcal{E}$ for which $B \subset A \subset C$ and $\nu(C - B) = 0$. Clearly $\pi_C \le \pi_A \le \pi_B$, and if $\pi_C < \pi_B$, then $\int_0^{\pi_B} I_B \circ X_s\, ds = 0$ while $\int_0^{\pi_C} I_C \circ X_s\, ds > 0$. It follows that $I_{C-B} \circ X_s > 0$ on a set of positive Lebesgue measure, and hence that $\int_0^\infty e^{-s} I_{C-B} \circ X_s\, ds > 0$. Taking expectations,

$$P\{\pi_B \ne \pi_C\} \le P\left\{ \int_0^\infty e^{-s} I_{C-B} \circ X_s\, ds > 0 \right\} = 0,$$

since otherwise $\nu(C - B) > 0$. Since π_A is caught between π_B and π_C, it follows that $P\{\pi_A = \pi_B\} = 1$. But by (i), π_B is an $\mathcal{F}_{t+}^\circ$-stopping time, so evidently π_A is an $(\mathcal{F}_t)$-stopping time. □

This has some consequences in potential theory. Suppose that (X_t) is a Hunt process. Then functions such as $P^x\{T_A < \infty\}$, where A is an analytic set, are nearly Borel measurable in x (§3.4, Theorem 6.) This is a fairly deep fact. The corresponding result for π_A is much easier, and the conclusion is stronger.

Proposition 6.8. *Let $A \in \mathcal{E}^{\sim}$ and let $p \geq 0$. Then $f(x) \equiv E^x\{e^{-p\pi_A}\}$ is p-excessive, and therefore nearly Borel measurable. If $A \in \mathcal{E}$, then f is Borel measurable.*

Proof. We have seen that π_A is a stopping time. Moreover, we claim that it is even a perfect terminal time: $\pi_A \leq t + \pi_A \circ \theta_t$, with equality if $t < \pi_A$, and $\pi_A = \lim_{t\downarrow 0}(t + \pi_A \circ \theta_t)$. The first inequality is evident from the fact that $\pi_A = \inf\{t : L_t > 0\}$ where $L_t = \int_0^t I_A \circ X_s\, ds$, and the second follows from the fact that $t \mapsto L_t$ is continuous.

Now $e^{-pt} P_t f(x) = e^{-pt} E^{X_t}\{e^{-p\pi_A}\}$ which is $= E^x\{e^{-p(t+\pi_A\circ\theta_t)} \mid \mathcal{F}_t\} \leq E\{e^{-p\pi_A}\}$, so f is p-supermedian, and thanks to Hunt's Lemma (see Theorem 9.4.8 of the *Course*) the right-hand side converges to $f(x)$ as $t \to 0$ since $\pi_A = \lim_{t\downarrow 0}(t + \pi_A \circ \theta_t)$. Thus f is in fact p-excessive.

Finally, if $A \in \mathcal{E}$, $e^{-p\pi_A}$ is $\mathcal{F}^\circ$-measurable. But then if Y is $\mathcal{F}^\circ$-measurable, $x \mapsto E^x\{Y\}$ is Borel. (Indeed, if $Y = g(X_t)$ for some bounded Borel g, $E^x\{Y\} = P_t g(x)$, which is Borel. A similar equation holds for Y of the form $g_1(X_{t_1}) \ldots g_n(X_{t_n})$, and a monotone class argument shows $x \mapsto E^x\{Y\}$ is $\mathcal{F}^\circ$-measurable for all bounded $\mathcal{F}^\circ$-measurable Y.) □

6.3. General Theory

We will need some basic results from P.-A. Meyer's general theory of processes in the sequel. Their thrust is that for nice processes, if something interesting ever happens, it happens at a stopping time. These are quite general theorems, and even a partially complete treatment of them would require new machinery. However, we will only need some special cases which we can state with what we already know. Proving them is another question, and we will simply say that the interested reader can find them, and much more, in Dellacherie and Meyer [2].

Let $\boldsymbol{E}$ be a locally compact metric space, and let $\mathcal{E}$ be the set of all Borel subsets of $\boldsymbol{E}$. Let $(\Omega, \mathcal{F}, P)$ be a complete probability space, and $(\mathcal{F}_t)$ a right-continuous, augmented filtration. The basic result is a special case of Meyer's Section Theorem. This has many useful consequences; we will give several.

Theorem 6.9. *(Section Theorem) Let $\{X_t, t \in \boldsymbol{T}\}$ be a process with values in $\boldsymbol{E}$, adapted to $(\mathcal{F}_t)$. Let $A \in \mathcal{E}$ and suppose that (X_t) has right-continuous (resp. left-continuous) sample paths. If $\varepsilon > 0$, then there exists a stopping time (resp. predictable time) T with the property that $X_T \in A$ a.s. on $\{T < \infty\}$ and $P\{T < \infty\} \geq P\{\exists t \geq 0 : X_t \in A\} - \varepsilon$.*

Here are two useful corollaries. The first is immediate, and it easily implies the second.

Corollary 6.10. *Let $\{X_t, t \in \boldsymbol{T}\}$ be an adapted process with values in $\mathbf{E}$. Let $A \in \mathcal{E}$ and suppose that either*

(a) *(X_t) has right-continuous sample paths and $P\{X_T \in A\} = 0$ for all stopping times T; or*

(b) *(X_t) has left-continuous sample paths and $P\{X_T \in A\} = 0$ for all predictable times T.*
Then $P\{\exists\, t \geq 0 : X_t \in A\} = 0$.

Definition 6.11. *Two stochastic processes $\{X_t,\ t \in \boldsymbol{T}\}$ and $\{Y_t\ :\ t \in \boldsymbol{T}\}$ are* **indistinguishable** *if $P\{\exists\, t \in \boldsymbol{T}\ :\ X_t = Y_t\ \forall t \geq 0\} = 1$.*

Corollary 6.12. *Let $X = \{X_t,\ t \in \boldsymbol{T}\}$ and $Y = \{Y_t,\ t \in \boldsymbol{T}\}$ be two adapted right-continuous (resp. left-continuous) processes with values in $\mathbf{E}$. Let f and g be Borel measurable functions on $\mathbf{E}$. If $P\{f(X_T) = g(Y_T)\} = 1$ for all finite stopping times (resp. predictable times) T, then $f(X_.)$ and $g(Y_.)$ are indistinguishable.*

Proof. Just apply Corollary 6.10 to the process $\{(X_t, Y_t),\ t \in \boldsymbol{T}\}$ and the set $A = \{(x, y) \in \mathbf{E} \times \mathbf{E} : f(x) \neq g(y)\}$. □

Here are two more consequences of the Section Theorem. The first is immediate, and it easily implies the second.

Theorem 6.13. *Let $\{X_t,\ t \in \boldsymbol{T}\}$ be an adapted right-continuous (resp. left-continuous) process with state space $\mathbf{E}$. If $A \in \mathcal{E}$ has the property that $\{t : X_t \in A\}$ is a.s. countable, then there exists a sequence of stopping times (resp. predictable times) $\{T_n, n = 1, 2, \dots\}$ such that $\{I_{\{t:\, X_t \in A\}},\ t \in \boldsymbol{T}\}$ and $\{I_{\{t:\, t = T_n,\ \text{some } n\}}, t \in \boldsymbol{T}\}$ are indistinguishable.*

Theorem 6.14. *If $\{X_t, t \in \boldsymbol{T}\}$ is an adapted process with values in $\mathbf{E}$ which is either left- or right-continuous, then there is a sequence of stopping times $\{T_n,\ n = 1, 2, \dots\}$ such that the set $\{t\ :\ X_{t-} \neq X_t$ or X_{t-} doesn't exist$\}$ is contained in the set $\{t\ :\ t = T_n,\ \text{some } n\}$.*

This leads to a useful criterion for the continuity of stochastic processes. We will limit ourselves to processes of the form $\psi(X_t)$, where (X_t) is either right or left continuous. Basically, we can reduce sample-function continuity to a question of continuity along a sequence of stopping (or predictable) times. We will give a slightly refined form, which only involves the convergence of the *expectations*. The proof re-cycles the transfinite induction argument of Theorem 6 of §3.4.

Theorem 6.15. *Let $\{X_t, t \in \boldsymbol{T}\}$ be a right-continuous (resp. left-continuous) adapted process with state space $\mathbf{E}$. Let ψ be a bounded real-valued Borel function on $\mathbf{E}$. Then a necessary and sufficient condition that $t \mapsto \psi(X_t)$ be* *right-continuous (resp. left-continuous) is that if* (T_n) *is a decreasing sequence of bounded stopping times (resp. increasing sequence of uniformly bounded predictable times) with limit* T,

$$\lim_{n\to\infty} E\{\psi(X_{T_n})\} = E\{\psi(X_T)\}. \tag{6.5}$$

If $\psi(X_t)$ *is right-continuous (resp. left continuous), a necessary and sufficient condition that it be rcll (resp. lcrl) is that if* (T_n) *is an increasing sequence of uniformly bounded stopping times (resp. decreasing sequence of bounded predictable times), then* $\lim_{n\to\infty} E\{\psi(X_{T_n})\}$ *exists.*

Proof. The condition is evidently necessary. We will show it is sufficient. It is clearly enough to show it on a bounded interval $[0, N]$. We assume by truncation if necessary that all stopping times in the following take their values in $[0, N]$. The proof is different in the two cases. Let us do the right-continuous case first. Let $Y_t = \psi(X_t)$.

Let $S < N$ be a stopping time and $r \in \boldsymbol{Q}$. Let $\Lambda_r = \{Y_S < r < \limsup_{t\downarrow S} Y_t\}$. The filtration is right continuous and complete, so $\Lambda_r \in \mathcal{F}_S$. Thus the time S_{Λ_r} defined by

$$S_{\Lambda_r} = \begin{cases} S & \text{on } \Lambda_r \\ N & \text{on } \Lambda_r^c \end{cases}$$

is a stopping time. Let $p = P\{S_{\Lambda_r} < N\}$. Suppose that $p > 0$. Let $S' \le N$ be a stopping time which is strictly greater than S_{Λ_r} on the set $\{S_{\Lambda_r} < N\}$. On Λ_r the set $Z \equiv \{t \in (S, S') : X_t \in \psi^{-1}(r, \infty)\}$ is non-empty, so by the Section Theorem, for any $\varepsilon > 0$ there is a stopping time $T \le N$ such that $T < N \Longrightarrow X_T \in Z$, and $P\{T < N\} \ge p - \varepsilon$, so that

$$P\{S_{\Lambda_r} < T < S'; Y_T > r\} \ge p - \varepsilon. \tag{6.6}$$

Let us define a sequence (T_n) of stopping times decreasing to S_{Λ_r}. Let T_1 satisfy (6.6) with $S' = N$ and $\varepsilon = p/2$. By induction, let T_n satisfy (6.6) with $S' = T_{n-1} \wedge (S_{\Lambda_r} + 1/n)$ and $\varepsilon = p2^{-n}$. Then as $Y_{S_{\Lambda_r}} < r < Y_{T_n}$,

$$\liminf_{n\to\infty} E\{Y_{T_n} I_{\{T_n<\infty\}}\} \ge pr > E\{Y_{S_{\Lambda_r}} I_{\{S_{\Lambda_r}<\infty\}}\}.$$

This contradicts (6.5), hence $p = 0$. Thus for any stopping time S and $r \in \boldsymbol{Q}$, $P\{Y_x < r < \limsup_{t\downarrow S} Y_t\} = 0$. This argument also applies to the lim inf and it follows that (Y_t) is a.s. right continuous at any stopping time.

Now we do a transfinite induction, as in §3.4 Theorem 6: let $T_0 = 0$, and define $T_{n+1} = \inf\{t : T_n < t \le N,\ |Y_t - Y_{T_n}| > \varepsilon\}$. It may happen that $\lim_n T_n < N$. If so, define $T_\omega = \lim_n T_n$ and continue with $T_{\omega+1}, T_{\omega+2}, \ldots$, through the countable ordinals: if β is an ordinal and T_β is defined, $T_{\beta+1} = \inf\{T : T_\beta < t \le N,\ |Y_t - Y_{T_\beta}| > \varepsilon\}$, and if β is a limit ordinal, $T_\beta = \sup_{\alpha<\beta} T_\alpha$. The process is right continuous at each stopping time, so $T_\beta < T_{\beta+1}$ a.s. on $\{T_\beta < N\}$. Thus, on $\{T_{\beta+1} < N\}$, $T_0 < T_1 < \ldots < T_\beta < T_{\beta+1} < \ldots$. Since there can be at most countably many disjoint open intervals, T_β is eventually equal to N for some countable ordinal, and hence there exists a countable ordinal β for which $P\{T_\beta = N\} = 1$.

Any $t \in \boldsymbol{T}$ falls in some interval $[T_\beta, T_{\beta+1})$, and $|Y_t - Y_{T_\beta}| \le \varepsilon$. It follows that if s is in the same interval, $|Y_t - Y_s| \le 2\varepsilon$, so we have for all $t \in \boldsymbol{T}$ that $\limsup_{s\downarrow t} |Y_s - Y_t| \le 2\varepsilon$, uniformly for $t \in \boldsymbol{T}$. By letting $\varepsilon \to 0$ through a sequence, we see that with probability one, $t \mapsto Y_t$ is right continuous.

Once we know the process is right continuous, we note that it will have left limits on $(0, N)$ if and only if for each $\varepsilon > 0$ the above stopping times satisfy $\lim_{n\to\infty} T_n \ge N$. If $\lim_{n\to\infty} T_n < N$ for some $\varepsilon > 0$, then it is clear that $\lim_{n\to\infty} Y_{T_n}$ fails to exist on a set of positive measure, for then $|Y_{T_n} - Y_{T_{n+1}}| \ge \varepsilon$. This does not quite imply the second statement; we leave it to the reader to modify the stopping times to make sure the expectations also fail to converge.

That takes care of the right-continuous case. Now suppose that (X_t) is left-continuous. Suppose without loss of generality that the bounded function ψ takes values in $[0, 1]$. Then so does Y_t. Let $D = \{t_i,\ i = 1, 2, \ldots\}$ be an essential separability set for (Y_t). Consider the process

$$Y_{mn}(t) = Y_{t_m} I_{\{t_m < t \le t_m + 2^{-n}\}}.$$

This is left continuous and adapted. We combine the (Y_{mn}) in a single process by setting

$$\hat{Y}_t = (Y_{mn}(t),\ m, n = 1, 2, , \ldots).$$

This has values in the space $[0, 1]^{\mathbb{N}^2}$, which is compact and metrizable in the product topology. As each coordinate is left-continuous, so is $(\hat{Y}_t)$. Denote an element $y \in [0, 1]^{\mathbb{N}^2}$ by $y = (y_{mn})$. Define functions on $[0, 1]^{\mathbb{N}^2}$ by

$$\bar{g}_n(y) = \sup_m (y_{mn}), \qquad \underline{g}_n(y) = \inf_m (y_{mn}),$$

$$\bar{h}(y) = \inf_n \bar{g}_n(y), \qquad \underline{h}(y) = \sup_n \underline{g}_n(y).$$

Then $\bar{g}_n(\hat{Y}_t) = \sup\{Y_s : s \in D, t - 2^{-n} \le s < t\}$ and $\bar{h}(\hat{Y}_t) = \limsup_{s\uparrow t, s\in D} Y_s$. Because D is an essential separability set, $\bar{g}_n(Y_t)$ and $\underline{g}_n(Y_t)$ equal respectively the essential sup and essential inf over $[t - 2^n, t)$, and we have with probability one that

$$\bar{h}(\hat{Y}_t) = \operatorname{ess}\limsup_{s\to t-} Y_s, \quad \underline{h}(\hat{Y}_t) = \operatorname{ess}\liminf_{s\to t-} Y_s \quad \forall\, t \in \boldsymbol{T}.$$

Now if $\bar{h}(\hat{Y}_t) = \underline{h}(\hat{Y}_t) = Y_t$ for all t, then $Y_t = \operatorname{ess}\lim_{s\to t-} Y_t$ for all t, which implies that (Y_t) is left continuous by Theorem 6.2. Thus if (Y_t) is not left continuous, at least one of the sets $\{t \le N : Y_t < \bar{h}(\hat{Y}_t)\}$ and

$\{t \leq N : Y_t > \underline{h}(\hat{Y}_t)\}$ is non-empty with strictly positive probability. Say it is the former. Then there is $r \in \boldsymbol{Q}$ such that $P\{\exists t : Y_t < r < \bar{h}(\hat{Y}_t)\} > 0$. Noting that $(X_t, \hat{Y}_t)$ is left continuous on $\boldsymbol{E} \times [0,1]^{N^2}$, we can apply Theorem 6.9: there is a predictable time T for which $(X_T < \hat{Y}_t) \in \{(x, y) : \psi(x) < r, \bar{h}(y) > r\}$ a.s. on $T \leq N$, and $P\{T \leq N\} = p > 0$. (The theorem gives us a possibly unbounded stopping time, but we truncate it by N so that $T \leq N$.) In other words, $P\{Y_T < r, \bar{h}(\hat{Y}_T) > r\} = p > 0$.

It follows that if S is any stopping time with $S < T$ on $\{T \leq N\}$, the set $\{t : S < t < T,\ Y_t > r\}$ has an accumulation point at T, so the Section Theorem 6.9 again implies that there is a predictable time T' such that $S < T' \leq T$ and $X_{T'} > r$ a.s. on $\{T' < T\}$, and $P\{T' < T\} \geq P\{T \leq N\} - \varepsilon$.

Now T is predictable, so there exists a sequence (S_n) announcing it. If we apply the above remarks to each S_n with, say, $\varepsilon = p4^{-n}$, we get a sequence (T_n') of predictable times with $T_n' \leq T$, $Y_{T_n'} > r$ if $T_n' < T$, and $P\{T_n' < T\} \geq p(1 - 4^{-n})$. We can make it an increasing sequence by setting $T_n = \max(T_1', \dots, T_n')$. Since $Y_{T_n} > r$ and $Y_T < r$ on $\{T_n < T\ \forall n\}$, it is clear that $\liminf_n E\{Y_{T_n}\} > E\{Y_T\}$. But this contradicts the hypothesis, so evidently $p = 0$.

We leave the statement about right limits to the reader. □

Exercises

1. Show that the essentially open sets of $\boldsymbol{R}$ form a topology.
2. Here is another definition of essential limits: $\operatorname{ess\,lim}_{s \to t} f(s) = L$ if for each $\varepsilon > 0$ there exists an essentially open set G containing t such that $|f(s) - L| < \varepsilon$ if $s \in G$. Show the two definitions coincide.
3. Prove (ii) of Proposition 6.2 without using Lebesgue measurability by noticing that for each t there is $\delta > 0$ such that for a.e. s, $|f(s) - \varphi(s)| < \varepsilon$ and $|\varphi(t) - \varphi(s)| < \varepsilon$. Conclude by a covering argument that for a.e. s on the line, $|f(s) - \varphi(s)| < \varepsilon$, and hence that $f(s) = \varphi(s)$ a.e.. Conclude that f must be Lebesgue measurable.
4. Let $A \subset \boldsymbol{R}$ be any subset, and let A' be the set of all accumulation points in the essential topology. Show that A' is closed and perfect in the ordinary topology. (A is perfect if all elements of A are accumulation points of A.)
5. There is a related topology, called the **approximate topology** which can be defined as follows: a point t is an accumulation point of a set $A \subset \boldsymbol{R}$ if for all $\epsilon > 0$, $m^*\{A \cap (t, t + \varepsilon)\}/\varepsilon \to 1$ as $\varepsilon \downarrow 0$. Show that a Lebesgue measurable function on $\boldsymbol{R}$ is approximately continuous at a.e. t.
6. Show that the approximate topology is finer than the essential topology.
7. Let us weaken the definition of separability slightly for the approximate topology by replacing "essential limit" by "approximate limit" and "$=$" by "$\leq$" in Definition 6.4, and call the result "approximate separability."

Show that a measurable process is approximately separable. Does there exist a separability set for which there is equality?

8. Let (X_t) be a Hunt process on $\boldsymbol{E}$. We say a set $A \subset \boldsymbol{E}$ is open in the **essential fine topology** if for each $x \in \boldsymbol{E}$ there is a universally measurable set $B \subset A$ such that $P^x\{\pi_{B^c} = 0\} = 0$. Verify that this is a topology which is finer than the fine topology, and show that
   (a) the essential fine closure of a set $A \subset \boldsymbol{E}$ is finely closed;
   (b) an essential finely continuous function is finely continuous.

Notes on Chapter 6

§6.1. The essential topology is closely related to the approximate topology. The properties given here, such as essential separability and its application to stochastic processes, were developed in Chung and Walsh [1]. The proof of essential separability given here is new. Doob [4] showed that there is a class of topologies, including the approximate topology, which share similar properties. An overview from a different perspective, along with some applications including the essentially fine topology, can be found in Walsh [1].

§6.3. This section contains a minimal account of the Section Theorem and some of its consequences. These are at the core of Meyer's Théorie Générale des Processus. In the interest of simplicity, we only consider right-continuous and left-continuous processes, instead of optional and predictable processes. While this is sufficient for our needs, the interested reader should look up the complete account, which can be found in Dellacherie [2] and in Dellacherie and Meyer [2]. The ideas of optionality and predictability are well worth knowing, and they give the theory both greater generality and greater elegance.

(It is an ironic historical fact that much of the "Théorie Générale" was developed in order to study Markov processes and potential theory, but the theory itself became in essence a theory of martingales. For some time it seemed a lucky coincidence that many of the theorems in potential theory had martingale counterparts. It eventually became apparent that this was no coincidence, for in a certain sense, martingale theory itself is the proper setting and potential theory is a special case.)

Chapter 7

Markov Chains: a Fireside Chat

7.1. Basic Examples

A (continuous-time) Markov chain is a Markov process with a countable discrete state space. We will always assume it has stationary transition probabilities. This is an interesting class of processes in its own right, but it is also a useful test-bed for theorems: it is large enough to exhibit most of the interesting behavior found in general Markov processes, and the setting (discrete space) is simple enough to avoid inessential technical complications, and – above all – to show the effects and limits of any hypotheses one is tempted to make.

We won't study Markov chains in depth, but we'll use them to test the limits of Hunt processes. Hunt processes were invented to handle potential theories, not to describe all Markov processes, so there is no a priori reason to think that they should include all Markov chains. Indeed they do not, but the reasons for this have to do with some interesting behavior, involving what is called "boundary theory." This will drive the succeeding chapters, in which we will see that a small change in the basic hypotheses leads to the theory of Ray processes, which does indeed encompass Markov chains as well as the Hunt and Feller processes we have studied.

We will not prove anything in this section. We will just chat about some results. Derivations and detailed proofs can be found in Chung [2], and, for the boundary theory, in Chung [3].

We can assume that the state space is a subset of the positive integers, or, exceptionally, all integers. Let us first consider a Markov chain $\{X_t, t \in \boldsymbol{T}\}$ on the finite state space $\boldsymbol{E} = \{1, \dots, n\}$. The classical way to describe a Markov chain is by its **transition probabilities**. For $i, j \in \boldsymbol{E}$, let

$$p_{ij}(t) = P\{X_t = j \mid X_0 = i\} = P\{X_{s+t} = j \mid X_s = i\}.$$

These are the transition probabilities. The matrix of derivatives at zero is often called the ***Q*-matrix** (q_{ij}), where:

$$q_{ij} = p'_{ij}(0), \ i, j \in \boldsymbol{E}.$$

Let

$$\lambda_i = -q_{ii}, \text{ and } \bar{p}_{ij} = \frac{q_{ij}}{\lambda_i}.$$

In the general theory, the transition semigroup of (X_t) is $P_t(i,\cdot) \stackrel{\text{def}}{=} \sum_j p_{ij}(t)\delta_j(\cdot)$ and (q_{ij}) is its infinitesimal generator.

Note that the state space is discrete and compact, so that (P_t) is a Feller semigroup, and the Markov chain itself has a version which is strongly Markov and right-continuous. In particular, it is a Hunt process.

There is a simple description of the sample paths in this case. If $X_0 = i$, then X_t stays at i for an exponential time with parameter $\lambda_i > 0$, then jumps to a different state j, where it stays for a further exponential length of time with parameter λ_j, then jumps to a third state, and so on. The sample path is a step-function with integer values.

More precisely, define the jump times by induction: $\tau_0 = 0$ and for $n \geq 1$,

$$\tau_{n+1} = \inf\{t > \tau_n : X_t \neq X_{\tau_n}\}.$$

Then $\{X_{\tau_n},\ n = 1, 2, \dots\}$ is a discrete-time Markov chain with one-step transition matrix $(\bar{p}_{ij})$:

$$\bar{p}_{ij} = P\{X_{\tau_{n+1}} = j \mid X_{\tau_n} = i\} = \frac{q_{ij}}{\lambda_i}, \quad i \neq j.$$

If we condition on the value of X_{τ_n}, or equivalently, since (X_t) is a strong Markov process, if we condition on $\mathcal{F}_{\tau_n}$, then the inter-jump times are exponential random variables:

$$P\{\tau_{n+1} - \tau_n > t \mid X_{\tau_n} = i\} = e^{-\lambda_i t}.$$

(X_{τ_n}) is called the *jump chain* and the τ_n are the *jump times*.

Conversely, given a one-step transition probability matrix $(\bar{p}_{ij})$ and strictly positive reals λ_i, $i = 1, \dots n$, there is a Markov chain with these $(\bar{p}_{ij})$ and (λ_i).

The description of the process is less simple when the state space is infinite. Then (P_t) need not be a Feller semigroup, for it is possible that $p_{ij}(t)$ does not tend to zero as $j \to \infty$. We will look at three examples, each further from the framework of Hunt processes than the one before.

EXAMPLE 7.1. Let $E = \{0,\ 1,\ 2, \dots\}$ and consider the process which has $\bar{p}_{i,i+1} = 1$, $\bar{p}_{ij} = 0$ if $j \neq i + 1$, and let $\lambda_i = 2^i$. The corresponding chain is known as a pure-birth process. If $X_0 = 0$, then $X_{\tau_1} = 1$, $X_{\tau_2} = 2$, $X_{\tau_3} = 3, \dots, X_{\tau_n} = n, \dots$, so that the process increases in jumps of one and tends to infinity as n does. The inter-jump time $\tau_{n+1} - \tau_n$ is exponential with parameter 2^n, so that if $\tau_\infty = \lim_n \tau_n$, then $E\{\tau_\infty\} = \sum_n E\{\tau_{n+1} - \tau_n\} = \sum_n 2^{-n} < \infty$. Thus $\tau_\infty < \infty$ a.s. so that $X_t \to \infty$ as $t \uparrow \tau_\infty$; τ_∞ is the time of the first infinity, or the **first blow-up time** of (X_t).

A question arises: what happens after τ_∞? Nothing we have said so far determines this. Many things can happen. The simplest is that (X_t) is killed at τ_∞, so $X_t = \partial$ for $t \geq \tau_\infty$. The lifetime is then $\zeta = \tau_\infty$. This is called the **minimal chain.** The minimal chain is not a Hunt process, for its lifetime is finite, and it fails to have a left limit there. It is, however, a standard process: a standard process differs from a Hunt process in that it need not have left limits at its lifetime. This does not exhaust the possibilities: there are many other ways to continue the process after τ_∞.

This poses a general problem: find all chains which continue the minimal chain. (In other words, find all Markov chains which agree with the minimal chain until τ_∞.) This problem is dealt with in Chung [3]; we will just look at some examples here.

One such continuation, pointed out by Doob (see Theorem 6, II.19 of Chung [2]) is this: after the blow-up, set $X_{\tau_\infty} = 0$ and restart. With probability one, it will blow up again, in which case it will restart at 0 again. Each time it blows up, it restarts at zero. Figure 7.1 shows a typical sample path.

This process is not a Hunt process, and is not even a standard process, for it fails to have left limits at the blow-up times, which are finite. Thus quasi-left continuity fails.

The point at infinity is a boundary point of the state space, but it should really be considered as point of the state space itself, for there is a moderate Markov property (§2.4) attached to it: when the process approaches ∞, it jumps to the origin a.s. The point at infinity is what is called a *branching point* (see §8.2) of this process.

This chain approaches infinity continuously, but it is also possible for a chain to jump directly to infinity. Indeed, if we may anticipate the discussion of reversal in Chapter 10, one can reverse the sense of time in a Markov chain – i.e. run it backward – and the result is also a Markov chain. If we do this to the above chain, we see that the reversed chain in some sense jumps from the state space to infinity, and then re-enters the state space continuously, as in (b) of Fig. 7.2 below. This figure also shows several other possible behaviors of a stable Markov chain at infinity. It is possible for a chain to both approach and leave infinity continuously, as in (c), and it is also possible to have behavior as in (d) and (e), in which, to adapt the terminology of §3.4, the point at infinity is regular for itself, i.e. regular for the set $\{\infty\}$. In boundary theory, such a point is called **sticky** because the process finds it hard to leave it.

The following example shows how a sticky boundary point can occur, and why it is analogous to some familiar phenomena in diffusion processes.

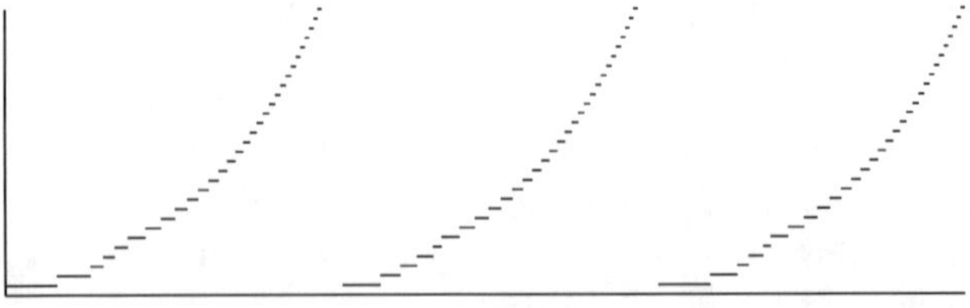

Fig. 7.1. Doob's continuation of the minimal chain.

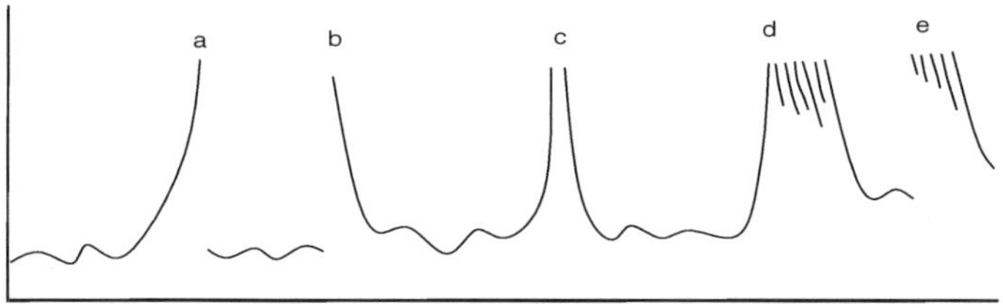

Fig. 7.2. Different types of boundary behavior for a stable Markov chain.

EXAMPLE 7.2. Let $\boldsymbol{E} = \{0, \pm 1, \pm 2, \ldots\}$ and let the jump-chain start at the origin and have transition probabilities $\bar{p}_{01} = \bar{p}_{0,-1} = 1/2$ and for $i \geq 1$, $\bar{p}_{i,i+1} = (n+1)/2n$, $\bar{p}_{i,i-1} = (n-1)/2n$, $\bar{p}_{ij} = 0$ if $j \neq i \pm 1$. Let it have the symmetric transition probabilities for negative i. Set $\lambda_0 = 1$ and $\lambda_i = i^4$ for $i \neq 0$.

One can show that this chain will eventually converge to either infinity or minus infinity at a finite time τ_∞. Once again, we can kill it at τ_∞ to get the minimal chain, but it is also possible to continue the chain to get a process with the property that τ_∞ is a.s. a limit point of times t for which $\liminf_{s \to t, s \in \boldsymbol{Q}} |X_s| = \infty$. (This is by no means obvious, but it can be shown. The reader will see immediately that there is a problem in defining the value of the process at such times! That is why we use convergence along rational values, where we can define the value of X_t unambiguously.)

Notice the absolute value signs above. If we look at the signed values of X_t, we see that τ_∞ is a limit points of times at which the process takes arbitrarily large negative values, and also of points at which it takes arbitrarily large positive values. In particular,

$$\limsup_{t \to \tau_\infty,\, t \in \boldsymbol{Q}} X_t = \infty, \qquad \liminf_{t \to \tau_\infty,\, t \in \boldsymbol{Q}} X_t = -\infty.$$

Both infinity and minus infinity are boundary points of the state space. The process can approach either one as $t \uparrow \tau_\infty$. Immediately afterward, it oscillates wildly between plus and minus infinity. This means that there is no satisfactory way to define X_{τ_∞} to make it a strong Markov process, much less a Hunt process. Both points are sticky; however, like Siamese twins, they are somehow joined together.

This behavior seems puzzling until we consider the process $Y_t \equiv 1/X_t$. As X_t approaches infinity, Y_t approaches zero from the right. As X_t approaches minus infinity, Y_t approaches zero from the left. If Y_t oscillates around zero, then X_t swings wildly between plus and minus infinity.

It is by no means obvious that such a process exists, but in fact it can be constructed from a standard Brownian motion, (B_t). Essentially, one gets (Y_t) by interpolating the Brownian motion: the t for which $Y_t = \pm 1/n$ correspond to times s when $B_s = \pm 1/n$; then we let $X_t = 1/Y_t$. So the wild behavior of

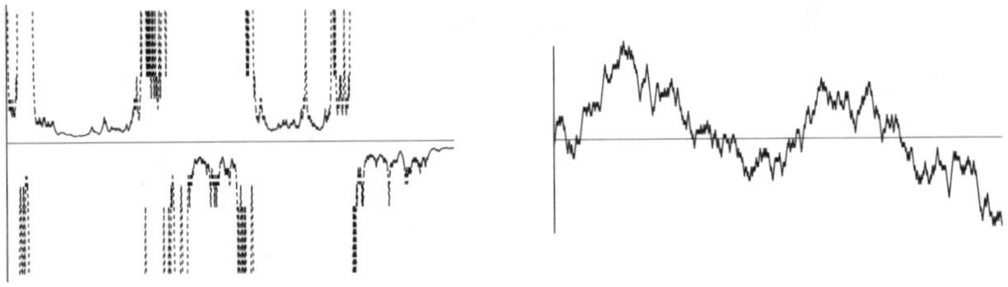

Fig. 7.3. The right-hand graph is of the modified diffusion Y_t; the left-hand graph is of the Markov chain $X_t = 1/Y_t$.

(X_t) at infinity actually comes from continuous behavior of (Y_t) at zero. The diagram above gives a very rough picture of the paths of (X_t) and (Y_t). We will not go into the details here. Suffice it to say that the above can be done rigorously with an important technique called time-changing. Example 7.3 below can be constructed with the same technique. However, it is outside the scope of this book. (See Sharpe, [1] VIII 65 or Blumenthal and Getoor [1] V (2.11).)

In Example 7.2 the infinities of X_t form an uncountable set, albeit one of Lebesgue measure zero. How do we define X_t at its infinities? Secondly, is it possible to salvage the strong Markov property? Since infinities may occur at stopping times, these two questions are related.

In this particular case, we see from the construction that one can define (Y_t) for all t by continuity, and the result is a strong Markov process on the space $\{0, 1/2, 1/3, \ldots\}$. The origin in this space corresponds to both plus and minus infinity for X_t. We conclude that we can define (X_t) to be a strong Markov process on the *one-point compactification* of its state space. But this is a very particular case, and we cannot expect this trick to work in any generality.

As a hint of things to come, note that the word "compactification" just entered the discussion.

EXAMPLE 7.3. Our final example may seem even more pathological than the foregoing. In each of the preceding examples, all states were "stable," i.e. q_{ii} was finite. It is possible, however, that some – or even all – of the states are "instantaneous" i.e. $q_{ii} = -\infty$. The sample path behavior changes radically. States are no longer holding points. If the process starts in an instantaneous state, then it will hit infinitely many other states in any time interval $(0, \delta)$.

Chung [2] gives an example of such a chain, due to Blackwell. We will outline another example here, one which is derived from a relatively nice, continuous strong Markov process, by simply transforming the state space. (A strong Markov process with continuous sample paths is a **diffusion**. A diffusion on some interval $(a, b) \subset \boldsymbol{R}$ is **regular** if for every $x \in (a, b)$ $P^x\{T_{(x,\infty)} = T_{(-\infty,x)} = 0\} = 1$.)

Feller and McKean [1] give an example of a regular diffusion on $[0, \infty)$ which spends all its time on the rationals. That is, they show the existence of a diffusion process $\{Z_t, t \in \boldsymbol{T}\}$ on the positive half line, for which the Lebesgue measure of the total time that (Z_t) spends on the set of irrational numbers is almost surely zero, and for any fixed $t > 0$, $P\{Z_t \in \boldsymbol{Q}\} = 1$. (As we remarked above, this diffusion can be constructed as a time-change of Brownian motion, but we will just accept its existence here.)

Now $\{Z_t,\ t \in \boldsymbol{Q}^+\}$ is also a Markov process, and its values are a.s. in $\boldsymbol{Q}$. Let (r_n) be an enumeration of $\boldsymbol{Q}^+$. Define a one-to-one function h from $\boldsymbol{Q}^+$ to $\boldsymbol{N}$ by $h(r_n) = n$. If we map (Z_t) by h, we will get another Markov process $\{h(Z_t),\ t \in \boldsymbol{Q}^+\}$ which has state space $\boldsymbol{N}$, To extend the parameter set to $\boldsymbol{T}$, we take the usual lower-semi-continuous version (Chung [2]) by setting $X_t = \liminf_{s \downarrow t,\ s \in \boldsymbol{Q}} h(Z_s)$. Then $\{X_t, t \in \boldsymbol{T}\}$ is a Markov chain, and except for a set of t of Lebesgue measure zero, $X_t = h(Z_t)$.

A moment's thought should convince the reader that all states of (X_t) are unstable. This follows from the regularity of (Z_t): if $X_t = n$, then $Z_t = r_n$, and as (Z_t) is regular, it hits infinitely many rationals in any interval $(t, t + \delta)$, which means that (X_t) hits infinitely many integers in the same time interval. This gives us a process which has the rather strange behavior: $\limsup_{s \to t} X_s = \infty$ for *all* t. It can not be strongly Markov: not only has it no nice continuity properties, but it is even undefined at an uncountable set of t. Question: is there any way to salvage the strong Markov property?

We know where (X_t) comes from, so we know it is a diffusion at heart. While it seems rather much to hope for, we ask: "Can we map (X_t) back to the original diffusion (Z_t) *without knowing* h? Surprisingly, the answer is "yes," as we will see in the next chapter. We will return to all of these examples in §9.1.

Notes on Chapter 7

In spite of their innocent appearance, Markov chains are a good source of nasty counterexamples. Many more can be found in Chung [2], including a different example, due to Blackwell, of a chain whose states are all instantaneous. The Feller-McKean diffusion and its equivalence to a Markov chain comes from Feller and McKean [1].

Chapter 8

Ray Processes

8.1. Ray Resolvents and Semigroups

Let $\boldsymbol{F}$ be a compact separable metric space with topological Borel field $\mathcal{E}$ and let $C(\boldsymbol{F})$ be the space of continuous functions on $\boldsymbol{F}$. The resolvent of a semigroup was defined in §2.1. It is the Laplace transform of the semigroup, and as such is often smoother and easier to handle. It can be be defined on its own, without an underlying semigroup. Let $\{U_p,\ p > 0\}$ be a **Markov resolvent** on $\boldsymbol{F}$, which means that for each $x \in \boldsymbol{F}$ and $p > 0$, $pU_p(x, \cdot)$ is a probability measure on the Borel sets of $\boldsymbol{F}$, and U_p satisfies the resolvent equation (see §3.2 Prop. 7):

$$U_p - U_q = (q - p)U_p U_q, \quad p, q > 0. \tag{8.1}$$

We will need two formulas on resolvents and their derivatives (see Getoor [1]): if f is a bounded Borel function on $\boldsymbol{F}$,

$$\frac{d^n}{dp^n} U_p f = (-1)^n n!(U_p)^{n+1} f, \tag{8.2}$$

$$\frac{d^n}{dp^n} pU_p f = (-1)^{n+1} n!(U_p)^n (I - pU_p) f. \tag{8.3}$$

Since resolvents are inherently smoother than semigroups (see Exercise 1) smoothness restrictions on them can be less onerous than the same restrictions on the semigroup. In particular, if we require that the resolvent, rather than the semigroup, have the Feller property, we get a surprisingly far-reaching extension of the Feller process, called a Ray process. A Feller semigroup takes $C(\boldsymbol{F}) \longrightarrow C(\boldsymbol{F})$. A Ray resolvent does the same, but for the resolvent, not the semigroup.

Let $p \geq 0$. A function $f \in \mathcal{E}_+^{\sim}$ is **p-supermedian** if for all $q > 0$, $qU_{p+q} f \leq f$. This is closely related to superaveraging, defined in §2.1: a superaveraging function is supermedian, but the converse need not be true.

Let $\mathcal{S}$ be the class of continuous 1-supermedian functions, i.e. the class of all positive continuous functions f on $\boldsymbol{F}$ such that $pU_{p+1} f \leq f$ for all $p > 0$. Let $\mathcal{S} - \mathcal{S} = \{f - g : f, g \in \mathcal{S}\}$ be the set of all differences of functions in $\mathcal{S}$. Then a Ray resolvent is a Feller resolvent with an additional separation property.

Definition 8.1. *A resolvent* (U_p) *on* $\boldsymbol{F}$ *is a* **Ray resolvent** *if*

(i) $\forall\, p > 0,\ U_p : C(\boldsymbol{F}) \mapsto C(\boldsymbol{F})$;
(ii) $\mathcal{S}$ *separates points of* $\boldsymbol{F}$.

The fundamental theorem on Ray resolvents is due to D. Ray. The original theorem only involved one semi-group, but there are actually two which are relevant, one left-continuous and one right-continuous. They are closely related and can easily be derived from each other, but one governs a right-continuous strong Markov process and the other, a left-continuous moderate Markov process, so it seems worthwhile to include them from the outset. Here then is (a trivial extension of) Ray's Theorem. We follow a proof of P.-A. Meyer.

Theorem 8.2. *Let* (U_p) *be a Ray resolvent. There exist two Borel measurable Markov semigroups* (P_t^0) *and* (P_t^1) *on* $\boldsymbol{F}$, *each having* (U_p) *as a resolvent, such that for all* $f \in C(\boldsymbol{F})$ *and* $x \in \boldsymbol{F}$

(i) $P_0^0 f(x) = f(x)$ *and* $t \mapsto P_t^0 f(x)$ *is left-continuous on* $(0, \infty)$;
(ii) $t \mapsto P_t^1 f(x)$ *is right-continuous on* $[0, \infty)$;
(iii) $P_t^1 = P_0^1 P_t^0 = P_t^0 P_0^1$.

Another way of stating (i) and (ii) is that for each $x \in \boldsymbol{F}$, $t \mapsto P_t^i(x, \cdot)$ is left-continuous if $i = 0$ and right-continuous if $i = 1$ in the vague topology of measures on $\boldsymbol{F}$. Since both semigroups have the same resolvent, which is to say the same Laplace transform, the P_t^i must be equal for a.e. t. Since one is left-continuous and one is right-continuous, $P_t^0 f(x) = P_t^1 f(x)$ except at (at most) countably many discontinuities.

The resolvents have the Feller property, but the semigroups (P_t^i) may not; that is, they may not take continuous functions into continuous functions. The most we can say is that they are Borel measurable.

Proof. The class of 1-supermedian functions contains constants and is closed under sums and finite minima, so $\mathcal{S} - \mathcal{S}$ is a lattice containing constants and separating points, hence it is dense in $C(\boldsymbol{F})$ by the Stone-Weierstrass theorem. Fix a point $x \in \boldsymbol{F}$ and a function $f \in \mathcal{S}$, and let

$$g(p) = (I - pU_{p+1}) f(x). \tag{8.4}$$

Since f is in $\mathcal{S}$, $g(p) \geq 0$ for all $p \geq 0$, and from (8.2) and (8.3),

$$(-1)^n \frac{d^n}{dp^n} g(p) \geq 0.$$

Thus g is completely monotone, and hence by a theorem of Bernstein, is the Laplace transform of a measure ν (which depends on x and f, of course):

$$g(p) = \int_0^\infty e^{-pt}\, \nu(dt),$$

where $\nu([0,\infty)) = g(0) = f(x)$ and $\nu(\{0\}) = \lim_{p\to\infty} g(p) = f(x) - \lim_{p\to\infty} pU_{p+1} f(x)$. We now define

$$P_t^0 f(x) = e^t\,(f(x) - \nu([0,t)))\,, \tag{8.5}$$

$$P_t^1 f(x) = e^t\,(f(x) - \nu([0,t]))\,. \tag{8.6}$$

Note that $t \mapsto P_t^0 f(x)$ is left-continuous and $t \mapsto P_t^1 f(x)$ is right-continuous. It is immediate that the $P_t^i f$ are positive, and that $e^{-t} P_t^i f(x) \le f(x)$. The remainder of the proof consists in verifying that the two are indeed Markov semigroups and correspond to the resolvent U_p.

Does $U_p f(x)$ equal $\int_0^\infty e^{-pt}\, P_t^i f(x)\,dt$? Since the $P_t^i f(x)$ differ at at most countably many t, it is enough to verify this for $i = 0$. Then

$$\begin{aligned}
\int_0^\infty e^{-pt} P_t^0 f(x)\,dt &= \int_0^\infty e^{-pt}\, e^t\,[f(x) - \nu([0,t))]\,dt \\
&= \frac{1}{p-1} f(x) - \int_0^\infty dt\, e^{-(p-1)t} \int_{[0,t)} \nu(ds) \\
&= \frac{1}{p-1}\,(f(x) - g(p-1)) \\
&= \frac{1}{p-1} f(x) - \frac{1}{p-1}\big(f(x) - (p-1)U_p f(x)\big) \\
&= U_p f(x).
\end{aligned}$$

The measure ν depends linearly on f so that $f \mapsto P_t^i f(x)$ is linear on $\mathcal{S}$, hence on $\mathcal{S} - \mathcal{S}$. If $f \in \mathcal{S} - \mathcal{S}$, and $f \ge 0$, then (8.2) tells us that $(-1)^n \frac{d^n}{dp^n} U_p f(x)$ is positive for $n \ge 1$. Now $U_p f$ is positive (since f is) so $U_p f(x)$ is completely monotone. Thus there is a measure μ on $[0,\infty)$ such that

$$U_p f(x) = \int_0^\infty e^{-pt}\, \mu(dt);$$

but we have just seen this is

$$= \int_0^\infty e^{-pt}\, P_t^0 f(x)\,dt.$$

It follows that $P_t^0 f(x)$ is positive for a.e. t; but it is left-continuous in t, so it is positive for all $t > 0$, and for $t = 0$ as well, since $P_t^0 f(x) = f(x) \ge 0$.

Moreover, for each $p > 0$,

$$\frac{1}{p} = U_p 1(x) = \int_0^\infty e^{-pt} \, P_t^0 1(x) \, dt,$$

so that $P_t^0 1(x) = 1$ for a.e. t, hence for all t. In fact, we can say more. If $f \in \mathcal{S}$, $\|f\| - f$ is positive and in $\mathcal{S} - \mathcal{S}$, so that $P_t^0 f(x) \le \|f\|$. Thus $f \mapsto P_t^0 f(x)$ is a positive bounded linear functional on $\mathcal{S} - \mathcal{S}$, and hence on $C(\boldsymbol{F})$. We denote the corresponding measure by $P_t^0(x, \cdot)$. As $P_t^0 1 = 1$, it is a probability measure for each $t \ge 0$. The same holds for $i = 1$; we need only replace left continuity by right continuity above.

Next, we claim that $x \mapsto P_t^i f(x)$ is Borel measurable for $f \in C(\boldsymbol{F})$. Let

$$\mathcal{G} = \left\{ g \in C([0, \infty]) : x \mapsto \int_0^\infty e^{-t} g(t) P_t^i f(x) \, dt \text{ is Borel measurable} \right\}.$$

Constant functions and functions of the form e^{-pt} are in $\mathcal{G}$, since then the integral equals $cU_1 f(x)$ and $U_{p+1} f(x)$ respectively, which are continuous. Sums and products of such functions are also in $\mathcal{G}$, so that $\mathcal{G}$ contains an algebra which is dense (Stone-Weierstrass) in $C([0, \infty])$. But $\mathcal{G}$ is closed under uniform convergence, so $\mathcal{G} = C([0, \infty])$. Then $P_{t_0}^i f(x) = \lim \int_0^\infty e^{-t} g_n(t) P_t^i f(x) \, dt$, where g_n is positive, continuous, satisfies $\int_0^\infty g_n(t) \, dt = e^{t_0}$ and has support in $(t_0 - 1/n, t_0)$ if $i = 0$ or in $(t_0, t_0 + 1/n)$ if $i = 1$. (If $i = 0$, the case $t = 0$ is, as always, special, but then $P_0^0 f(x) = f(x)$.)

Next, we verify that the (P_t^i) are semigroups. It is enough to show that

$$P_s^i P_t^i f(x) = P_{s+t}^i f(x) \tag{8.7}$$

for $s \ge 0, t \ge 0$, and $f \in \mathcal{S}$. If $i = 0$ and either s or t is zero, this is immediate, since $P_0^0 = I$. In general, for each x, $t \mapsto P_{s+t}^i f(x)$ and $t \mapsto P_t^i f(x)$ are left-continuous if $i = 0$ and right-continuous if $i = 1$, hence the same is true of $t \mapsto P_s^i P_t^i f(x)$. Thus it is enough to verify the Laplace transform of (8.7). The Laplace transform of the left hand side is

$$\begin{aligned} \int_0^\infty e^{-pt} \, P_s^i P_t^i f(x) \, dt &= \int P_s^i(x, dy) \int_0^\infty e^{-pt} \, P_t^i f(y) \, dt \\ &= P_s^i U_p f(x). \end{aligned} \tag{8.8}$$

The transform of the right hand side of (8.7) is

$$\int_0^\infty e^{-pt} \, P_{s+t}^i f(x) \, dt. \tag{8.9}$$

Again, both (8.8) and (8.9) are either left or right-continuous in s, depending on i. Thus it is enough to consider their Laplace transforms in s.

The transform of (8.8) is

$$\int_0^\infty e^{-qs} P_s^i U_p f(x)\, ds = U_q U_p f(x),$$

while the transform of (8.9) is

$$\int_0^\infty ds\, e^{-qs} \int_0^\infty e^{-pt} P_{s+t}^i f(x)\, dt.$$

Following the calculations in §3.2 Proposition 7, this is

$$\frac{1}{q-p}\left[U_p f(x) - U_q f(x)\right] = U_q U_p f(x)$$

by the resolvent equation. Thus (P_t^i) is a semigroup, $i = 1,\ 2$.

Finally, (iii) follows easily: $P_t^1 f(x) = P_t^0 f(x)$ for a.e. t, so that if $f \in \mathcal{S}$, $P_t^1 f(x) = \lim_{s\downarrow t} P_s^1 f(x) = \lim_{s\downarrow t} P_s^0 f(x) = \lim_{s\downarrow t} P_0^0 P_s^0 f(x) = P_0^0 P_t^1 f(x)$, and the other equality is similar. This completes the proof. □

8.2. Branching Points

Let us make a slight change in notation, and denote the right-continuous semigroup by (P_t) instead of (P_t^1), and the left-continuous semigroup by $(\bar{P}_t)$ instead of (P_t^0).

Now $\bar{P}_0$ is the identity ($\bar{P}_0(x, \{x\}) = 1$) but P_0 may not be. If we had to point to a single fact and say "Here is the key to Ray processes," that would be it. There are many contenders for that honor, but most of them are ultimately consequences of this. It leads to branching points.

Definition 8.3. *A point $x \in \boldsymbol{F}$ is a* **branching point** *if $P_0(x, \{x\}) < 1$. If x is a branching point, then $P_0(x, \cdot)$ is the* **branching measure** *at x. The set of branching points is denoted by B, and the set of non-branching points is $D = \boldsymbol{F} - B$.*

This is a new idea: branching points do not occur in Feller and Hunt processes. It will be important below. The reason for this will emerge in good time, but for the moment, let us just say that we will have no occasion to regret the loss of the quasi-left continuity which was so important in understanding the sample paths of Feller and Hunt processes.

Let us define some particular classes of branching points for later use.

Definition 8.4. *A branching point x is* **degenerate** *if $P_0(x, \cdot)$ is a point mass. Denote the set of degenerate branching points by B_d, and for $x \in B_d$, let $k(x)$ be the unique $y \in \boldsymbol{F}$ such that $P_0(x, y) = 1$.*

Let ∂ be a distinguished point in $\boldsymbol{F}$.

Definition 8.5. *A branching point x is a* **fork** *if there is a point $y \in \boldsymbol{F}$ and a coefficient $p(x) \in (0, 1)$ such that $P_0(x, \cdot) = p(x)\delta_\partial(\cdot) + (1 - p(x))\delta_y(\cdot)$. The set of forks is denoted by Π.*

A degenerate branching point branches directly to a single point in D. A fork branches to two, one of which is distinguished in advance. (One might think a fork should be a branching point whose branching measure is supported by a doubleton. However, we have a specific use in mind where ∂ will indeed be distinguished: it is the cemetery.)

Then B, B_d and Π are Borel sets, and the associated mappings are also Borel. Consider B, for instance.

The set $C(\boldsymbol{F})$ is separable, hence so is $\mathcal{S}$. Thus there is a countable dense subset $\mathcal{S}_{\boldsymbol{F}} \subset \mathcal{S}$, and, as $\mathcal{S} - \mathcal{S}$ is dense in $C(\boldsymbol{F})$, so is $\mathcal{S}_{\boldsymbol{F}} - \mathcal{S}_{\boldsymbol{F}}$. Note that $f \in \mathcal{S} \Longrightarrow P_0 f \le f$, so that x is a branching point if and only if $P_0 f(x) < f(x)$ for some $f \in \mathcal{S}$, or, equivalently, for some $f \in \mathcal{S}_{\boldsymbol{F}}$. In fact, if (f_n) is an enumeration of $\mathcal{S}_{\boldsymbol{F}}$, then $f_0 \stackrel{\text{def}}{=} \sum_n 2^{-n} \frac{f}{1+\|f\|}$ is a bounded continuous supermedian function, and

$$B = \{x : P_0 f_0(x) < f_0(x)\}.$$

It follows that B is a Borel set. Moreover, as $P_0 f_0 = f_0$ in D, and $P_0 = P_0 P_0$, $0 = P_0(f_0 - P_0 f_0)$, hence $P_0(x, B) = 0$ for all x, including $x \in B$. Since $P_t = P_t P_0$, it follows that

$$P_t(x, B) = 0 \text{ for all } x \in \boldsymbol{F} \text{ and } t \ge 0. \tag{8.10}$$

Similarly, one can see that B_d is Borel by considering $P_0(f^2)(x) - (P_0 f(x))^2$, which is zero for all $f \in C(\boldsymbol{F})$ if and only if $P_0(x, \cdot)$ is a point mass. Thus the set of degenerate branching points is the intersection of B with the Borel set on which $P_0(f^2)(x) - (P_0 f(x))^2 > 0$ for some f in a countable dense subset of $C(\boldsymbol{F})$. It follows that k is Borel, since for any Borel $A \subset \boldsymbol{F}$, $B_d \cap \{x : k(x) \in A\} = B_d \cap \{x : P_0(x, A) = 1\}$, which is Borel since $P_0(x, \cdot)$ is. Π is similar. Note that

$$x \in B_d \Longrightarrow P_s(x, \cdot) = P_s(k(x), \cdot),\ s \ge 0. \tag{8.11}$$

8.3. The Ray Processes

There are two semigroups, hence there are two Markov processes, one for each. The following theorem tells us that one is a right-continuous strong Markov process (X_t) corresponding to the semigroup (P_t); the other is a left continuous moderate Markov process $(\bar{X}_t)$ corresponding to $(\bar{P}_t)$. They are closely related: $X_t = \lim_{s \downarrow\downarrow t} \bar{X}_s$ and $\bar{X}_t = \lim_{s \uparrow\uparrow t} X_s$. This question could have come up with Feller and Hunt processes in Chapters 2 and 3, but it was not important then. It is the new phenomenon, branching points, which makes the left-continuous process significant.

Theorem 8.6. *For each probability measure μ on $\boldsymbol{F}$, there exist two processes, $\{X_t^0,\ t \in \boldsymbol{T}\}$ and $\{X_t^1,\ t \in \boldsymbol{T}\}$, which satisfy*

(i) *(X_t^0) has left-continuous sample paths which have right limits, it has initial distribution μ, and it is moderately Markov with transition semigroup (P_t^0).*
(ii) *(X_t^1) has right-continuous sample paths which have left limits, it has initial distribution μP_0, and it is strongly Markov with transition semigroup (P_t^1).*
(iii) *With probability one, $X_t^1 = \lim_{s\downarrow\downarrow t} X_t^0$, $\forall t \in \boldsymbol{T}$.*

Proof. Let us adjoin a point $\{0-\}$ to the parameter set $\boldsymbol{R}^+$ to give us the parameter set $\boldsymbol{R}^{\pm} = \{0-\} \cup [0, \infty)$. (The reason for this is that we are trying to define two processes on the same space, and one is the left limit of the other, so we need the "left limit" at zero.) Let μ be a probability measure on F. Then, (see §1.2) there exists a Markov process $\{Y_t,\ t \in \boldsymbol{R}^{\pm}\}$ which admits (P_t) as transition semigroup and has absolute distribution μP_t. In particular, the distribution of Y_0 is μP_0, which does not equal μ if μ charges the set B of branching points, and the transition probability from time $t = 0-$ to $t = 0$ is $P_0(\cdot, \cdot)$. Now Y is right-continuous in probability at any $t \geq 0$ for if $f \in C(\boldsymbol{F})$,

$$E\{(f(Y_{t+s}) - f(Y_t))^2 \mid Y_t\} = P_s f^2(Y_t) + f(Y_t)^2 - 2f(Y_t)P_s f(Y_t).$$

As s decreases to zero this goes to zero on the set $\{Y_t \in D\}$. But this set has probability $\mu P_t(D) = 1$.

If $f \in \mathcal{S}$, f is 1-supermedian so $\{e^{-t} f(Y_t),\ t \geq 0\}$ is a supermartingale. Thus it has left and right limits everywhere along the rationals. This is true simultaneously for all f in a countable dense subset $\mathcal{S}_{\boldsymbol{F}}$ of $\mathcal{S}$. Since $\mathcal{S}_{\boldsymbol{F}} - \mathcal{S}_{\boldsymbol{F}}$ is dense in $C(\boldsymbol{F})$, the restriction of Y_t to the rationals has left and right limits everywhere in the topology of $\boldsymbol{F}$ a.s..

Define processes

$$X_t = \lim_{\substack{s\downarrow\downarrow t \\ s\in\boldsymbol{Q}}} Y_s \qquad \text{if } t \geq 0,$$

$$\bar{X}_t = \begin{cases} \lim\limits_{\substack{s\uparrow\uparrow t \\ s\in\boldsymbol{Q}}} Y_s & \text{if } t > 0 \\ Y_{0-} & \text{if } t = 0. \end{cases}$$

Since Y was right-continuous in probability, X_t will itself be a right continuous version of Y. $\bar{X}_t$ will be left-continuous, and it will equal X_t at each point of continuity. Therefore $P\{\bar{X}_t = Y_t\} = 1$ if (X_t) is a.s. continuous at t, which will be true at all but at most a countable set of t. Thus there will be a countable dense subset $S \in \boldsymbol{R}^+$ such that $P\{X_t = \bar{X}_t = Y_t,\ \forall t \in S\} = 1$. Notice that for all $t \geq 0$, $X_t = \lim_{s\downarrow\downarrow t,\ s\in S} \bar{X}_s$, which implies (iii).

Let us denote the natural filtrations by $\bar{\mathcal{F}}_t = \sigma\{\bar{X}_s,\ s \le t\}$ and $\mathcal{F}_t = \bar{\mathcal{F}}_{t+}$, completed with respect to the underlying probability measure on the space. Let T be a finite stopping time relative to the filtration $(\mathcal{F}_t)$ (resp. a predictable time relative to $(\bar{\mathcal{F}}_t)$) and let $\mathcal{G}$ be $\mathcal{F}_{T+}$ (resp. $\bar{\mathcal{F}}_{T-}$), let $Z_t = X_t$ (resp $Z_t = \bar{X}_t$), and let $Q_t = P_t$ (resp. $Q_t = \bar{P}_t$). Suppose we can show that for each $p > 0$ and $f \in C(\boldsymbol{F})$ that

$$E\left\{\int_0^\infty e^{-ps} f(Z_{T+s})\,ds \mid \mathcal{G}\right\} = U_p f(Z_T) \quad \text{a.s.} \tag{8.12}$$

Then we claim that for each $t > 0$,

$$E\{f(Z_{T+t}) \mid \mathcal{G}\} = Q_t f(Z_T) \quad \text{a.s.} \tag{8.13}$$

To see this we repeat an argument we used before. Let $\mathcal{H}$ be the class of h in $C([0, \infty])$ for which

$$E\left\{\int_0^\infty e^{-t} h(t) f(Z_{T+t})\,dt \mid \mathcal{G}\right\} = \int_0^\infty e^{-t} h(t) Q_t f(Z_T)\,dt \quad \text{a.s.} \tag{8.14}$$

Then $\mathcal{H}$ is linear, closed under uniform convergence, and contains the constants, as well as functions of the form e^{-pt} by (8.12). This forms an algebra, so by the Stone-Weierstrass Theorem, $\mathcal{H}$ is dense in, and therefore contains, $C[0, \infty]$. Now let h run through a sequence of positive functions (h_n) with $\int h_n(s)\,ds = e^{t_0}$ and $\text{supp}(h_n) \subset (t_0, t_0 + 1/n)$ (resp. $\text{supp}(h_n) \subset (t_0 - 1/n, t_0)$) and remark that we can go to the limit on both sides of (8.14) to get (8.13).

It remains to verify (8.12) for stopping times (resp. predictable times). By the Markov property, if $s \in S$,

$$E\left\{\int_0^\infty e^{-pt} f(Z_{s+t})\,dt \mid \mathcal{F}_s\right\} = U_p f(X_s), \quad \text{a.s.}$$

Note that $Z_s = X_s = \bar{X}_s = Y_s$ for $s \in S$, so this equation holds for X, $\bar{X}$, and Y as well as for Z. By the usual extension, if T is a stopping time with values in S (so that it takes on only countably many values) we have

$$E\left\{\int_0^\infty e^{-pt} f(Z_{T+t})\,dt \mid \mathcal{F}_T\right\} = U_p f(Z_T), \quad \text{a.s.} \tag{8.15}$$

If T is an arbitrary finite stopping time, we can approximate it with S-valued stopping times $T_n > T$ such that $T_n \ge T_{n+1} \downarrow T$. Since $T_n + t \downarrow T + t$ for each $t \ge 0$, $Z_{T_n+t} \to X_{T+t}$ and the right hand side of (8.15) converges to $U_p f(X_T)$ while the left hand side converges boundedly to the conditional expectation given $\cap_n \mathcal{F}_{T_n} = \mathcal{F}_{T+}$ as desired, giving (8.12). Thus X_t is a strong Markov process.

On the other hand, suppose T is predictable. Let T_n be a sequence of stopping times which announces T: $T_n < T_{n+1} < T$ a.s. on $\{T > 0\}$ and $T_n \uparrow T$ a.s.

Note that *we can even assume that the T_n take values in S.* (See Exercise 12 in §1.3.) Thus (8.15) holds for the T_n. Now let $n \to \infty$ and notice that as $T_n \uparrow\uparrow T$, $Z_{T_n+t} \to \bar{X}_{T+t}$ – it is the left limit which comes into play here – and, as $U_p f$ is continuous the right hand side of (8.15) converges to $U_p f(\bar{X}_T)$, while the conditional expectation on the left hand side converges boundedly to $E\{\int_0^\infty e^{-pt} f(\bar{X}_{T+t}) \mid \bigvee_n \mathcal{F}_{T_n}\}$ which is what we wanted to prove, since $\bigvee_n \mathcal{F}_{T_n} = \mathcal{F}_{T-}$. This shows that $(\bar{X}_t)$ is moderately Markov. □

8.4. Jumps and Branching Points

We gave the left and right-continuous processes equal billing in the previous two sections, but now we will concentrate on the right-continuous process (X_t). The left-continuous moderate Markov process will just be called (X_{t-}).

The division of optional times into predictable and totally inaccessible times will be useful. A stopping time T is **totally inaccessible** if for any predictable time S, $P\{0 < S = T < \infty\} = 0$. (See §3.1, Exercise 6.) For instance the statement that a Hunt process is quasi-left-continuous simply means that its discontinuities can only occur at totally inaccessible times.

Are Ray processes quasi-left-continuous? In general, no. If there are branching points, the process can have jumps, and in fact it *must* jump when it converges to a branching point. This means that quasi-left continuity fails. This is less of a loss than it might seem. Branching points help explain how the process jumps. If $X_{T-} \in B$ at some predictable time T, then, as we said, X will be discontinuous, but in fact we know the exact distribution of the jump: it is given by the branching measure associated to that branching point. So if we know the left limit X_{T-}, we know the distribution of the right limit X_T as well. One might say that with a Ray process, the left hand *does* know what the right hand is doing.

Let $(\mathcal{F}_t)$ be the filtration of (X_t), augmented as usual. It is complete and right-continuous (§2.3, Theorem 4 and its Corollary). We might point out that the key to this right-continuity is well-hidden in the proof of Theorem 8.6. It is that little "+" in the statement "Let $\mathcal{G}$ be $\mathcal{F}_{T+}$.")

Proposition 8.7. *Let T be a predictable time. Then on $\{0 < T < \infty\}$ the conditional distribution of X_T given $\mathcal{F}_{T-}$ is $P_0(X_{T-}, \cdot)$.*

Proof. The left-continuous process X_{t-} is moderately Markov, so that for each $f \in C(\boldsymbol{F})$, $t > 0$ and bounded random variable $\xi \in \mathcal{F}_{T-}$, $E\{\xi f(X_{T+t-})\} = E\{\xi \bar{P}_t f(X_{T-})\}$. Now as $t \downarrow 0$, $X_{T+t-} \to X_T$ while $\bar{P}_t f \to P_0 f$. The convergence is bounded, so we can take the limit and conclude that $E\{f(X_T) \mid \mathcal{F}_{T-}\} = P_0 f(X_{T-})$, which does it. □

Thus the process is continuous at a predictable time if its left-hand limit is in D, and it is discontinuous if its left-hand limit is in B, in which case the distribution of its landing place is given by the branching measure. Let us go deeper. Recall that we defined polar and semi-polar sets in §3.5.

Proposition 8.8.

(i) *The set B of branching points is polar for (X_t) and semi-polar for (X_{t-}).*

(ii) *Let $Z = \{t : X_{t-} \in B\}$. Then $Z \subset \{t : X_{t-} \neq X_t\}$, and furthermore, Z is contained in a countable union of graphs of predictable times.*

Proof. (i) (X_t) is right-continuous so by Theorem 6.9, if it hits B with positive probability, there is an optional T such that $P\{X_T \in B\} > 0$. If $f \in C(\boldsymbol{F})$, the strong Markov property at T gives

$$f(X_T) = E\{f(X_T) \mid \mathcal{F}_{T+}\} = P_0 f(X_T) \text{ a.s.}$$

But this can't hold for all f since $P_0(x, \{x\}) = 0$ on B. The fact that B is semi-polar for (X_{t-}) follows immediately from the fact that the process jumps whenever $X_{t-} \in B$, and there are only countably many jumps, so X_{t-} can be in B for only countably many t.

(ii) The fact that $X_{t-} \neq X_t$ whenever $X_{t-} \in B$ follows from (i): with probability one, X_t is never in B. Since Z is contained in the set of jumps, it is a.s. countable, and (X_{t-}) is left-continuous in t, so the second conclusion follows from Theorem 6.13. □

We will complete the general description of the jumps of a Ray process in Theorem 8.13, but we will first need to gather some facts about martingales on Ray filtrations.

8.5. Martingales on the Ray Space

Let $(\mathcal{F}_t)$ be the augmented filtration generated by (X_t). There is an intimate connection between the discontinuities of the process, filtration, and of the martingales on that filtration. Indeed, when the process jumps, the filtration is likely to jump too: before the jump, the filtration may not know the landing point; after, it does. Further, discontinuities in the filtration are signaled by discontinuities in martingales, while continuity properties in filtrations imply corresponding properties of martingales. For instance, right-continuity of the filtration implies that martingales have right-continuous versions. And finally, to complete the circle, the continuity of martingales is intimately connected with the continuity of the Ray process. Theorem 8.10 highlights this connection.

The following lemma is a direct consequence of Propositions 8.7 and 8.8 (ii).

Lemma 8.9. *Let μ be an initial measure. Then for P^μ-a.e. ω, $X_{t-}(\omega) \in B_d \Longrightarrow X_t(\omega) = k(X_{t-}(\omega))$ for all $t > 0$, where $k(x)$ is defined in Definition 8.4.*

Theorem 8.10. *Let $\{M_t, \mathcal{F}_t,\ t \in \boldsymbol{T}\}$ be a right continuous martingale on $(\Omega, \mathcal{F}, P^\mu)$. For P^μ-a.e. ω, $t \mapsto M_t(\omega)$ is continuous wherever $t \mapsto X_t(\omega)$ is, and it is also continuous at all t for which $X_{t-}(\omega) \in B_d$.*

Thus if (X_t) is continuous at t, so are all martingales, and they are also continuous when (X_{t-}) encounters a degenerate branching point, even though the process itself jumps at such a time.

Let us define the rcll Ray process on the canonical space Ω of rcll functions $\omega : \boldsymbol{R}^+ \mapsto \boldsymbol{F}$ by $X_t(\omega) = \omega(t),\ \omega \in \Omega$. The **shift operator** θ_t is defined by $\theta_t\omega(s) = \omega(s+t)$, $s, t \geq 0$.

To prove Theorem 8.10, notice that it is enough to prove it on $[0, t_0]$ for any $t_0 > 0$. Since $M_t = E\{M_{t_0} \mid \mathcal{F}_t\}$, it is enough to prove it for martingales of the form $M_t = E\{Z \mid \mathcal{F}_t\}$ for $Z \in L^1(\Omega, \mathcal{F}, P^\mu)$, and it is even enough to prove it for a dense set of Z, since if $E\{|Z_n - Z|\} \to 0$, $\sup_{t \leq t_0} |E\{Z_n - Z \mid \mathcal{F}_t\}| \to 0$ by Doob's submartingale maximal inequality. Here is one natural choice of Z. Let $p_1, p_2, \dots, p_n > 0$, $f_1, \dots f_n \in C(\boldsymbol{F})$, and put

$$Z = \prod_{i=1}^{n} \int_0^\infty e^{-p_i t} f_i(X_t)\, dt. \tag{8.16}$$

These random variables are bounded and they are also total in $L^1(\Omega, \mathcal{F}, P^\mu)$ for any μ. Indeed, we know that random variables of the form $\prod_{i=1}^n f_i(X_{t_i})$ are total, for they generate $\mathcal{F}$, and they can be approximated in turn by linear combinations of such Z. To see this, note that finite linear combinations of the functions $e^{-pt},\ p \geq 0$ form an algebra which separates points on the line and contains constants (case $p = 0$), so by the Stone-Weierstrass Theorem, it is dense in $C[0, \infty]$. Let $g_n \in C^+[0, \infty]$ have support in $[t, t + 1/n]$ and satisfy $\int e^{-t} g_n(t)\, dt = 1$. (X_t) is right-continuous, so $\int e^{-s} g_n(s) f_i(X_s)\, ds \to f_i(X_t)$. We can approximate g_n uniformly by h_n of the form $\sum_{j=1}^m a_j e^{-p_j s}$, so the above integral can be approximated arbitrarily well by $\int e^{-s} h_n(s) f_i(X_s)\, ds = \sum_j a_j \int e^{-(p_j+1)s} f_i(X_s)\, ds$, which is a linear combination of the Z. Thus they are total.

We can compute the conditional expectations explicitly. Denote the random variable in (8.16) by $Z = Z(p_1, \dots, p_n; f_1, \dots, f_n)$. By splitting $\int_0^\infty = \int_0^t + \int_t^\infty$, we can write $E\{Z(p_1, \dots, p_n; f_1, \dots, f_n) \mid \mathcal{F}_t\}$ as a sum of terms of the form

$$\left(\prod_{i=1}^{p} \int_0^t e^{-r_i s} h_i(s)\, ds\right)\left(E\left\{\prod_{j=1}^{n-p} \int_t^\infty e^{-q_j s} g_j(X_s)\, ds \mid \mathcal{F}_t\right\}\right). \tag{8.17}$$

Here, $h_i(s) = f_k(s)$ for some k, $g_j = f_i$ for some i, and the r_i and q_j are chosen from $p_1, \dots, p_n$. The first factor is clearly continuous in t and if we apply the Markov property, we see the second can be written

$$E^{\mu}\{e^{-(q_1+\ldots+q_m)}Z(q_1,\ldots,q_m;g_1,\ldots,g_m)\circ\theta_t \mid \mathcal{F}_t\}$$
$$= e^{-(q_1+\ldots+q_m)}E^{X_t}\{Z(q_1,\ldots,q_m;g_1,\ldots,g_m)\}. \tag{8.18}$$

Thus we need the following:

Lemma 8.11. *The function* $x \mapsto E^x\{Z(q_1,\ldots,q_m;g_1,\ldots,g_m)\}$ *is continuous.*

Proof. Let $J_n(x;\, p_1,\ldots p_n;\, f_1\ldots f_m) = E^x\{Z(q_1,\ldots,q_m;g_1,\ldots,g_m)\}$. We will show by induction that J_n is continuous in x. It is if $n=1$ since $U_p f$ is continuous if f is. Suppose it is for $n-1$. Consider

$$\frac{1}{t}\Big(J_n(x) - e^{-(p_1+\ldots+p_n)t}P_t J_n(x)\Big)$$
$$= \frac{1}{t}E^x\left\{\prod_{i=1}^n \int_0^\infty e^{-p_i s} f_i(X_s)\,ds - \prod_{i=1}^n \int_t^\infty e^{-p_i s} f_i(X_s)\,ds\right\}.$$

Let $t \to 0$. This converges boundedly to

$$\sum_{i=1}^n f_i(x)J_{n-1}(x;\; p_1,\ldots,\hat{p}_j,\ldots,p_n;\; f_1,\ldots,\hat{f}_j,\ldots,f_n) \stackrel{\text{def}}{=} G_n(x),$$

where the hat over a variable means it is suppressed. Then G_{n-1} is continuous by the induction hypothesis, hence so is $J_n = U_{p_1+\ldots+p_n}G_n$, for U_p is a Ray resolvent. This proves the lemma. □

The proof of Theorem 8.10 is now immediate: by (8.17) and (8.18) $E\{Z \mid \mathcal{F}_t\}$ is the product of a continuous function of t with a continuous function of X_t, and is therefore continuous wherever (X_t) is. It is also continuous if $X_{t-} = x \in B_d$ by Lemma 8.9, since then $E^x\{Z\} = E^{h(x)}\{Z\}$. Linear combinations of such Z are dense in L^1, so we are done. □

8.6. A Feller Property of P^x

This section is a digression, but it follows so directly from the preceding that we could not resist including it. A Ray resolvent takes $C(\boldsymbol{F})$ into $C(\boldsymbol{F})$, but its semigroup P_t may not. That is, the Ray semigroup may not be a Feller semigroup, for a Feller process has no branching points. Nevertheless, the measures on path space do have a certain Feller property.

As before, we take the probability space Ω to be the rcll functions from $\boldsymbol{R}^+$ to $\boldsymbol{F}$, with $X_t(\omega) = \omega(t)$. This space is often given the Skorokhod topology, but we will give it a simpler and more familiar one, namely the topology of convergence in (Lebesgue) measure: $\omega_n \to \omega$ if the functions $\omega_n(t)$ converge in measure to $\omega(t)$, or, equivalently, if $X_.(\omega_n) \to X_.(\omega')$ in measure. This topology is metrizable, and a compatible metric is given by

$$d(\omega, \omega') = \sum_n 2^{-n} \int_0^\infty e^{-t} |f_n(X_t(\omega)) - f_n(X_t(\omega'))| \, dt,$$

where (f_n) is a sequence of continuous 1-supermedian functions which separate points of $\boldsymbol{F}$ and have values in $[0, 1]$. Let $\mathcal{P}(\Omega)$ be the space of all probability measures on Ω with the topology of weak convergence: $\mu \to \nu$ iff $\mu(h) \to \nu(h)$ for all continuous h on Ω.

Note that the Borel field of Ω is exactly the (uncompleted) sigma field $\mathcal{F}^o \equiv \sigma\{X_t,\ t \geq 0\}$. This is clear in one direction: the Borel field is generated by d, which is $\mathcal{F}^o$-measurable. In the other direction, the functions $\omega \mapsto \int_a^b f(X_s(\omega))\, ds$ are continuous on Ω if f is continuous, hence X_s itself is Borel measurable, and $\mathcal{F}^o$ is contained in the Borel field, so they are equal.

Theorem 8.12. *The function $x \mapsto P^x$ is continuous from $\boldsymbol{F}$ to $\mathcal{P}(\Omega)$.*

Proof. We first show that $\{P^x,\ x \in \boldsymbol{F}\}$ is tight. (In fact, the entire family (P^μ) is tight.) Let $I_n^k = [(k-1)2^{-n}, k2^{-n}]$, $k = 1, 2, \dots, 2^n$ be the dyadic subintervals of $[0, 1]$. Let $D_{n,j}^k$ be the number of strict upcrossings of I_n^k, by the right-continuous supermartingale $e^{-t} f_n(X_t)$. ("Strict" means that the process goes from strictly below to strictly above the interval. See the *Course* for upcrossings.) By Doob's upcrossing inequality, for any initial measure μ,

$$E^\mu\{D_{nj}^k\} \leq 2^n.$$

Choose constants A_{nj}^k large enough so that $\sum_{k,n,j} \frac{2^n}{A_{nj}^k} < \infty$. Fix $\varepsilon > 0$. The set

$$K \stackrel{\text{def}}{=} \{\omega : D_{nj}^k \leq \varepsilon^{-1} A_{nj}^k,\ \forall\, k, n, j\}$$

satisfies $P^\mu\{K\} > 1 - \varepsilon$. We will show that K is compact in the topology of convergence in measure, which will imply that the family (P^μ), and in particular, (P^x), is tight.

Before proving this, let us show that it implies the theorem. If $y \to x$ in $\boldsymbol{F}$, the family (P^y) is tight so by Prokhorov's theorem it has at least one limit point in $\mathcal{P}(\Omega)$. Now linear combinations of the random variables Z of (8.16) form an algebra of bounded continuous functions on Ω which separate points and contain the constants. Thus they are dense in $C(K)$ for each compact K (Stone-Weierstrass) so two measures ν and ν' on Ω are equal if $\nu(Z) = \nu'(Z)$ for all such Z. But the functions $y \mapsto E^y(Z) \equiv P^y(Z)$ are continuous by Lemma 8.11, so that the only possible limit point of the P^y as $y \to x$ is P^x. Thus $x \mapsto P^x$ is continuous.

It remains to verify that K is compact. Let $(\omega_m) \subset K$ and note that $f_j(\omega_m(t)) \in [0,1]$ by hypothesis, so, by taking a subsequence if necessary, we may assume that for fixed j and t, $f_j(\omega_m(t))$ converges to a limit; and by taking a further diagonal subsequence we may assume that this holds simultaneously for all j and all rational $t \geq 0$. The f_j are continuous and separate points of $\boldsymbol{F}$, so that that for all rational $t \geq 0$, $\omega_n(t)$ converges in the compact $\boldsymbol{F}$ to a limit, say $\bar{\omega}(t)$. Now $\bar{\omega}(t)$ is defined for rational t, so its upcrossings are defined, and, because we used strict inequalities for the upcrossings, we see that $D^k_{nj}(\bar{\omega}) \leq A^k_{nj}$ for all n, j, k. This means that the function $f_j(\bar{\omega}(t))$, t rational, has finitely many upcrossings of each dyadic interval, and thus has left and right limits everywhere. This being true for all j, we conclude that

$$\omega_\infty(t) \stackrel{\text{def}}{=} \lim\{\bar{\omega}(s) : s \in \boldsymbol{Q},\ s > t,\ s \to t\}$$

exists, is rcll, and hence, as ω_∞ has no more upcrossings than $\bar{\omega}$, that $\omega_\infty \in K$.

Finally, we must show that a subsequence of (ω_n) converges in measure to ω_∞. We will prove this for the case where Ω has values in $[0,1]$. The general case then follows by applying this to $f_j(\omega(t))$, $j = 1, 2, \ldots$ and taking a diagonal subsequence.

Let $\varepsilon = 2^{-q}$ and let $A(\omega_m, \varepsilon)$ be the (finite) set of points $t_0 = 0$, $t_{k+1} = \inf\{t > t_k : |e^{-t}(\omega_m(t) - e^{-t_k}\omega_m(t_k)| > \varepsilon\}$. Between t_k and t_{k+1} there can be at most one upcrossing or one downcrossing of each interval I^k_q by any $e^{-t} f_j(\omega_m(t))$, and hence there can be at most finitely many, say N_ε, total points in $A(\omega_m, \varepsilon)$, independent of m. We can thus find a diagonal subsequence (ω'_m) for which $A(\omega'_m, \varepsilon)$ converges to a set $B(\varepsilon)$. (This does not use anything delicate about set-convergence: for fixed ε the sets $A(\omega_m, \varepsilon)$ have at most N_ε points, so there is a subsequence for which, first, the number of points, and second, the smallest, next smallest, etc. points of $B(\omega_m, \varepsilon)$ all converge in the usual sense.) There can be at most N_ε points in $A(\varepsilon)$. Now take a further diagonal sequence to get convergence simultaneously for all $\varepsilon = 2^{-q}$.

If $t \notin A(\varepsilon)$ and if m is large enough, there exists $\delta > 0$ such that the distance from t to $A(\omega'_m, \varepsilon)$ is $> \delta$. Thus the ω'_m have oscillation $\leq 2\varepsilon$ on $[t-\delta, t+\delta]$. But the same must be true in the limit, so that the oscillation of ω_∞ is $\leq 2\varepsilon$ on $[t-\delta, t+\delta]$. Thus if $r \in \boldsymbol{Q} \cap [t-\delta, t+\delta]$, and m is large enough so that $|\omega_m(r) - \omega_\infty(r)| < 2\varepsilon$,

$$|\omega_m(t) - \omega_\infty(t)| \leq |\omega_m(t) - \omega_m(r)| + |\omega_m(r) - \omega_\infty(t)|$$
$$+ |\omega_\infty(r) - \omega_\infty(t)| \leq 6\varepsilon.$$

Thus $\omega_m \to \omega_\infty$ off the countable set $\cup_q B(2^{-q})$, which is better than convergence in measure. □

8.7. Jumps Without Branching Points

Let us look at those jumps of a Ray process which are not associated with branching points. Not surprisingly, these all occur at totally inaccessible times.

What is surprising is that jump times are the *only* totally inaccessible times. Any optional time at which the process is continuous is actually predictable. This, together with Proposition 8.7, gives the full connection between jumps and stopping times.

Theorem 8.13.

(i) *Let T be a stopping time with the property that $X_{T-} \in D$ and $X_{T-} \neq X_T$. Then T is totally inaccessible.*
(ii) *Let T be a stopping time with the property that (X_t) is P^μ-a.s. continuous at T on the set $\{0 < T < \infty\}$. Then T is predictable.*

Proof. (i) Let S be a predictable time and $f \in C(\boldsymbol{F})$. By Proposition 8.7,

$$E\{(f(X_S) - f(X_{S-}))^2 \mid \mathcal{F}_{S-}\} = P_0 f^2(X_{S-}) - 2f(X_{S-})P_0 f(X_{S-}) + f^2(X_{S-}).$$

Now $P_0(x, \{x\}) = 1, \forall x \in D$, so this vanishes a.s. on $\{X_{S-} \in D\}$, and it does so simultaneously for a countable dense set in $C(\boldsymbol{F})$. This implies that X_t is a.s. continuous at S, and hence that $P\{0 < T < \infty,\ S = T\} = 0$.

(ii) Suppose that $E\{T\} < \infty$. (If it is not, we can consider $T \wedge N$ instead and then let $N \to \infty$.) Consider the process

$$V_t \equiv E\{T - (t \wedge T) \mid \mathcal{F}_t\} = E\{T \mid \mathcal{F}_t\} - t \wedge T.$$

Then V_t is a positive supermartingale and has a right-continuous version, so we can assume it is right-continuous. Thus if V_s or V_{s-} ever vanishes, $V_t = 0$ for all $t > s$. If $t \geq 0$ and $\Lambda = \{V_t \leq 0, T > t\}$, then

$$\int_\Lambda V_t\, dP^\mu = \int_\Lambda (T - t)\, dP^\mu.$$

The integrand in the second integral is strictly positive, while that in the first is negative, so evidently $P^\mu\{\Lambda\} = 0$. Thus $V_t > 0$ a.s. for each rational $t < T$, and hence for all $t < T$.

Moreover, X_t is continuous at T, hence by Proposition 8.10, so is the martingale $E\{T \mid \mathcal{F}_t\}$. Thus V is continuous at T, and $V_T = 0$. Let $T_n = \inf\{t : V_t \leq 1/n\}$. Then T_n is optional, $T_n < T$ on $\{T > 0\}$, and $T_n \to T$. Therefore T is predictable. □

The answer to the the question of quasi-left continuity is now clear: it holds on D. The reader can verify that:

Corollary 8.14. *(Quasi-left continuity) Let (T_n) be an increasing sequence of optional times with limit T. Then on the set $\{0 < T < \infty\}$, $\lim X_{T_n} = X_T$ if and only if $\lim_n X_{T_n} \in D$.*

8.8. Bounded Entrance Laws

Consider the Ray process $\{X_t, t > 0\}$ defined for strictly positive t. Can it always be extended to $t = 0$? Another way of posing the question is to ask if there is any way to enter the space other than by starting at a point of $\boldsymbol{F}$. This kind of question is associated with entrance boundaries and entrance laws; it has received a lot of attention.

Definition 8.15. *An* **entrance law** *is a family* $\{\mu_t, t > 0\}$ *of measures on* $\boldsymbol{F}$ *such that* $\mu_t P_s = \mu_{s+t}$ *for each* $s, t > 0$.

Note that the total mass of the entrance law is constant. We say it is **bounded** if this mass is finite. We do not lose any generality by assuming that the mass is one, so that each μ_t is a probability measure. (There are also *unbounded* entrance laws. However, that subject is deeper, and we will not enter into it.)

If μ is an initial measure, $(\mu P_t)_{t>0}$ is an entrance law. Are there any others? For Ray processes, the answer is "no." The space is already complete, and we get nothing new from entrance laws.

Theorem 8.16. *For each bounded entrance law* μ_t *there exists a unique finite measure* μ *supported by* D *such that* $\mu_t = \mu P_t$ *for all* $t > 0$.

Proof. If $f \in \mathcal{S}$, $t \mapsto e^{-t} P_t f$ is decreasing, so there is a finite limit as $t \to 0$. The same is true for $t \mapsto \int f(x)\, \mu_t(dx)$. Thus the limit also exists for $f \in \mathcal{S} - \mathcal{S}$, which is dense in $C(\boldsymbol{F})$. Now $\boldsymbol{F}$ is compact, so this means that as $t \to 0$, (μ_t) converges in the weak topology of measures. Let μ be the limit.

Now let $f \in \mathcal{S}$ and consider

$$\int U_p f(x)\, \mu_t(dx) = \int_0^\infty ds\, e^{-ps} \int f(x)\, \mu_{t+s}(dx).$$

Let $t \to 0$ and use the fact that $U_p f$ is continuous to see that the left hand side tends to $\int U_p f(x)\, \mu(dx)$. On the right-hand side, note that $t \mapsto \int f(x)\, \mu_{t+s}(dx)$ is monotone, and converges to $\int f(x)\, \mu_s(dx)$ for a.e. s, hence by dominated convergence

$$\int_0^\infty e^{-ps} \mu P_s f\, ds = \int_0^\infty ds\, e^{-ps} \int \mu_s(dx)\, f(x)\, ds.$$

Invert the Laplace transform and let f run through the total set $\mathcal{S}$ to see that $\mu P_s = \mu_s$.

Now notice that $\mu P_t = \mu P_0 P_t$ so that μ_t converges weakly to μP_0. But $\mu = \lim_{t \to 0} \mu_t = \mu P_0$, so μ is supported by D. Uniqueness is evident. □

8.9. Regular Supermedian Functions

We say a supermedian function f is **regular** if, for each initial measure μ, $t \to f(X_{t-})$ is P^μ-a.e. left-continuous. A supermedian function which is right-continuous along the paths of (X_t) is excessive, so regular supermedian functions are the left-handed version of excessive functions. (This is consistent with the definition of regular excessive functions which we will encounter in §13.14.) In the next section we will need supermedian functions which have continuity properties for a process with a fixed initial distribution, and we will use "left-regular" and "right-regular" to distinguish them – "right-regular supermedian" is not quite equivalent to "excessive" in that context – but for the moment, we will just use the one word "regular."

This definition is not local, in that it does not involve a topology on the space. If we had a strong Markov dual, which would bring with it a cofine topology, then regularity would be equivalent to continuity in the cofine topology. However, if there is no cofine topology, we do not know how to define it without involving the sample paths.

The differences between supermedian, excessive, and regular supermedian functions can be seen in the trivial special case where X_t is uniform motion to the right on the line. In that case, a supermedian function is positive and decreasing, an excessive function is positive, decreasing and right-continuous, a regular supermedian function is positive, decreasing and left-continuous and a regular excessive function is decreasing and continuous.

Proposition 8.17. *The p-potential of a function is both p-excessive and regular p-supermedian.*

Proof. Let $f \geq 0$ be a measurable. Then $U_p f$ is p-excessive, so we need only check regularity. Let T be a finite predictable time. By (8.12)

$$U_p f(X_{T-}) = E\left\{\int_0^\infty e^{-ps} f(X_{T+s})\, ds \mid \mathcal{F}_{T-}\right\}.$$

If (T_n) is an increasing sequence of uniformly bounded predictable times with limit T, then $\int_0^\infty e^{-ps} f(X_{T_n+s})\, ds \to \int_0^\infty e^{-ps} f(X_{T+s})\, ds$ whenever the integrands on the left are finite, so by bounded convergence,

$$\begin{aligned}\lim_{n\to\infty} E\{U_p f(X_{T_n-})\} &= \lim_{n\to\infty} E\left\{\int_0^\infty e^{-ps} f(X_{T_n+s})\, ds\right\}\\ &= E\left\{\int_0^\infty e^{-ps} f(X_{T+s})\, ds\right\} = E\{U_p f(X_{T-})\}.\end{aligned}$$

It follows by Theorem 6.15 that $t \mapsto U_p f(X_{t-})$ is left-continuous, hence $U_p f$ is regular. □

The remainder of the section deals only with the left-continuous process, so, to avoid continually writing X_{t-}, for this section only, we let (X_t) denote the left-continuous moderate Markov Ray process. We define it on the canonical space Ω of left-continuous functions from $[0, \infty)$ to $\boldsymbol{F}$ by $X_t(\omega) = \omega(t)$.

The debut D_A of a set A was defined in §1.5. A slight extension:

Definition 8.18. *If $A \subset F$, then the* **debut of A after t** *is*

$$D_A^t = \inf\{s \geq t : X_s \in A\},$$

with the usual convention that $\inf \emptyset = \infty$.

Here are some of its elementary properties for left-continuous paths:

1° $D_A^t = t + D_A \circ \theta_t$.
2° $t \mapsto D_A^t$ is increasing, and if $s < t$ and $D_A^s \geq t$, then $D_A^s = D_A^t$.
3° If A is closed, $t \mapsto D_A^t$ is left-continuous.
4° If $A_n \supset \bar{A}_{n+1} \supset A \equiv \cap_n A_n$, then $D_{A_n}^t \to D_A^t$.

These are easily seen by noting that if $Z_A = \{t : X_t \in A\}$, then $Z_A \circ \theta_s = \{t : X_{s+t} \in A\}$ and $D_A^t = \inf\{s : Z_A \cap [t, s] \neq \emptyset\}$. Then 1° and 2° are clear, and 3° follows from the fact that if A is closed and (X_t) is left-continuous, then Z_A is closed under increasing limits. In 4°, clearly $D_A^t \geq \lim_n D_{A_n}^t$ since the $D_{A_n}^t$ increase in n. Then either $D_{A_n}^t \to \infty$, in which case $D_A^t = \infty$, or the limit is finite, in which case $X_{D_A^t} = \lim_n X_{D_{A_n}^t} \in \cap_n \bar{A}_n = A$, so $D_A^t \leq \lim_n D_{A_n}^t$, which implies equality.

Proposition 8.19. *A decreasing limit of regular p-supermedian functions is a regular p-supermedian function.*

Proof. Let (f_n) be a sequence of p-supermedian functions decreasing to f. Then $f_n(x) \geq pU_{p+q} f_n(x)$ for $q > 0$, and we can go to the limit by monotone convergence as $n \to \infty$ to see that f is p-supermedian.

Let (T_n) be an increasing sequence of uniformly bounded predictable times with limit T. Since $\{e^{-pT_n} f(X_{T_n}),\ n = 1, 2, \dots\}$ is a positive supermartingale with an $\mathcal{F}_{T-}$-measurable final element $e^{-pT} f(X_T)$, $\lim_n E\{e^{-pT_n} f(X_{T_n})\} \geq E\{e^{-pT} f(X_T)\}$. On the other hand, f_k is regular, so $f_k(X_T) = \lim_n f_k(X_{T_n})$, and $\lim_n E\{e^{-pT_n} f(X_{T_n})\} \leq \lim_n E\{e^{-pT_n} f_k(X_{T_n})\} = E\{e^{-pT} f_k(X_T)\}$. Let $n \to \infty$ to see $\lim_n E\{e^{-pT_n} f(X_{T_n})\} \leq E\{e^{-pT} f(X_T)\}$. It follows that $\lim_n E\{e^{-pT_n} f(X_{T_n})\} = E\{e^{-pT} f(X_T)\}$ for all increasing uniformly bounded sequences of predictable times, hence $t \mapsto f(X_t)$ is left-continuous by Theorem 6.15. □

Proposition 8.20. *Let $p > 0$ and let $K \subset F$ be closed. The functions ψ_K and ϕ_K defined for the left-continuous process (X_t) by*

$$\psi_K(x) = E^x\{e^{-pD_K}\} \quad \textit{and} \quad \phi_K(x) = P^x\{D_K < \infty\}$$

are regular p-supermedian and regular supermedian respectively.

Proof. Apply the moderate Markov property at a finite predictable time T:

$$e^{-pT}\psi_K(X_T) = e^{-pT}E\{e^{-pD_K\circ\theta_T} \mid \mathcal{F}_{T-}\} = E\{e^{-pD_K^T} \mid \mathcal{F}_{T-}\}$$

by 1°. Since $D_K^T \geq D_K$

$$E^x\{e^{-pT}\psi_K(X_T)\} = E^x\{e^{-pD_K^T}\} \leq E^x\{e^{-pD_K}\} = \psi_K(x),$$

so ψ_K is supermedian. To see it is regular, let (T_n) be an increasing sequence of uniformly bounded predictable times with limit T and note that

$$\lim_n E\{\psi_K(X_{T_n})\} = \lim_n E\{e^{-p(D_K^{T_n}-T_n)}\}.$$

By 3°, $D_K^{T_n} \to D_K^T$, so this is

$$= E\{e^{-p(D_K^T-T)}\} = E\{e^{-pD_K\circ\theta_T}\} = E\{\psi_K(X_T)\}.$$

By Theorem 6.15, this implies that $t \mapsto \psi_K(X_t)$ is left-continuous, hence ψ_K is regular. The proof for ϕ_K is similar. □

8.10. Ray-Knight Compactifications: Why Every Markov Process is a Ray Process at Heart

In his original article, Ray suggested that every Markov process could be embedded in a Ray process. While his proposed embedding scheme had a subtle flaw, he was right. Frank Knight patched it with the seemingly-innocuous Lemma 8.24 below. This turned out to be the golden key to embeddings.

Indeed, any time-homogeneous Markov process on a nice state space can be modified slightly to become a Ray process. "Slightly" is relative; the modification may be fairly extensive in some sense – after all, it turns an ordinary Markov process into a strong Markov process, and it may involve a complete change of topology on the underlying space – but the sample paths are changed at most on Lebesgue null-sets, and if the original process was a right continuous strong Markov process, it is completely unchanged.

This means that Ray processes are canonical in some sense. Theorems about them have implications for all Markov processes. A word of warning: Ray processes may be canonical, but the modifications are not. There is considerable latitude, and one can often tailor them to fit the situation. See e.g. Example 8.56. Meanwhile, let us see how to construct them.

We will discuss the compactification of a Markov process with a fixed initial distribution, as opposed to the more common setting of a family of Markov processes, one for each initial distribution. This means that we can't assume that there is a process starting from each point, and consequently we can't be sure the transition function forms a semigroup. This makes things harder – we will deal with this problem below in §8.13 – but it is dictated by some applications, such as time-reversal, where one is given a single, immutable process, and must construct the transition semigroup from scratch. But then, one important use of compactification is exactly to construct semigroups.

Lemma 8.21. *Let Λ be a set and let Γ be a countable set of mappings of $\Lambda \times \Lambda$ into Λ. If G is a countable subset of Λ, then the smallest set $H \subset \Lambda$ satisfying*

(i) $G \subset H$
(ii) $\forall h \in \Gamma,\ h(H \times H) \subset H$

is countable.

Proof. Let $h_1, h_2, \dots$ be a sequence such that for each n, $h_n \in \Gamma$ and each $h \in \Gamma$ occurs infinitely often in the sequence. Let $C_0 = C$ and, by induction, let $C_{n+1} = C_n \cup h_{n+1}(C_n \times C_n)$. Then $H = \cup_n C_n$ works. □

Let $(\Omega, \mathcal{F}, P)$ be a probability space, $(\mathcal{F}_t)_{t \geq 0}$ a filtration of subsets of $\mathcal{F}$, and let $(\boldsymbol{E}, \mathcal{E})$ be locally compact separable metric space and its Borel field. Let $\bar{\boldsymbol{E}}$ be the one-point compactification of $\boldsymbol{E}$. Consider the following.

Hypothesis 8.22. (MP) *There is a Markov resolvent $(U_p)_{p>0}$ on $(\boldsymbol{E}, \mathcal{E})$ and a measurable stochastic process $\{X_t,\ t \in \boldsymbol{T}\}$ adapted to $(\mathcal{F}_t)$ such that for all $t \geq 0$, all $p > 0$ and all $f \in C(\bar{\boldsymbol{E}})$,*

$$E\left\{\int_0^\infty e^{-ps} f(X_{t+s})\,ds \mid \mathcal{F}_t\right\} = U_p f(X_t) \ \ a.s. \tag{8.19}$$

A few remarks are in order. Hypothesis (MP) implies that (X_t) is a single time-homogeneous Markov process, so its initial distribution is fixed.

We will only use $\bar{\boldsymbol{E}}$ as a crutch: it is just a convenient way to get a separable family of uniformly continuous functions on $\boldsymbol{E}$. Let $(U_p)_{p>0}$ be a Markov resolvent on $(\boldsymbol{E}, \mathcal{E})$, i.e. it satisfies the resolvent equation identically, $x \mapsto U_p(x, A)$ is $\mathcal{E}$-measurable for all $A \in \mathcal{E}$, and $U_p(x, \boldsymbol{E}) = 1/p$. (If we have a sub-Markovian resolvent we know how to extend it to a Markovian resolvent by adding a point ∂ to the state space for nefarious purposes.) For the moment we do not assume any continuity conditions on the process (X_t), but only that $(\omega, t) \mapsto X_t(\omega)$ is $\mathcal{F} \times \mathcal{L}$ measurable, where $\mathcal{L}$ denotes the Lebesgue sets on the line.

A set of functions Λ is **inf-stable** if $f, g \in \Lambda \Longrightarrow f \wedge g \in \Lambda$, and it is **U – stable** if $f \in \Lambda,\ p > 0 \Longrightarrow U_p f \in \Lambda$.

Definition 8.23. *If G is a set of positive functions, then $\mathcal{S}(G)$ is the smallest positive inf-stable, U-stable cone which contains G.*

Lemma 8.24. *(Knight) Let $G \subset \mathcal{S}$. Then $\mathcal{S}(G) \subset \mathcal{S}$ and if G is separable in the sup-norm on $\boldsymbol{E}$, so is $\mathcal{S}(G)$.*

Proof. $\mathcal{S}$ is itself a positive inf-stable U-stable cone containing G, and therefore it contains $\mathcal{S}(G)$, which is the smallest such.

Suppose G is separable. Let Γ be the set of mappings

$$(f, g) \mapsto f + g$$

$$(f, g) \mapsto f \wedge g$$

$$(f, g) \mapsto pf,\ 0 < p \in \boldsymbol{Q}$$

$$(f, g) \mapsto U_p f,\ 0 < p \in \boldsymbol{Q},$$

of $\mathcal{S}(G) \times \mathcal{S}(G) \mapsto \mathcal{S}(G)$. Note that all these mappings are continuous in the sup-norm. Let G_1 be a countable dense subset of G and let $\mathcal{S}'(G_1)$ be the smallest superset of G_1 which is stable under all mappings $h \in \Gamma$. $\mathcal{S}'(G_1)$ is countable by Lemma 8.21, and by continuity, its closure $\overline{\mathcal{S}'(G_1)}$ is still stable under all $h \in \Lambda$, and is also stable under U_p for all $p > 0$, and under multiplication by all real $p > 0$, not just rational p. Thus $\overline{\mathcal{S}'(G_1)}$ is an inf-stable, U-stable cone containing G_1, hence $\overline{\mathcal{S}'(G_1)} \supset \mathcal{S}(G_1)$. But $G \subset \bar{G}_1 \subset \overline{\mathcal{S}'(G_1)} \Longrightarrow \mathcal{S}(G) \subset \overline{\mathcal{S}'(G_1)}$, and the latter is separable. □

Theorem 8.25. *Assume Hypothesis (MP) and that there is a family G of bounded 1-supermedian functions which separate points of $\boldsymbol{E}$ and is itself separable (sup-norm). Then there exists a compact metric space $\boldsymbol{F}$ and a Ray resolvent $(\hat{U}_p)_{p>0}$ on $\boldsymbol{F}$ such that*

(i) *$\boldsymbol{E}$ is a dense Borel subset of $\boldsymbol{F}$;*
(ii) *each $h \in \mathcal{S}(G)$ is the restriction to $\boldsymbol{E}$ of a unique function in $C(\boldsymbol{F})$;*
(iii) *$\mathcal{S}(G)$ (extended to $\boldsymbol{F}$ by continuity) is total in $C(\boldsymbol{F})$;*
(iv) *for each $x \in \boldsymbol{E}$ and $A \in \mathcal{E}$, $\hat{U}_p(x, A) = U_p(x, A)$.*

Furthermore there is a right-continuous Ray process $\{\hat{X}_t,\ t \in \boldsymbol{T}\}$ on $\boldsymbol{F}$ such that

(v) *$\hat{X}_t$ admits $(\hat{U}_p)$ as resolvent;*
(vi) *for a.e. t, $P\{\hat{X}_t = X_t\} = 1$.*

In particular, the right-continuous regularization (in $\boldsymbol{F}$) of (X_t) is a Ray process.

Proof. We may suppose without loss of generality that $1 \in G$. $\mathcal{S}(G)$ is separable by Lemma 8.24. We metrize $\boldsymbol{E}$ to make all $f \in \mathcal{S}(G)$ uniformly continuous, as follows. Let H be a countable dense (sup-norm) subset of $\mathcal{S}(G)$. $G \subset \mathcal{S}(G)$, so $\mathcal{S}(G)$ separates points of $\boldsymbol{E}$, hence so does H. Write $H = \{h_n\}$ and define the metric ϱ by

$$\varrho(x, y) = \sum_n 2^{-n}(1 + \|h_n\|)^{-1}|h_n(y) - h_n(x)|, \; x, y \in \boldsymbol{E}. \tag{8.20}$$

Let $\boldsymbol{F}$ be the completion of $\boldsymbol{E}$ in this metric. Since ϱ is bounded, $\boldsymbol{F}$ is compact. The injection map of $\boldsymbol{E}$ into $\boldsymbol{F}$ is one-to-one and Borel measurable, so $\boldsymbol{E}$ is a Borel subset of $\boldsymbol{F}$. Each function h_n is uniformly continuous in the metric ϱ, and can therefore be extended to $C(\boldsymbol{F})$. H is dense in $\mathcal{S}(G)$, so the extension is unique.

Denote the extension of h by $\hat{h} \in C(\boldsymbol{F})$, and let $\hat{\mathcal{S}}(G) = \{\hat{h} : h \in \mathcal{S}(G)\}$. We can extend the resolvent to $\boldsymbol{F}$ by continuity: for $h \in \mathcal{S}(G)$, define $\hat{U}_p\hat{h} = \widehat{U_p h}$. This is continuous on $\boldsymbol{F}$. The resolvent equation extends to $\hat{U}_p$. Indeed, apply the resolvent equation to $h \in \mathcal{S}(G)$. Then $U_p f \in \mathcal{S}(G)$, so it extends by continuity to $\boldsymbol{F}$; it follows that for each $p > 0$, $q > 0$ and $f \in \mathcal{S}(G)$, $U_p h - U_q h = (p - q)U_p U_q h$; all terms extend to $\boldsymbol{F}$ by continuity and $\hat{U}_p\hat{h} - \hat{U}_q\hat{h} = (p - q)\hat{U}_p\hat{U}_q\hat{h}$.

Now $\mathcal{S}(G)$ separates points of $\boldsymbol{E}$, hence by construction, its extension $\hat{\mathcal{S}}(G)$ separates points of $\boldsymbol{F}$ as well. Moreover, if $h \in \mathcal{S}(G)$, h is 1-supermedian, so $pU_{p+1}h \le h$ on $\boldsymbol{E}$. But both h and $U_{p+1}h$ are in $\mathcal{S}(G)$, so the inequality also extends to $\boldsymbol{F}$ by continuity, implying that $\hat{h}$ is continuous and 1-supermedian on $\boldsymbol{F}$. Thus $\hat{\mathcal{S}}(G)$ is a collection of continuous 1-supermedian functions which contains constants, separates points and is closed under finite minima. Then the Stone-Weierstrass theorem implies that $\hat{\mathcal{S}}(G) - \hat{\mathcal{S}}(G)$ is dense in $C(\hat{\boldsymbol{F}})$. Thus $\hat{U}_p f$ is defined and continuous on $\boldsymbol{F}$ for all $f \in C(\boldsymbol{F})$, and the continuous 1-supermedian functions separate points of $\boldsymbol{F}$. Thus $(\hat{U}_p)_{p>0}$ is a Ray resolvent on $\boldsymbol{F}$ which satisfies (i) – (iv).

Let us now construct the Ray process. Let $h \in H$. Since h is 1-supermedian, $e^{-t}h(X_t)$ is a supermartingale which, while not necessarily separable, has essential left and right limits at all t with probability one by Corollary 6.5.

This is true simultaneously for all $h_n \in H$; as $\boldsymbol{F}$ is compact and the h_n are continuous and separate points of $\boldsymbol{F}$, we conclude that X_t itself has essential limits in the topology of $\boldsymbol{F}$. Thus define $\hat{X}_t = \text{ess}\lim_{s\downarrow t} X_s$. From Proposition 6.2

(a) $t \mapsto \hat{X}_t$ is right-continuous in $\boldsymbol{F}$;
(b) for a.e. ω, the set $\{t : \hat{X}_t \neq X_t\}$ has Lebesgue measure zero;
(c) the set $\{t : P\{\hat{X}_t \neq X_t\} < 1\}$ has Lebesgue measure zero.

It remains to verify that $(\hat{X}_t)$ is a Ray process. Now for a.e. t, $\hat{X}_t = X_t$ a.s., so that if we ignore a set of t of measure zero, $(\hat{X}_t)$ is a Markov process with resolvent U_p, and therefore with resolvent $\hat{U}_p$, since $\hat{U}_p = U_p$ on $\boldsymbol{E}$.

By (a), $t \mapsto U_p h(\hat{X}_t)$ is right-continuous for h in H, and hence for all $h \in C(\boldsymbol{F})$, for $\hat{H} - \hat{H}$ is dense in $C(\boldsymbol{F})$. Now we need only repeat the argument that led to (8.13) to conclude that $(\hat{X}_t)$ is a right-continuous Markov process (no exceptional t-set) having a Ray resolvent, and is therefore a Ray process. □

Thus, if we ignore a null-set of t, an ordinary Markov process has a standard modification which is a Ray process. However, we should be careful. The original topology may not survive the compactification: $\boldsymbol{E}$ is a subset of $\boldsymbol{F}$, but not necessarily a subspace. (This is not too surprising. The intrinsic topology for a Markov process is the fine topology, which is usually quite different from the metric topology of the state space, and it is only the fine topology which – for a strong Markov process, at least – is preserved by the embedding.) Thus, the embedding is not really a compactification of the original topology. Nevertheless, it is called **Ray-Knight compactification,** and we use the same name for the embedding of the state space or the process, or both.

The Ray-Knight compactification is not canonical, for it depends on the initial set G of 1-supermedian functions we choose. This gives a flexibility which is useful in applications, but it makes it hard to sum up Ray-Knight compactifications in a single theorem.

Let us introduce some abbreviations to shorten the upcoming discussion. A process whose sample paths are right-continuous and have left limits at all $t > 0$ a.s. will be called **rcll** while a process whose sample paths are a.s. left-continuous and have right limits at all t will be called **lcrl**.

Theorem 8.25 puts nearly minimal conditions on the Markov process. If we start with more, we should get more. In particular, if we start with a strong or moderate Markov process, we might hope to embed it directly, without exceptional sets. The next theorem shows that we can.

We introduce two new hypotheses, one for strong Markov processes and one for moderate Markov processes.

Hypothesis 8.26. (SMP) *Hypothesis (MP) holds and for all $f \in \mathcal{E}_b$ and all $p > 0$, both $t \mapsto X_t$ and $t \mapsto U_p f(X_t)$ are rcll.*

Hypothesis 8.27. (MMP) *Hypothesis (MP) holds and for all $f \in \mathcal{E}_b$ and all $p > 0$, both $t \mapsto X_t$ and $t \mapsto U_p f(X_t)$ are lcrl.*

We assume as usual that the probability space $(\Omega, \mathcal{F}, P)$ is complete and that the filtration $(\mathcal{F}_t)$ is augmented. Under (SMP) the filtration is right-continuous (see Theorem 4 of §2.3).

Lemma 8.28. *Let $G \subset \mathcal{S}$. Assume Hypothesis (SMP) (resp. Hypothesis (MMP)). If $t \mapsto f(X_t)$ is rcll (resp. lcrl) for all $f \in G$, the same holds for all $f \in \mathcal{S}(G)$.*

Proof. If f and g are bounded and measurable, and if $t \mapsto f(X_t)$ and $t \mapsto g(X_t)$ are a.s. rcll (resp. lcrl), so are $t \mapsto f(X_t) + g(X_t)$, $f(X_t) \wedge g(X_t)$ and

$U_p f(X_t)$, the last courtesy of Hypotheses (SMP) and (MMP). Thus the set of functions f for which $t \mapsto f(X_t)$ is rcll (resp. lcrl) is linear, inf-stable and U-stable, and hence contains $\mathcal{S}(G)$. □

Definition 8.29. *A set $A \in \boldsymbol{E}$ is X*-**polar** *if $P\{\exists t : X_t \in A\} = 0$.*

Note that (X_t) has a fixed initial distribution, so that this is weaker than the usual notion of polarity, in which the probability that (X_t) hits A is zero for all possible initial distributions.

Theorem 8.30. *Suppose that Hypothesis (SMP) (resp. (MMP)) holds and that there exists a family G of bounded 1-supermedian functions which separates points and is itself separable, such that for each $f \in G$, $t \mapsto f(X_t)$ is rcll (resp. lcrl). Then there is a Ray-Knight compactification $\boldsymbol{F}$ of $\boldsymbol{E}$ and a Ray process $(\hat{X}_t)$ such that $P\{\exists t : X_t \neq \hat{X}_t\} = 0$ (resp. $P\{\exists t : X_t \neq \hat{X}_{t-}\} = 0$). In particular, the process $\{X_t,\ t \in \boldsymbol{T}\}$ is, considered as a process on $\boldsymbol{F}$, a rcll strong Markov process (resp. lcrl moderate Markov process).*

Proof. Let $\boldsymbol{F}$ be the compactification and $(\hat{X}_t)$ the Ray process guaranteed by Theorem 8.25. By Lemma 8.28, $t \mapsto f(X_t)$ is rcll (resp. lcrl) for all $f \in \mathcal{S}(G)$, and therefore for all the h_n in (8.20). These define the topology of $\boldsymbol{F}$, so that if (x_j) is a sequence in $\boldsymbol{E}$, then $x_j \to x_0$ in the topology inherited from $\boldsymbol{F}$ iff $h_n(x_j) \to h_n(x_0)$ for all n. It follows that (X_t) is rcll (resp. lcrl) in the topology of $\boldsymbol{F}$. Moreover, Fubini's theorem tells us that $X_t = \hat{X}_t$ for a.e. t. Under Hypothesis (SMP), both (X_t) and $(\hat{X}_t)$ are right-continuous, hence we conclude $X_t \equiv \hat{X}_t$. Since $(\hat{X}_t)$ is a strong Markov process, so is (X_t). Under Hypothesis (MMP), the same reasoning tells us that $X_t \equiv \hat{X}_{t-}$, which is a moderate Markov process by Theorem 8.6. □

We considered a single process with a fixed initial distribution above. However, the usual setting for Markov processes involves a family of processes, one for each possible initial distribution. Moreover, one commonly starts from the semigroup, not the resolvent. Let us recast these results in a setting which we can apply directly to Hunt, Feller, and standard processes.

Let $(P_t)_{t \geq 0}$ be a Markov semigroup on $\boldsymbol{E}$. For the sake of applications, we will loosen the measurability requirements slightly. We assume that if $f \in b\mathcal{E}$, $P_t f \in b\mathcal{E}^{\sim}$. Here are the "usual hypotheses" so often encountered in the literature.

Hypothesis 8.31. (UH)

(i) *For each probability measure μ on $\boldsymbol{E}$ there exists a Markov process $\{X_t,\ t \in \boldsymbol{T}\}$ on $\boldsymbol{E}$ which admits (P_t) as a semigroup, has initial distribution μ, and has sample paths which are a.s. right-continuous.*

(ii) *Let f be a p-excessive function. Then f is nearly-Borel measurable, and for a.e. ω, $t \mapsto f(X_t)$ is right-continuous on $\boldsymbol{T}$.*

Hypothesis (UH) implies that (X_t) is a strong Markov process, and that its natural filtration $(\mathcal{F}_t)_{t\geq 0}$, suitably augmented, is right-continuous. (In fact, one of the achievements of Hunt's theory was to show that (UH) holds for Hunt processes. As it happens, this makes the rest of the theory work, so it is customary to simply take it as a hypothesis. We reassure the reader that we will not use this to define yet another class of Markov processes, but only to show that all such can be transformed into Ray processes.) Most of the usual suspects – Feller, Hunt, and standard processes – satisfy Hypothesis (UH). Ray processes, interestingly enough, do not, since (i) precludes branching points.

Theorem 8.32. *Assume Hypothesis (UH). For each initial law μ on $\boldsymbol{E}$, the process $\{X_t, t \geq 0\}$ on $\boldsymbol{E}$ with initial measure μ is, considered as a process on $\boldsymbol{F}$, a Ray process admitting semigroup $(\hat{P}_t)$.*

Lemma 8.33. *If the resolvent separates points of $\boldsymbol{E}$, there exists a countable family of 1-potentials which separates points of $\boldsymbol{E}$.*

Proof. If $x \neq y$, there is $p > 0$ and $f \in \mathcal{E}_b$ for which $U_p f(x) \neq U_p f(y)$. By the resolvent equation, if $g = f + (1-p)U_p f$, $U_1 g = U_p f$, so $U_1 g(x) \neq U_1 g(y)$. Thus U_1 itself separates points, so if (f_n) is a countable dense subset of $C^+(\bar{\boldsymbol{E}})$, then $(U_1 f_n)$ is a separating set of 1-potentials. □

Proof. (of Theorem 8.32.) There is a technical problem: the resolvent (U_p) of (P_t) is only universally measurable, not Borel measurable as before. We leave it to the reader to verify that we can still apply Theorem 8.30. Notice that $\boldsymbol{E}$ itself has no branching points, since by (i) of (UH) we can take μ to be a point mass at any point of $\boldsymbol{E}$. Then the resolvent $(U_p)_{p>0}$ of (P_t) separates points of $\boldsymbol{E}$, since if f is continuous on $\bar{\boldsymbol{E}}$, take $\mu = \delta_x$ in (i) to see that $\lim_{p\to\infty} pU_p f(x) = f(x)$. By Lemma 8.33 there is a countable separating set G of 1-potentials. Apply Theorem 8.30 with this G and note that the resulting compactification is independent of the initial measure μ. □

8.11. Useless Sets

We continue the last section, in which a process satisfying (UH) is embedded in a Ray process on the compact space $\boldsymbol{F}$ via a Ray-Knight compactification. The embedding used to prove Theorem 8.32 was rather special, but it is in fact the most useful type, and deserves a name.

Let $\mathcal{U} = \{U_1 f,\ f \geq 0 \text{ Borel}\}$ be the set of 1-potentials of functions, and let $\mathcal{U}_b$ be the set of bounded elements of $\mathcal{U}$. A compactification generated by a separable set $G \subset \mathcal{U}_b$ is a **$\mathcal{U}$-compactification**. $\mathcal{U}$-compactifications have some nice properties. Here is one.

Lemma 8.34. *The set of degenerate branching points in a $\mathcal{U}$-compactification is empty.*

Proof. If $G \subset \mathcal{U}_b$, it is easily seen that $\mathcal{S}(G)$ consists of potentials of positive functions and finite infima of such potentials; this remains true when $\mathcal{S}(G)$ is extended by continuity to $\hat{\mathcal{S}}(G)$ on $\boldsymbol{F}$. If x is a degenerate branching point and $y = k(x)$, then $y \neq x$ and $U_p f(x) = U_p f(y)$ for every f and $p > 0$, hence $h(x) = h(y)$ for every $h \in \hat{\mathcal{S}}(G)$. But $\hat{\mathcal{S}}(G)$ is total in $C(\boldsymbol{F})$, so $h(x) = h(y)$ for every $h \in C(\boldsymbol{F})$, hence $x = y$, a contradiction. □

Let us consider some other exceptional sets which may be introduced by compactifications. We say that a set $K \subset \boldsymbol{F}$ is **μ-useless** if with P^μ probability one, neither (X_t) nor $(\hat{X}_{t-})$ ever hits K. We say K is **useless** if it is μ-useless for all μ on $\boldsymbol{E}$. (Notice that μ sits on $\boldsymbol{E}$, *not* $\boldsymbol{F}$.) The following corollaries of Theorem 8.32 concern a (possibly non-$\mathcal{U}$) compactification $\boldsymbol{F}$ of $\boldsymbol{E}$.

Corollary 8.35. *The set of $x \in \boldsymbol{F}$ for which $P_0(x, \boldsymbol{F} - \boldsymbol{E}) > 0$ is useless.*

Proof. Theorem 8.32 implies that for $y \in \boldsymbol{E}$, the debut $D_{\boldsymbol{F}-\boldsymbol{E}}$ is P^y-a.s. infinite. Let $C = \{x \in \boldsymbol{F} : P_0(x, \boldsymbol{F} - \boldsymbol{E}) > 0\}$. If $y \in \boldsymbol{E}$ and if C is not P^y-polar for (X_{t-}), there exists a predictable time T, not identically infinite, for which $X_{T-} \in C$ on $\{T < \infty\}$. Then

$$P\{X_T \in \boldsymbol{F} - \boldsymbol{E}\} = E\{P_0(X_{T-}, \boldsymbol{F} - \boldsymbol{E})\} > 0.$$

This contradicts the fact that $D_{\boldsymbol{F}-\boldsymbol{E}}$ is a.s. infinite. □

Corollary 8.36. *The set of points in $\boldsymbol{F} - \boldsymbol{E}$ which are not branching points is useless.*

Indeed, for such a point, the measure $P_0(x, \cdot)$ is not carried by $\boldsymbol{E}$.

Corollary 8.37. *The set of $x \in \boldsymbol{F}$ such that the measures $U_p(x, \cdot)$ are not carried by $\boldsymbol{E}$ is useless.*

Indeed, if $P_0(x, \cdot)$ is carried by $\boldsymbol{E}$, then so is $U_p(x, \cdot)$, since $U_p = U_p P_0$.

Corollary 8.38. *If for each $x \in \boldsymbol{E}$ the measure $U_p(x, \cdot)$ is absolutely continuous with respect to a measure ξ, the set of $x \in \boldsymbol{F}$ for which $U_p(x, \cdot)$ is not absolutely continuous with respect to ξ is useless.*

Indeed, if $P(x, \cdot)$ is carried by $\boldsymbol{E}$, $U_p(x, \cdot) = (P_0 U_p)(x, \cdot)$ is absolutely continuous with respect to ξ.

Remark 8.39. $\mathcal{U}$-compactifications introduce a minimal number of branching points. Others may introduce many more. In particular, the set of degenerate branching points need not always be useless; in fact, it may even be useful, as in Example 8.56 below.

8.12. Hunt Processes and Standard Processes

Let us see how Hunt processes and their standard cousins survive compactification. The state space of a Hunt process is locally compact. From Theorem 8.30 we know that it can be embedded in a larger compact space – with a possible change of topology – in such a way that the Hunt process is embedded in a one-to-one way. Thus, in a sense, the Hunt process is already a Ray processes. However, although the process embeds without change, the same is not true of its left limits. A priori, the left limits in $\boldsymbol{F}$ could be branching points. So the emphasis in this section will be on the relations between the original processes and the branching points.

A Ray process is a Hunt process if it is quasi-left continuous. This holds if the set of branching points is empty, or, more generally, if it is polar for (X_{t-}). For embedded processes, the only relevant initial measures are those which sit on the original space $\boldsymbol{E}$. Thus an embedded process is a Hunt process if the set of branching points is useless.

Proposition 8.40. *For a $\mathcal{U}$-compactification of a Hunt process (X_t), B is useless. Thus (X_t) is also a Hunt process when considered as a process on $\boldsymbol{F}$. In particular, for P^μ-a.e. ω, $\lim_{s\uparrow\uparrow t} X_s$ exists in both the topologies of $\boldsymbol{E}$ and of $\boldsymbol{F}$, the two limits are equal, and $\{X_{t-},\ t>0\}$ is a left continuous moderately Markov Ray process with values in $\boldsymbol{E}$.*

Proof. Suppose X_t is a Hunt process on $\boldsymbol{E}$ and that $(\hat{X}_t)$ is the corresponding Ray process. Let X_{t-} denote the left limit of X *in the topology of* $\boldsymbol{E}$ and let $\hat{X}_{t-}$ denote the left limit of $\hat{X}$ *in the topology of* $\boldsymbol{F}$. Now $X_t = \hat{X}_t$ for all t, so that the proposition will be proved if we show that $X_{t-} = \hat{X}_{t-}$ for all t. Indeed, $\boldsymbol{E}\cap B = \emptyset$ and X_{t-} never hits B, hence B is useless. The rest follows from the known properties of $(\hat{X}_{t-})$. Consider a predictable time T which is announced by a sequence (T_n) of optional times. Then $\lim_n X_{T_n} = X_T$ a.s. in the topology of $\boldsymbol{E}$, so that $\hat{X}_T$ is $\mathcal{F}_{T-}$-measurable. On the other hand, $\lim_n X_{T_n} = \hat{X}_{T-}$ also exists in the topology of $\boldsymbol{F}$. As $(\hat{X}_{t-})$ is a moderate Markov process, the conditional distribution of X_T given $\mathcal{F}_{T-}$ is $P_0(\hat{X}_{T-},\,\cdot\,)$. In order that X_T be $\mathcal{F}_{T-}$-measurable, $P_0(X_{T-},\,\cdot\,)$ must be a point mass. (For $P_0 f(X_{T-}) = f(X_T)$, so, squaring, $(P_0 f(X_{T-}))^2 = f^2(X_T) = P_0 f^2(X_{T-})$ for a countable dense set of f, and the point mass assertion follows from Jensen's inequality.) Since there are no degenerate branching points, $\hat{X}_{T-}$ must be a non-branching point a.s., so that $\hat{X}_{T-} = X_T$. But this means that $\hat{X}_{T-} = X_{T-}$ since both equal X_T.

Now both $(\hat{X}_{t-})$ and (X_{t-}) are left-continuous in their respective topologies, so that the fact that they are a.s. equal at each predictable time implies that $P^\mu\{X_{t-} = \hat{X}_{t-},\ \forall\, t>0\} = 1$. □

Remark 8.41. In other words, if the original process is a Hunt process, so is its compactification. As long as $X_0 \in \boldsymbol{E}$, (X_{t-}) will not encounter branching points. This does not imply that there are no branching points. The compactification may introduce some, but they can't be reached from $\boldsymbol{E}$. If the Ray process starts with an initial measure on $\boldsymbol{F} - \boldsymbol{E}$ it may well encounter them.

Ever since the publication of the book by Blumenthal and Getoor, standard processes have been the central class of Markov processes in probabilistic potential theory. They are an extension of Hunt processes, in which quasi-left continuity holds everywhere except possibly at the lifetime. More exactly, a standard process is a process satisfying Hypothesis (UH) whose state space has a distinguished absorbing state ∂, and a lifetime $\zeta = \inf\{t : X_t = \partial\}$, and whose paths have left limits and are quasi-left continuous *except* possibly at the lifetime ζ : for each increasing sequence (T_n) of optional times with limit T, (X_{T_n}) converges to X_T in the topology of $\boldsymbol{E}$ a.s. on $\{T < \zeta\}$. But $X_{\zeta-}$ may not exist.

Standard processes arise in a natural way when a Hunt process is killed at some random time. We will see that in a sense, *all* standard process arise this way.

As before, we will denote the left-hand limits in the topology of $\boldsymbol{E}$ by X_{t-} and the left-hand limits in the topology of $\boldsymbol{F}$ by $\hat{X}_{t-}$. (There is no need to distinguish between X_t and $\hat{X}_t$, since they are identical by Theorem 8.32, but we have to be very careful with the left limits.) Recall from §8.2 that Π is the set of forks.

Theorem 8.42. *Consider a $\mathcal{U}$-compactification of a standard process (X_t). Let μ a probability distribution on $\boldsymbol{E}$. Then*

(i) *for P^μ-a.e. ω, for all $t < \zeta$, either $X_{t-} = \hat{X}_{t-}$, or else $\hat{X}_{t-} \in \Pi$ and $X_{t-} = X_t$.*
(ii) *The set $B - \Pi$ is useless.*

Proof. Consider the set $\{0 < t < \zeta : X_{t-} \neq \hat{X}_{t-},\ \hat{X}_{t-} \notin B\}$, which is a subset of $H \equiv \{0 < t : X_{t-}$ *doesn't exist or* $X_{t-} \neq \hat{X}_{t-}\}$. This is a predictable set which is contained in the union of a countable number of graphs of predictable times. If T is predictable, we have by the moderate Markov property that $\hat{X}_{t-} \notin B \Longrightarrow \hat{X}_{T-} = X_T$, while $T < \zeta \Longrightarrow X_{T-} = X_T$ by the standard character of (X_t). Thus $T \notin H$, hence for P^μ-a.e. ω, H is empty. Thus for P^μ-a.e. ω, for all $t < \zeta(\omega)$, either $\hat{X}_{t-} = X_{t-}$ or $\hat{X}_{t-} \in B$ and $X_{t-} = X_t$.

This proves (i) with Π replaced by B, and (i) will follow as stated if we prove (ii), which we will now do. The argument is similar to that which showed that B was useless for a Hunt process, but is somewhat more delicate.

Let $T > 0$ be a stopping time such that $\hat{X}_{T-} \in B$ P^μ-a.e. on $\{T < \infty\}$. Let $\mathcal{G}$ be the sigma field generated by $\mathcal{F}_{T-}$ and the set $\Lambda = \{T < \zeta\} = \{X_T \neq \partial\}$.

Let $f \in C(\boldsymbol{F})$ with $f(\partial) = 0$. With the convention that $\frac{0}{0} = 0$ we have

$$E^{\mu}\{f(X_T)I_{\{T<\infty\}} \mid \mathcal{G}\} = \frac{E^{\mu}\{f(X_T)I_{\{T<\infty\}}I_{\Lambda} \mid \mathcal{F}_{T-}\}}{P^{\mu}\{T<\infty, \Lambda \mid \mathcal{F}_{T-}\}} I_{\Lambda} + \frac{E^{\mu}\{f(X_T)I_{\{T<\infty\}}I_{\Lambda^c} \mid \mathcal{F}_{T-}\}}{P^{\mu}\{T<\infty, \Lambda^c \mid \mathcal{F}_{T-}\}} I_{\Lambda^c}.$$

Let $h = I_{\boldsymbol{F}-\partial}$. The second term vanishes since $f(\partial) = 0$, and we can write the first in terms of P_0, to see this is

$$= \frac{P_0(fh)(\hat{X}_{T-})}{P_0h(\hat{X}_{T-})} I_{\{T<\zeta\}}.$$

On the other hand, X is quasi-left continuous at T on the set $\{T < \zeta\}$, so $X_{T-} = X_T$, and the above is

$$= \frac{P_0(fh)(X_T)}{P_0h(X_T)} I_{\{T<\zeta\}} = f(X_T)I_{\{T<\zeta\}},$$

this last because P_0 is the identity on D. Define

$$Qf(x) = \frac{P_0(fh)(x)}{P_0h(x)}.$$

Thus $f(X_T)I_{\{T<\zeta\}} = Qf(\hat{X}_{T-})I_{\{T<\zeta\}}$. Square both sides and compare:

$$Q(f^2)(\hat{X}_{T-})I_{\{T<\zeta\}} = (Qf(\hat{X}_{T-}))^2 \, I_{\{T<\zeta\}}.$$

This is true P^{μ}-a.s. for f in a countable dense subset of $C(\boldsymbol{F})$, and we conclude from Jensen's inequality that if $\lambda(A) = P^{\mu}\{\hat{X}_{T-} \in A\}$, that for λ-a.e. $x \in B$, $Q(x, \cdot)$ is either a unit mass or the zero measure. But if Q is the zero measure, x must branch directly to ∂, and is therefore degenerate. There are no degenerate branching points, so Q must be a unit mass. We conclude that for λ-a.e. $x \in B$, that P_0 is of the form

$$P_0(x, \cdot) = p(x)\delta_{\partial}(\,\cdot\,) + (1 - p(x))\delta_y(\,\cdot\,)$$

for some $y \in \boldsymbol{E}$, where $p(x) = P_0(x, \partial)$ of course. Now a priori it is possible that either $y = \partial$ or $p(x) = 0$ or 1. However any of these possibilities implies that x is a degenerate branching point which branches either to ∂ or y, and this can not happen. We conclude that x is a fork so that $X_{T-} \in \Pi$ on $\{T < \zeta\}$.

Now the set $\{t : \hat{X}_{t-} \in B\}$ is contained in the union of graphs of countably many predictable times, and we have just seen that if T is predictable, $X_{T-} \notin B - \Pi$ a.s.. Thus $(\hat{X}_{t-})$ never hits $B - \Pi$. Neither does (X_t), since it never hits B. Thus $B - \Pi$ is μ-useless for any μ which sits on $\boldsymbol{E}$, and is hence useless, as claimed. □

Remark 8.43. We see from this theorem that there are three ways the standard process can be killed. First, it can converge directly to ∂. Since ∂ is not a branching point (it's absorbing), (X_t) will hit it and stick. (∂ was isolated in the topology of $\boldsymbol{E}$, but it might not be isolated in $\boldsymbol{F}$.) A second possibility is that the process jumps directly from a non-branching point to ∂. This jump will be at a totally inaccessible time by Theorem 8.13. Finally, the process may approach a fork, say x. In that case, it is killed with probability $p(x)$, and jumps to some other point $y \in \boldsymbol{E}$ with probability $1 - p(x)$. Notice that if only the first two kinds of killing are present, (X_t) is actually a Hunt process, for it will be quasi-left continuous at all times. If the third kind is present, then the process will still be quasi-left continuous up to time ζ, since the only possible times that quasi-left continuity could fail are when $\hat{X}_{t-}$ is at a fork, say x, and jumps directly to a point y. But by (i), $X_{t-} = X_t$ then, so the process is actually continuous.

This itself does not quite prove that a standard process is a killed Hunt process, but one can construct a Hunt process from (X_t) by extending its lifetime as follows: each time $\hat{X}_{t-}$ is at a fork x which branches to y, say, the process either jumps to y or it is killed. If the process is killed, we extend its lifetime by setting $X_t = y$ instead, and then let it continue independently according to its transition function until the next time it approaches a fork, when we continue the process again. The resulting extension will be a Hunt process – for the third type of killing has been eliminated – and of course we can recover the original process by killing the extension at the forks. An actual proof of this is rather long, but the idea is clear, and we leave the task of carrying it out as an exercise.

8.13. Separation and Supermedian Functions

This section is somewhat technical. Theorems 8.25 and 8.30 require a set of 1-supermedian functions which separate points of $\boldsymbol{E}$ and have some additional continuity properties. The problem is: does such a set always exist, and if so, how do we find it? It is easy if the resolvent itself separates points – then the 1-potentials are enough by Lemma 8.33 – but it isn't if it doesn't. If the resolvent doesn't separate points – and it may well not – we need more supermedian functions. We will show that under Hypotheses (SMP) or (MMP), we can find them. Here is the main theorem. Compatibility is defined below.

Definition 8.44. *A function f on $\boldsymbol{E}$ is* **right-regular for** X *if $t \mapsto f(X_t)$ is a.s. right-continuous. It is* **left-regular for** X *if $t \mapsto f(X_t)$ is a.s. left-continuous.*

We emphasize "for X" above since (X_t) is a Markov process with a fixed initial distribution, and the continuity property may well be true for some initial distributions and false for others.

Theorem 8.45. *Suppose that Hypothesis (SMP) (resp. (MMP)) holds relative to a compatible resolvent. Then there exists a countable family* (φ_n) *of bounded Borel 1-supermedian functions such that*

(i) (φ_n) *separates points of* $\mathbf{E}$;
(ii) *for all* n, φ_n *is right-regular (resp. left-regular) for* X.

The proof proceeds by a sequence of lemmas. The functions we want are associated with first hitting times. We will do some unashamed bootstrapping: we first reduce to a quotient space in which the resolvent separates points, compactify the quotient space, and use the first-hitting times of the resulting Ray process to construct the functions we need. In what follows, we assume Hypothesis (SMP) (resp. (MMP)). The main difficulty comes from the branching points, which, in this setting, can be defined as follows.

Definition 8.46. *A point* $x \in \mathbf{E}$ *is a* **non-branching point** *if, for all* $f \in C(\bar{\mathbf{E}})$,

$$\lim_{p\to\infty} pU_p f(x) = f(x).$$

It is a **branching point** *otherwise. The set of branching points is denoted* B.

As before, the process does not often encounter branching points.

Lemma 8.47.

(i) *Under (SMP),* B *is* X*-polar.*
(ii) *Under (MMP),* $m\{t : X_t \in B\} = 0$ *a.s.*

Proof. Let T be a stopping (resp. predictable) time at which (X_t) is a.s. right-continuous. Let $f \in C(\bar{\mathbf{E}})$. Then

$$f(X_T) = \lim_{s\downarrow T} f(X_s) = \lim_{n\to\infty} \frac{1}{n}\int_0^\infty e^{-ns} f(X_{T+s})\,ds.$$

By Hunt's Lemma and Hypothesis (SMP) (resp. (MMP)),

$$\begin{aligned} f(X_T) &= E\{f(X_T) \mid \mathcal{F}_T \text{ (resp. } \mathcal{F}_{T-})\} \\ &= \lim_{p\to\infty} E\left\{\int_0^\infty pe^{-ps} f(X_{t+s})\,ds \mid \mathcal{F}_T \text{ (resp. } \mathcal{F}_{T-})\right\} \qquad (8.21) \\ &= \lim_{p\to\infty} pU_p f(X_T). \end{aligned}$$

(To be completely correct, we should let $p \to \infty$ through rational values and note that pU_p is continuous in t.) This holds a.s. simultaneously for a countable dense subset of $C(\bar{\mathbf{E}})$, and we conclude that X_T is a.s. a non-branching point. Under (SMP) (X_t) is right-continuous, so this holds for every finite stopping time T. It follows from Corollary 6.10 that B is X-polar, proving (i).

Under (MMP), however, (X_t) is left-continuous, not right-continuous, so (8.21) holds at times when (X_t) is continuous. In particular, it holds for $T \equiv t$ if t is a.s. a continuity point of (X_t), and it implies that $P\{X_t \in B\} = 0$. Then (ii) follows by an application of Fubini's theorem. □

There are some properties that we might expect of a resolvent satisfying (SMP) or (MMP) which may not – quite – hold. For instance, Lemma 8.47 suggests that $U_p(\cdot, B)$ should vanish identically, but in fact, it is quite possible that there are exceptional points where it is strictly positive. But this only happens on a small (X-polar) set and it is possible to tweak the resolvent to eliminate it without endangering (SMP) or (MMP). Let us consider *compatible* resolvents.

Definition 8.48. *A resolvent $(U_p)_{p>0})$ on $\mathbf{E}$ is* **compatible** *if*

(i) *for all $x \in \mathbf{E}$, and $p > 0$, $U_p(x, B) = 0$; and*
(ii) *for all $x \in \mathbf{E}$, the vague limit $\lim_{p\to\infty} pU_p(x, \cdot)$ exists.*

As it happens, there is no loss in assuming from the start that we are dealing with a compatible resolvent, as the following lemma shows.

Lemma 8.49. *Assume (SMP) (resp. (MMP)). Then there is a compatible resolvent $(\hat{U}_p)_{p>0}$ which also satisfies (SMP) (resp. (MMP)) and which differs from $(U_p)_{p>0}$ on at most an X-polar set.*

Proof. Suppose that $N_0 \in \mathcal{E}$ satisfies

$$m\{t : X_t \in N_0\} = 0 \text{ a.s.} \tag{8.22}$$

where m is Lebesgue measure. Then the set $N' = \{x : U_1(x, N_0) > 0\}$ is X-polar. Indeed, from (8.22), if T is any stopping (resp. predictable) time,

$$U_1(X_T, N_0) = E\left\{\int_0^\infty e^{-t} I_{N_0}(X_{T+t})\, dt \mid \mathcal{F}_T \ (resp.\ F_{T-})\right\} = 0$$

since the integrand vanishes a.s.. Thus $P\{X_T \in N'\} = 0$ for all such T, hence by Theorem 6.10 N' is X-polar.

Let $N_0 = B$. Then (8.22) holds by Lemma 8.47. Set $N_{n+1} = N_n \cup \{x : U_1(x, N_n) > 0\}$, n=0, 1,.... By induction, N_n satisfies (8.22) for all n, hence so does $C_0 \stackrel{\text{def}}{=} \cup_n N_n$. Let $C' = \{x : U_1(x, C) > 0\}$. Then $B \cup C' \subset C_0 \in \mathcal{E}$ and C' is X-polar.

There is one other exceptional set to deal with. It is possible that the set $M \stackrel{\text{def}}{=} \{x : \lim_{p\to\infty} pU_p(x, \cdot)$ does not exist in $\mathbf{E}\}$ is non-empty, where the limit is in the vague topology of measures on $\bar{\mathbf{E}}$. M is necessarily a subset of B. Note that as the paths have right limits under either (SMP) or (MMP), for any stopping (resp. predictable) time T and $f \in C(\bar{\mathbf{E}})$, $\lim_{s\downarrow\downarrow 0} f(X_{T+s})$ exists.

If we write (8.21) with t replaced by T and $\mathcal{F}_t$ replaced by $\mathcal{F}_T$ (resp. $\mathcal{F}_{T-}$), we see that $\lim_{p\to\infty} pU_p f(X_T)$ exists a.s.. This being so for a countable dense subset (f_n), we conclude that the vague limit $\lim_{p\to\infty} pU_p(X_T, \cdot)$ exists a.s.. The limit is a priori a measure on $\bar{\boldsymbol{E}}$, but in fact, as all limit points of the paths are in $\boldsymbol{E}$, it must sit on $\boldsymbol{E}$ a.s. Thus with probability one, $P\{X_T \in M\} = 0$ for all such T, hence M is X-polar. Set $C = C' \cup M$. Then C is also X-polar.

We now modify the resolvent. Choose a point $x_0 \in \boldsymbol{E} - C_0$ and set

$$\hat{U}_p(x, \cdot) = \begin{cases} U_p(x, \cdot) & \text{if } x \in \boldsymbol{E} - C \\ U_p(x_0, \cdot) & \text{if } x \in C. \end{cases} \tag{8.23}$$

This makes each point in C a branching point which branches directly to x_0. Since $U_1(x, \cdot)$ – and hence $U_p(x, \cdot)$ for all p – doesn't charge C, $U_pU_q = \hat{U}_p\hat{U}_q$ on $\boldsymbol{E} - C$, and also on C by construction. Since (U_p) satisfies the resolvent equation, this means that $(\hat{U}_p)$ does too. Moreover, the vague limit $\lim_{p\to\infty} pU_p(x, \cdot)$ exists in $\boldsymbol{E}$ for all x. Indeed, the limit is δ_x if $x \in \boldsymbol{E} - C$ and δ_{x_0} if $x \in C$.

Finally, C is X-polar so that $\hat{U}_p f(X_t) \equiv U_p(X_t)$, so if the latter is rcll (resp. lcrl), so is the former. □

Now we will deal with the remaining thorny problem: the resolvent may not separate points. If it doesn't, we will reduce the problem to a quotient space in which the resolvent does separate points. Let (f_n) be a countable dense (sup-norm) subset of $\mathcal{S}(U_1(C^+(\bar{\boldsymbol{E}})))$. Such a set exists by Knight's Lemma since $C^+(\bar{\boldsymbol{E}})$, and therefore $U_1(C^+(\bar{\boldsymbol{E}}))$, are separable. Define $d(x, y)$ by

$$d(x, y) = \sum_{n=1}^{\infty} 2^{-n}(1 + \|f_n\|)^{-1}|f_n(y) - f_n(x)|, \quad x, y \in \boldsymbol{E}.$$

This will be a metric if the resolvent separates points. Otherwise it is a pseudo-metric and generates an equivalence relation by

$$x \sim y \iff d(x, y) = 0 \iff U_p(x, \cdot) = U_p(y, \cdot)\ \forall\, p > 0.$$

Let $\boldsymbol{E}^*$ be the set of equivalence classes of elements of $\boldsymbol{E}$ and let $h : \boldsymbol{E} \mapsto \boldsymbol{E}^*$ map each x to its equivalence class. Then $\boldsymbol{E}^*$ is a metric space if we give it the metric $d^*(\xi, \eta) \stackrel{\text{def}}{=} d(h^{-1}(\xi), h^{-1}(\eta))$, $\xi, \eta \in \boldsymbol{E}^*$. Define a process on $\boldsymbol{E}^*$ by $X_t^* \stackrel{\text{def}}{=} h(X_t)$. Let $\mathcal{E}^*$ be the topological Borel field of E^* and note that h is a measurable map from $(\boldsymbol{E}, \mathcal{E})$ to $(\boldsymbol{E}^*, \mathcal{E}^*)$. Define $U_p^*(\xi, A)$ by $U_p^*(h(x), A) = U_p(x, h^{-1}(A))$. As $U_p(\cdot, A)$ is constant on equivalence classes, we can write this unambiguously as $U_p^*(\xi, A) = U_p(h^{-1}(\xi), h^{-1}(A))$. Under the hypothesis (SMP) (resp. (MMP)), $t \mapsto f_n(X_t)$ is a.s. right-continuous (resp. left-continuous).

Notice that if $x \neq y$ are non-branching points, there exists $f \in C(\bar{E})$ such that

$$\lim_{p\to\infty} pU_p f(x) = f(x) \neq f(y) = \lim_{p\to\infty} pU_p f(y),$$

so that the resolvent separates x and y, and $d(x, y) > 0$. Consequently, we have

Lemma 8.50.

(i) *If $x \neq y$ and $x \sim y$, one of the two is a branching point.*
(ii) *h is one-to-one from $\mathbf{E} - B$ into $\mathbf{E}^*$.*

Lemma 8.51. *Under (SMP) (resp. (MMP)), $(U_p^*)_{p>0}$ is a resolvent on $\mathbf{E}^*$ which separates points and (X_t^*) is a Markov process which satisfies (SMP) (resp. (MMP)) relative to it.*

Proof. If $f \in b\mathcal{E}^*$ then

$$U_p^* f(\xi) = U_p(f \circ h)(h^{-1}(\xi)); \tag{8.24}$$

this is clear if $f = I_A$, $A \in \mathcal{E}^*$ since then $U_p^* f(\xi) = U_p(h^{-1}(\xi), h^{-1}(A)) = U_p(f \circ h)(h^{-1}(\xi))$. (8.24) follows for simple f by linearity and hence for all f by a monotone class argument. The resolvent equation follows from this and a little manipulation:

$$\begin{aligned} U_p^* U_q^* f(\xi) &= U_p^* \left[(U_q(f \circ h)) \circ h^{-1}\right](\xi) \\ &= U_p \left[(U_q(f \circ h)) \circ h^{-1} \circ h\right](h^{-1}(\xi)) \\ &= U_p \left[U_q(f \circ h)\right](h^{-1}(\xi)) \\ &= \frac{1}{q-p} \left[U_p(f \circ h) - U_q(f \circ h)\right](h^{-1}(\xi)) \\ &= \frac{1}{q-p} \left[U_p^* f(\xi) - U_q^* f(\xi)\right]. \end{aligned}$$

where we used the fact that (U_p) satisfies the resolvent equation. Now let us check the Markov property. If $f \in \mathcal{E}_b^*$,

$$E\left\{\int_0^\infty e^{-ps} f(X_{t+s}^*)\, ds \mid \mathcal{F}_t\right\} = E\left\{\int_0^\infty e^{-ps} f \circ h(X_{t+s})\, ds \mid \mathcal{F}_t\right\}.$$

By (MP) and (8.24) this is

$$= U_p(f \circ h)(X_t) = U_p^* f(X_t^*).$$

To check (SMP) (resp. (MMP)) we must show that $t \mapsto U_p^* f(X_t)$ is rcll (resp. lcrl) for each $f \in \mathcal{E}_b^*$. But

$$U_p^* f(X_t^*) = U_p(f \circ h)(h^{-1}(X_t^*)) = U_p(f \circ h)(X_t),$$

which is rcll by (SMP) (resp. lcrl by (MMP).) This is true in particular for the functions which define the metric, so that $t \mapsto X_t^*$ is itself rcll (resp. lcrl.) $\square$

Under either (SMP) or (MMP), Lemma 8.33 implies that (X_t^*) and (U_p^*) satisfy the hypotheses of Theorem 8.30, so there exists a Ray-Knight compactification $(\boldsymbol{F}^*, \bar{U}_p)$ of $(\boldsymbol{E}^*, U_p^*)$. Let $(\bar{X}_t)$ be the corresponding right-continuous Ray process, let $(\bar{P}_t)$ be its semigroup and let $(\bar{P}^\mu)$ be the distribution of $(\bar{X}_t)$ with initial distribution μ.

So: (X_t) and (X_t^*) are fixed processes with fixed initial distributions. $(\bar{X}_t)$ is a Ray process which can have different initial distributions, but in the rcll case, if $\bar{X}_0 = X_0^*$, then $\bar{X}_t \equiv X_t^*$. In other words, (X_t^*) is a realization of $(\bar{X}_t)$ for its particular initial distribution. (In the lcrl case, (X_t^*) is a realization of $(\bar{X}_{t-})$.) However, the topologies of $\boldsymbol{E}$ and $\boldsymbol{E}^*$ may differ: the map h is measurable, but not necessarily continuous.

Penetration times were defined in §6.2. Define

Definition 8.52. *Let $t \geq 0$, $A \in \mathcal{E}$. The* **first penetration time of A after t** *is*

$$\pi_A^t = \inf\{s \geq t : m\{t \leq u \leq s : X_u \in A\} > 0\},$$

where m is Lebesgue measure.

Lemma 8.53. *Let $A \subset \boldsymbol{E}$ be open and $K \subset \boldsymbol{E}$ be closed. Then there exist 1-supermedian Borel functions ϕ_A and ψ_K on $\boldsymbol{E}$ such that*

(i) *under (SMP), ϕ_A is right-regular and for all finite stopping times T,*

$$\phi_A(X_T) = E\left\{e^{-(D_A^T - T)} \mid \mathcal{F}_T\right\} \quad a.s. \tag{8.25}$$

(ii) *under (MMP), ψ_K is left-regular, and for all finite predictable times T,*

$$\psi_K(X_T) = E\left\{e^{-(D_K^T - T)} \mid \mathcal{F}_{T-}\right\} \quad a.s. \tag{8.26}$$

where D_K^t is the debut of K after t.

Proof. Let $A^* = h(A - B)$. Since h is one-to-one on $\boldsymbol{E} - B$ (Lemma 8.50) the general theory tells us $A^* \in \mathcal{E}^*$. Consider the Ray process $(\bar{X}_t)$ and define

$$\bar{\phi}_{A^*}(\xi) = \bar{E}^\xi\{e^{-\bar{\pi}_{A^*}}\}$$

where $\bar{E}^\xi$ is the expectation given that $\bar{X}_0 = \xi$, and $\bar{\pi}_{A^*}$ is the first penetration time of A^* by $(\bar{X}_t)$. This is $\mathcal{E}^*$-measurable and 1-excessive for $(\bar{P}_t)$. Define

ϕ_A on $\boldsymbol{E}$ by

$$\phi_A(x) = \begin{cases} \bar{\phi}_{A^*}(h(x)) & \text{if } x \in \boldsymbol{E} - (A \cap B) \\ 1 & \text{if } x \in A \cap B. \end{cases}$$

Now U_p puts no mass on B, so

$$pU_{p+1}\phi_A(x) = p\bar{U}_{p+1}\bar{\phi}_{A^*}(h(x)) \le \bar{\phi}_{A^*}(h(x)) \le \phi_A(x),$$

so ϕ_A is 1-supermedian. We claim that $\phi_A = 1$ on A. This is true by definition on $A \cap B$. Otherwise, note that A is open so if $x \in A - B$, then $pU_p(x, A) \to 1$ as $p \to \infty$. Moreover, $U_p(x, B) = 0$ so, as $A - B \subset h^{-1}(A^*) \subset A \cup B$ (Lemma 8.50) we have $U_p(x, A) = U_p(x, h^{-1}(A^*))$. But this equals $U_p^*(h(x), A^*) = \bar{U}_p(h(x), A^*)$, so $p\bar{U}_p(h(x), A^*) \to 1$, and evidently $\bar{P}^{h(x)}\{\bar{\pi}_{A^*} = 0\} = 1$. It follows that $\phi_A(x) = \bar{\phi}_A(h(x)) = 1$ as claimed.

Let us show (8.25). Under (SMP), (X_t) is rcll and B is X-polar by Lemma 8.47 so $D_A^T = \pi_A^T = \pi_{A-B}^T$ a.s. and, as $X_t^* = h(X_t)$ this equals $\bar{\pi}_A^T$, which is the first penetration time of A^* after T by (X_t^*). As (X_t^*) equals the Ray process $(\bar{X}_t)$,

$$E\left\{e^{-(D_A^T - T)} \mid \mathcal{F}_T\right\} = E\left\{e^{-(\pi_{A^*}^T - T)} \mid \mathcal{F}_T\right\} = \bar{\phi}_{A^*}(X_T^*).$$

If $X_T \in \boldsymbol{E} - A \cap B$, this equals $\phi_A(X_T)$, while if $X_T \in A \cap B$, $D_A^T = T$, and both sides of (8.25) equal one.

The right-regularity of ϕ_A follows immediately, since, as (X_t) never hits B, $\phi_A(X_t) \equiv \bar{\phi}_A(X_t^*)$, and this last is right-continuous and has left limits since $\bar{\phi}_A$ is $\bar{P}_t$-1-excessive.

Now (8.25) also holds for predictable T under (MMP) with $\mathcal{F}_T$ replaced by $\mathcal{F}_{T-}$, but the above argument needs to be modified since (X_t) is now lcrl, and the left-continuous process may hit B. However, it spends zero (Lebesgue) time there, so if T is predictable, we still have $D_A^T = \pi_A^T$ on the set $\{X_T \notin A\}$. Thus the argument is valid on $\{X_T \notin A\}$, while if $X_T \in A$, $D_A^T \equiv T$ and $\phi_A(X_T) = 1$, so (8.25) holds a.e. with $\mathcal{F}_T$ replaced by $\mathcal{F}_{T-}$.

For $K \in \boldsymbol{E}$ closed, let A_n be open neighborhoods of K such that $A_n \supset \bar{A}_{n-1}$ and $K = \cap_n A_n$. Then $\lim_n D_{A_n}^T = D_K^T$. (See §8.9 for the properties of debuts for left-continuous processes.)

Define $\psi_K(x) = \lim_{n\to\infty} \phi_{A_n}(x)$. This is 1-supermedian, since it is a decreasing limit of 1-supermedian functions, and, by (8.25)

$$\begin{aligned} \psi_K(X_T) &= \lim_n \phi_{A_n}(X_T) = \lim_n E\left\{e^{-(D_{A_n}^T - T)} \mid \mathcal{F}_{T-}\right\} \\ &= E\left\{e^{-(D_K^T - T)} \mid \mathcal{F}_{T-}\right\} \end{aligned}$$

where the last equality comes from Hunt's Lemma. Thus ψ_K satisfies (8.26).

To show ψ_K is left-regular, we let (T_n) be an increasing sequence of uniformly bounded predictable times with limit T. For a closed K, $t \mapsto D_K^t$ is left-continuous. By (8.26) and the bounded convergence theorem

$$\lim_n E\left\{\psi_K(X_{T_n})\right\} = \lim_n E\left\{e^{-(D_K^{T_n}-T_n)}\right\} = E\left\{e^{-(D_K^T-T)}\right\}.$$

This implies that $t \mapsto \psi_K(X_t)$ is left-continuous by Theorem 6.15. □

We can now prove Theorem 8.45.

Proof. (of Theorem 8.45) There exists a countable set of 1-potentials, say $(U_1 f_n)$, which separates all pairs (x, y) which are not equivalent, i.e. for which $d(x, y) > 0$. If there are equivalent points, we use Lemma 8.53 to find a countable family of regular 1-supermedian functions which separate them, as follows.

If $x \sim y$, one of the two, say x, must be a branching point. Let (A_n) be an open base of the topology of $\bar{E}$, and choose n such that $x \in A_n$, $y \in A_n^c$. Then $\phi_{A_n}(x) = 1$, but, as x is a branching point, if A_n is small enough, $\lim_{n\to\infty} pU_p(x, A) < 1$. Since $U_p(y, \cdot) = U_p(x, \cdot)$, $\lim_{n\to\infty} pU_p(y, A) < 1$ too, which implies that $\bar{P}^y\{\pi_{A_n^*} > 0\} > 0$. Thus $\phi_{A_n}(y) < 1$ while $\phi_{A_n}(x) = 1$.

It follows that (ϕ_{A_n}) is a set of regular 1-supermedian functions which separate all pairs of equivalent points. Thus they, together with the $(U_1 f_n)$, separate points of $\boldsymbol{E}$.

Note that the same argument shows that x and y are separated by the $\psi_{\bar{A}_n}$ as well. Thus $(U_1 f_n) \cup (\phi_{A_n})$ and $(U_1 f_n) \cup (\psi_{\bar{A}_n})$ are countable right regular and left regular separating sets under (SMP) and (MMP) respectively. □

8.14. Examples

EXAMPLE 8.54. Let $\boldsymbol{E}$ be the lazy crotch, i.e. the set of the three rays A, B, and C pictured below in Fig. 8.1. Let (X_t) be uniform motion to the right on each, except that when it reaches the origin, it tosses a (figurative) coin, and goes on to ray B or ray C with probability one-half each. This is a Markov process, but it is not strongly Markov, since there is no way to define the transition probability at the origin to satisfy the Markov property at both stopping times T_B and T_C, the first times the process enters the interior of B and C respectively.

Let us see what a Ray-Knight compactification does. Any positive function which decreases in the direction of motion will be supermedian. Choose any two such functions which separate points ($\pi/2 - \arctan x$ and $\pi/2 + 0 \wedge \arctan y$ will do, for instance) and compactify. Then the compactification $\boldsymbol{F}$ will be homeomorphic to the second figure below. Notice that the origin has split into three points, represented by the three circles. When the Ray process on $\boldsymbol{F}$ reaches the hollow circle, it jumps to one of the two solid circles with

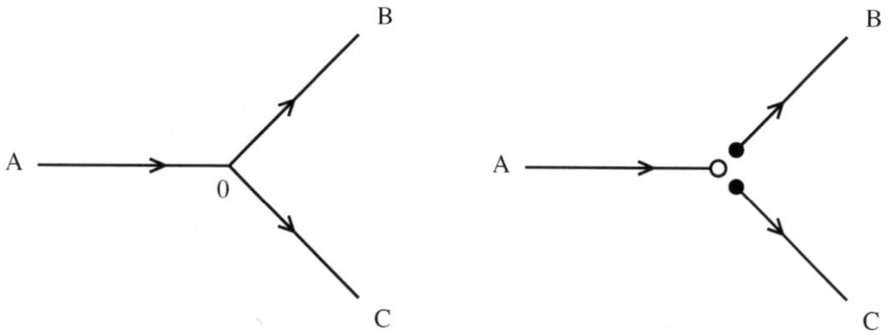

Fig. 8.1. The lazy crotch and its Ray compactification.

probability 1/2 each, and then continues its motion on that ray. Thus the hollow circle is a branching point, with branching measure sitting on the two solid circles. There is no longer a problem at either T_A or T_B, and the process is easily seen to be strongly Markov.

EXAMPLE 8.55. Useful counter-examples in probabilistic potential theory are often trivial, such as uniform motion. Here is one which is even more trivial than most, but may help us understand the uniqueness – or the lack of it – in $\mathcal{U}$-compactifications. Let $\boldsymbol{E} = [-1, 1]$, and consider the trivial process $X_t \equiv X_0$, with resolvent $U_p(x, \cdot) = p^{-1}\delta_x(\cdot)$. Then $pU_p f = f$ for all f, and all positive functions are not only excessive functions, but potentials. If $\boldsymbol{E}$ has the usual topology, (X_t) is a Ray process. That doesn't prevent us from compactifying it, however. If we start with $f_1(x) \equiv x + 1$, and $G = \{f_1\}$, the compactification is $\boldsymbol{E}$ again. If we start with $G = \{f_1, f_2\}$, where $f_2(x) = I_{\{x \geq 0\}}$, then the the point 0 splits into two points, 0_- and 0_+, say, and the Ray-Knight compactification is homeomorphic to $[-1, 0_-] \cup [0_+, 1]$. This is a second $\mathcal{U}$-compactification.

One might object that in the second compactification, f_2 is not continuous, but our definition of $\mathcal{U}$-compactification did not require that. And indeed, if we define a $\mathcal{U}_c$-compactification to be a $\mathcal{U}$-compactification based on *continuous* f, then one can show that $\mathcal{U}_c$-compactifications are unique in the sense that they are homeomorphic. But then this example shows that $\mathcal{U}_c$-compactifications depend on the original topology of $\boldsymbol{E}$, for we could easily have given $\boldsymbol{E}$ a topology in which f_2 was continuous without affecting the continuity properties of the paths.

EXAMPLE 8.56. Consider the classical Dirichlet problem on the unit disk: find a function h which is harmonic on the open unit disk $\boldsymbol{E}$ in the plane, continuous on its closure, and equal to a given continuous function f on the boundary. This has the elegant probabilistic solution given in §4.4 (1). Let (X_t) be Brownian motion on $\boldsymbol{E}$, and let ζ be the first time X exits $\boldsymbol{E}$. Then

$$h(x) = E^x\{f(X_\zeta)\}. \tag{8.27}$$

To translate this to the general probabilistic setup, add an isolated point ∂ for use as a cemetery, and re-define (X_t) to be Brownian motion for $t < \zeta$, and equal to ∂ for $t \geq \zeta$. Its state space is now $\boldsymbol{E} \cup \partial$. This is a fundamental example of a standard process which is not a Hunt process. The original process took values in the plane and $X_{\zeta-}$ was in the unit circle. But the new process takes values in $\boldsymbol{E} \cup \partial$, which does not contain the circle, so $X_{\zeta-}$ simply does not exist in this space.

Let us now $\mathcal{U}$-compactify the new space $\boldsymbol{E} \cup \partial$. Notice that as z approaches points of the unit circle, $pU_p(z, \cdot)$ approaches the unit mass at ∂, which equals $pU_p(\partial, \cdot)$. Thus the compactified space $\boldsymbol{F}$ is isomorphic to be the one-point compactification of the open unit disk, with ∂ playing the role of the one point. Up to its lifetime, the Ray process is Brownian motion on the unit disk as before. But in the topology of $\boldsymbol{F}$, as $t \to \zeta$, X_t will converge to ∂. Thus the compactified process is actually continuous, even at ζ , so $X_{\zeta-} \equiv X_\zeta \equiv \partial$. This is awkward: we've just lost the solution (8.27)!

Notice that this is entirely a question of $X_{\zeta-}$. In the plane, $X_{\zeta-}$ is a non-trivial random variable taking values in the unit circle. In $\boldsymbol{E}$, it simply does not exist. In the compact space $\boldsymbol{F}$, it is trivial, always being equal to ∂. We do not want this. We want to define $X_{\zeta-}$ so as to preserve (8.27).

The problem is that there were too few functions in G to separate points of the unit circle. We should have used a larger set.

If h is a positive harmonic function on $\boldsymbol{E}$ and zero on ∂, it is excessive (and therefore 1-supermedian) on $\boldsymbol{E} \cup \partial$. So we simply add enough harmonic functions to separate points of the unit circle. In the event, it is enough to let G consist of the three functions $h_1(x, y) \equiv 1$, $h_2(x, y) = 1 + x$ and $h_3(x, y) = 1 + y$, extended to vanish at ∂. These separate points of $\boldsymbol{E} \cup \partial$, and they separate points of the unit circle as well. When we compactify $\boldsymbol{E}$ with this scheme, the compactification can be identified with the closed unit disk plus the single isolated point ∂. The process (X_t) will still be Brownian motion for $t < \zeta$; at its lifetime, $X_{\zeta-}$ will be in the unit circle, while $X_\zeta = \partial$. Then (8.27) becomes

$$h(x) = E^x\{f(X_{\zeta-})\}. \tag{8.28}$$

Contrast this with (8.27): $X_{\zeta-}$, not X_ζ, figures. Indeed, X_ζ itself is trivial: $X_\zeta \equiv \partial$, since each point of the unit circle has become a degenerate branching point which branches directly to ∂. Thus we have preserved the probabilistic representation of harmonic functions at the (negligible!) price of introducing branching points.

Exercises

1. Let $X_t = (B_t, t)$ be space-time Brownian motion, where $\{B_t,\ t \geq 0\}$ is a standard Brownian motion in $\boldsymbol{R}^1$. Show that if $s \neq t$ the measures $P_s(x, \cdot)$ and $P_t(x, \cdot)$ are mutually singular as measures on $\boldsymbol{R}^2$, while the resolvent is absolutely continuous with respect to Lebesgue measure on $\boldsymbol{R}^2$.

2. If $f \in \mathcal{S}$, then $\hat{f} \overset{\text{def}}{=} P_0 f$ is 1-excessive. Careful: this uses the continuity of f, and is not necessarily true of all 1-supermedian functions.
3. Show that if T is a predictable time at which $P^\mu\{X_{T-} \in B - B_d\} > 0$, then there exists a martingale M which is P^μ-a.s. discontinuous on $\{X_{T-} \in B - B_d\}$.
4. Recall that if $A \in \mathcal{F}_T$, T_A is the optional time which equals T on A, $+\infty$ on A^c. For a stopping time T, let $A = \{0 < T < \infty,\ X_{T-} \in D,\ X_{T-} \neq X_T\}$. Show that for a Ray process, the totally inaccessible part of T is T_A, and the accessible part is T_{A^c}.
5. Let $\boldsymbol{E}$ be the first quadrant of the open upper half plane, and let X_t be translation to the right, parallel to the x axis in such a way that if it is at (x, y_0), it moves on the line $y = y_0$ at speed $1/y_0$ if $0 < x < 1$, and with speed 1 otherwise. Let G consist of the functions e^{-x} and e^{-y}, and compactify using G. Show that the original process is a Hunt process, but that the compactification includes the x axis as well, and that the points $(x, 0)$, $0 \le x < 1$ are degenerate branching points, all of whom branch directly to $(1, 0)$.
6. Consider the lazy crotch of Example 8.54, but let the process move to the left at unit speed instead of to the right. Verify that this is a Hunt process. Now kill it as follows: if the process comes from C, kill it with probability 1/2 as it passes the origin. If it comes from B, let it continue to A unscathed. (This implies adjoining a state ∂ to which the process jumps when it is killed. So be it.) Show that this is a standard process but not a Hunt process. Show that a $\mathcal{U}$-compactification $\boldsymbol{F}$ can be identified with the one-point compactification of the following: $\partial \cup A \cup B \cup C'$, where C' is a closed ray which is separated from $A \cup B$, and instead of connecting to zero, terminates in a fork, say c, which branches to ∂ with probability 1/2, and to the origin with probability 1/2. Describe the killing in terms of $\hat{X}_{t-}$ in $\boldsymbol{F}$.
7. Let (X_t) be a Feller process on $\boldsymbol{E}$. Show that if $\boldsymbol{F}$ is a $\mathcal{U}$-compactification of $\boldsymbol{E}$, then $\boldsymbol{F}$ is isomorphic to the one-point compactification of $\boldsymbol{E}$ (or to $\boldsymbol{E}$ itself if it is already compact) and the topology on $\boldsymbol{E}$ is that inherited from $\boldsymbol{F}$.
8. Construct an example of a Hunt process and a compactification of it which introduces a set of branching points which are not useless.
9. If $\boldsymbol{E}$ is a locally compact space, show that $\mathcal{S}(C_0^+(\boldsymbol{E}))$ is separable, where $C_0^+(\boldsymbol{E})$ is the set of continuous functions on $\boldsymbol{E}$ which have limit zero at infinity.
10. Let $\{B_t, t \ge 0\}$ be Brownian motion on the domain D, where D is the unit disk in $\boldsymbol{R}^2$ minus the positive x-axis. Consider the Ray-Knight compactification generated by the functions $f_1(x, y) = x$, $f_2(x, y) = y$, and $f_3(x, y) = \arg(z)$, where $\arg z$ is the argument of $z = x + iy$. Show that the compactification is homeomorphic to the closed unit disk.

Notes on Chapter 8

§8. Ray resolvents and Ray processes were developed by D. Ray in [1]. Ray found the exact hypotheses, and his theorem has needed no generalization: supermedian functions, not excessive functions, are the key. Getoor [1] and Sharpe [1] both give good, careful treatments of Ray processes under weaker measurability hypotheses than we do.

The first three sections of this chapter come from Ray [1], with the exception of the "left-handed" properties – a trivial addition – which was motivated by the appearance of the moderate Markov property in 1969; they were first pointed out in Walsh [2].

Proposition 8.7 and Theorems 8.13 and 8.10 summarize the properties of the jumps of a Ray process and their relation to martingales and filtrations. These have three sources: the behavior at the branching points is due to D. Ray [1]; the behavior at the non-branching points comes from Meyer's theorem that if a Hunt process is a.s. continuous at a stopping time, that stopping time must be predictable. The proof given here is from Chung and Walsh [2]. The relations between stopping times, right-continuity of filtrations, and continuity of martingales are typical results of the General Theory of Processes, and are taken from Walsh [2]. This all comes together in a remarkably neat way for Ray processes.

§8.6. This is from Meyer and Walsh [2].

§8.8. Theorem 8.16 was proved by Kunita and Watanabe in [3].

§8.9. This contains left-handed versions of some well-known theorems on excessive functions.

§8.10. D. Ray [1] suggested that almost any Markov process could be embedded in a Ray process. Unfortunately, the compactification he constructed was non-metrizable in general, and therefore not useful in many applications. Frank Knight [1] corrected this by proving Lemma 8.24, and that led directly to Theorem 8.25.

The embedding of Hunt and standard processes, and particularly, the fact that there is something to be learned from it, were pointed out in Meyer and Walsh [1]. §8.11 and §8.12 are taken from that article. The results extend to the general class of "right processes" with little change (see Sharpe [1]).

§8.13. This section is motivated by the need to compactify reverses of strong Markov processes. In that case, the excessive functions may not separate points, and it becomes necessary to find regular supermedian functions which do. The results are from Walsh [3].

Chapter 9

Application to Markov Chains

Since Markov chains have a relatively simple state space with a relatively simple (discrete) topology, one has to study them as chains qua chains. There is no freedom to add "harmless" simplifying assumptions. (Assume the chain is continuous... oops, it's constant!) Even right continuity is too much to ask, as we saw in Chapter 7. Thus one is forced to confront the perversities of the processes, and for this reason, many interesting facets of the behavior of general Markov processes were first discovered in continuous-time Markov chains. The subject remains a good touchstone for any purported general theory.

On a deeper level, it is immaterial whether or not we think that the discrete topology is simple. The key to the understanding of some of the wilder members of the breed is finding a topology – a different topology – more closely adapted to the process. This is where Ray-Knight compactifications enter.

We have seen some striking examples in Chapter 7, and now that we have the tools, we will attempt to explain them. We will do somewhat more than that. Although we have no intention of going deeply into the subject, we will first use the Ray-Knight compactifications to develop some of the basic path properties of chains, partially for their own interest, and partially to illustrate how compactifications can make some delicate analysis quite easy. After that, we will see how the examples fit in. Hopefully we will learn something about both Markov chains and Ray processes at the same time.

Let $\boldsymbol{E}$ be either $1, 2, \dots, N$ for some integer N, or $\boldsymbol{N}$ itself. Let $p_{ij}(t)$, $i, j \in \boldsymbol{E}, t \geq 0$, be real-valued measurable functions which satisfy:

(i) $p_{ij}(t) \geq 0$.

(ii) $\sum_{j \in \boldsymbol{E}} p_{ij}(t) \leq 1$, all $t \geq 0, i \in \boldsymbol{E}$.

(iii) $\sum_{j \in \boldsymbol{E}} p_{ij}(s) p_{jk}(t) = p_{ik}(s+t),\ s, t \geq 0,\ i, k \in \boldsymbol{E}$.

(iv) $p_{ii}(0) = 1$ for $i \in \boldsymbol{E}$.

The (p_{ij}) are the *transition probabilities* of a Markov chain. They were introduced earlier, in §1.2, Example 1. Their basic properties are derived in

Chung [2] II.1–3, and can be summarized:

1° $t \mapsto p_{ij}(t)$ is continuous.

2° In fact, $t \mapsto p_{ij}$ is even differentiable for all $t > 0$, and at $t = 0$, $q_i \stackrel{\text{def}}{=} \lim_{t \downarrow 0} t^{-1}(p_{ii}(0) - p_{ii}(t))$ exists in the extended reals: $0 \leq q_i \leq \infty$. If q_i is finite, the $p_{ij}(t)$ are continuously differentiable, and we set $q_{ij} = p'_{ij}(0)$, $j \neq i$.

3° $p_{ij}(t)$ is either strictly positive for all t or identically zero.

4° $p_{ii}(0) = 0$ or 1.

We excluded the zero value in (iv) since if $p_{ii}(0) = 0$, the state is trivial: $p_{ii}(\cdot) \equiv 0$ by 3°, and $p_{ji}(\cdot) \equiv 0$ by 1° and (iii). Thus the chain will never hit i, and we can safely drop it from the state space.

The (p_{ij}) form a sub-Markov semigroup; if the sum in (ii) is strictly less than one for some i, we can make (p_{ij}) into a Markov semigroup as usual by adding a cemetery state ∂ to $\boldsymbol{E}$ to get $\boldsymbol{E}_\partial \equiv \boldsymbol{E} \cup \{\partial\}$, and extend the semigroup to $\boldsymbol{E}_\partial$ by setting $p_{i\partial}(t) = 1 - \sum_{j \in \boldsymbol{E}} p_{ij}(t)$ and $p_{\partial\partial}(t) = 1$. The resolvent is then defined on $\boldsymbol{E}_\partial$ by

$$U_p(i, \{j\}) = \int_0^\infty e^{-pt} p_{ij}(t)\, dt.$$

9.1. Compactifications of Markov Chains

There is a canonical way of compactifying chains. Let $\mathcal{G}$ be the set of functions $\{I_{\{i\}}, i \in \boldsymbol{E}_\partial\}$, and let $G = \{U_1 f, f \in \mathcal{G}\}$. Then G separates points of $\boldsymbol{E}_\partial$ by 1° and (iv), so it generates a compactification via Theorem 8.25. Then there is a compact metric space $\boldsymbol{F}$ and a Ray resolvent $(\bar{U}_p)$ on $\boldsymbol{F}$ such that $\boldsymbol{E}_\partial$ is a dense subset (tho not necessarily a subspace) of $\boldsymbol{F}$. The space $\boldsymbol{F}$ is determined (up to homeomorphism) by

(F1) $\boldsymbol{F}$ is a compact metric space;

(F2) $\boldsymbol{E}_\partial$ is a dense subset of $\boldsymbol{F}$;

(F3) for each $j \in \boldsymbol{E}_\partial$ the mapping $i \mapsto U_1(i, \{j\})$ extends to a function in $C(\boldsymbol{F})$;

(F4) the (extended) functions in (F3) separate points of $\boldsymbol{F}$.

Remark 9.1. This procedure will always give a Ray-Knight compactification of a Markov chain. However, we exaggerated when we said that G was canonical. It is only *nearly* canonical. There are cases in which we may want to add extra 1-supermedian functions to it. For example we may want to separate some boundary points, as we did in Example 8.56, or we may simply wish to make the compactification more intuitive. See Remark 9.13 and the Exercises at the end of this chapter.

Let $\bar{P}_t$ be the corresponding Ray semigroup. Then, by uniqueness of the Laplace transform, $\bar{P}_t(i, \{j\}) = p_{ij}(t)$, $i, j \in \boldsymbol{E}_\partial$, $t \geq 0$.

Let $\{\bar{X}_t, t \geq 0\}$ be the corresponding rcll Ray process. If $\bar{X}_0 = i \in \boldsymbol{E}_\partial$, $U_p(i, \boldsymbol{F} - \boldsymbol{E}_\partial) = 0$, so $X_t \in \boldsymbol{E}_\partial$ for a.e. t. Then it is not hard to see the following:

1° $x, y \in \boldsymbol{F}$: then $x = y \Longleftrightarrow \bar{U}_p f(x) = \bar{U}_p f(y)$, for all $f \in \mathcal{G}$;
2° $x_n \to x \Longleftrightarrow \bar{U}_1 f(x_n) \to \bar{U}_1 f(x)$, for all $f \in \mathcal{G}$;
3° $\bar{U}_p f(x) = U_p f(x)$ for all $f \in \mathcal{G}$, $p > 0$;
4° if $\bar{X}_0 = X_0$, then with probability one, $\bar{X}_t = X_t$ for (Lebesgue) a.e. t.

The center of interest is not the Ray process $(\bar{X}_t)$ but the raw – i.e. uncompactified – chain (X_t) which takes values in the one-point compactification $\boldsymbol{E}_\partial \cup \{\infty\}$. A priori, the process (X_t) may have extremely irregular sample paths, so the first task is to define a nice version of it. Rather than define (X_t) from the transition probabilities, we can use $(\bar{X}_t)$. Define $h(x)$ on $\boldsymbol{F}$ by

$$h(x) = \begin{cases} x & \text{if } x \in \boldsymbol{E}_\partial \\ \infty & \text{otherwise,} \end{cases}$$

and set

$$X_t = h(\bar{X}_t).$$

We need the point at infinity to define the process in a reasonable way at all times t. Now $(\bar{X}_t)$ is rcll in the topology of $\boldsymbol{F}$, but this topology has a tenuous relation to the discrete topology of $\boldsymbol{E}_\partial$, where the sample paths of (X_t) are not guaranteed to have any regularity properties at all.

Two things are clear. First, since $\sum_{j \in \boldsymbol{E}_\partial} p_{ij}(t) = 1$ for all t and $i \in \boldsymbol{E}_\partial$, then $\bar{X}_0 \in \boldsymbol{E}_\partial \Longrightarrow \bar{X}_t = X_t$ a.s. for each t. Next, thanks to Fubini's theorem, the set of t for which $X_t \neq \bar{X}_t$ has Lebesgue measure zero a.s. Thus (X_t) is a Markov process with the transition probabilities $(p_{ij}(t))$ inherited from $(\bar{X}_t)$. (But, careful, it does not necessarily have either the strong or the moderate Markov property.) That means that (X_t) is a Markov chain. Moreover, if we stick to essential limits and first penetration times as opposed to first hitting times, we can treat (X_t) and $(\bar{X}_t)$ interchangeably – the two will have the same essential limits and penetration times.

9.2. Elementary Path Properties of Markov Chains

We will use the Ray-Knight compactification to derive some of the sample path properties of Markov chains. The following lemma holds for both (X_t) and $(\bar{X}_t)$.

Lemma 9.2. *Each point in $\boldsymbol{E}_\partial$ is regular for itself. Moreover, if $i \in \boldsymbol{E}_\partial$, and π_i is the first penetration time of $\{i\}$, $P^i\{\pi_i = 0\} = 1$.*

Proof. For each $t > 0$, $P^i\{\pi_i < t\} = P^i\{t^{-1}\int_0^t I_{\{i\}}(\bar{X}_s)\,ds > 0\} = E\{t^{-1}\int_0^t I_{\{i\}}(X_s)\,ds\} = t^{-1}\int_0^t p_{ii}(s)\,ds \to 1$ as $t \to 0$ since by (iv), $p_{ii}(s) \to 0$ as $s \to 0$. □

Theorem 9.3. *With probability one, in the topology of $\boldsymbol{E}_\partial$ the path $s \mapsto X_s$ can have at most one finite essential limit point from the right and one finite essential limit point from the left at any t.*

Proof. If $s \mapsto X_s$ had two finite essential limit points from the right at some time t, so would $s \mapsto \bar{X}_s$. But this can't happen since $(\bar{X}_s)$ is right continuous. Similarly, it can't have two limit points from the left since $(\bar{X}_s)$ has left limits. □

Proposition 9.4. *$\{X_t,\ t \geq 0\}$ is a Markov chain with values in the one-point compactification $\boldsymbol{E}_\partial \cup \{\infty\}$. Its paths are lower-semicontinuous from the right, and for each optional T, $X_T = \liminf_{t\downarrow 0} X_{T+t}$ a.s. on $\{X_T \in \boldsymbol{E}_\partial\}$.*

Proof. We showed the Markov property above. Let us consider the continuity. Let d be the metric on $\boldsymbol{F}$. If $x \in \boldsymbol{F} - \boldsymbol{E}_\partial$, $d(x, j) > 0$ for all $j \in \boldsymbol{E}_\partial$, hence the distance from x to $\{1, \dots, N\}$ is strictly positive for any N. By right-continuity, if $\bar{X}_t \notin \{1, \dots, N\}$, then $\bar{X}_s \notin \{1, \dots, N\}$ for s in some interval $(t, t+\delta)$, $\delta > 0$. Now either $X_s = \bar{X}_s$ or $X_s = \infty$. In either case, $X_s \notin \{1, \dots, N\}$ for all $s \in (t, t+\delta)$. It follows that if $t < \zeta$ and $X_t \notin \{1, \dots, N\}$, then $\liminf_{s\downarrow t} X_s \geq N$. Thus for all $t < \zeta$, $X_t \leq \liminf_{s\downarrow t} X_s$.

Let T be optional and apply the strong Markov property on $\{X_T = i\} = \{\bar{X}_T = i\}$. Since $(\bar{X}_t)$ is strongly Markov, Lemma 9.2 implies that i is an essential limit point of $\bar{X}_{T+t}$ – and therefore of X_{T+t} – as $t \downarrow 0$. Thus $X_T \geq \operatorname{ess\,lim\,inf}_{t\downarrow 0} X_{T+t} \geq \liminf_{t\downarrow 0} X_{T+t}$, which implies equality. □

Corollary 9.5. *Let T be optional. Then (X_t) satisfies the strong Markov property on the set $\{X_T \in \boldsymbol{E}_\partial\}$.*

Proof. Suppose T is optional with respect to the filtration $(\bar{\mathcal{F}}_t)$ generated by $(\bar{X}_t)$ (which contains the filtration generated by (X_t)). Then $(\bar{X}_t)$ satisfies the strong Markov property at T so that for $j \in \boldsymbol{E}_\partial$, $P\{X_{T+t} = j \mid \bar{\mathcal{F}}_T\} = P\{h(\bar{X}_{T+t}) = j \mid \bar{\mathcal{F}}_T\} = P\{h(\bar{X}_{T+t}) = j \mid \bar{X}_T\}$. The set $\{\bar{X}_T \in \boldsymbol{E}_\partial\}$ is a union of sets $\{\bar{X}_T = i\}$, and it is enough to prove the strong Markov property on each of these. But for $i \in \boldsymbol{E}_\partial$, $X_T = i \iff \bar{X}_T = i$, so the above probability is just $\bar{P}^i\{\bar{X}_t = j\} = p_{ij}(t)$, which is what we wanted to prove. □

Remark 9.6. Corollary 9.5 appears more useful than it really is, for the time when one most wants to use the Markov property is when the process equals infinity... which is exactly when the strong Markov property fails.

Chung [9] uses a different lower-semicontinuous version (X_t^+), which can be defined in terms of (X_t) by $X_t^+ \stackrel{\text{def}}{=} \operatorname{ess\,lim\,inf}_{s\downarrow t} X_s$. By Proposition 9.4, the two are a.s. equal at optional times, and therefore at fixed times, so that (X_t^+) is also a Markov chain with the same transition probabilities, and moreover, Corollary 9.5 also applies to (X_t^+). However, in spite of the fact that they are equal a.s. at optional times, they need not be identical. See Exercise 2.

Note that the process (X_t) is a function of a right-continuous adapted process, so the section theorem applies to it. However, the section theorem evidently does not always apply to (X_t^+) – if it did, $\bar{X}_t$ and X_t^+ would be identical. For aficionados, (X_t^+) is an example of a process which is progressively measurable but not necessarily well-measurable.

9.3. Stable and Instantaneous States

The above results are not at all straightforward without the aid of compactifications. This is evident if one re-examines Example 7.3 of §7. That particular Markov chain satisfies $\limsup_{s\downarrow t} = \infty$ for *all* t. The sample paths are wildly discontinuous. (The *essential* lim sups are also infinite, so this is not just a matter of exceptional null sets.)

Let π_i^* be the first penetration time of $\boldsymbol{F} - \{i\}$. The paths of $(\bar{X}_t)$ are rcll, so this equals its first hitting time of $\boldsymbol{F} - \{i\}$. (We use the first penetration time because it is the same for (X_t) and $(\bar{X}_t)$.) By the Blumenthal zero-one law (§2.3 Theorem 6), $\bar{P}^i\{\pi_i^* = 0\} = 0$ or 1.

Definition 9.7. *A state i is* **stable** *if $\bar{P}^i\{\pi_i^* = 0\} = 0$, and it is* **instantaneous** *if $\bar{P}^i\{\pi_i^* = 0\} = 1$. A chain is* **stable** *if all its states are stable.*

Noting that $(\bar{X}_t)$ is right continuous, and that the first hitting and first penetration times of $\boldsymbol{F} - \{i\}$ are equal, we see that

$$\{\pi_i^* > t\} = \{\forall\, n : \bar{X}_{\frac{t}{n}} = \bar{X}_{\frac{2t}{n}} = \ldots = \bar{X}_t = i\}$$

so that

$$\bar{P}^i\{\pi_i^* > t\} = \lim_{n\to\infty} \bar{P}^i\{\bar{X}_{\frac{t}{n}} = \bar{X}_{\frac{2t}{n}} = \ldots = \bar{X}_t = i\} = \lim_{n\to\infty} p_{ii}(\tfrac{1}{n})^n.$$

But now $q_i = \lim_{t\downarrow 0} t^{-1}(1 - p_{ii}(t))$ so if q_i is finite, this is

$$= \lim_{n\to\infty} (1 - \tfrac{q_i t}{n})^n = e^{-q_i t}.$$

It is easy to see that this also holds if $q_i = \infty$, since then the above limit is zero. We have proved

Proposition 9.8. $q_i < \infty \iff$ *i is stable;*

$q_i = \infty \iff$ *i is instantaneous.*

If i is stable and $X_0 = i$, π_i^* is an exponential random variable with parameter q_i.

We now look at the sets of constancy of (X_t) for both stable and instantaneous states. For $i \in \boldsymbol{E}_\partial$ let $S_i(\omega) = \{t : X_t(\omega) = i\}$.

Theorem 9.9. *Let $i \in \boldsymbol{E}_\partial$ be a stable state. With probability one the set S_i is the union of a locally finite set of left-semi-closed intervals, where "locally finite" means that there are a.s. only finitely many of them in any finite time interval. If i is instantaneous, S_i is nowhere-dense, and if $X_0 = i$, then X_t hits infinitely many other states in each time interval* $(0, \delta)$, $\delta > 0$.

Proof. The sets S_i and $\bar{S}_i \stackrel{\text{def}}{=} \{t : \bar{X}_t = i\}$ are identical, so it is enough to prove it for the latter. (The bar in $\bar{S}_i$ just means that it is associated with $\bar{X}$, not that it is closed!) Let $T_i = \inf\{t > 0 : \bar{X}_t = i\}$ and $T_i^* = \inf\{t > 0 : \bar{X}_t \neq i\}$. First suppose that i is stable. Define a sequence of optional times by

$$\tau_1 = T_i, \quad \sigma_1 = \tau_1 + T_i^* \circ \theta_{\tau_1},$$

and by induction, if we have defined $\tau_1 \leq \sigma_1 \leq \ldots \leq \tau_{n-1} \leq \sigma_{n-1}$, define

$$\tau_n = \sigma_{n-1} + T_i \circ \theta_{\sigma_{n-1}}, \quad \sigma_n = \tau_n + T_i^* \circ \theta_{\tau_n}.$$

By right continuity, $\bar{X}_{\tau_n} = i$ a.s. on $\{\tau_n < \infty\}$. On the other hand, $\bar{X}_{\sigma_n} \neq i$ a.s., for by the definition of σ_n, if $\bar{X}_{\sigma_n} = i$, then σ_n must be a limit point from the right of the set $\{t : \bar{X}_t \neq i\}$, which would imply that $T_i^* \circ \theta_{\sigma_n} = 0$. This contradicts the strong Markov property: σ_n is optional and i is stable, so $\bar{X}_{\sigma_n} = i \Longrightarrow \bar{P}\{T_i^* \circ \theta_{\sigma_n} = 0 \mid \bar{X}_{\sigma_n}\} = \bar{P}^i\{T_i^* = 0\} = 0$, a contradiction. But $(\bar{X}_t)$ is right-continuous, so if $\bar{X}_{\sigma_n} \neq i$, evidently $\tau_{n+1} > \sigma_n$. Thus we have $\tau_1 < \sigma_1 < \ldots < \sigma_{n-1} < \tau_n < \sigma_n < \ldots$ as long as the times are finite.

The proof for the stable case will be finished if we show that $\lim_n \tau_n = \infty$, for then it will follow that $\bar{S}_i = \cup_n[\tau_n, \sigma_n)$, and the union is locally finite.

Notice that on the set $\{\tau_n < \infty\}$, $\sigma_n - \tau_n$ is independent of $\bar{\mathcal{F}}_{\tau_n}$ and has the same distribution as T_i^*, which is exponential (q_i). But $\tau_n \geq \sum_{k=1}^{n-1}(\sigma_k - \tau_k)$, and the the sum of a series of i.i.d. exponential random variables is a.s. infinite. Therefore $\lim_n \tau_n = \infty$.

Now suppose i is instantaneous, and $X_0 = \bar{X}_0 = i$. Then 0 is an essential limit point of $\bar{S}_i$. Therefore, it cannot be a limit point of $\bar{S}_j$ for any other j by Theorem 9.3.

Let $0 \leq a < b$. We claim that $[a, b]$ cannot be contained in the closure of $\bar{S}_i$. If $\bar{X}_a \neq i$, then by right-continuity in $\boldsymbol{F}$, there is $\varepsilon > 0$ such that $X_s \neq i$ for $s \in [a, a + \varepsilon)$, and we are done. If $\bar{X}_{\underline{a}} = i$, then since i is instantaneous, a is a limit from the right of t for which $\bar{X}_t \neq i$. By right continuity, for each such t, there is an interval $[t, t + \epsilon)$ on which $\bar{X}_t \neq i$. Thus there exist open sub-intervals of $[a, b]$ on which $\bar{X}_t \neq i$, which verifies the claim. This is true simultaneously for all rational $a < b$, and it follows that $\bar{S}_i$, and therefore S_i, is nowhere dense. □

The following corollary may help visualize the paths of a stable chain: they are rcll in the extended reals. This doesn't preclude infinite values, but it at least implies that the infinities are nowhere dense. This only holds for stable chains. Chains with instantaneous states can have dense infinities, as we will see below.

Corollary 9.10. *The sample paths of a stable chain are rcll in the one-point compactification of* $\boldsymbol{E}_\partial$.

Proof. This is clear from Theorem 9.9 if $\boldsymbol{E}$ is finite. Suppose $\boldsymbol{E} = \boldsymbol{N}$. The set $\{t : X_t \in \boldsymbol{E}\}$ is a countable union of intervals of the form $[s, t)$ on which (X_t) is constant by Theorem 9.9, so (X_t) is right-continuous on this set. The complement of $\{t : X_t \in \boldsymbol{E}\}$ is $\{t : X_t = \infty\}$. Since the paths of (X_t) are lower semi-continuous from the right, $X_t = \infty \Longrightarrow X_t \le \liminf_{s\downarrow t} X_s \Longrightarrow \lim_{s\downarrow t} X_s = X_t$, which proves right-continuity.

To see that left limits exist, note that if (X_s) has no finite limit points as $s \uparrow t$, then the left-hand limit exists in the extended reals: $\lim_{s\uparrow t} X_s = \infty$. Suppose there exists a finite left limit point, say i. Then there exists a sequence (u_n) such that $u_n < t$, $u_n \to t$, and $X_{u_n} = i$. By Theorem 9.9 each u_n is contained in an interval $[s_n, t_n)$ on which $X_s = i$. These intervals are locally finite, so there can be at most finitely many distinct $s_n < t$, hence there exists n for which $t \in [s_n, t_n]$, and $\lim_{s\uparrow t} X_s = i$ exists. □

A chain with only finitely many states is stable. (Indeed, $X_0 = i$, so 0 is an essential limit point of S_i by Lemma 9.2. It cannot be an essential limit point of S_j for any other state j by Theorem 9.3, so $T_i^* > 0$.)

The sample paths are step functions. Once we have an infinite state space, however, the sample-function behavior becomes more complicated, even if we restrict ourselves to stable chains. It is time to look at this.

9.4. A Second Look at the Examples of Chapter 7

We gave three examples of Markov chains in Chapter 7 which were emphatically not Hunt processes, and whose sample paths ranged from somewhat irregular to frankly bizarre. However, if we compactify them, we will see that this apparent irregularity is a sham: beneath their colorful exteriors they are highly respectable Ray processes in disguise.

EXAMPLE 9.11. Example 7.1 concerned a pure-birth process which blows up in finite time: the jumps $\tau_1 < \tau_2 < \dots$ satisfied $\tau_\infty \stackrel{\text{def}}{=} \lim_{n\uparrow\infty} \tau_n$ is a.s. finite. Moreover, it returns to the state space immediately after its blow-up. Let us generalize the return slightly. Let ν be a probability measure on $\boldsymbol{E}_\partial$, and suppose that $P\{X_{\tau_\infty} = i \mid \mathcal{F}_{\tau_\infty -}\} = \nu(\{i\})$. (If $\nu(\{\partial\}) = 1$, this is the minimal chain. Otherwise, its paths look much like Fig. 7.1.)

Then X_{τ_∞} takes values in $\boldsymbol{E}_\partial$, and the sample path has an infinite discontinuity at τ_∞. Thus the process blows up and then, in Doob's words, "makes its ghostly return to the state space."

Let $\boldsymbol{F}$ be the compactification of $\boldsymbol{E}_\partial$, as in §9.1.

Lemma 9.12. *Suppose ν is not a point mass. Then there exists a unique point $\xi \in \boldsymbol{F}$ such that $\bar{X}_{\tau_\infty-} = \xi$ a.s.. Moreover ξ is a branching point with branching measure ν.*

Proof. Remember that $\bar{X}_t = X_t$ if $X_t \in \boldsymbol{E}_\partial$ and hence, for a.e. t. Let $f \in \mathcal{G}$ and consider the moderate Markov property of $(\bar{X}_t)$ at τ_∞:

$$\bar{U}_1 f(X_{\tau_\infty-}) = E\Big\{\int_0^\infty e^{-t} f(\bar{X}_{\tau_\infty+t})\,dt \mid \mathcal{F}_{T_{\infty-}}\Big\}.$$

The process jumps back into the space with distribution ν, independent of $\mathcal{F}_{\tau_\infty-}$ so we can calculate this: it is

$$= \int_E \bar{U}_1 f(x)\nu(dx) = \nu\bar{U}_1 f.$$

This is independent of $\bar{X}_{\tau_\infty-}$. But functions of the form $\bar{U}_1 f$ separate points of $\boldsymbol{F}$, so this determines $\bar{X}_{\tau_\infty-}$, which means that $\bar{X}_{\tau_\infty-}$ is a constant, say ξ.

Now ν is not a point mass, and the $U_1 f$ separate points, so that there is no $i \in \boldsymbol{E}_\partial$ such that $\bar{U}_1 f(i) = \nu\bar{U}_1 f$ for all $f \in \mathcal{G}$. Thus $\xi \notin \boldsymbol{E}_\partial$. Since the process converges to ξ but does not hit it ($X_{\tau_\infty} \in \boldsymbol{E}_\partial$ and $\xi \notin \boldsymbol{E}_\partial$) it must be a branching point, and its branching measure is clearly ν. □

Remark 9.13. The point ξ can be identified with the point at infinity, and in fact $\boldsymbol{F}$ is homeomorphic to the one-point compactification of E_∂. (See the Exercises.) We needed the fact that ν was not a point mass to show that $\xi \notin E_\partial$. If ν were a point mass at i_0, say, it would turn out that the compactification identified ξ with i_0 instead of ∞! This may seem strange, but in spite of that, it is not particularly interesting – that is, it is not intrinsic, but is just an artifact of the particular compactification we used. It can be avoided by adding another supermedian function to G to separate ξ and i_0. We leave this as an exercise.

Remark 9.14. Note that if we define $P_t(\infty, \cdot) = \nu P_t(\cdot)$, (X_t) is both a strong and moderate Markov process on the one-point compactification of $\boldsymbol{E}_\partial$. This was actually the first place that the moderate Markov property was found (K.L. Chung [9].)

EXAMPLE 9.15. Example 7.2 of Chapter 7 differs from Example 9.11 in that its state space is $\boldsymbol{E} \stackrel{\text{def}}{=} \{\pm 1, \pm 2, \cdots\}$ so that both positive and negative infinities are possible, and both occur. We gave away the origin of this chain when we first described it – it is an approximation of an inverse Brownian motion – but let us ignore that for the minute, and see what its Ray-Knight compactification looks like.

Recall the following facts about the sample path behavior of (X_t).

1° $\tau_\infty < \infty$ a.s.
2° $\bar{X}_t = X_t$ if $t < \tau_\infty$.
3° τ_∞ is a limit from the right of points s for which $\limsup_{u\downarrow s} X_u = \infty$ and of s for which $\liminf_{u\downarrow s} X_u = -\infty$.

Lemma 9.16. *F is homeomorphic to the one-point compactification of E_∂. The point at infinity is a non-branching point which is regular for itself, and $(\bar{X}_t)$ is continuous at τ_∞.*

Proof. The key is that the process jumps from each point to a nearest neighbor, so that if $X_0 = i > 0$, (X_t) hits all $n > i$ before τ_∞. Similarly, if $X_0 = i < 0$, (X_t) hits all $n < i$ before τ_∞. Consequently, if $X_0 = i > 0$, $f \in \mathcal{G}$,

$$\lim_{t\uparrow\tau_\infty} U_1 f(X_t) = \lim_{n\to\infty} U_p f(n),$$

so

$$\lim_{t\uparrow\infty} \bar{U}_1 f(\bar{X}_t) = \lim_{n\to\infty} U_1 f(n) = \bar{U}_1 f(\bar{X}_{\tau_\infty-}).$$

Thus $\bar{U}_1 f(X_{\tau_\infty-})$ is deterministic. Since $\{\bar{U}_1 f,\ f \in \mathcal{G}\}$ separates points of F, there exists $\xi_+ \in F$ such that $\bar{X}_{\tau_\infty-} = \xi_+$. By symmetry, there exists $\xi_- \in F$ such that if $X_0 < 0$, $X_{\tau_\infty-} = \xi_-$, and $\bar{U}_1 f(\xi_-) = \lim_{n\to\infty} \bar{U}_1 f(n)$.

But now $(\bar{X}_t)$ is right-continuous at τ_∞ so by 3° above,

$$\begin{aligned}\bar{U}_1 f(X_{\tau_\infty}) &= \lim_{n\to\infty} U_1 f(n) = \bar{U}_1 f(X_{\xi+})\\ &= \lim_{n\to-\infty} U_1 f(n) = \bar{U}_1 f(\xi-).\end{aligned}$$

Therefore $\xi_+ = \xi_- \overset{\text{def}}{=} \xi$ and $\bar{U}_1 f(\xi) = \lim_{|n|\to\infty} U_p f(n)$. It follows that any neighborhood of ξ must contain $\{n : |n| \geq N\}$ for some N.

We leave it as an exercise to show that $\xi \notin E_\partial$. Since E_∂ is dense in F, there are no other possible limit points: if $x \in F - E_\partial$, then any neighborhood of x must intersect all neighborhoods of ξ, hence $x = \xi$. Thus $F = E_\partial \cup \{\xi\}$, and neighborhoods of ξ are complements of finite subsets of E_∂, so ξ is indeed the Alexandroff point.

Now τ_∞ is a limit point from the right of t at which $\lim_{s\downarrow t} X_t = \infty$, which means it is a right limit point of t for which $\bar{X}_t \notin E_\partial$, and, as there is only one point in $F - E_\partial$, it is a limit point of t for which $\bar{X}_t = \xi$. By the strong Markov property, then, ξ must be regular for itself. □

Remark 9.17.

(1) A boundary point which is regular for itself is called **sticky** in the literature.

(2) It is instructive to map $\boldsymbol{F}$ into $\boldsymbol{F}' = \{0, \pm 1, \pm 1/2, \pm 1/3 \ldots\} \cup \{\partial\}$ by $n \mapsto 1/n$ and $\infty \mapsto 0$. This is again a homeomorphism if we give $\boldsymbol{F}'$ the usual Euclidean topology. Let $\bar{X}'_t = 1/\bar{X}_t$ with the convention that $1/\infty = 0$, and note that $\tau_\infty = \inf\{t : \bar{X}'_t = 0\}$. Then $\bar{X}'_t$ is continuous at τ_∞. So in fact $(\bar{X}'_t)$ is a well-behaved Markov process which is continuous at what originally looked like a wild discontinuity.

(3) We can now look back at the origin of this example: $(\bar{X}'_t)$ is just a Markov chain approximation of the original Brownian motion of Example 7.3.

EXAMPLE 9.18. Let us recall the third example of Chapter 7. (Z_t) was the Feller-McKean diffusion on $\boldsymbol{R}$. It was mapped into a Markov chain (X_t) with state space $\boldsymbol{N}$ by $X_t = h(Z_t)$, where $h(r_n) = n$ for some enumeration (r_n) of $\boldsymbol{Q}^+$. Let (P^Z_t) and (U^Z_p) be the semigroup and resolvent of (Z_t) respectively. We have for each $t > 0$ and $x \in \boldsymbol{R}$ that $P_t(x, \boldsymbol{Q}) = 1$, so that $U^Z_p(x, \boldsymbol{R} - \boldsymbol{Q}) = \int_0^\infty e^{-pt} P^Z_t(x, \boldsymbol{R} - \boldsymbol{Q})\, dt = 0$. Now $X_t = h(X_t)$ so if $Z_0 = r_i$, then

$$P\{X_t = j \mid X_0 = i\} = P\{Z_t = r_j \mid Z_0 = r_i\} = P^Z_t(r_i, \{r_j\}),$$

hence the transition probabilities of (X_t) are given by $p_{ij}(t) = P^Z_t(r_i, \{r_j\})$ and the resolvent is $U_1(i, \{j\}) = U^Z_1(r_i, \{r_j\})$.

Note that $x \mapsto U^Z(x, \{r_j\})$ is continuous. Indeed, (Z_t) is a Hunt process, so that $t \mapsto U^Z_1(X_t, \{r_j\})$ is right-continuous. Suppose $Z_0 = x$. (Z_t) is regular, so that in all intervals $(0, \delta)$, (Z_t) hits both (x, ∞) and $(-\infty, x)$. It follows that

$$U^Z_1(x, \{r_j\}) = \lim_{t \downarrow 0} U^Z_1(X_t, \{r_j\}) = \lim_{y \downarrow 0} U^Z_1 f(y) = \lim_{y \uparrow 0} U^Z_1 f(y).$$

Since x is arbitrary, $U^Z_1(\cdot, \{r_j\})$ is continuous, as claimed.

Lemma 9.19. *$\boldsymbol{F}$ is homeomorphic to the one-point compactification $\bar{\boldsymbol{R}}$ of the real line, and $\boldsymbol{N} \subset \boldsymbol{F}$ is homeomorphic to $\boldsymbol{Q}$ under the map $n \mapsto r_n$.*

Proof. Map $\boldsymbol{N} \mapsto \boldsymbol{Q}$ in the one-point compactification of $\boldsymbol{R}$ by $n \mapsto h^{-1}(n) = r_n$. We must check that $\bar{\boldsymbol{R}}$ satisfies (F1) – (F4). We only need to deal with U_p for $p = 1$.

First, (F1) and (F2) clearly hold. Next, with the above identification, $U_1(i, \{j\})$ can be written $U_1(r_i, \{r_j\} = U^Z_1(r_i, \{r_j\})$. But the latter is the restriction to $\boldsymbol{Q}$ of a function which is continuous on $\boldsymbol{R}$. Set $\bar{U}_1(x, \{r_j\}) =$

$U_1^Z(x, \{r_j\})$ for $x \in \boldsymbol{R}$. Noting that $U_1^Z(x, \{r_j\}) \to 0$ as $|x| \to \infty$, define $\bar{U}_1(\infty, \{r_j\}) = 0$. Then $\bar{U}_1(\cdot, \{r_j\})$ is continuous on $\bar{\boldsymbol{R}}$, which verifies (F3).

Finally, $\bar{U}_1$ separates points of $\bar{\boldsymbol{R}}$ since it separates points of $\boldsymbol{R}$ and ∞ is the only point at which $\bar{U}_1(x, \{r_j\}) = 0$ for all h. □

Remark 9.20. It follows that under the embedding, the Ray process $\bar{X}_t$ on $\boldsymbol{F}$ is exactly (Z_t). The diffusion never reaches the point at infinity, so ∞ is useless in $\boldsymbol{F}$, and can be ignored. Thus, up to a homeomorphism, the Ray-Knight compactification gives us the original diffusion back again! The Ray-Knight compactification, like the Princess' kiss, turns this strange Markov chain back into the handsome diffusion it sprang from.

Notice that there are no branching points in $\boldsymbol{F}$: all points are even regular for themselves. This includes the point at infinity, which turns out to be an absorbing point.

Exercises

1. Show that if $X_t^+ = i$, then $X_t = i$. Conclude that in general, X_t and X_t^+ can only differ on the set $\{t : X_t \in \boldsymbol{E},\ X_t^+ = \infty\}$.
2. Show in Example 9.18 that if t is the start of an excursion of the diffusion from r_i, that $X_t = i$ while $X_t^+ = \infty$. Thus, in general the two versions can differ only when $X_t \in \boldsymbol{E}$ and $X_t^+ = \infty$.
3. Show that if all states are stable, $X_t^+ \equiv X_t$.
4. In Example 9.11, let ν be a point mass at $i_0 \in \boldsymbol{E}$.
   (i) Show that the compactification identifies the point at infinity with i_0 or, equivalently, that for all n, $\{i_0,\ n,\ n+1, \dots\}$ is a neighborhood of i_0 in the topology of $\boldsymbol{F}$.
   (ii) Let $g(i) = E^i\{e^{-\tau_\infty}\}$. Show that g is 1-supermedian and that if it is added to G (see §9.1), then the resulting compactification $\boldsymbol{F}$ will be isomorphic to the one-point compactification of $\boldsymbol{E}_\partial$, and the point at infinity will be a degenerate branching point of $\boldsymbol{F}$.

   Hint: see the following exercise.
5. Let (X_t) be a stable Markov chain on $\boldsymbol{N}$ with well-ordered jump times $\tau_1 < \tau_2 < \dots \uparrow \tau_\infty$. Suppose that $\tau_\infty < \infty$ a.s., and that (X_t) has the property that for any increasing sequence (n_j) of integers, with probability one, X_t hits infinitely many n_j before τ_∞. Let $\boldsymbol{F}$ be a $\mathcal{U}$-compactification of $\boldsymbol{N}$. (ii) $\boldsymbol{F} = \boldsymbol{N}$ (setwise, not as topological spaces) and there exists a unique $i_0 \in \boldsymbol{N}$ such that for each $n \in \boldsymbol{N}$, $\{i_0,\ n,\ n+1, \dots\}$ is an $\boldsymbol{F}$-neighborhood of i_0. This last happens iff $P\{X_{\tau_\infty} = i_0\} = 1$.

   Hint: the proof of Lemma 9.16 may help. See the previous exercise for the intuition.

Notes on Chapter 9

§9. The aim of this chapter is partly to illustrate a typical application of Ray processes and compactifications, and it is partly to introduce the reader to some delicate and beautiful processes which were once at the cutting edge of research in probability. Classical proofs of the theorems in §9.2 and §9.3 can be found in part II of Chung [9], along with their history and references. These are well-known basic results in Markov chain theory. (We should qualify that since it might be hard these days to find someone who knows them all. However, there was a period in which the study of continuous time Markov chains was one of the most vibrant branches of probability. These theorems were indeed well-known then, and their proof added to more than one reputation.) The easy proofs one gets with Ray-Knight compactifications should indicate the power of the method.

The properties 1°–4° of the transition probabilities are non-trivial. The continuity is due to Doob and the differentiability was proved by D. Austin.

§9.1: Ray-Knight compactifications were undoubtedly first used for Markov chains, but we cannot trace this back before 1967 in Kunita-Watanabe [3]. It was realized early on that some form of compactification was needed, but work concentrated on the Martin boundary for some time. Chung [10] gives an account of this. See also T. Watanabe's article [1].

§9.2: Theorem 9.3 is due to J. L. Doob and Corollary 9.5 is due to K. L. Chung.

§9.3 Theorem 9.9 is due to P. Lévy for the stable states, and to J. L. Doob for instantaneous states. Corollary 9.10 was discovered independently by A. O. Pittenger and J. B. Walsh [6].

Chapter 10

Time Reversal

The Markov property has an elegant informal statement: the past and the future are conditionally independent given the present. This is symmetric in time: if $\{X_t, t \geq 0\}$ is a Markov process and if we reverse the sense of time by setting, say, $Y_t = X_{1-t}$, then $\{Y_t,\ 0 \leq t \leq 1\}$ is also a Markov process. However (Y_t) will usually not have stationary transition probabilities, even if (X_t) does. To get a homogeneous process we must reverse from a *random* time, say τ, to get $Y_t = X_{\tau-t}$. If τ is well-chosen – and that is not a hard choice, as we will see – (Y_t) will be a Markov process with stationary transition probabilities.

We will treat the reversal question for Ray processes. This is no restriction, since the result for other Markov processes follows from a Ray-Knight compactification. We will only reverse from the process lifetime. Once again, this is a small restriction, since a double-reversal argument (see Theorem 12.3) shows that any possible reversal time is the lifetime of some sub-process of X.

Let $(U_p)_{p>0}$ be a Ray resolvent on a compact metric space $\boldsymbol{F}$, with corresponding right-continuous semigroup $(P_t)_{t\geq 0}$. We assume that $\boldsymbol{F}$ has a distinguished point ∂ which is isolated in the topology of $\boldsymbol{F}$. Let $\boldsymbol{E} = \boldsymbol{F} - \{\partial\}$; $\boldsymbol{E}$ is also compact. Let $\mathcal{E}$ be the topological Borel field of $\boldsymbol{F}$.

Let Ω be the canonical space of functions from $[0, \infty)$ to $\boldsymbol{F}$ which are right-continuous, have left limits, and admit a lifetime $\zeta(\omega)$: if $\omega \in \Omega$, then $\omega(t) \in \boldsymbol{E}$ if $t < \zeta(\omega)$, and $\omega(t) = \partial$ if $t \geq \zeta(\omega)$. We suppose that the Ray process $\{X_t,\ t \geq 0\}$ is defined canonically on $(\Omega, \mathcal{F}, P)$: $X_t(\omega) = \omega(t)$, $t \geq 0$, $\omega \in \Omega$.

Let $(\tilde{X}_t)$ be the reversal of (X_t) from its lifetime:

$$\tilde{X}_t(\omega) = \begin{cases} X_{\zeta-t}(\omega) & \text{if } 0 < t \leq \zeta(\omega) \\ \partial & \text{if } t > \zeta(\omega) \text{ or } \zeta(\omega) = \infty. \end{cases} \tag{10.1}$$

The original process (X_t) is right-continuous, so its reverse, $(\tilde{X}_t)_{t>0}$, is left-continuous. We can not expect $(\tilde{X}_t)$ to be strongly Markov – it may well not be – but we will see that it has the natural form of the strong Markov property for left-continuous processes, the moderate Markov property.

The transition function of $(\tilde{X}_t)$ depends on the initial distribution of (X_t), so we will fix an initial distribution μ of X_0. Then the distribution of X_t is $\mu_t \stackrel{\text{def}}{=} \mu P_t$, $t \geq 0$, and the distribution of $\tilde{X}_t$ is $K_t(A) \stackrel{\text{def}}{=} P\{\tilde{X}_t \in A\}$. We assume that $P^{\mu}\{\zeta < 00\} > 0$.

Let us define a measure G by

$$G(A) = \mu U_0(A) = \int_0^\infty \mu_t(A)\, dt.$$

Note that G may not be finite, or even sigma finite, but we shall see that it does have certain finiteness properties.

If λ is a measure on $(\boldsymbol{E}, \mathcal{E})$, let us introduce the notation

$$\langle f, g \rangle_\lambda = \int f(x) g(x)\, \lambda(dx).$$

The purpose of this chapter is to prove the following theorem.

Theorem 10.1. *Let $\{X_t, t \geq 0\}$ be a Ray process with initial distribution μ. Suppose that $P^{\mu}\{\zeta < \infty\} > 0$. Then the reverse process $\{\tilde{X}_t, t > 0\}$ is a left-continuous moderate Markov process with left-continuous moderate Markov semigroup $(\tilde{P}_t)_{t \geq 0}$ which satisfies $\tilde{P}_0(x, \cdot) = \delta_x(\cdot)$, and for $f, g \in C(\boldsymbol{E})$ and $t > 0$,*

$$\langle f, \tilde{P}_t g \rangle_G = \langle g, P_t f \rangle_G. \tag{10.2}$$

Moreover, its absolute distributions K_t satisfy, for all bounded, continuous and integrable functions k on $[0, \infty)$,

$$\int_0^\infty k(t) K_t f\, dt = \int_{\boldsymbol{E}} E^x\{k(\zeta)\} f(x) G(dx). \tag{10.3}$$

Example 10.2. To see why the strong Markov property may not hold for the reverse, consider the example in Fig. 10.1 below, in which (X_t) starts with probability one-half in rays B and C, and performs uniform motion to the right on all three rays. This is strongly Markov, since the motion is deterministic. On the other hand, the reverse is exactly the process of Example 8.54, so it is not strongly Markov. However, it is easy to see that it is moderately Markov.

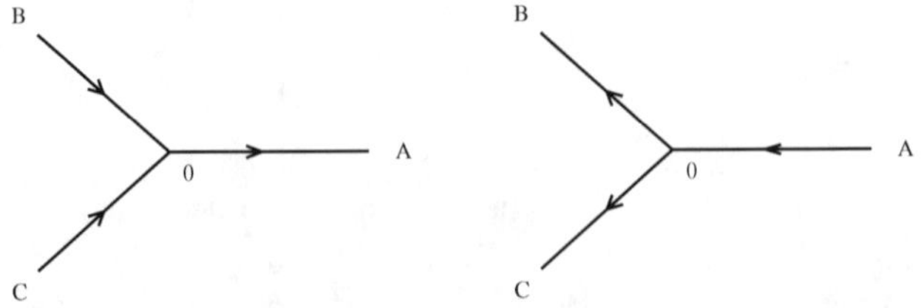

Fig. 10.1. The forward and reverse processes of Example 10.2.

For each x, let $L(x,t) \stackrel{\text{def}}{=} P_t(x,\{\partial\})$ be the distribution function of the lifetime ζ. We make the following hypotheses.

Hypothesis 10.3. *(H1) For each $x \in \boldsymbol{E}$, $t \mapsto L(x,t)$ is absolutely continuous with density $\ell_t(x)$;*

(H2) $t \mapsto \ell_t(x)$ is equicontinuous in x, $x \in \boldsymbol{E}$.

We will first prove the theorem under Hypotheses (H1) and (H2), and then we will completely eliminate them at the end. The proof is in several steps. We first find the finite-dimensional distributions of the reverse process, and use these to construct a loose transition function. This may not be a semigroup, however. In order to construct a true transition semigroup $(\tilde{P}_t)$ and resolvent $(\tilde{U}_p)$ we first construct approximate versions $(\hat{P}_t^1)$ and $(\hat{U}_p^1)$. We then improve them in stages – $(\hat{U}_p^2)$ will be a better version of the resolvent – in order to arrive at the desired transition function and resolvent.

Lemma 10.4. *For each $x \neq \partial$, and $s, t \geq 0$,*

$$P_t \ell_s(x) = \ell_{s+t}(x). \tag{10.4}$$

Proof. If $0 < u < v$,

$$\begin{aligned}\int_0^t \ell_{s+r}(x)\,dr &= P_{s+t}(x,\partial) - P_s(x,\partial) = P_s\big(P_t(x,\partial) - P_0(x,\partial)\big)\\ &= \int P_s(x,dy)\int_0^t \ell_r(y)\,dr = \int_0^t P_s\ell_r(x)\,dr.\end{aligned}$$

This holds for each t, hence $P_s\ell_r(x) = \ell_{s+r}(x)$ for a.e. r. But $\ell_t(x)$ is continuous in t by $(H2)$, so this holds for all r. □

Define a measure K_s on $\boldsymbol{E}$ by

$$K_s(A) = \int_0^\infty \mu P_t(I_A \ell_s)\,dt, \quad A \in \mathcal{E}.$$

By Fubini,

$$K_s(\boldsymbol{E}) = \int_0^\infty \mu P_t(\ell_s)\,dt = \mu\int_0^\infty \ell_{s+t}\,dt = \mu P\{s \leq \zeta < \infty\} \leq 1.$$

Thus K_s is a bounded measure. If we extend it to $\boldsymbol{F}$ by $K_s(\{\partial\}) = 1 - K(\boldsymbol{E})$, it is a probability measure. Note that $K_s(dx) = \ell_s(x)G(dx)$.

Let $\varrho(x) = \int_0^\infty e^{-t}\ell_t(x)\,dt = E^x\{e^{-\zeta}; \zeta < \infty\}$. Then $\varrho(x) \geq 0$, and $\varrho(x) > 0$ if $P^x\{\zeta < \infty\} > 0$. By hypothesis, then, $\varrho(x)$ is not identically 0 and

$$0 < \int_{\boldsymbol{E}} \varrho(x) G(dx) \leq 1. \tag{10.5}$$

In particular, we see that G is sigma-finite on $\{x \in \boldsymbol{E} : P^x\{\zeta < \infty\} > 0\}$. We can now give our first reversal result.

Theorem 10.5. *Under hypotheses* $(H1)$ *and* $(H2)$, *the absolute distribution of* $\tilde{X}_s$ *is* K_s, *and the joint distribution of* $\tilde{X}_s$ *and* $\tilde{X}_t$, $0 < s \leq t$ *is*

$$P\{\tilde{X}_s \in A, \tilde{X}_t \in B\} = \int_B G(dx) \int_A P_{t-s}(x, dy)\ell_s(y), \tag{10.6}$$

where $A, B \in \mathcal{E}$. *More generally, if* $0 < t_1 \leq \ldots \leq t_n$ *and* $A_1, \ldots, A_n \in \mathcal{E}$,

$$P\{\tilde{X}_{t_j} \in A_j,\ j = 1, \ldots, n\}$$
$$= \int_{A_n} G(dx_n) \int_{A_{n-1}} P_{t_n - t_{n-1}}(x_n, dx_{n-1}) \ldots \int_{A_1} P_{t_2 - t_1}(x_2, dx_1)\ell_{t_1}(x_1). \tag{10.7}$$

Proof. The assertion about K_s follows upon taking $t = s$ and $B = A$ in (10.6). Equations (10.6) and (10.7) are proved by essentially the same argument, so we will just prove (10.6), and leave it to the reader to apply the argument to (10.7).

The idea is simple once we bring ζ into the picture: write

$$P\{\tilde{X}_s \in A, \tilde{X}_t \in B\} = \int_0^\infty P\{X_{u-s} \in A, X_{u-t} \in B, \zeta \in du\}, \tag{10.8}$$

and express this in terms of the transition probabilities. Let f and g be continuous, $0 \leq f \leq 1$, $0 \leq g \leq 1$, and set $\zeta_n = \frac{k}{2^n}$ if $\zeta \in \left[\frac{k-1}{2^n}, \frac{k}{2^n}\right)$, and $\zeta_n = \infty$ if $\zeta = \infty$. Then $\zeta_n \downarrow \zeta$, hence $X_{\zeta_n - t} \to X_{\zeta - t}$. Let $\alpha_k = \frac{k}{2^n}$. Then

$$\begin{aligned}
E\{f(\tilde{X}_s)g(\tilde{X}_t)\} &= E\{f(X_{\zeta-s})g(X_{\zeta-t})\} \\
&= \lim_{n\to\infty} E\{f(X_{\zeta_n-s})g(X_{\zeta_n-t})\} \\
&= \lim_{n\to\infty} \sum_{k>2^n t} E\big\{f(X_{\alpha_k-s})g(X_{\alpha_k-t}), \zeta \in [\alpha_k, \alpha_{k+1})\big\}.
\end{aligned}$$

Now $\alpha_k - t \leq \alpha_k - s$, so the kth summand is

$$E\left\{g(X_{\alpha_k-t})f(X_{\alpha_k-s})\int_0^{2^{-n}} \ell_{s+u}(X_{\alpha_k-s})\,du\right\}$$

which can be written

$$\int_0^{2^{-n}} \mu P_{\alpha_k-t}(gP_{t-s}(f\ell_{s+u}))\,du,$$

hence the above is

$$= \lim_{n\to\infty} \sum_{k>2^n t} \int_0^{2^{-n}} \mu P_{\alpha_k - t}(g P_{t-s}(f \ell_{s+u}))\, du.$$

Define a function h_n by

$$h_n(\alpha_k + u) = \mu P_{\alpha_k - t}(g P_{t-s}(f \ell_{s+u})),\ k \geq 2^n t,\ u \geq 0,$$

and write the above sum as

$$= \lim_{n\to\infty} \sum_{k>2^n t} \int_0^{2^{-n}} h_n(\alpha_k + u)\, du = \lim_{n\to\infty} \int_t^\infty h_n(\tau)\, d\tau. \tag{10.9}$$

But now for each $\tau > t$,

$$\lim_{n\to\infty} h_n(\tau) = \mu P_{\tau - t}(g P_{t-s}(f \ell_s)).$$

Since f and g are positive and bounded by one,

$$h_n(\alpha_k + u) \leq \mu P_{\alpha_k - t} P_{t-s} \ell_u = \mu \ell_{\alpha_k + u},$$

by Lemma 10.4. But this is integrable:

$$\int_t^\infty \mu \ell_v\, dv \leq \int \mu(dx) \int_0^\infty \ell_v(x)\, dv \leq 1$$

since (ℓ_s) is the density of ζ. Thus we can go to the limit in (10.9) by dominated convergence to see it is

$$\begin{aligned} &= \int_t^\infty \mu P_{u-t}(g P_{t-s}(f \ell_s))\, du && (10.10)\\ &= G(g P_{t-s}(f \ell_s)). \end{aligned}$$

We have shown that if f and g are positive continuous functions on $\boldsymbol{E}$,

$$E\{f(\tilde{X}_s) g(\tilde{X}_t)\} = \int_{\boldsymbol{E}} G(dx) g(x) \int_{\boldsymbol{E}} P_{t-s}(x, dy) f(y) \ell_s(y).$$

A monotone class argument shows that this holds for $f = I_A$, $g = I_B$, which implies (10.6). □

10.1. The Loose Transition Function

Definition 10.6. *A family* $\big(\hat{P}_t(x, \cdot), x \in \boldsymbol{E}, t > 0\big)$ *of sub-Markovian kernels on* $(\boldsymbol{E}, \mathcal{E})$ *is a* **loose transition function** *for the process* $\{\tilde{X}_t, t > 0\}$ *if for each* $s > 0, t > 0$ *and* $f \in C(\boldsymbol{E})$,

$$E\{f(\tilde{X}_{s+t}) \mid \tilde{X}_u, u \leq s\} = \hat{P}_t f(X_s) \text{ a.s.} \tag{10.11}$$

Remark 10.7. 1. The process $(\tilde{X}_t)$ jumps to ∂ at its lifetime and stays there, so if its loose transition function $\hat{P}_t$ is only defined on $\boldsymbol{E}$, we extend it to $\boldsymbol{F}$ by setting

$$\hat{P}_t(x, \{\partial\}) = 1 - \hat{P}_t(x, \boldsymbol{E}) \text{ and } \hat{P}_t(\partial, \{\partial\}) = 1. \tag{10.12}$$

We use the word "loose" because it may not form a semigroup. If $(\tilde{X}_t)$ has a loose transition function, it is evidently a Markov process.

A loose transition function $\hat{P}_t$ is characterized by a duality relation: $\hat{P}_t$ is the adjoint of P_t relative to the measure G. Indeed:

Lemma 10.8. *Let $(\hat{P}_t)_{t\geq 0}$ be a family of sub-Markovian kernels on $(\boldsymbol{E}, \mathcal{E})$ such that $\hat{P}_0 = I$. Then $(\hat{P}_t)$ is a loose transition function for $(\tilde{X}_t)$ if and only if for all $t > 0$ and positive Borel functions f and g on $\boldsymbol{E}$,*

$$\langle f, P_t g\rangle_G = \langle \hat{P}_t f, g\rangle_G. \tag{10.13}$$

Proof. Let $0 < s < t$ and $g,\ h \in C(\boldsymbol{E})$ and let $f = \ell_s h$. We calculate $E\{h(\tilde{X}_s), g(\tilde{X}_t)\}$ in two ways. First, by (10.6), it is

$$= \int G(dx) g(x) P_{t-s}(h\ell_s)(x) = \langle P_{t-s} f, g\rangle_G.$$

On the other hand, if $(\hat{P}_s)$ is a loose transition function, the distribution of $\tilde{X}_s$ is $dK_s = \ell_s dG$ so the expectation is also

$$= \int K_s(dx)\, h(x) \hat{P}_{t-s} g(x) = \int G(dx)\, f(x) \hat{P}_{t-s} g(x) = \langle f, \hat{P}_{t-s} g\rangle_G.$$

Thus a loose transition function is dual to (P_t). Conversely, to show that $(\hat{P}_t)$ is a loose transition function for $(\tilde{X}_t)$, it is enough to show that $(\hat{P}_t)$ generates the finite-dimensional distributions of $(\tilde{X}_t)$ by the usual formula (see (3) of §1.2.) Suppose that $(\hat{P}_t)$ satisfies the duality relation (10.13). We will treat the case $n = 3$; the argument extends easily to the general case, but we will leave the induction to the interested reader. (Note that the argument above, read from bottom to top, verifies that it generates the bivariate distributions.)

Let $t_1 > 0$, $t_2 = t_1 + a$, $t_3 = t_2 + b$, where $b, c > 0$, and let f_1, f_2, and $f_3 \in C(\boldsymbol{E})$. Then if we write out (10.7) in operator notation for this case, we see that

$$\begin{aligned} E\left\{\prod_{j=1}^{3} f_j(\tilde{X}_j)\right\} &= \int G(dx) f_3 P_b(f_2 P_a(f_1 \ell_{t_1}))(x) \\ &= \langle f_3, P_b f_2 P_a(f_1 \ell_{t_1})\rangle_G \\ &= \langle \hat{P}_b f_3, f_2 P_a(f_1 \ell_{t_1})\rangle_G \\ &= \langle f_2 \hat{P}_b f_3, P_a(f_1 \ell_{t_1})\rangle_G \\ &= \langle \hat{P}_a f_2 \hat{P}_b f_3, f_1 \ell_{t_1}\rangle_G \end{aligned}$$

But $\ell_{t_1} dG = dK_{t_1}$ so this is

$$= \iiint K_{t_1}(dx_1) f_1(x_1) \hat{P}_a(x_1, dx_2) f_2(x_2) \hat{P}_b(x_2, dx_3) f_3(x_3),$$

which is the formula we wanted to prove. □

We will now show the existence of a loose dual transition function. It is clear that if it exists, $\hat{P}_t(x, A)$ has to be a version of the Radon-Nikodym derivative $P\{\tilde{X}_s \in dx, \tilde{X}_{s+t} \in A\}/P\{\tilde{X}_s \in dx\}$. However, the strong Markov property of $(\tilde{X}_t)$ requires an extremely good version of this Radon-Nikodym derivative, so we shall proceed step-by-step. We first construct a version of $\tilde{P}_t(x, A)$ which is measurable in (x, t) and a measure in A.

Proposition 10.9. *Under Hypothesis 10.3 there exists a family* $(\hat{P}_t^1)_{t \geq 0}$ *of sub-Markovian kernels on* $(\boldsymbol{E}, \mathcal{E})$ *which is a loose transition function for* $(\tilde{X}_t)$.

Proof. We will construct a family satisfying (10.13). Define

$$k(x) = \int_0^\infty e^{-s} \ell_s(x) ds.$$

By (H2), $k(x) = 0 \Longleftrightarrow \ell_s(x) = 0 \;\; \forall s \geq 0$. By Lemma 10.4, $P_t k = \int_0^\infty e^{-s} P_t \ell_s \, ds = e^t \int_t^\infty e^{-s} \ell_s \, ds \leq e^t k$, so that k is 1-excessive with respect to (P_t). Define a measure K on $\boldsymbol{E}$ by $K(A) = \int_0^\infty e^{-s} K_s(A) \, ds = \int_A k(x) G(dx)$.

Since $K_s(\boldsymbol{E}) \leq 1$ for each s, $K(\boldsymbol{E}) \leq 1$. For $t > 0$ define

$$\Pi_t(A, B) = \int_A G(dx) \int_B P_t(x, dy) k(y).$$

Then

$$\Pi_t(A, \boldsymbol{E}) \leq \int_A G(dx) \, e^t k(x) \leq e^t K(A),$$

while as $G P_t \leq G$,

$$\Pi_t(\boldsymbol{E}, B) \leq \int_B G(dy) \, k(y) = K(B).$$

Thus both $\Pi_t(A, \cdot)$ and $\Pi_t(\cdot, B)$ are absolutely continuous with respect to K. Rather than defining $\hat{P}_t^1(x, \cdot)$ as a measure, we will define it as a linear functional on $C(\boldsymbol{E})$.

Let D_0 be a countable dense subset of $C(\boldsymbol{E})$, and let D be the smallest class of functions which contains D_0 and is closed under addition and multiplication by rationals. Then (Lemma 8.21) D is a countable dense subset of $C(\boldsymbol{E})$. For $f \in D$, let

$$\Pi_t(f, B) = \int f(x) \Pi_t(dx, B).$$

Let $L(f, y)$ be a version of the Radon-Nikodym derivative $\Pi_t(f, dy)/K(dy)$ which is $\mathcal{E} \times \mathcal{B}$-measurable in (y, t). We claim there is a set $N \in \mathcal{E}$ with $K(N) = 0$ such that if $y \in \boldsymbol{E} - N$,

(a) $L(f, y) \geq 0$ if $f \in D$, $f \geq 0$;
(b) $L(cf, y) = cL(f, y)$ if $f \in D$, $c \in \boldsymbol{Q}$;
(c) $L(f + g, y) = L(f, y) + L(g, y)$, $f, g \in D$;
(d) $|L(f, y)| \leq \|f\|$ if $f \in D$,

where $\|f\|$ is the supnorm of f. The proofs of these are all similar and all trivial. For instance, for (b), if $B \in \mathcal{E}$,

$$\int_B L(cf, y)\, K(dy) = \Pi_t(cf, B) = c\Pi_t(f, B) = \int_B cL(f, y)\, K(dy),$$

so we can choose B to be $\{y : L(cf, y) > cL(f, y)\}$ or $\{y : L(cf, y) < cL(f, y)\}$ to see that (b) holds K-a.e. for each pair $c \in \boldsymbol{Q}$ and $f \in D$. Since there are only countably many such pairs, (b) holds K-a.e. simultaneously for all of them.

But now, if $y \in \boldsymbol{E} - N$, (d) tells us that we can extend $L(y, \cdot)$ from D to $C(E)$ by continuity, and (a)–(d) will hold for all $f \in C(E)$ and $c \in \boldsymbol{R}$.

Thus $L(\cdot, y)$ is a positive linear functional on $C(\boldsymbol{E})$, hence it defines a measure $L(A, y)$ on $\mathcal{E}$ with total mass ≤ 1. We define

$$\hat{P}_t^1(y, A) = L(A, y) \ \text{ if } y \in \boldsymbol{E} - N,\ A \in \mathcal{E},\ t > 0; \tag{10.14}$$

$$\hat{P}_t^1(y, A) = 0 \ \text{ if } y \in N,\ t > 0; \tag{10.15}$$

$$\hat{P}_0^1(y, \cdot) = \delta_y(\cdot). \tag{10.16}$$

Thus $\hat{P}_t^1$ is a kernel on $(\boldsymbol{E}, \mathcal{E})$, and, of course, $\hat{P}_t^1 f(y) = L(f, y)$ for $f \in C(\boldsymbol{E})$. It remains to verify (10.13).

$$\begin{aligned}\langle \hat{P}_t^1 f, g\rangle_G &= \int \hat{P}_t^1 f(x) g(x)\, G(dx)\\ &= \int L(f, x) g(x) G(dx).\end{aligned}$$

Now $\Pi_t(f, dx) = L(f, x)\, K(dx) = L(f, x)\, G(dx)/k(x)$:

$$\begin{aligned}&= \int \Pi_t(f, dx)\frac{g(x)}{k(x)}\\ &= \Pi_t\left(f, \frac{g}{k}\right)\\ &= \iint f(x)\, G(dx) P_t(x, dy)\, \frac{g(y)}{k(y)} k(y)\\ &= \langle f, P_t g\rangle_G.\end{aligned}$$

Thus $(\hat{P}_t^1)$ is dual to (P_t), so by Lemma 10.8 it is a loose transition function. □

10.2. Improving the Resolvent

The reverse process is actually moderately Markov, but not relative to the loose transition function of the last section. It is ... well ... too loose. We need to tighten it. We will follow the path of least resistance, which, in this case, is to regularize the resolvent, since it is marginally smoother than the transition function. It comes to the same thing in the end, since once we find the resolvent, we construct the semigroup by a Ray-Knight compactification. We will switch back and forth between the forward and reverse processes as we do this. We are still working under Hypothesis 10.3. For $x \in \boldsymbol{F}$, define a measure

$$\hat{U}_p(x, \cdot) = \int_0^\infty e^{-pt} \hat{P}_t(x, \cdot)\, dt.$$

This is a measure on $\boldsymbol{F}$, since $\hat{P}_t^1$ is defined on $\boldsymbol{F}$ by (10.12), and for each p and x, $\hat{U}_p^1(x, \boldsymbol{F}) = 1/p$. Then for any bounded measurable f on $\boldsymbol{F}$,

$$\hat{U}_p^1 f(x) = \int_0^\infty e^{-ps} \hat{P}_t^1 f(x)\, dt. \tag{10.17}$$

Let $\tilde{\mathcal{F}}_t = \sigma\{\tilde{X}_s, s \le t\}$ be the natural filtration of $(\tilde{X}_t)$ completed as usual by adding all null sets of $\mathcal{F}$ to $\tilde{\mathcal{F}}_0$. Since $\hat{P}_t$ is a loose transition function,

$$\hat{P}_s^1 f(X_t) = E\{f(\tilde{X}_{t+s}) \mid \mathcal{F}_t\}.$$

If we integrate this (and we have to be careful because we are using Fubini's theorem with conditional expectations) we see that

$$\hat{U}_p^1 f(\tilde{X}_t) = E\Big\{\int_0^\infty e^{-ps} f(\tilde{X}_{t+s})\, ds \mid \tilde{\mathcal{F}}_t\Big\}. \tag{10.18}$$

Lemma 10.10. *If $f \in \mathcal{E}_b^+$, $\{e^{-pt}\hat{U}_p^1 f(\tilde{X}_t), \tilde{\mathcal{F}}_t, t \ge 0\}$ is a supermartingale, and both $t \mapsto \hat{U}_p^1 f(\tilde{X}_t)$ and $t \mapsto \hat{U}_p^1 f(X_t)$ have essential left and right limits at all t a.s. For each $\varepsilon > 0$ and $N > 0$, $p \mapsto \hat{U}_p^1 f(x)$ is uniformly equicontinuous on $[\varepsilon, \infty)$ for $x \in \boldsymbol{E}$ and $\|f\| \le N$.*

Proof. The supermartingale property follows from §2.1 Proposition 2, and the existence of essential limits for $(\tilde{X}_t)$ follows directly from Corollary 6.5. But $X_t = \tilde{X}_{\zeta - t}$, so essential right (resp. left) limits of $(\tilde{X}_t)$ at t translate into essential left (resp. right) limits of (X_t) at $\zeta - t$, so $t \mapsto \hat{U}_p^1 f(X_t)$ also has essential left and right limits. The last statement follows from (10.17) and the fact that the derivative of a Laplace transform has an easy bound in terms of $\|f\|$. □

The above proof highlights the usefulness of the essential limit. Not only does it exist under rather weak conditions, it is invariant under time reversal.

This is not true of, say, limits along rationals. (After reversal, the time parameter is $\zeta - t$, and one can no longer tell for which t $\zeta - t$ is rational!)

If $(\tilde{X}_t)$ is moderately Markov, it should be left-continuous along the resolvent, but Lemma 10.10 only promises essential limits. This calls for regularization. Left continuity for $(\tilde{X}_t)$ corresponds to right continuity for (X_t) so we will use the original process to do the smoothing. Recall that μ is the initial distribution of the original process (X_t). Let (f_n) be a countable dense subset of $C(\boldsymbol{F})$. Define

$$\hat{U}_p^2 f_n(x) = \liminf_{r\to\infty} rU_r(\hat{U}_p^1 f_n)(x). \tag{10.19}$$

Note that we use U_r above, not $\hat{U}_r$. It is clear that $\hat{U}_p^2 f_n(x)$ is measurable in (x, p). We establish its properties in a sequence of lemmas.

Lemma 10.11. *Let $g \in b\mathcal{E}$ have the property that with P^μ-probability one, $t \mapsto g(X_t)$ has essential left and right limits at all t. Define*

$$\bar{g}(x) = \liminf_{r\to 0} rU_r g(x). \tag{10.20}$$

Then

(i) *there is a μ-polar set K such that for $x \in \boldsymbol{E} - K$, the limit in* (10.20) *exists;*
(ii) *$t \mapsto \bar{g}(X_t)$ is right continuous and has left limits P^μ-a.s.;*
(iii) *$\bar{g}(X_t) = \operatorname{ess\,lim}_{s\to t+} g(X_s)$ for all t, P^μ-a.s.;*
(iv) *$\bar{g}(X_t) = g(X_t)$ for a.e. t, P^μ-a.s..*

Proof. Let T be a stopping time. Then

$$\bar{g}(X_T) = \liminf_{r\to\infty} U_r g(X_T) = \liminf_{r\to\infty} E\left\{\int_0^\infty re^{-rt} g(X_{T+t})\,dt \mid \mathcal{F}_T\right\}.$$

Let $Z_t \stackrel{\text{def}}{=} \operatorname{ess\,lim}_{s\to t+} g(X_s)$ for $t \geq 0$. This is defined for all t by hypothesis, and by Proposition 6.2 it is right-continuous and equals $g(X_t)$ for a.e. t. By the Blumenthal zero-one law, Z_T is $\mathcal{F}_T$-measurable. Thus by bounded convergence, the above is

$$= E\{Z_T \mid \mathcal{F}_T\} = Z_T.$$

Now (Z_t) is right-continuous and adapted, $(\bar{g}(X_t))$ is a function of a right-continuous and adapted process, and the two are equal a.s. at each stopping time. By Corollary 6.12, the two processes are indistinguishable. Then (ii), (iii) and (iv) follow from Proposition 6.2.

To see (i), let $K = \{x : P^x\{\operatorname{ess\,lim}_{t\downarrow 0} g(X_t) \textit{ fails to exist}\} > 0\}$. The above argument shows that $P^\mu\{X_T \in K\} = 0$ for all stopping times T, which, again by Corollary 6.10, implies that with probability one (X_t) never hits K, so that K is μ-polar, as claimed. □

Lemma 10.12. *There is a μ-polar set K_1 such that for all $p > 0$ and $x \in \boldsymbol{F} - K_1$, (10.19) defines a measure $\hat{U}_p^2(x, \cdot)$ of mass $1/p$ on $\boldsymbol{F}$.*

Proof. Let

$$\Lambda = \{x \in \boldsymbol{E} : \lim_{r\to\infty} rU_r(\hat{U}_p^1 f_n)(x) \text{ exists for all } n \text{ and all rational } p > 0\}.$$

By Lemmas 10.10 and 10.11 $K_1 \stackrel{\text{def}}{=} \boldsymbol{F} - \Lambda$ is μ-polar (the lemmas apply to single n and p, but a countable union of μ-polar sets is μ-polar). Moreover, for fixed x, $rU_r\hat{U}_p^1(x, \cdot)$ is a measure of mass $1/p$ on the compact space $\boldsymbol{F}$. If the limit as $r \to \infty$ exists for a dense set of f_n, then the measures converge in the weak topology to a measure, also of mass $1/p$. This convergence holds for all rational p, but, as the measures are equicontinuous in p, it actually holds for all p. Thus, for $x \in \boldsymbol{F} - K_1$, $\hat{U}_p^2(x, \cdot)$ is a kernel on $\boldsymbol{F}$ of mass $1/p$. □

Lemma 10.13. *Let $f \in C(\boldsymbol{E})$, and $p > 0$.*

(i) *$t \mapsto \hat{U}_p^2 f(\tilde{X}_t)$ is a.s. left-continuous;*
(ii) *for all finite predictable times $T > 0$,*

$$\hat{U}_p^2 f(\tilde{X}_T) = E\left\{\int_0^\infty e^{-pt} f(\tilde{X}_{T+t})\,dt \mid \tilde{\mathcal{F}}_{T-}\right\}; \tag{10.21}$$

Proof. We prove this for positive f; the result follows by taking $f = f^+ - f^-$.

By Lemma 10.10, $\hat{U}_p^1 f(X_t)$ has essential left and right limits, so Lemma 10.11 (with $\bar{g}(x) = \hat{U}_p^1 f(x)$) implies that $t \mapsto \bar{U}_p^2 f(X_t)$ is P^μ-a.s. rcll. After reversing, $t \mapsto \hat{U}_p^2 f(\tilde{X}_t)$ is lcrl. Moreover, part (iv) of the same lemma implies that $\hat{U}_p^2 f(X_t) = \hat{U}_p^1 f(X_t)$ for a.e. $t < \zeta$, and hence by Fubini that $\hat{U}_p^2 f(\tilde{X}_t) = \hat{U}_p^1 f(\tilde{X}_t)$ a.s. for a.e. t.

We can choose a countable dense subset $D \subset \boldsymbol{R}^+$ such that $\hat{U}_p^2 f(\tilde{X}_t) = \hat{U}_p^1 f(\tilde{X}_t)$, $\forall t \in D$ a.s. By (10.18), if $T \equiv t \in D$

$$\hat{U}_p^2 f(\tilde{X}_T) = E\left\{\int_0^\infty e^{-ps} f(\tilde{X}_{T+s})\,ds \mid \tilde{\mathcal{F}}_T\right\}. \tag{10.22}$$

It follows by the usual argument (see e.g. (2) of §2.3) that (10.22) remains true if T is a stopping time for $(\tilde{X}_t)$ which takes values in the countable set D. If $T > 0$ is a predictable time, there exists a sequence (T_n) of stopping times announcing it, that is, $T_n < T$ a.s. and $T_n \uparrow T$. We can even assume that the T_n take values in D. (See Exercise 12, §1.3.) Then clearly

$$\int_0^\infty e^{-pt} f(\tilde{X}_{T_n+t})\,dt \longrightarrow \int_0^\infty e^{-pt} f(\tilde{X}_{T+t})\,dt,$$

and the convergence is bounded, while $\tilde{\mathcal{F}}_{T_n} \uparrow \tilde{\mathcal{F}}_{T-}$. By Hunt's Lemma (see Theorem 9.4.8 of the *Course*) we can take the limit as $n \to \infty$ on the two sides of (10.22) – for $\hat{U}_p^2 f(\tilde{X}_t)$ is left continuous – to see that it holds for T. □

Lemma 10.14. *There is a μ-polar set K_2 such that $(\hat{U}^2_p)_{p>0}$ satisfies the resolvent equation on $\boldsymbol{F} - K_2$.*

Proof. Let $f \in C_0(\boldsymbol{E})$, $p, q > 0$, and let $T > 0$ be a predictable time. The following is a disguised variant of the calculation in §3.2 Proposition 7. Note that from (10.21)

$$\hat{U}^2_p \hat{U}^2_q f(\tilde{X}_T) = E\left\{ \int_0^\infty e^{-ps} \hat{U}^2_q f(\tilde{X}_{T+s})\, ds \mid \tilde{\mathcal{F}}_T \right\}$$

and

$$E\{\hat{U}^2_q f(\tilde{X}_{T+s}) \mid \tilde{\mathcal{F}}_T\} = E\left\{ \int_0^\infty e^{-qu} f(\tilde{X}_{T+s+u})\, du \mid \tilde{\mathcal{F}}_T \right\},$$

so if Y is a bounded $\tilde{\mathcal{F}}_T$-measurable random variable,

$$\begin{aligned} E\{Y \hat{U}^2_p \hat{U}^2_q f(\tilde{X}_T)\} &= E\left\{ Y \int_0^\infty e^{-ps} \hat{U}^2_q f(\tilde{X}_{T+s})\, ds \right\} \qquad (10.23) \\ &= \int_0^\infty e^{-ps} E\{Y \hat{U}^2_q f(\tilde{X}_{T+s})\}\, ds \\ &= \int_0^\infty e^{-ps} E\left\{ Y \int_0^\infty e^{-qu} f(\tilde{X}_{T+s+u})\, du \right\} ds \\ &= E\left\{ Y \int_0^\infty \int_0^\infty e^{-ps-qu} f(\tilde{X}_{T+s+u})\, du\, ds \right\}. \end{aligned}$$

Let $v = s + u$ and integrate first over s. The last double integral becomes

$$\int_0^\infty \left[\int_0^v e^{(q-p)s}\, ds \right] e^{-qv} f(\tilde{X}_{T+v})\, dv$$

so (10.23) is

$$= (q - p)^{-1} E\{Y(\hat{U}^2_p f(\tilde{X}_T) - \hat{U}^2_q f(\tilde{X}_T))\}.$$

This is true for every bounded $\tilde{\mathcal{F}}_T$-measurable Y, so that for each predictable T and $f \in C(\boldsymbol{F})$,

$$\hat{U}^2_p f(\tilde{X}_T) - \hat{U}^2_q f(\tilde{X}_T) = (q - p) \hat{U}^2_p \hat{U}^2_q f(\tilde{X}_T) \ a.s. \qquad (10.24)$$

Thus if $N_{fpq} = \{x : \hat{U}^2_p f(x) - \hat{U}^2_q f(x) \neq (q - p)\hat{U}^2_p \hat{U}^2_q f(x)\}$ we have just shown that for a predictable time T, $P\{\tilde{X}_T \in N_{fpq}\} = 0$. Now $t \mapsto \tilde{X}_t$ is left continuous, so by Corollary 6.10, $P\{\exists t : \tilde{X}_t \in N_{fpq}\} = 0$. Thus as $X_t = \tilde{X}_{\zeta - t}$, $P\{\exists t : X_t \in N_{fpq}\} = 0$ too, i.e. N_{fpq} is μ-polar. It follows that if K_2 is the union of N_{fpq} over rational p, q, and a countable dense subset $(f_n) \subset C_0(\boldsymbol{F})$, then K_2 is also μ-polar, and the resolvent equation holds on $\boldsymbol{F} - (K_1 \cup K_2)$ for each f_n and each rational p and q. But all terms are continuous in p and q, and the resolvent is also continuous in f (supnorm) so the resolvent equation holds identically in $\boldsymbol{F} - (K_1 \cup K_2)$. □

Lemma 10.15. *Let K be a μ-polar set. Then there exists a μ-polar set $\hat{K} \supset K$ such that if $x \in \boldsymbol{E} - \hat{K}$, $\hat{U}^2_p(x, \hat{K}) = 0$.*

Proof. For any μ-polar set M let $M' = \{x : \hat{U}^2_p(x, M) > 0\}$. If T is a predictable time $\hat{U}^2_p(\tilde{X}_T, M) = E\{\int_0^\infty e^{-pt} I_M(\tilde{X}_{T+t})\, dt \mid \mathcal{F}_{T-}\} = 0$ since, as X_t (and therefore $\tilde{X}_t$) never hits M, the integrand vanishes identically. Then $P\{\tilde{X}_T \in M'\} = 0$ for all predictable T. It follows that $(\tilde{X}_t)$, and hence (X_t), never hits M', so M' is μ-polar. Now let $M_1 = K$, $M_2 = M_1 \cup M'_1, \dots M_{k+1} = M_k \cup M'_k$. This is a sequence of μ-polar sets, and $\hat{K} \stackrel{\text{def}}{=} \cup_n M_n$ is the desired set. □

Now we can prove the existence of a proper resolvent for $(\tilde{X}_t)$. Let $\mathcal{E}_b$ and $\mathcal{E}_b^+$ denote the bounded Borel functions and the positive bounded Borel functions respectively.

Proposition 10.16. *There exists a compatible Markov resolvent $(\tilde{U}_p)_{p>0}$ on $\boldsymbol{F} = \boldsymbol{E} \cup \partial$ such that $(\tilde{X}_t)$ and $(\tilde{U}_p)$ satisfy Hypothesis (MMP).*

Proof. Let K_1 and K_2 be the μ-polar sets of Lemmas 10.12 and 10.14. By Lemma 10.15 there is a μ polar set $\hat{K}$ which contains $K_1 \cup K_2$ and has the property that for each $x \in \boldsymbol{E} - \hat{K}$, and each $p > 0$, $\hat{U}^2_p(x, \hat{K}) = 0$. Then $(\hat{U}^2_p)$ satisfies the resolvent equation on $\boldsymbol{F} - \hat{K}$, and we can change its values on $\hat{K}$ without affecting the resolvent equation on $\boldsymbol{F} - \hat{K}$. Thus we define

$$\tilde{U}_p(x, \cdot) = \begin{cases} \hat{U}^2_p(x, \cdot) & \text{if } x \in \boldsymbol{F} - \hat{K} \\ \dfrac{1}{p}\delta_\partial(\cdot) & \text{if } x \in \hat{K}. \end{cases} \tag{10.25}$$

Then $(\tilde{U}_p)$ satisfies the resolvent equation on $\boldsymbol{F} - \hat{K}$ by Lemma 10.14, and it satisfies it trivially on $\hat{K}$, so it satisfies it on $\boldsymbol{F}$. $\tilde{U}_p$ and $\hat{U}^2_p$ differ only on the μ-polar set $\hat{K}$, so for any $f \in C(\boldsymbol{E})$, $\tilde{U}_p f(\tilde{X}_t)$ and $\hat{U}^2_p f(\tilde{X}_t)$ are indistinguishable. Thus (10.21) holds for $\tilde{U}_p$, as well as for $\hat{U}^2_p$, at least if $f \in C(\boldsymbol{E})$. But a monotone class argument extends it to $f \in \mathcal{E}_b^+$.

To see that $t \mapsto \tilde{U}_p f(\tilde{X}_t)$ is lcrl, let (T_n) be a uniformly bounded sequence of predictable times for $(\tilde{X}_t)$ which increase to a limit T. Then by the above extension of (10.21)

$$\lim_{n\to\infty} E\{\tilde{U}_p f(\tilde{X}_{T_n})\} = \lim_{n\to\infty} E\left\{\int_0^\infty e^{-pt} f(\tilde{X}_{T_n+t})\, dt\right\}.$$

Since $T_n \uparrow T$ and $(\tilde{X}_t)$ is left-continuous, this is

$$= E\left\{\int_0^\infty e^{-pt} f(\tilde{X}_{T+t})\, dt\right\} = E\{\tilde{U}_p f(\tilde{X}_T)\}.$$

This implies that $t \mapsto \tilde{U}_p f(\tilde{X}_t)$ is left continuous by Theorem 6.15. Thus $(\tilde{X}_t)$ and $(\tilde{U}_p)$ satisfy (MMP). Finally, Lemma 8.49 assures us that we may assume without loss of generality that $(\tilde{U}_p)$ is compatible. □

10.3. Proof of Theorem 10.1

Suppose that Hypotheses (H1) and (H2) hold. (Recall that we used these to construct the original reverse transition function, so everything above is subject to them.) Then Proposition 10.16 tells us that $(\tilde{X}_t)$ and $(\tilde{U}_p)$ satisfy Hypothesis (MMP). By Lemma 8.49 we may assume without loss of generality that $(\tilde{U}_p)$ is compatible. Theorem 8.45 guarantees the existence of a separating set of supermedian functions, so that the hypotheses of Theorem 8.30 hold, and it follows that there exists a Ray-Knight compactification and an associated left continuous Ray semigroup $(\tilde{P}_t)$. The restriction of that semigroup to $\boldsymbol{E} \cup \partial$ is a left-continuous moderate Markov semigroup on $\boldsymbol{E} \cup \partial$, and $(\tilde{X}_t)$ is a left-continuous moderate Markov process relative to it.

The absolute distribution of $\tilde{X}_s$ is given by $K_s(dx) = \ell_s(x)\, G(dx)$, so that $\int_0^\infty k(s) f(\tilde{X}_s)\, ds = \int_0^\infty \int_{\boldsymbol{E}} k(s)\ell_s(x) f(x) G(dx)\, ds$. Integrate first over s and note that $\int_0^\infty k(s)\ell_s(x)\, ds = E^x\{k(\zeta)\}$, to see that (10.3) holds for each positive integrable function k.

Thus Theorem 10.1 is proved under Hypothesis 10.3. It remains to show it is unnecessary.

10.4. Removing Hypotheses (H1) and (H2)

Let (X_t) be a process on $\boldsymbol{F}$ with lifetime ζ. Let us modify it by adding three states, ∂_1, ∂_2, and ∂_3, and three exponential random variables with parameter $p > 0$, ξ_1, ξ_2, and ξ_3, which are independent of each other and of the process (X_t). Then set

$$X'_t = \begin{cases} X_t & \textit{if } 0 \le t < \zeta \\ \partial & \textit{if } \zeta \le t < \zeta + \xi_1 \\ \partial_1 & \textit{if } \zeta + \xi_1 \le t < \zeta + \xi_1 + \xi_2 \\ \partial_2 & \textit{if } \zeta + \xi_1 + \xi_2 \le t < \zeta + \xi_1 + \xi_2 + \xi_3 \\ \partial_3 & \textit{if } t \ge \zeta + \xi_1 + \xi_2 + \xi_3. \end{cases}$$

Thus the process (X'_t) is the same as (X_t) until its lifetime, but instead of remaining at ∂, it stays there an exponential length of time, then jumps to ∂_1, and from there to ∂_2, and finally to ∂_3, which becomes the new cemetery. This modification preserves the Markov property after ζ, and (X'_t) remains a rcll strong Markov process which satisfies the conditions of the theorem. Its new lifetime, the first hit of ∂_3, is $\zeta' = \zeta + \xi_1 + \xi_2 + \xi_3$. This now satisfies (H1) and (H2). To see this, let $\tau \equiv \xi_1 + \xi_2 + \xi_3$. Then τ is independent of the original process (X_t) and has density $p^3 t^2 e^{-pt}/2$. If the distribution of ζ given $X_0 = x$ is $L(x, dt)$, then the corresponding density of $\zeta' = \zeta + \tau$ is $\ell_t(x) \equiv (p^3/2)\int_0^t L(x, ds)(t-s)^2 e^{-p(t-s)}$; this is continuously differentiable, bounded by $p^2/2$; its derivative is bounded by $2p/e^2$, independent of x. Thus it satisfies (H1) and (H2). Let $\tilde{X}'_t = X'_{\zeta - t}$ be the reverse of (X'_t) from its lifetime. By what we have shown above, it is a moderate Markov process relative to the moderate Markov semigroup $(\tilde{P}_t)$. It starts at ∂_2, jumps to ∂_1, then ∂, and then enters the state space at time τ. Now τ is the first hitting time of $\boldsymbol{E}$, so it is a stopping time, and $\tau + \epsilon$ is a predictable time for any $\epsilon > 0$, so by the moderate Markov property of $(\tilde{X}'_t)$, $\{\tilde{X}'_{\tau+\varepsilon+t},\ t \geq 0\}$ is a moderate Markov process with semigroup $(\tilde{P}_t)$. But $\tilde{X}'_{\tau+\varepsilon+t} = \tilde{X}_{t+\varepsilon}$, and it follows easily that $\{\tilde{X}_t,\ t > 0\}$ is a moderate Markov process on $\boldsymbol{E} \cup \partial$ admitting transition semigroup $(\tilde{P}_t)$.

To verify (10.3), notice that it holds for (X'_t). Let the parameter p of the exponential random variables ξ_i tend to infinity, so that $\tau = \zeta' - \zeta = \xi_1 + \xi_2 + \xi_3$ tends to zero in distribution. This means that $X'_t \to X_t$ in distribution. If $f \in C(\boldsymbol{F})$, $f(\tilde{X}'_t) \to f(\tilde{X}_t)$ in distribution as well. As k is bounded, continuous and integrable, and f is bounded and continuous, we conclude that $E\{\int_0^\infty k(t) f(\tilde{X}'_t)\,dt\} \to E\{\int_0^\infty k(t) f(\tilde{X}_t)\,dt\}$, so that (10.3) holds for $(\tilde{X}_t)$. This finishes the proof of Theorem 10.1.

Notes on Chapter 10

Some history is in order. The fact that a Markov process reversed in time is still a Markov process has been known since the beginnings of the subject. So has the fact that this reversal seldom has stationary transition probabilities. While some of the reversal properties of specific processes like Brownian motion and the Ornstein-Uhlenbeck process are well-known, the general phenomenon remained ill-understood.

P.-A. Meyer, writing in 1967 [2], in the appendix to Ch. XII, summarizes the problems of reversal as follows: "Nous avons vu au n° 3 que le retournement du temps préserve le caractère markovien des processus. Malheureusement, il ne préserve le plus souvent aucune autre propriété intéressante: par exemple, supposons que l'on effectue le renversement du sens du temps sur un processus (X_t) admettant une fonction de transition homogène dans le temps: on ignore si le processus obtenu admet une fonction de transition; même si celle-ci existe, elle ne sera en générale pas homogène dans le temps. Enfin, la fonction de transition du processus retourné (si elle existe) ne dépend pas seulement de la fonction de transition (P_t) de (X_t), mais aussi de la loi d'entrée de ce processus. Il s'agit donc d'une operation peu satisfaisante."

Meyer is talking about reversal from a fixed time. While he wrote several articles within a year which showed that reversal was an operation *beaucoup plus satisfaisante* than this indicates, it is a fair summary of the situation at that time. Probabilists were at best ambiguous about the subject. Kai Lai Chung, speaking of the 1960s, said [private communication] "The problem of *true* reversal, namely path by path, was at the time regarded either as 'obvious' or 'impossible and unnecessary.'"

There is one case in which reversal works without complications: if the original process is itself a stationary process, its reversal from a fixed time is a Markov process whose transition function is easily found. Kolmogoroff [1, 2], showed that the forward and reverse processes were in duality, although these papers were analytic and did not involve explicit path-by-path reversal. Later, E. Nelson [1] showed that this held in much greater generality.

If a process is not stationary, its reverse from a fixed time will only have stationary transition probabilities in relatively trivial cases, and the strong Markov property is unavailable.

In hindsight, one can see that the problem has three levels:

(i) when are the transition probabilities stationary?
(ii) Is the reverse a strong Markov process?
(iii) What is the connection with duality?

To answer (i), one generally must reverse from a random time, not a fixed time. The modern, path-by-path theory of reversal dates from Hunt's seminal 1960 paper [4]. Hunt reversed a discrete time Markov chain from a last exit time to get a Markov chain with stationary transition probabilities. While it was clear that his ideas must hold in greater generality, the passage from discrete to continuous was delicate, and centered on the strong Markov property.

Chung showed in [9] that one could reverse a stable continuous-time Markov chain from its first infinity, using the density of the lifetime to calculate the reverse transition probabilities. The strong Markov property then followed by general Markov chain theory.

The first general answer to (ii) was by Ikeda, Nagasawa and Sato in two papers; See Ikeda, Nagasawa and Sato [1], Nagasawa [1]. The former reversed from the lifetime, and the latter generalized the time of reversal from a last exit time to a co-optional time – called an L-time – of a continuous-time Markov Processes. The key element was the assumption that there was a strong Markov dual process. This guaranteed that the right-continuous reverse was strongly Markov. At that time, the relation between duality and time reversal was not much better understood than it had been in Kolmogorov's time. It was clear that the relation was close, but the duality hypothesis postulated the existence of a strong Markov dual, and one had to ask if the reverse would still be strongly Markov without that. Counter-examples showed there were simple situations where it wasn't.

Then Chung and Walsh [3] introduced the moderate Markov property – which had been first observed at the first infinity of a Markov chain by K. L. Chung [10] – and showed that even without duality, the raw (i.e. left-continuous) reverse of a right-continuous strong Markov process is a moderate Markov process. The reversal was from the process lifetime, but it had become clear by then that this was in fact the general case. They proved the existence of transition probabilities for the reverse, but didn't show they formed a semigroup. P.-A. Meyer gave a different proof of the theorem in [4], using a deep result of Mokobodzki, and constructed a transition semigroup for the reverse.

This did not imply that the right-continuous reverse was strongly Markov, though, which re-raised question (iii) of the relation between reversal and duality. Walsh [7] showed in generality that it was possible to compactify the state space in such a way that both the forward and right continuous reverse process were strongly Markov, but this didn't answer the question if both were strongly Markov in the *original* topology. This was important since Hunt's memoir [5] showed that the most delicate results in potential theory required duality hypotheses. Smythe and Walsh took this up in [1]; the results are recounted in Chapter 15.

The definition of the moderate Markov property and Theorem 10.1 are from Chung and Walsh [3], with the exception of the existence of the semi-group, due to P.-A. Meyer [4]. The construction of the transition semigroup follows Walsh [3].

Chapter 11

h-Transforms

Let us introduce an important class of transformations of Markov processes, known variously as "h-path processes," "superharmonic transforms" or, as we will call them, "h-transforms." They were introduced by J. L. Doob in his study of the boundary limits of Brownian motion. They have an intuitive interpretation: they have to do with conditioning the process on its behavior at its lifetime.

We will need a class of processes large enough to be closed under these transforms, which means a slight relaxation of our hypotheses. In particular, we will not be able to assume that the sample paths have left limits at their lifetimes, since an h-transform may fail to have a left limit at the lifetime, even if the original process has one. Other than that, the processes will satisfy Hypothesis (SMP).

As before, we let $\boldsymbol{E}_\partial$ denote a locally compact metric space $\boldsymbol{E}$ with an isolated "cemetery point" ∂ adjoined. We let $\bar{\boldsymbol{E}}$ be the one-point compactification of $\boldsymbol{E}$. (This is merely for convenience: $\boldsymbol{E}$ is the proper state space and $\bar{\boldsymbol{E}}$ will only be used to provide a convenient set of uniformly continuous functions on $\boldsymbol{E}$.) Let $\mathcal{E}$ and $\mathcal{E}^{\sim}$ be the Borel sets of $\boldsymbol{E}$ and their universal completion respectively.

Let (P_t) be a Borel measurable semigroup on $\boldsymbol{E}_\partial$. We assume that for each probability measure μ on $\boldsymbol{E}_\partial$, there exists a Markov process $\{X_t, t \in \boldsymbol{T}\}$ with state space $\boldsymbol{E}_\partial$ and transition semigroup (P_t), satisfying:

(i) $P\{X_0 \in A\} = \mu P_0(A)$;
(ii) (X_t) is a strong Markov process;
(iii) there exists a random variable ζ, $0 \le \zeta \le \infty$, such that

$$X_t(\omega) \in \boldsymbol{E} \ \text{if}\ t < \zeta(\omega) \quad \text{and} \quad X_t(\omega) = \partial \ \text{if}\ t \ge \zeta(\omega);$$

(iv) $t \mapsto X_t(\omega)$ is right continuous on $[0, \infty)$ and has left limits in $\boldsymbol{E}$ on $(0, \zeta)$.

If we compare this with §3.1, we see that we have dropped three properties of Hunt processes: we do not assume that $X_{\zeta-}$ exists, we do not assume quasi-left continuity, and we do not assume that $P^x\{X_0 = x\} = 1$ for all x. Otherwise, the setting is that of Chapter 3, and we shall adopt that notation without further mention.

It will be convenient to use the **canonical probability space**: let Ω be the space of functions $\omega : \boldsymbol{R}_+ \mapsto \boldsymbol{E}_\partial$ which are right continuous, admit a lifetime ζ, and have a left limit except possibly at ζ. We can define $\{X_t, t \geq 0\}$ on this probability space by $X_t(\omega) = \omega(t)$, $\omega \in \Omega$. This space comes equipped with the shift operator θ_t (see §1.2 (6)): $\theta_t\omega(s) = \omega(s+t)$.

Let $\mathcal{F}_t^0$ be the natural filtration on Ω: $\mathcal{F}_t^0 = \sigma\{X_s, 0 \leq s \leq t\}$. Set $\mathcal{F}^0 = \cup_t \mathcal{F}_t^0$; and let $\mathcal{F}$ and $\mathcal{F}_t$ be the augmented right-continuous sigma fields.

11.1. Branching Points

As we saw with Ray processes, the only possible values of $P_0(x, \{x\})$ are zero and one. As with Ray processes, we define

Definition 11.1. *A point* $x \in \boldsymbol{E}$ *is a* **branching point** *if* $P(x, \{x\}) = 0$, *and is a* **non-branching** *point otherwise. The set of branching points is denoted by* B.

By and large, branching points have the same properties they had for Ray processes. First, $B \in \mathcal{E}$. Indeed, if (f_n) is a countable set of functions dense in $C(\bar{\boldsymbol{E}})$, then $B = \cup_n\{x : |f_n(x) - P_0 f_n(x)| > 0\} \in \mathcal{E}$. Note also that for all $x \in \boldsymbol{E}$, $P_0(x, B) = 0$. Indeed, for any n and $x \in \boldsymbol{E}$, apply the strong Markov property at $T \equiv 0$:

$$f_n(X_0) = E^x\{f_n(X_0) \mid \mathcal{F}_{0+}\} = P_0 f_n(X_0) \text{ a.s..}$$

It follows that with P^x-probability one, $X_0 \in \{x : P_0 f_n(x) = f_n(x), \forall n\} = \boldsymbol{E} - B$. Thus $0 = P^x\{X_0 \in B\} = P_0(x, B)$. More generally,

Proposition 11.2. *B is a Borel measurable polar set.*

Proof. If T is a stopping time, then by the strong Markov property at T,

$$P^x\{X_T \in B\} = E^x\{P_0(X_T, B)\} = 0.$$

This implies that B is polar. □

11.2. h-Transforms

Let h be an excessive function, and let $\boldsymbol{E}_h = \{x : 0 < h(x) < \infty\}$.

Definition 11.3.

$$_hP_t(x, dy) = \begin{cases} \dfrac{h(y)}{h(x)} P_t(x, dy) & \text{if } x \in \boldsymbol{E}_h \\ 0 & \text{if } x \in \boldsymbol{E} - \boldsymbol{E}_h. \end{cases} \tag{11.1}$$

Remark 11.4. 1° If $f \in \mathcal{E}^+$, then ${}_hP_t f(x) = \frac{I_{E_h}(x)}{h(x)} P_t(fh)(x)$, where we make the convention that $0 \cdot \infty = 0$. This is nearly Borel measurable.

2° ${}_hP_t(x, \boldsymbol{E} - \boldsymbol{E}_h) = 0$ for all $x \in \boldsymbol{E}_h$ and $t \geq 0$.

This is true by definition if $x \in \boldsymbol{E} - \boldsymbol{E}_h$. If $x \in \boldsymbol{E}_h$, $P_t h(x) \leq h(x) < \infty$, so ${}_hP_t(x, \{h = \infty\}) = \frac{1}{h(x)} \int_{\{h=\infty\}} P_t(x, dy)h(y) \leq \frac{1}{h(x)} P_t h(x) \leq 1$. But the integral can only be 0 or ∞, hence it must vanish. Thus ${}_hP_t(x, \boldsymbol{E} - \boldsymbol{E}_h) = {}_hP_t(x, \{h = 0\}) = \frac{1}{h(x)} \int_{\{h=0\}} P_t(x, dy)h(y) = 0$.

Proposition 11.5. *$({}_hP_t)$ is a sub-Markov semigroup on $\boldsymbol{E}$.*

Proof. If $x \in \boldsymbol{E}_h$, then ${}_hP_t(x, \boldsymbol{E}) = \frac{1}{h(x)} \int_{\boldsymbol{E}} P_t(x, dy)h(y) = \frac{1}{h(x)} P_t h(x) \leq 1$, while if $x \notin \boldsymbol{E}_h$, ${}_hP_t(x, \boldsymbol{E}) = 0$. In either case, ${}_hP_t(x, \boldsymbol{E}) \leq 1$.

To check the semigroup property, let $f \in \mathcal{E}$, $0 \leq f \leq 1$, and let $s, t \geq 0$. If $x \notin \boldsymbol{E}_h$, ${}_hP_t f(x) = 0$ for all t, and the semigroup property is trivial. If $x \in \boldsymbol{E}_h$,

$$
\begin{aligned}
{}_hP_s({}_hP_t f)(x) &= \frac{1}{h(x)} \int_{\boldsymbol{E}} P_s(x, dy)h(y) \frac{I_{\boldsymbol{E}_h}(y)}{h(y)} P_t(fh)(y) \\
&= \frac{1}{h(x)} \int_{\boldsymbol{E}_h} P_s(x, dy) P_t(fh)(y).
\end{aligned}
$$

But $P_s(x, \{h = \infty\}) = 0$ while, as $0 \leq f \leq 1$, $P_t(fh)(y) \leq P_t h(y) \leq h(y)$, which vanishes where $h = 0$, so we can integrate over all of $\boldsymbol{E}$ instead of $\boldsymbol{E}_h$:

$$
= \frac{1}{h(x)} \int_{\boldsymbol{E}} P_s(x, dy) P_t(fh)(y) = \frac{1}{h(x)} P_{s+t}(fh)(x) = {}_hP_{s+t} f(x).
$$

□

We set ${}_hP_t(\partial, \{\partial\}) = 1$ and ${}_hP_t(x, \{\partial\}) = 1 - {}_hP_t(x, \boldsymbol{E})$ as usual to extend ${}_hP_t$ to a Markov semigroup on $\boldsymbol{F}$.

Remark 11.6. 1° Note that we have shown that

$$
\int_{\boldsymbol{E}} {}_hP_s(x, dy) {}_hP_t f(y) = \frac{I_{\boldsymbol{E}_h}(x)}{h(x)} \int_{\boldsymbol{E}} P_s(x, dy) P_t(hf)(y). \tag{11.2}
$$

2° The set of branching points for ${}_hP_t$ is $B \cup (\boldsymbol{E} - \boldsymbol{E}_h)$. Indeed, if $x \in \boldsymbol{E} - \boldsymbol{E}_h$, then x is a degenerate branching point which branches directly to ∂. If $x \in \boldsymbol{E}_h$, however, x is a branching point for ${}_hP_t$ iff it is a branching point for P_t, for ${}_hP_0(x, \{x\}) = \frac{1}{h(x)} P_0(x, \{x\})h(x) = P_0(x, \{x\})$.

A function f on $\boldsymbol{E}$ is ***h*-excessive** if it is excessive for $({}_hP_t)$. h-excessive functions are essentially ratios of excessive functions of the form $\frac{u}{h}$, as the following shows.

Proposition 11.7. *If v is h-excessive, then $v = 0$ on $\boldsymbol{E} - \boldsymbol{E}_h$, and there exists an excessive function u such that $u = hv$ on $\{h < \infty\}$. Conversely, if u is excessive and if v satisfies*

(i) $v = 0$ *on* $\boldsymbol{E} - \boldsymbol{E}_h$;
(ii) $u = vh$ *on* $\boldsymbol{E}_h$,

then v is h-supermedian, and ${}_hP_0 v$ is h-excessive.

Remark 11.8. If $\boldsymbol{E} - \boldsymbol{E}_h = \emptyset$, then Proposition 11.7 simplifies: v is h-excessive iff $v = \frac{u}{h}$ for some excessive function u.

Proof. If v is h-excessive, $v = \lim_{t\downarrow 0} {}_hP_t v$, which vanishes on $\boldsymbol{E} - \boldsymbol{E}_h$. Let

$$\hat{u}(x) = \begin{cases} h(x)v(x) & \text{if } h(x) < \infty \\ \infty & \text{if } h(x) = \infty \end{cases}$$

(where by convention $0 \cdot \infty = 0$.) We claim that $\hat{u}$ is supermedian, i.e. $\hat{u} \geq P_t\hat{u}$. This is clear on $\{h = \infty\}$, and also on $\{h = 0\}$, where $\hat{u} = P_t\hat{u} = 0$. On the other hand, if $x \in \boldsymbol{E}_h$,

$$P_t\hat{u}(x) = \int_{\boldsymbol{E}} P_t(x, dy)h(y)v(y) = h(x)_h P_t v(x) \leq h(x)v(x) = \hat{u}(x). \tag{11.3}$$

Let $u(x) = \lim_{t\downarrow 0} P_t\hat{u}(x)$ be the excessive regularization of $\hat{u}$. By (11.3), if $x \in \boldsymbol{E}_h$,

$$u(x) = \lim_{t\downarrow 0} h(x)_h P_t v(x) = h(x)v(x),$$

so (i) and (ii) hold.

Conversely, suppose u is excessive and v satisfies (i) and (ii). If $x \in \boldsymbol{E} - \boldsymbol{E}_h$, both $v(x)$ and ${}_hP_t v(x)$ vanish for all t, while if $x \in \boldsymbol{E}_h$, ${}_hP_t v(x) = \frac{1}{h(x)} \int P_t(x, dy)h(y)v(y)$. But $hv = u$ on $\boldsymbol{E}_h$, so that

$${}_hP_t v(x) = \frac{1}{h(x)} \int_{\boldsymbol{E}_h} P_t(x, dy)u(y) \leq \frac{1}{h(x)} P_t u(x) \leq \frac{u(x)}{h(x)} = v(x).$$

Thus v is h-supermedian. Now let $t \to 0$. If $x \notin B$, the fact that $\boldsymbol{E}_h$ is fine open implies that for $x \in \boldsymbol{E}_h$, $\int_{\boldsymbol{E}_h} P_t(x, dy)u(y) \to u(x)$, hence ${}_hP_t v(x) \uparrow v(x)$.

In general, since ${}_hP_t(x, \cdot)$ does not charge B,

$$\lim_{t\downarrow 0} {}_hP_t v(x) = \lim_{t\downarrow 0} {}_hP_0({}_hP_t v)(x) = {}_hP_0 v(x),$$

since ${}_hP_t v(x) \uparrow v(x)$. □

Note that if v is h-excessive and u is the excessive function we constructed above, then $\boldsymbol{E}_v \subset \boldsymbol{E}_u$. There may not be equality, but one can show that $\boldsymbol{E}_u - \boldsymbol{E}_v$ is in the set of points which are regular for $\boldsymbol{E}_v$ but not in $\boldsymbol{E}_v$ itself, and is therefore semi-polar.

11.3. Construction of the h-Processes

We have two Markov semigroups on $\boldsymbol{E}_\partial$, (P_t) and $({}_hP_t)$. We want to compare the corresponding processes. We will first construct a realization of both processes on the canonical space $\bar{\Omega}$ of *all* functions $\omega : \boldsymbol{R}_+ \mapsto \boldsymbol{E}_\partial$, then show that the processes have right-continuous versions, and finally transfer the processes to the canonical space Ω of right-continuous functions.

Let $\bar{X}_t$ be the canonical process on $\bar{\Omega}$, i.e. $\bar{X}_t(\omega) = \omega(t)$. Let $\bar{\mathcal{F}}^0 = \sigma\{\bar{X}_t,\ t \geq 0\}$ and $\bar{\mathcal{F}}_t^0 = \sigma\{\bar{X}_s,\ s \leq t\}$. For each x we construct measures $\bar{P}^x$ and ${}_h\bar{P}^x$ on $\bar{\mathcal{F}}_0$: if $A_1, \dots, A_n \in \mathcal{E}$ are subsets of $\boldsymbol{E}_h$ and $0 \leq t_1 < \dots < t_n$, put $\Lambda = \{\bar{X}_{t_1} \in A_1, \dots, \bar{X}_{t_n} \in A_n\}$. Then set

$$\bar{P}^x\{\Lambda\} = \int_{A_1} P_{t_1}(x, dx_1) \int_{A_2} P_{t_2 - t_1}(x_1, dx_2) \dots \int_{A_n} P_{t_n - t_{n-1}}(x_{n-1}, dx_n), \tag{11.4}$$

and define ${}_hP^x\{\Lambda\}$ by the same formula, with P_t replaced by ${}_hP_t$. By induction on n using (11.2), we see that ${}_h\bar{P}^x\{\Lambda\}$ equals

$$\frac{I_{\boldsymbol{E}_h}(x)}{h(x)} \int_{A_1} P_{t_1}(x, dx_1) \int_{A_2} P_{t_2 - t_1}(x_1, dx_2) \dots \int_{A_n} P_{t_n - t_{n-1}}(x_{n-1}, dx_n) h(x_n). \tag{11.5}$$

Extend both $\bar{P}^x$ and ${}_h\bar{P}^x$ to $\bar{\mathcal{F}}^0$ as usual. Let $\bar{E}^x$ and ${}_h\bar{E}^x$ be the corresponding expectation operators. Set $t = t_n$, and rewrite (11.5) as

$${}_h\bar{P}^x\{\Lambda,\ \bar{X}_t \in \boldsymbol{E}\} = \frac{I_{\boldsymbol{E}_h}(x)}{h(x)} \bar{E}^x\{\Lambda;\ h(\bar{X}_t)\}. \tag{11.6}$$

Note that (11.6) extends to all $\Lambda \in \bar{\mathcal{F}}_t^0$ – both sides are measures – and in particular,

(a) ${}_h\bar{P}^x\{\bar{X}_t \in \boldsymbol{E}\} = \dfrac{I_{\boldsymbol{E}_h}(x)}{h(x)} P_t h(x);$

(b) $$\begin{aligned} {}_h\bar{P}^x\{\lim_{\substack{s \downarrow t' \\ s \in \boldsymbol{Q}}} \bar{X}_s = \bar{X}_{t'}, X_t \in \boldsymbol{E}\} &= \frac{I_{\boldsymbol{E}_h}(x)}{h(x)} \bar{E}^x\{\lim_{\substack{s \downarrow t' \\ s \in \boldsymbol{Q}}} \bar{X}_s = \bar{X}_{t'}; h(\bar{X}_t)\} \\ &= \frac{I_{\boldsymbol{E}_h}(x)}{h(x)} P_t h(x), \end{aligned}$$

since the original process was right continuous. Similarly, since the original process is free of oscillatory discontinuities on $(0, \zeta)$, we can take Λ to be the event on which $\bar{X}_s$ has both right and left limits along the rationals of $(0, t)$ to see that for $x \in \boldsymbol{E}_h$,

(c) ${}_h\bar{P}^x\{\Lambda, \bar{X}_t \in \boldsymbol{E}\} = \dfrac{1}{h(x)} P_t h(x).$

But this is ${}_hP^x\{X_t \in \boldsymbol{E}\}$ by (a), so evidently for ${}_hP_t^x$-a.e. ω, $s \mapsto \bar{X}_s(\omega)$ has both left and right limits along the rationals of $(0, t]$, so long as $\zeta(\omega) > t$, hence it has both left and right limits along the rationals on $(0, \zeta(\omega))$. Set

$$\hat{X}_t(\omega) = \lim_{\substack{s \downarrow t \\ s \in \mathbf{Q}}} \bar{X}_s(\omega).$$

Then $t \mapsto \hat{X}_t$ is a.s. right continuous with a lifetime $\hat{\zeta}(\omega) = \inf\{t \in \mathbf{Q} : \bar{X}_t = \partial\}$, it has left limits on $(0, \zeta)$, and for each fixed $s \geq 0$, ${}_h\bar{P}^x\{\hat{X}_s = \bar{X}_s\} = 1$ by (b).

We can now pull back both ${}_h\bar{P}^x$ and $\bar{P}^x$ from probabilities on $(\bar{\Omega}, \bar{\mathcal{F}}^0)$ to probabilities ${}_hP^x$ and P^x on $(\Omega, \mathcal{F}^0)$, the canonical space of right-continuous functions. Thus (11.6) holds in Ω as well as in $\bar{\Omega}$.

Denote the process we have just defined by (X_t^h). This is the ***h*-transform** of (X_t). A close look at the above proof reveals that we can not assume that (X_t^h) has a limit at its lifetime, even if (X_t) does.

Theorem 11.9. *(X_t^h) is a right-continuous strong Markov process which has left limits except possibly at ζ. If T is a stopping time with respect to $(\mathcal{F}_{t+})$, $\Lambda \in \mathcal{F}_{T+}$ and $x \in \mathbf{E}_h$, then*

$$ {}_hP^x\{\Lambda; \zeta > T\} = \frac{1}{h(x)} E^x\{\Lambda; h(X_T)\}. \tag{11.7}$$

Proof. Note that (11.7) extends (11.6) in two ways. The extension to the augmented σ-field $\mathcal{F}$ is technical but straightforward. Consider it done. In order to extend it from a fixed time t to a stopping time T, let us define discrete stopping times $T_n = k2^{-n}$ if $(k-1)2^{-n} \leq T < k2^{-n}$ $n = 1, 2, \ldots$ and $T_n = \infty$ if $T = \infty$. Then $\{T = k2^{-n}\} \in \mathcal{F}_{k2^{-n}}$ and so is $\Lambda \cap \{T_n = k2^{-n}\}$, so by (11.6) ${}_hP^x\{\Lambda; T_n = k2^{-n}, \zeta > k2^{-n}\} = \frac{1}{h(x)} E^x\{\Lambda; T_n = k2^{-n}; h(X_{k2^{-n}})\}$ so that the left-hand side of (11.6) with T replaced by T_n becomes

$$\begin{aligned} {}_hP_t^x\{\Lambda; \zeta > T_n\} &= \sum_{n=1}^{\infty} {}_hP_t^x\{\Lambda; T_n = k2^{-n}, \zeta > k2^{-n}\} \\ &= \sum_{n=1}^{\infty} E^x\{\Lambda; T_n = k2^{-n}; h(X_{T_n})\} \\ &= \frac{1}{h(x)} E^x\{\Lambda; h(X_{T_n})\}. \end{aligned}$$

Thus (11.7) holds with T replaced by T_n. Let $n \to \infty$. $T_n \downarrow T$, so $\{\zeta > T_n\} \uparrow \{\zeta > T\}$, and ${}_hP_t^x\{\Lambda; \zeta > T_n\} \to {}_hP_t^x\{\Lambda; \zeta > T\}$. Furthermore, $h(X_{T_n}) \to h(X_T)$ and $E^x\{\Lambda; h(X_{T_n})\} = E^x\{\Lambda; E\{h(X_{T_n}) \mid \mathcal{F}_T\}\}$, which increases to $E^x\{\Lambda; h(X_T)\}$ since h is excessive. Thus we can pass to the limit on both sides as $n \to \infty$ to get (11.7).

It remains to verify the strong Markov property for (X_t^h). Let $\Lambda_1 \in \mathcal{F}_{T+}$ and let $\Lambda_2 = \{X_{T+t_1} \in A_1, \ldots, X_{T+t_n} \in A_n\}$, where $0 \le t_1 < \ldots < t_n$ and $A_1, \ldots, A_n$ are Borel sets in $\boldsymbol{E}$. By (11.7),

$$
\begin{aligned}
{}_hP^x\{\Lambda_1 \cap \Lambda_2; \zeta > T + t_n\} &= \frac{I_{E_h}(x)}{h(x)} E^x\left\{\Lambda_1 \cap \Lambda_2; h(X_{T+t_n})\right\} \\
&= \frac{I_{E_h}(x)}{h(x)} E^x\left\{\Lambda_1; E^x\{\Lambda_2; h(X_{T+t_n}) \mid \mathcal{F}_{T+}\}\right\}.
\end{aligned}
$$

Now (X_t) itself is strongly Markov, so by the strong Markov property at time T, this is

$$
\begin{aligned}
&= \frac{I_{E_h}(x)}{h(x)} E^x\left\{\Lambda_1; E^{X_T}\{I_{\Lambda_2} \circ \theta_T h(X_{t_n})\}\right\} \\
&= \frac{I_{E_h}(x)}{h(x)} E^x\left\{\Lambda_1; h(X_T)\frac{I_{E_h}(X_T)}{h(X_T)} E^{X_T}\{I_{\Lambda_2} \circ \theta_T h(X_{t_n})\}\right\} \\
&= \frac{I_{E_h}(x)}{h(x)} E^x\left\{\Lambda_1; h(X_T)\,{}_hE^{X_T}\{I_{\Lambda_2} \circ \theta_T; \zeta > t_n\}\right\} \\
&= {}_hE^x\left\{\Lambda_1, \zeta > T; {}_hE^{X_T}\{I_{\Lambda_2} \circ \theta_T; \zeta > t_n\}\right\},
\end{aligned}
$$

which implies the strong Markov property for $({}_hP_t)$. □

11.4. Minimal Excessive Functions and the Invariant Field

Let $\{u_\alpha,\ \alpha \in J\}$ be a family of excessive functions, where J is a parameter set, and let ν be a finite measure on J. Suppose that $(\alpha, x) \mapsto u_\alpha(x)$ is $\mathcal{J}^\nu \times \mathcal{E}$ - measurable, where $\mathcal{J}^\nu$ is the σ-field of ν-measurable subsets of J. Consider the function

$$
h(x) = \int u_\alpha \nu(d\alpha). \tag{11.8}
$$

Then

1° $h \in \mathcal{E}^+$ (by Fubini's Theorem).

2° h is excessive. (It may be $\equiv \infty$.)

Indeed, $P_t h(x) = \int P_t u_\alpha(x)\nu(d\alpha) \le \int u_\alpha(x)\nu(dx) = h(x)$. As $t \downarrow 0$, $P_t u_\alpha \uparrow u_\alpha$, so $P_t h \uparrow h$ by monotone convergence.

Proposition 11.10. *Define h by (11.8) and write* ${}_\alpha P = {}_{u_\alpha}P$. *Then, for* $\Lambda \in \mathcal{F}$ *and* $x \in \boldsymbol{E}_h$:

$$
{}_hP^x\{\Lambda\} = \frac{1}{h(x)} \int u_\alpha(x)\,{}_\alpha P^x\{\Lambda\}\nu(d\alpha). \tag{11.9}
$$

Proof. If $\Lambda \in \mathcal{F}_t$ and $x \in \boldsymbol{E}_h$,

$$ {}_hP^x\{\Lambda; \zeta > t\} = \frac{1}{h(x)} E^x\{\Lambda; h(X_t)\} = \frac{1}{h(x)} \int E^x\{\Lambda; u_\alpha(X_t)\} \nu(d\alpha) $$

by Fubini's Theorem. Now if $u_\alpha(x) = 0$, then $E^x\{\Lambda; u_\alpha(X_t)\} = 0$ too, and, since $h(x) < \infty$, $\nu\{\alpha : u_\alpha(x) = \infty\} = 0$. Thus the above is

$$ = \frac{1}{h(x)} \int u_\alpha(x)_\alpha P^x\{\Lambda; \zeta > t\} \nu(d\alpha). $$

Thus (11.9) holds for all t and all $\Lambda \in \mathcal{F}_t$ and therefore for all $\Lambda \in \mathcal{F}$. □

Let $A \in \mathcal{E}$ and let u be excessive. Recall the operator P_A defined in §3.4. We call $P_A u$ the **reduite of u over A**:

$$ P_A u(x) = E^x\{u(X_{T_A})\}, $$

where, as usual, $T_A = \inf\{t > 0 : X_t \in A\}$.

Proposition 11.11. *If $A \in \boldsymbol{E}$ and u is excessive, and if $u(x) < \infty$, then*

$$ P_A u(x) = u(x)_u P^x\{T_A < \infty\}. \tag{11.10} $$

Proof. Since $P_A u \le u$, (11.10) holds if $u(x) = 0$. If $x \in \boldsymbol{E}_u$, then

$$ E^x\{u(X_{T_A})\} = E^x\{T_A < \zeta; u(X_{T_A})\} = u(x)_u P^x\{T_A < \infty\}. $$ □

Definition 11.12. *An excessive function h is* **minimal** *if, whenever u and v are excessive and $u + v = h$, both u and v are proportional to h.*

Definition 11.13. *The* **invariant field** *$\mathcal{I}$ is the set of all $\Lambda \in \mathcal{F}$ which have the property that for all t,*

$$ \theta_t^{-1}\{\Lambda \cap \{\zeta > 0\}\} = \Lambda \cap \{\zeta > t\}. $$

It is not hard to check that $\mathcal{I}$ is a σ-field. It is connected with the behavior of X at its lifetime. Events such as $\{\zeta = \infty\}$ and $\{X_{\zeta-} \in A\}$ are invariant, but in general $\{X_t \in A\}$ is not.

Minimal excessive functions are connected with potentials. For Brownian motion in $\boldsymbol{R}^3$, for instance, $f(x) \equiv 1/|x - y|$, where $y \in \boldsymbol{R}^3$ is fixed, is a minimal excessive function.

Heuristically, an h-transform can be thought of as the process (X_t) conditioned to have a given behavior at ζ. This is particularly transparent when h is minimal as we will see in Theorem 11.21 below.

The set $\{h = \infty\}$ can be a nuisance. The following lemma and its corollary will occasionally help us handle it.

Lemma 11.14. *Let h and u be excessive functions such that $u(x) = h(x)$ on the set $\{h < \infty\}$. Then there is an excessive function v such that $h = u + v$.*

Remark 11.15. This can happen. Consider uniform translation to the right on the positive x-axis. Let $h(x) = \infty$ if $x < 0$ and $h(x) = 1$ if $x \geq 0$, and let $u \equiv 1$. Both f and u are excessive, $u = h$ on $\{h < \infty\}$, but $u \neq h$. In this case, $v(x) = \infty$ if $x < 0$ and $v(x) = 0$ if $x \geq 0$. This is typical: the function v can always be chosen to takes on just the two values 0 and ∞.

Proof. Let $A = \{x : h(x) = \infty,\ u(x) < \infty\}$, and let $k(x) = P^x\{T_A < \infty\}$. Put $C = \{x : k(x) > 0\}$ and define

$$v(x) = \begin{cases} \infty & \text{if } x \in C \\ 0 & \text{otherwise.} \end{cases}$$

Notice that $u = h$ on A^c. (If $x \in A^c$ and $h(x) < \infty$, then $u(x) = h(x)$ by hypothesis. The other possibility is $u(x) = h(x) = \infty$.) Thus A^c is fine closed, so A is fine open. It follows that k is excessive and $k = 1$ on A, so $A \subset C$. Note that $k(x) > 0 \Longrightarrow h(x) = \infty$, so that $h(x) = u(x) + v(x)$ for all x. To see that v is excessive, note that $C^c = \{x : k(x) = 0\}$ is stochastically closed, so that, in particular, $P_t(x, C) = 0$ if $x \in C^c$. Thus $v(x) = P_t v(x) = 0$ if $x \in C^c$. Since $v \equiv \infty$ on C, it is clear that $v \geq P_t v$ there. Finally, C is fine open, so that $x \in C \Longrightarrow P_t(x, C) > 0$ for small t, hence for small t, $P_t v(x) = \infty = v(x)$. □

The following corollary is immediate.

Corollary 11.16. *Suppose h is a minimal excessive function and that $\boldsymbol{E}_h \neq \emptyset$. If u is excessive and $u = h$ on $\{h < \infty\}$, then $u \equiv h$.*

Proposition 11.17. *If h is excessive and $\Lambda \in \mathcal{I}$, then the function $u(x) \stackrel{\text{def}}{=} {}_hP^x\{\Lambda;\ \zeta > 0\}$ is h-excessive. If h is minimal, u is constant on $\boldsymbol{E}_h$. Finally, h is minimal if and only if $\mathcal{I}$ is trivial under ${}_hP^\mu$ for all initial measures μ on $\boldsymbol{E}_h$.*

Proof. We may as well assume that $\boldsymbol{E}_h \neq \emptyset$, since the proposition is empty if $\boldsymbol{E}_h$ is. Let $x \in \boldsymbol{E}_h$. Then

$${}_hP_t u(x) = {}_hE^x\{\theta_t^{-1}(\Lambda \cap \{\zeta > 0\})\} = {}_hE^x\{\Lambda;\ \zeta > t\} \leq u(x);$$

when $t \downarrow 0$, $\{\zeta > t\} \uparrow \{\zeta > 0\}$, so that ${}_hP_t u \uparrow u$, and u is excessive.

Next, let $\Lambda \in \mathcal{I}$ and define a second excessive function by $v(x) = {}_hP^x\{\Lambda^c;\ \zeta > 0\}$. By Proposition 11.7 there are excessive $\hat{u}$ and $\hat{v}$ such that $\hat{u} = \hat{v} = 0$ on $\{h = 0\}$ and $\hat{u} = hu$, $\hat{v} = hv$ on $\{h < \infty\}$. We claim that $\hat{u} + \hat{v} = h$. If $x \in \boldsymbol{E}_h - B$, then ${}_hP^x\{\zeta > 0\} = 1$, so $\hat{u}(x) + \hat{v}(x) = h(x)$.

But in fact this is also true if $x \in B \cap \{h < \infty\}$, since then $P_0(x, \cdot)$ does not charge $B \cup \{h = \infty\}$, so that $h(x) = P_0 h(x) = P_0(\hat{u} + \hat{v})(x) = \hat{u}(x) + \hat{v}(x)$. Thus $\hat{u} + \hat{v} = h$ on $\{h < \infty\}$. Since $\boldsymbol{E}_h \neq \emptyset$, Corollary 11.16 tells us that $\hat{u} + \hat{v} = h$ everywhere, as claimed. But now h is minimal, so both $\hat{u}$ and $\hat{v}$ are proportional to h, and $\hat{u} = ah$ for some a. Then $ah(x) = u(x)h(x)$ for all $x \in \boldsymbol{E}_h$, hence $u(x) = a$. Thus u is constant on $\boldsymbol{E}_h$.

If u is not constant on $\boldsymbol{E}_h$, then $\hat{u}$ and $\hat{v}$ are not proportional, so h is not minimal. Moreover, if, say, $u(x) \neq u(y)$, and $\mu = (1/2)(\delta_x + \delta_y)$, then ${}_hP^\mu(\Lambda \cap \{\zeta > 0\}) = (1/2)(u(x) + u(y))$, which is strictly between zero and one, hence $\mathcal{I}$ is not trivial under ${}_hP^\mu$. If h is minimal, on the other hand, then for any $\Lambda \in \mathcal{I}$, u will be constant. We claim this implies that $\mathcal{I}$ is trivial.

Indeed, if $\Lambda \in \mathcal{I}$ is such that $\Lambda \subset \{\zeta > 0\}$, let $\Gamma \in \mathcal{F}_t$ be contained in $\{\zeta > t\}$. Then for an initial measure μ,

$$\begin{aligned} {}_hP^\mu\{\Lambda \mid \Gamma\} &= {}_hP^\mu\{\Lambda \cap \{\zeta > t\} \mid \Gamma\} = {}_hP^\mu\{\theta_t^{-1}(\Lambda \cap \{\zeta > 0\}) \mid \Gamma\} \\ &= {}_hE^\mu\{u(X_t) \mid \Gamma\}. \end{aligned}$$

Now $u(X_t)$ is constant, say $u(X_t) = a$ on $\{\zeta > t\} \subset \Gamma$. Thus ${}_hP^\mu\{\Lambda \mid \Gamma\} = a = {}_hP^\mu\{\Lambda\}$.

It follows that Λ and Γ are independent. But events of the form Γ generate $\mathcal{F}$. Therefore Λ is independent of $\mathcal{F}$, therefore of itself, and therefore has probability either zero or one. □

11.5. Last Exit and Co-optional Times

Define the **last exit time** L_A from a set $A \subset \boldsymbol{E}$ by

$$L_A = \sup\{t \geq 0 : X_t \in A\},$$

with the convention that the sup of the empty set is zero. If $A \in \mathcal{E}$, $\{L_A < \zeta\} \in \mathcal{I}$. If h is minimal excessive, we say a set $A \in \mathcal{E}$ is ***h*-thin** if

$${}_hP^x\{L_A < \zeta\} = 1, \ x \in \boldsymbol{E}_h.$$

Proposition 11.18. *If $A \in \mathcal{E}$, h is minimal, and $\boldsymbol{E}_h \neq \emptyset$, then A is h-thin iff $P_A h \neq h$.*

Proof. Since $\{L_A < \zeta\} \in \mathcal{I}$ and since $\mathcal{I}$ is trivial under ${}_hP^x$ by Proposition 11.17, ${}_hP^x\{L_A < \zeta\} = 0$ or 1. Note that $T_A = \infty \iff L_A = 0$, and $L_A \geq T_A$ if T_A is finite. If $x \in \boldsymbol{E}_h - B$,

$$h(x)_hP^x\{L_A < \zeta\} \geq h(x)P^x\{T_A = \infty\} = h(x) - P_A h(x).$$

Thus if $P_A h(x) < h(x)$ for some $x \in \boldsymbol{E}_h$, ${}_hP^x\{L_A < \zeta\}$ is strictly positive and therefore identically one. Conversely, if $P_A h(x) = h(x)$ for all $x \in \boldsymbol{E}_h$, then $P_A h \equiv h$ by Corollary 11.16, and ${}_hP^x\{T_A < \infty\} = 1$ for all $x \in \boldsymbol{E}_h$. But now, if $L_A < \zeta$, there exits $t \in \boldsymbol{Q}$ such that $L_A < t < \zeta$, and for this t, $T \circ \theta_t = \infty$. Thus if (r_n) is an enumeration of the rationals,

$$ {}_hP^x\{L_A < \zeta\} \le \sum_n {}_hP^x\{T_A \circ \theta_{r_n} = \infty, \zeta > r_n\} = 0. \qquad \square $$

Definition 11.19. *A minimal excessive function h has a* **pole** *at $y \in \boldsymbol{E}$ if, whenever V is a neighborhood of y, V^c is h-thin.*

Remark 11.20. This could be called "h-thin at the lifetime," or "h-thin at the pole," to distinguish it from the thinness defined in §3.5, which concerns thinness at zero. The two ideas concern thinness at the two ends of the sample path, 0 and ζ. However, the terminology is well-established at both ends, and there is no danger of confusion.

Theorem 11.21. *Suppose h is a minimal excessive function and $\boldsymbol{E}_h \neq \emptyset$. Then h has a pole at y iff ${}_hP^x\{X_{\zeta-} = y\} = 1$, $\forall x \in \boldsymbol{E}_h$. If h does not have a pole at y, ${}_hP^x\{X_{\zeta-} = y\} = 0$.*

Remark 11.22. This implies that a minimal excessive function can have at most one pole. Note that we have not assumed that $X_{\zeta-}$ exists.

Proof. $\{X_{\zeta-} = y\} \in \mathcal{I}$, so ${}_hP^x\{X_{\zeta-} = y\} \equiv 0$ or 1 on $\boldsymbol{E}_h$. Suppose h has a pole at y. Let V be a neighborhood of y. Then, as V^c is h-thin, there is a random $\tau < \zeta$ such that $X_t \in V$ a.s. if $\tau < t < \zeta$. This is true simultaneously for a neighborhood base (V_n) of y, which implies that $\lim_{t \uparrow \zeta} X_t = y$.

Conversely, if ${}_hP^x\{X_{\zeta-} = y\} = 1$, then for any neighborhood V of y, ${}_hP^x\{L_{V^c} < \zeta\} = 1$, so V^c is h-thin, and h has a pole at y.

Finally, if y is not a pole of h, ${}_hP^x\{X_{\zeta-} = y\} < 1$, hence it is identically zero. $\square$

There is a connection between last exit times and h-transforms which we will continually exploit. In fact, it extends to more general times, called *co-optional* times.

Definition 11.23. *A random variable L, $0 \le L \le \infty$, is* **co-optional** *if for all $t \ge 0$, $L \circ \theta_t = (L - t)^+$.*

Lemma 11.24. *Let L be a co-optional time and define*

$$ c_L(x) = P^x\{L > 0\}. $$

Then c_L is excessive.

Proof. $P_t c_L(x) = P^x\{L \circ \theta_t > 0\} = P^x\{(L - t)^+ > 0\}$. This is dominated by $P^x\{L > 0\} = c_L(x)$. Clearly $P_t c_L \to c_L$ as $t \to 0$. $\square$

Define X^L, the **process killed at L** by

$$X_t^L = \begin{cases} X_t & \text{if } t < L \\ \partial & \text{if } t \geq L. \end{cases}$$

Remark 11.25. It is readily verified that the last exit time L_A is co-optional and that $c_{L_A}(x) = P^x\{T_A < \infty\}$. According to the following theorem, the process killed at its last exit from A is a c_{L_A}-transform.

Theorem 11.26. *Let L be a co-optional time. Then X^L is a c_L-transform of X.*

Proof. If $f \in \mathcal{E}^+$ and $\Gamma \in \mathcal{F}_t$,

$$\begin{aligned} E^x\{\Gamma;\, f(X_t^L)\} &= E^x\{\Gamma;\, f(X_t), L > t\} \\ &= E^x\{\Gamma;\, f(X_t), L \circ \theta_t > 0\} \\ &= E^x\{\Gamma;\, f(X_t)c_L(X_t)\}; \end{aligned}$$

or, if we replace f by f/c_L:

$$E^x\{\Gamma; c_L^{-1}(X_t^L) f(X_t^L)\} = E^x\{\Gamma;\, f(X_t)\}.$$

Thus

$$\begin{aligned} E^x\{\Gamma;\, f(X_{t+s}^L)\} &= E^x\{\Gamma;\, f(X_{t+s})c_L(X_{t+s})\} \\ &= E^x\{\Gamma;\, P_s(fc_L)(X_t)\} \\ &= E^x\{\Gamma; c_L^{-1}(X_t^L) P_s(fc_L)(X_t^L)\}, \end{aligned}$$

i.e. (X_t^L) is a Markov process with transition function $c_L^{-1}(x)P_t(x, dy)c_L(dy)$. □

Remark 11.27. The interpretation of the h-transformation as conditioning is particularly transparent for minimal excessive functions. Suppose that h is a minimal excessive function with pole y. Theorem 11.21 suggests the following: (X_t^h) is the process (X_t) conditioned to converge to to the pole of h at its lifetime. There is more to this, however, for there is a relationship with the lifetime itself. A better description of the h-transform is this: it is the original process conditioned to converge to its pole, and then killed at its last exit from the pole. (If h blows up at its pole, then the first hitting and last exit times are one and the same. However, if h is bounded, as it would be for Brownian motion on a bounded interval of the line, the first hitting time and last exit times are different, and one can check the above description by a direct calculation.) See the exercises for another look at the idea of conditioning.

We occasionally want to take an h-transform of another h-transform. The result is again an h-transform in the following sense.

Theorem 11.28. *Let h be excessive and let v be h-excessive. Then there exists an excessive function u such that $u = vh$ on $\{h < \infty\}$, and ${}_uP_t(x, \cdot) = {}_v({}_hP_t)(x, \cdot)$ except possibly on the semi-polar, h-polar set of points in $\{h = \infty\}$ which are not regular for $\{h = \infty\}$.*

Proof. The difficulty in the proof is technical: we must deal with the set $\{h = \infty\}$. Let u_1 be the excessive function constructed in Proposition 11.7 such that $u_1 = vh$ on $\{h < \infty\}$. To define u, first let

$$u_2(x) = \begin{cases} u_1(x) & \text{if } h(x) < \infty \\ \infty & \text{if } h(x) = \infty. \end{cases}$$

Note that u_2 is supermedian: $u_2 \geq P_t u_2$ is clear on $\{h = \infty\}$, where u_2 is infinite, and if $h(x) < \infty$, $P_t(x, \{h = \infty\}) = 0$, so $P_t u_2(x) = P_t u_1(x) \leq u_1(x) = u_2(x)$.

Then set $u(x) = \lim_{t \downarrow 0} P_t u_2(x)$ to get an excessive function. From the above, $u = u_1 = vh$ on $\{h < \infty\}$. Note that if $h(x) = \infty$, then the process from x satisfies $h(X_t) = \infty$ until the first time it hits the set $\{h < \infty\}$, and is finite from there on. Thus, if x is regular for $\{h = \infty\}$, $P_t(x, \{h = \infty\}) > 0$ for small enough t, hence $P_t u_2(x) = \infty$, hence $u(x) = u_2(x) = \infty$.

Let N be the set of points in $\{h = \infty\}$ which are not regular for $\{h = \infty\}$. Then N is a semi-polar set which is also h-polar, and, if we trace our way back through the definitions, we see that $\boldsymbol{E}_u$ contains $\boldsymbol{E}_v$, and they differ by at most the set N. Since $\boldsymbol{E}_v \subset \boldsymbol{E}_h \cup N$, we can write

$${}_v({}_hP_t)(x, dy) = \frac{I_{\boldsymbol{E}_v}(x)}{v(x)h(x)} P_t(x, dy) v(y) h(y). \tag{11.11}$$

$${}_uP_t(x, dy) = \frac{I_{\boldsymbol{E}_u}(x)}{u(x)} P_t(x, dy) u(y), \tag{11.12}$$

The expressions in (11.11) and (11.12) are equal except possibly if $x \in N$. □

11.6. Reversing h-Transforms

We will investigate the general problem of the time-reversal of an h-transform under duality hypotheses in §13. However, there is one useful general result that we can give here.

The transition function of a time-reversed process depends on the initial measure, so the transition function of the reverse of an h-transform depends on both h and the initial measure. These two can balance, as the following proposition shows. The corollary reveals an interesting aspect of time reversal from co-optional times.

Proposition 11.29. *Let h be an excessive function and μ a probability measure which sits on $\mathbf{E}_h$, such that h is μ-integrable. Let (X_t) have initial measure μ, and let (X_t^h) have initial measure ν, where $\nu(dx) = Ch(x)\mu(dx)$ and C is chosen to make ν a probability measure. Then the reverses of (X_t) and (X_t^h) from their respective lifetimes, where finite, have the same transition probabilities.*

Proof. Let P_t and $\tilde{P}_t$ be the transition probabilities of (X_t) and its reverse $(\tilde{X}_t)$. By Lemma 10.8 we need only show that ${}_hP_t$ and $\tilde{P}_t$ satisfy the duality relation (10.2) with μ replaced by ν. But then $G\mu$ is replaced by $G^h\nu$, where

$$G^h\nu(A) = C\int_A \int_{\mathbf{E}_h} h(z)\mu(dz)\left[\int_0^\infty \frac{1}{h(z)} P^x(z, dx) I_{\mathbf{E}_h}(x)ds\right] \tag{11.13}$$

$$= C\int_A \mu(dz)U_0(hI_{\mathbf{E}_h})(z), \tag{11.14}$$

so that $G^h\nu(dz) = Ch(z)I_{\mathbf{E}_h}(z)$, leading to

$$\langle f, \tilde{P}_t g\rangle_{G^h\nu} = C\langle fhI_{\mathbf{E}_h}, \tilde{P}_t g\rangle_G = C\langle g, P_t(fhI_{\mathbf{E}_h})\rangle_G.$$

Now $P_t fhI_{\mathbf{E}_h}(x) = h(x)P_t f(x)$ so this is

$$= C\langle gh, {}_hP_t f\rangle_G = \langle g, {}_hP_t f\rangle_{G^h\nu}.$$

□

Apply this to the process killed at a co-optional time L. Suppose (X_t) has initial measure μ. The process is immediately killed at time zero on the set $\{L = 0\}$, which may have positive probability. Indeed, if $X_0 = x$, the probability it is killed at zero is $P^x\{L = 0\} = 1 - c_L(x)$, so the initial measure is not $\mu(dx)$ but $c_L(x)\mu(dx)$. Thus it follows that the processed reversed from the co-optional time has the same transition probabilities as the original reverse.

Informally, then, the transition probabilities of the reverse from a co-optional time are independent of the co-optional time. Formally, we have

Corollary 11.30. *Let (X_t) be a Ray process with initial distribution μ. Let L be a co-optional time for (X_t), and let $c_L(x) = P^x\{L > 0\}$. Suppose μ sits on the set $\{x : c_L(x) > 0\}$. Then the reverse of (X_t) from its lifetime ζ, where finite, and the reverse of (X_t) from L, where finite, have the same transition probabilities.*

Remark 11.31.

1. The hypothesis that (X_t) is a Ray process is purely to simplify the statement. The extension to more general processes follows by a Ray-Knight compactification. In particular, the results above hold without change for the more general "right processes" and therefore for all the Markov processes we will be considering later.

2. The result is intuitive, once one realizes that co-optional times are stopping times for the reverse. Indeed, the canonical example of a co-optional time is a last-exit time, which is a first-hitting time for the reverse. We will pursue this point further in Chapter 12.

Exercises

1. Show that last exit times and lifetimes are co-optional. Show that if L is co-optional, so is $L - t$ for any $t > 0$. Show that the minimum and maximum of two co-optional times are co-optional.
2. Let (X_t) be Brownian motion on the interval $(0, 1)$, killed when it hits either end point. Let $y \in (0, 1)$ and let $h(x)$ be the continuous function which satisfies $h(0) = h(1) = 0, h(y) = 1$, and h is linear on the intervals $(0, y)$ and $(y, 1)$. (So h is a tent-function.) Show that h is a minimal excessive function with pole y. Show that the h-transform of X is Brownian motion which is conditioned to hit y, and is killed at its last exit from y. Show that (X_t^h) hits y infinitely often. Show that its lifetime is a totally inaccessible stopping time for (X_t^h).
3. Let $\{f_\alpha, \ \alpha \in I\}$ be a family of minimal excessive functions such that f_α has pole y_α. Let μ be a measure on I, and, assuming the necessary measurability, define $h(x) = \int f_\alpha(x)\nu(dx)$. Show that for $x \in A, \ C \in \mathcal{E}$, that the h-transform satisfies
$$_hP^x\{X_{\zeta-} \in C\} = h(x)^{-1} \int_{\{\alpha : y_\alpha \in C\}} f_\alpha(x)\mu(dx).$$
4. (Continuation) Suppose that $h(x) \equiv 1$ in the previous exercise. Find the distribution of $X_{\zeta-}$, given that $X_0 = x$, and justify the statement that the u_α transform is the original process conditioned on $X_{\zeta-} = y_\alpha$.

Notes on Chapter 11

The h-transform was created by Doob in order to put Brelot's relative potential theory [2] in a probabilistic setting. Relative potential theory concerns the ratios f/h, where h is a fixed strictly-positive harmonic function, and f is a harmonic or superharmonic function. They were first conceived in Doob [6] and made their definitive entrance in Doob [5]. (Due to the vagaries of waiting times, the second actually appeared before the first.)

The history of h-transforms from that time on is simple: the important results are all in Doob [5]. This may still be the best reference on the subject, since it presents the theory in its purest form, for Brownian motion in a domain in $\boldsymbol{R}^n$. It contains the relation with conditioning, the connection with the Martin boundary – which we will study in Chapter 14 – the Dirichlet problem and its relation to the invariant field (called the "stochastically ramified boundary" in Doob [6].) The theory has been adapted and generalized to various situations, but the adaptation is invariably straightforward. The main thing needed to generalize his treatment to our setting is some care with the sets where the excessive functions are infinite or zero. These cause no problem with Brownian motion, but may in the general case. The only important result in the chapter which is not in Doob [5] is Theorem 11.26, which was proved in Meyer, Smythe and Walsh [1].

The h-transform has a natural generalization to transforms by positive multiplicative functionals which are not necessarily monotone. This is done in Sharpe [1]. Indeed, $M_t = h(X_t)/h(X_0)$ is a multiplicative functional. The h-transform remains the most important such transform, but the generalization sheds light on the usual transformation by a (decreasing) multiplicative functional as in Blumenthal and Getoor [1]. This is associated purely with killing, while the general transformation, like the h-transform, involves a mixture of killing and conditioning.

Chapter 12

Death and Transfiguration: A Fireside Chat

There are some interesting topics which are relevant to our subject, but which will not be used in the sequel. While we don't want to let them pass unremarked, we don't want to go into great detail either. So we'll deal with them informally. We'll give the main definitions and theorems, and talk more about the intuition behind the proofs than the proofs themselves, even if this means resorting to sheer hand-waving in places. In short, we'll chat with the reader.

The subject is birth and death. The birth and death of processes, that is, their beginning and their end. But what we're about to show – the near identity of lifetimes, death times, birth times, reversal times, and their relation to transfiguration – has no theological connection whatsoever.

The reversal theorem (Theorem 10.1) only mentions reversing from process lifetimes. It is well known that there are other random reversal times, such as first hitting and last exit times, or, more generally, terminal times and co-optional times. However, the process lifetime suffices, because in a certain strong sense, it is the most general time of reversal: the above times are all lifetimes – or death times – of subprocesses.

Let $\{X_t,\ t \in \boldsymbol{T}\}$ be a strong Markov process on a locally compact separable metric space $\boldsymbol{E}$, which is right continuous and has left limits except possibly at its lifetime ζ.

Definition 12.1. *A random variable* β, $0 \le \beta \le \zeta$, *is a* **time of reversal** *for* (X_t) *if the process*

$$\tilde{X}_t^\beta \stackrel{\text{def}}{=} \begin{cases} X_{\beta-t} & \text{if } 0 < t \le \beta < \infty \\ \partial & \text{if } t > \beta \text{ or } \beta = \infty \end{cases}$$

is a left-continuous moderate Markov process.

Definition 12.2. *A process* (Y_t) *is a* **subprocess** *of* (X_t) *if there is a random variable* ξ, $0 \le \xi \le \zeta$, *such that*

$$Y_t = \begin{cases} X_t & \text{if } 0 \le t < \xi \\ \partial & \text{if } t \ge \zeta, \end{cases}$$

is a strong Markov process.

Note that we reverse from β only when it is finite, so that the actual time of reversal is β', where

$$\beta' \stackrel{\text{def}}{=} \begin{cases} \beta & \text{if } \beta < \infty \\ 0 & \text{if } \beta = \infty. \end{cases}$$

In terms of β',

$$\tilde{X}_t^{\beta} = \tilde{X}_t^{\beta'} = \begin{cases} X_{\beta'-t} & \text{if } 0 < t \le \beta \\ \partial & \text{if } t > \beta. \end{cases}$$

We let $\tilde{X}_t = \tilde{X}_t^{\zeta}$ denote the reverse from the lifetime. Then we have

Theorem 12.3. *A random variable β with values in $[0, \zeta]$ is a time of reversal if and only if β' is the lifetime of a sub-process of (X_t).*

We will give a basic idea of the proof, and then indicate how to make that idea rigorous.

One direction is clear: if β' is the lifetime of a subprocess, then Theorem 10.1 applies to show it is also a time of reversal.

The other direction has a simple heuristic proof by double reversal. If β is a time of reversal, then its reverse $(\tilde{X}_t^{\beta})$ is a Markov process which also has lifetime β'. We can reverse yet again from its lifetime to get another Markov process (Y_t) say, with stationary transition probabilities and lifetime β'; and $Y_t = X_t$ if $t < \beta'$. Thus (Y_t) is a subprocess of (X_t) with lifetime β', as claimed.

This argument is not rigorous, for while the first reversal works by definition, the second does not, since (X_t^{β}) is left continuous and in general is only moderately Markov, not strongly Markov, so that Theorem 10.1 doesn't apply. However, we can easily work around this by using a Ray-Knight compactification. Here is how to go about this.

By Theorem 8.30 we can find a compact space $\boldsymbol{F} \supset \boldsymbol{E}$ and a right continuous Ray process $(\hat{X}_t)$ on $\boldsymbol{F}$ such that $\hat{X}_t = \tilde{X}_{t+}$ and $\hat{X}_{t-} = \tilde{X}_t$. The limit $\tilde{X}_{t+}$ is taken in $\boldsymbol{F}$, but by the magic of Ray-Knight compactifications, the limit $\hat{X}_{t-}$ can be taken in the topology of either $\boldsymbol{E}$ or $\boldsymbol{F}$ – both topologies give the same limit. Since $(\hat{X}_t)$ is a right-continuous strong Markov process, it can be reversed from its lifetime. The reverse will be a left-continuous moderate Markov process. When we make it right-continuous, we get the process (Y_t), which is (X_t) killed at β'. There remains some work to verify that (Y_t) is in fact strongly Markov, but we will leave this aside. That finishes the proof.

There is a dual of death times, namely birth times.

Definition 12.4. *A random variable α, $0 \le \alpha \le \zeta$, is a* **birth time** *for (X_t) if the process $\{X_{\alpha+t},\ t > 0\}$ is a strong Markov process. A random variable β, $0 \le \beta \le \zeta$, is a* **death time** *for (X_t) if the process (X_t) killed at β is a strong Markov process.*

The most familiar example of a birth time is a stopping time. In this case, $(X_{\alpha+t})$ has the same transition probabilities as (X_t), but this is the exception. For other times, such as last exit times, the post-α process has a quite different transition probability. Thus the process starts a new and different life at α: it is transfigured. If we apply Theorem 12.3 to the reverse, we get

Corollary 12.5. *Suppose $\zeta < \infty$ a.s. If α is a random variable such that $\alpha \le \zeta$, then α is a birth time for (X_t) if and only if $\beta \equiv \zeta - \alpha$ is a death time for $(\tilde{X}_t)$.*

The proof is by reversal: if α is a birth time, the process $(X_{\alpha+t})$ can be reversed from its lifetime, which is $\beta = \zeta - \alpha$, and its reverse is $(\tilde{X}_t)$ killed at β. Thus β is the lifetime – and therefore the death time – of a subprocess of $(\tilde{X}_t)$. Conversely, if β is a death time for $(\tilde{X}_t)$, it is a time of reversal, and the (right-continuous) reverse of $(\tilde{X}_t)$ from β is $(X_{\alpha+t})$.

Examples

The theory of multiplicative functionals, as given for instance in Blumenthal and Getoor [1], provides a large family of death times. These are by and large randomized stopping times. Let us look at non-randomized times, i.e. random variables which are measurable with respect to the filtration generated by the process. We already know about stopping times and co-optional times. Let us introduce another dual pair, terminal times and co-terminal times. Recall the shift operator θ_t and the killing operator k_t.

Definition 12.6. *A stopping time T is a* **perfect exact terminal time** *if for each $\omega \in \Omega$ and $t < T(\omega)$ we have $T(\omega) = t + T(\theta_t\,\omega)$ and $T(\omega) = \lim_{t\downarrow 0} T(\theta_t\,\omega)$. A co-optional time L is a* **perfect exact co-terminal time** *if for each $\omega \in \Omega$ and $t > L(\omega)$, we have $L(\omega) = L(k_t\,\omega)$ and $L(\omega) = \sup_t L(k_t\,\omega)$.*

Many of the best known stopping times are terminal times. First hitting times are terminal times, but constant times, second-hitting times, and so forth are not. Co-terminal times are best thought of in terms of the reverse process $(\tilde{X}_t)$: if ζ is finite, and if $\tilde{T}$ is a stopping time for $(\tilde{X}_t)$, then $(\zeta - \tilde{T})^+$ is a co-optional time, and if $\tilde{T}$ is a terminal time for the reversed process, then $(\zeta - \tilde{T})^+$ is a co-terminal time. This characterizes co-terminal and co-optional times in the case where the lifetime is finite, but the given definitions work even when the lifetime is infinite.

Remark 12.7. The usual definitions of terminal and co-terminal times allow the defining relations to fail on exceptional null-sets. It is known that given an exact terminal or co-terminal time, one can always find one that is a.s. equal and "perfect," i.e. for which there are no exceptional null-sets.

Exactness refers to the second relation in the definitions; one can always regularize the time to make it exact. Therefore we will simply assume in what follows that our times are both perfect and exact, and say no more about it.

There is a duality between terminal and co-terminal times. (We will encounter this below in another guise in §13.6. The reader might find it interesting to compare the two.)

Proposition 12.8. *Let T be a terminal time. Then there exists an associated co-terminal time L_T, defined by $L_T = sup\{t > 0 : T \circ \theta_t < \infty\}$. Let $L \leq \zeta$ be a co-terminal time. Then there exists an associated terminal time T_L defined by $T_L = \inf\{t > 0 : L \circ k_t > 0\}$. Then L is the associated co-terminal time of T_L and T is the associated terminal time of L_T.*

We leave it to the reader to check that these definitions do what is claimed.

It is well-known that terminal times are death times. Indeed, the semigroup (P_t^T) of the process killed at a terminal time T is $P_t^T(x, A) = P^x\{X_t \in A,\ T > t\}$. According to Theorem 11.26, co-optional times are death times as well.

Since birth times correspond to killing times for the reverse process, we can see that stopping times and co-terminal times should be birth times. Indeed they are. In fact, we can go further and describe the semigroup of the process starting at a co-terminal time.

Let L be a co-terminal time, and let $T = T_L$ be the associated terminal time. If (P_t) is the transition function of (X_t), let (P_t^T) be the transition function of the process killed at T_L. Define a function $g(x) = P^x\{T = \infty\} = P^x\{L = 0\}$. Notice that this is invariant for the killed process, since

$$\begin{aligned} P_t^T g(x) &= P^x\{T \circ \theta_t = \infty,\ t < T\} = P^x\{T - t = \infty,\ t < T\} \\ &= P^x\{T = \infty\} = g(x). \end{aligned}$$

Let $K_t = {}_g P_t^T$ be the g-transform of the killed semigroup (P_t^T). Then

Theorem 12.9. *Let L be a co-terminal time. Then $\{X_{L+t},\ t > 0\}$ is strongly Markov with semigroup (K_t).*

If the lifetime is finite, a double-reversal argument shows that the post-L process is strongly Markov, but doesn't give the semigroup. In the general case, one proves the theorem by directly calculating the semigroup, using the properties of a co-terminal time. Notice that the transfigured process does not have the same semigroup as the original process. In fact the semigroup (K_t)

is given by a conditional probability:

$$K_t f(x) = \frac{1}{g(x)} P^T(fg)(x) = \frac{E^x\{f(X_t),\ T = \infty\}}{P^x\{T = \infty\}}.$$

Thus (K_t) is the semi-group conditioned on $T = \infty$. In the special case when L is the last exit time from a set A, then T is the first hit of A, and the post-L process has the same transition probabilities as the original process conditioned never to hit A.

Let's look at some other examples of birth and death times.

Suppose we kill the process twice in succession, first at a terminal time T, and then at a time L which is co-optional *for the killed process*. We get a sub-process whose lifetime is a "co-optional time before a terminal time." (For example, "The last exit from A before the first hit of B." Note that this time is neither a first hitting nor last exit time, since the process might or might not hit A again after its first hit of B.)

Similarly, if we re-start – transfigure – the process twice, first at a co-terminal time, and then at a stopping time for the re-started process, we obtain another process whose birth time is "a stopping time after a co-terminal time."

We can continue this scheme to get more birth and death times, but in fact we've already given the most interesting ones. Indeed, as Jacobsen and Pitman [1] discovered, these are the only birth and death times at which the Markov property holds. This is quite surprising, and their theorem is definitely worth a careful statement.

The Markov property in question is delicate – and perhaps we should not call it *the* Markov property, but rather, a particular Markov property.

To explain it, let us introduce the notion of the sigma field $\mathcal{F}_R$ for random variables $R \geq 0$ which may not be stopping times. We define $\mathcal{F}_R$ to be the sigma-field generated by $\mathcal{F}_S|_{\{S \leq R\}}$ where S ranges over all stopping times. This agrees with the usual definition if R is a stopping time, and it is easily verified that $R < S \Longrightarrow \mathcal{F}_R \subset \mathcal{F}_S$. Similarly, $\mathcal{F}_{R-}$ is the sigma field generated by $\mathcal{F}_S|_{\{S<R\}}$, where S ranges over all stopping times. Define $\mathcal{G}_R^-$ to be the sigma field generated by $\lim_{s\uparrow\uparrow R} f(X_s)$, where f ranges over all excessive functions, and, if R is a birth time, let $\mathcal{G}_R^+$ be the sigma field generated by $\operatorname{ess\,lim}_{s\downarrow\downarrow R} f(X_s)$, where f ranges over all functions supermedian *for the post-R process*. ($\mathcal{G}_R^-$ and $\mathcal{G}_R^+$ are the germ fields – technically, the invariant fields – just before and just after R, respectively, but one can think of them as $\sigma(X_{R-})$ and $\sigma(X_{R+})$ respectively.)

Definition 12.10. *A random variable α is a* **regular birth time** *if it is a birth time and if, for each bounded random variable ξ, $E\{\xi \circ \theta_\alpha\, I_{\{\alpha<\infty\}} \mid \mathcal{F}_\alpha\}$ is measurable with respect to $\sigma\{X_\alpha\}$. A random variable β is a* **regular death time** *if it is a death time and if, for each bounded random variable ξ, $E\{\xi \circ \theta_\beta\, I_{\{\beta<\infty\}} \mid \mathcal{F}_\beta\}$ is measurable with respect to $\mathcal{G}_\beta$.*

The following very interesting result was first proved by Jacobsen and Pitman for Markov Chains, then extended to continuous time in a sequence of papers by Pittenger, Sharpe, and Shih.

Theorem 12.11.

(i) *A random time β is a regular death time for (X_t) if and only if there exists a terminal time $T \geq \beta$ such that β is a co-terminal time for the process (X_t) killed at T.*
(ii) *A random time α is a regular birth time for (X_t) if and only if there exists a co-terminal time $L \leq \alpha$ such that $\alpha - L$ is a stopping time for the post-L process.*

Exercises

1. Let (X_t) be Brownian motion on $(-1, 1)$, killed when it hits the boundary. Let L_0 be the last exit time from the origin. Show that the post-L_0 process does not satisfy the strong Markov property at $t = 0$. Show, thus, that there exists a random variable ξ for which $E\{\xi \circ \theta_L \mid \mathcal{F}_L\}$ is not $\sigma(X_L)$-measurable. Find the sigma-field $\mathcal{G}^+(L)$.
2. Let (X_t) be a rcll process. Suppose that its lifetime ζ is finite. Let L be a co-optional time. Show that $(\zeta - L)^+$ is a stopping time for the reverse process $\tilde{X}_t = X_{\zeta - t}$.

Notes on Chapter 12

The fact that a time of reversal is the lifetime of a subprocess was implicit in Chung and Walsh [3], and may well have been known to Hunt. It was later proved in Walsh [3], where the double reversal argument was made rigorous. The argument there was somewhat more flexible than the one we presented here: it treated essentially Markov processes – processes in which the Markov property only holds for a.e. t – and constructed a Ray-Knight compactification for these which worked simultaneously for the forward and reverse processes. See also Walsh [7].

The fact that a co-optional time is a death time and a co-terminal time is a birth time comes from Meyer, Smythe and Walsh [1]. While there is no mention of reversal in that paper, the theorems were discovered via double-reversal arguments and, as P.-A. Meyer once said, "Knowing something is true is a tremendous hint."

The Markov property at the co-optional and co-terminal times is due to Getoor and Sharpe.

The remarkable characterization of regular birth and death times was discovered for discrete-time chains by Jacobsen and Pitman [1]. The extension to continuous-parameter processes presented considerable difficulty, and was done in a series of papers by Pittenger, Shih, and Sharpe. See Pittenger and Sharpe [1] for references to these, as well as a characterization of times which are simultaneously regular birth and death times.

Chapter 13

Processes in Duality

Consider a semiconductor. Think of it as a crystal lattice whose lattice sites can absorb and emit electrons. The electrons are lightly bound, and can easily migrate from site to site. A small bias will cause them to move. We call a lattice site with no electron a *hole*. Add enough electrons to the crystal to fill up about half the sites. A given lattice site will either have an electron or a hole. When an electron moves from site A to site B, we see two things: the hole B is filled by an electron, and A becomes a hole. A more dynamic description is that we see the electron jump from A to B, and, at the same time, we see the hole jump from B to A.

Thus there are two motions, that of the electrons and that of the holes. Without fussing too much about definitions, we can see that they are dual in some sense, and leave it at that. A net motion of electrons produces a negative current flowing in the direction of motion. But if we look at the holes instead, we see both a net movement and a positive current in the opposite direction.

Now suppose the electrons perform random walks in the crystal. The holes will then follow a dual random walk. Take a finer and finer grid, so that the random walks tend to a diffusion; then the holes' motions will tend to a dual diffusion.

The same description applies to other situations. In heat conduction in a lattice, for instance, heat energy, instead of electrons, can be transferred from site to site. While it is the energy which is being transferred, we visualize it as energetic particles which diffuse through the body carrying the heat energy with them. The holes have the dual interpretation: they take away the energy. Let's look at Kakutani's original discovery, the one which initiated the study of probabilistic potential theory, in that light. Let D be a domain in the plane filled with some heat conducting material, and let G be an open subset. Hold G at a temperature of 1°, and keep the temperature at the boundary of D at 0°. The problem is to find the temperature at an arbitrary point $z \in D$. Now G, being at a higher temperature, transfers thermal energy to nearby molecules, which then diffuse through the domain. We may think of G as emitting energetic Brownian particles. The temperature at z is just a measure of the local heat energy, so it is proportional to the number of these particles passing near z per unit time. Equivalently, it is proportional to the probability that a typical particle emitted from G passes near z. But recall the holes. If a particle starts from G, its dual, the hole, will end up in G. To find the probability that a particle from G passes near z, then, we can send a hole from z, and see if it ever hits G! This in fact was Kakutani's solution. He deduced that the temperature $T(z)$ at z was

$$T(z) = P^z\{\hat{X}_t \text{ ever hits } G\},$$

where $\hat{X}_t$ is the process corresponding to the motion of the holes. In his case, both X_t and $\hat{X}_t$ were Brownian motions, and it wasn't clear that two different processes were involved. But if the fluid were to move gently to the left, say, one would see a net drift to the left of the particles from G, and a corresponding drift of holes to the right. The processes of the particles and the holes would be different.

There is an irony in this. We probabilists have long claimed a superior intuition because we could study the paths of the particles. In fact, we were wrong: we didn't study the particles, we studied the holes!

13.1. Formal Duality

A close look at the physical description of the dual motions of the particles and holes above shows that both motions are going forward in time. But there is an intriguing ambiguity of time-sense: what is the "real" time-direction of the holes? Like the Escher prints in which water flows downwards in order to rise, or the ever-rising musical scales in which each note is a tone above the preceding but the pitch never rises more than an octave, this is something better appreciated by the observer than explained by the author.

Philosophical and physical questions aside, it is mathematically profitable to consider dual processes as a single process running in two opposing directions of time: one is a reflection of the other.

We are going to take this viewpoint and run with it, so to speak. This exposition offers two novelties. One is a question of viewpoint, one a question of preference: we will concentrate on the time-reversal aspect, and we will also let our processes have – indeed, glory in – branching points. Time reversal makes some apparently technical facts deeply intuitive; branching points initially complicate some statements and proofs, but once the preliminaries are over, they explain enough to show us what the usual assumptions – quasi-left-continuity in particular – add, and where they add it.

Let $\boldsymbol{E}$ be a locally compact separable metric space which is a subspace of a compact metric space $\bar{\boldsymbol{E}}$. The space $\bar{\boldsymbol{E}}$ will just serve to define a convenient set of uniformly continuous functions on $\boldsymbol{E}$; it can be taken to be the one-point compactification of $\boldsymbol{E}$. Let $\mathcal{E}$ be its be its Borel field. The sets of bounded measurable functions and positive measurable functions are denoted by $b\mathcal{E}$ and $\mathcal{E}^+$ respectively.

Definition 13.1. *Let ξ be a σ-finite Radon measure on $(\boldsymbol{E}, \mathcal{E})$. Two Markov resolvents $(U_p)_{p>0}$ and $(\hat{U}_p)_{p>0}$ on $(\boldsymbol{E}, \mathcal{E})$ are* **in duality relative to ξ** *if*

(i) $p > 0$ *and* $f,\ g \in \mathcal{E}^+ \Longrightarrow \int f(y) U_p g(y)\, \xi(dy) = \int \hat{U}_p f(x) g(x)\, \xi(dx)$;

(ii) *for all* $x, y \in \boldsymbol{E}$ *and* $p > 0$, $U_p(x, \cdot) \ll \xi(\cdot)$ *and* $\hat{U}_p(y, \cdot) \ll \xi(\cdot)$.

If only (i) holds, we say U_p and $\hat{U}_p$ are **in weak duality relative to ξ**.

Notation: $\langle f, g\rangle_\mu = \int f(x)g(x)\,\mu(dx)$. If $\mu = \xi$, we will often just write $\langle f, g\rangle$.

Consider two resolvents (U_p) and $(\hat{U}_p)$ which are in duality with respect to a σ-finite measure ξ. We say a function which is p-excessive with respect to U_p is ***p*-excessive**, while a function which is p-excessive with respect to $\hat{U}_p$ is ***p*-co-excessive**. The following theorem shows that we can choose extremely good versions of the densities of $U_p(x, \cdot)$ and $\hat{U}_p(x, \cdot)$ with respect to ξ.

Theorem 13.2. *Let (U_p) and $(\hat{U}_p)$ be resolvents in duality with respect to a σ-finite measure ξ. Then for each $p > 0$ there exists a function $u_p(x, y)$, $x, y \in \boldsymbol{E}$, such that*

(i) $u_p(\cdot, \cdot)$ *is* $\mathcal{E} \times \mathcal{E}$*-measurable*;

(ii) $x \mapsto u_p(x, y)$ *is p-excessive for each* $y \in \boldsymbol{E}$;

(iii) $y \mapsto u_p(x, y)$ *is p-coexcessive for each* $x \in \boldsymbol{E}$;

(iv) $U_p(x, dy) = u_p(x, y)\,\xi(dy)$ *and* $\hat{U}_p(y, dx) = u_p(x, y)\,\xi(dx)$.

Furthermore, if $0 < p \le q$, $u_p(x, y)$ *satisfies*

$$u_p(x, y) = u_q(x, y) + (q - p) U_p u_q(x, y) \tag{13.1}$$

$$= u_q(x, y) + (q - p) \hat{U}_q u_p(x, y), \tag{13.2}$$

where $U_p u_q(x, y) = \int U_p(x, dz) u_q(z, y)$ *and* $\hat{U}_q u_p(x, y) = \int \hat{U}_q(y, dz)\, u_p(x, z)$.

Remark 13.3. We have not assumed that either (U_p) or $(\hat{U}_p)$ is the resolvent of a strong Markov process. This theorem is about resolvents, not processes.

Proof. Write $\xi(dx) = dx$ for simplicity. Let us first show that (i)–(iv) imply (13.1) and (13.2). By the resolvent equation,

$$U_p = U_q + (q - p) U_p U_q,$$

and we see that for each x, (13.1) holds for ξ-a.e. y. By (iii), $u_p(x, \cdot)$ is p-coexcessive, so it holds for all y. In the same way, (13.2) holds for each x and y since by (ii), $u_p(x, y)$ is p-excessive in x for each y.

Now $U_p(x, dy) \ll \xi(dy)$ so that, by an argument of Doob, there exists a function $w_p(x, y) \in \mathcal{E} \times \mathcal{E}$ such that $U_p f(x) = \int w_p(x, y) f(y)\, dy$ for all $x \in \boldsymbol{E}$, $f \in b\mathcal{E}$.

This argument is both useful and well-known; here it is. Choose a strictly positive $h \in \mathcal{E}^+$ such that $\int h(x)\, dx = 1$, and let $\nu(dx) = h(x)\, dx$. Then ν is a reference probability measure. Let $\mathcal{G}_n$ be a finite partition of $\boldsymbol{E}$ of sets of diameter $\leq 1/n$. Suppose $\mathcal{G}_{n+1} \supset \mathcal{G}_n$. If $\mathcal{G}_n = \{G_{n1}, \dots, G_{nk_n}\}$, define $w_p^n(x, y)$ by

$$w_p^n(x, y) = \begin{cases} \dfrac{U_p(x,\ G_{nj})}{\nu(G_{nj})} & \text{if } y \in G_{nj} \text{ and } \nu(G_{nj}) > 0; \\ 0 & \text{if } \nu(G_{nj}) = 0. \end{cases}$$

For each fixed x, $\{w_p^n(x, \cdot),\ n = 1, 2, \dots\}$ is a positive martingale on the probability space $(\boldsymbol{E}, \mathcal{E}, \nu)$. It is uniformly integrable, since if w is any version of the Radon-Nikodym derivative $U_p(x, dy)/\nu(dy)$, then $w_p^n(x, \cdot) = E\{w \mid \mathcal{G}_n\}$. Since $\bigvee_n \mathcal{G}_n = \mathcal{E}$, $w_p^n(x, \cdot)$ converges a.e. to $E\{w \mid \bigvee_n \mathcal{G}_n\} = w$. Therefore, define

$$w_p(x, y) = h(y) \liminf_{n \to \infty} w_p^n(x, y).$$

Then $w_p(x, \cdot)$ is a version of $U_p(x, dy)/dy$, so it is ξ-integrable and $w_p(\cdot, \cdot)$ is $\mathcal{E} \times \mathcal{E}$-measurable since the same is true of w_p^n.

Now $w_p(x, y)$ is defined for all x and y, but it is only unique up to y-sets of measure zero, so we must smooth it in y. The limit should be p-coexcessive, so consider $\lim_{q \to \infty} q \hat{U}_{p+q} w_p(x, y)$. (If w_p were p-co-supermedian, this would be its p-coexcessive regularization.)

Fix x and $p > 0$ and set $w(y) = w_p(x, y)$. If $f \in b\mathcal{E}^+$, and $p, q > 0$,

$$\begin{aligned} \langle \hat{U}_{p+q} w, f \rangle = \langle w, U_{p+q} f \rangle &= \int w_p(x, y) U_{p+q} f(y)\, dy \\ &= U_p U_{p+q} f(x) = \frac{1}{q} \left(U_p f(x) - U_{p+q} f(x) \right), \end{aligned}$$

using duality, the fact that $w_p(x, \cdot)$ is the density of $U_p(x, \cdot)$, and the resolvent equation in turn. In other words,

$$\langle q \hat{U}_{p+q} w, f \rangle = \langle w, f \rangle - U_{p+q} f(x). \tag{13.3}$$

Since $f \geq 0$, this is $\leq \langle w, f \rangle$, and, as it holds for all such f, $q \hat{U}_{p+q} w(y) \leq w(y)$ for ξ-a.e. y. Moreover, since $U_{p+q} f(x) \leq \|f\|/(p+q) \to 0$ as $p \to \infty$,

$$\langle q \hat{U}_{p+q} w, f \rangle \to \langle w, f \rangle \text{ as } p \to \infty. \tag{13.4}$$

We claim that $q\hat{U}_{p+q}w(y)$ increases with q for *every* y. Indeed, if $r > q$,

$$r\hat{U}_{p+r}w(y) - q\hat{U}_{p+q}w(y) = (r-q)\hat{U}_{p+r}w(y) + q\big[\hat{U}_{p+r} - \hat{U}_{p+q}\big]w(y)$$
$$= (r-q)\big[\hat{U}_{p+r}w(y) - \hat{U}_{p+r}(q\hat{U}_{p+q}w)(y)\big]$$

using the resolvent equation. We know $q\hat{U}_{p+q}w \le w$ ξ-a.e., so this is

$$\ge (r-q)(\hat{U}_{p+r} - \hat{U}_{p+r})w(y) = 0.$$

Thus, let $u_p(x, y) = \lim_{q\to\infty} q\hat{U}_{p+q}w(y)$. Then $u_p(x, y) \le w(y)$ for a.e. y, hence it follows from (13.4) that $u_p(x, y) = w(y)$ for a.e. y. Moreover $u_p(x, \cdot)$ is p-coexcessive, being the increasing limit of the p-coexcessive functions $q\hat{U}_{p+q}w$. From (13.4) we have by monotone convergence that

$$\int u_p(x, y) f(y)\, dy = \langle w, f\rangle = U_p f(x),$$

so that $U_p(x, dy) = u_p(x, y)dy$. There are two things left to show:

(i) $\hat{U}_p(y, dx) = u_p(x, y)\, dx$;
(ii) $x \mapsto u_p(x, y)$ is p-excessive.

To show (i), let $f, g \in b\mathcal{E}^+$. Then

$$\langle \hat{U}_p f, g\rangle = \langle f, U_p g\rangle$$
$$= \int dx\, f(x) \int u_p(x, y) g(y)\, dy$$
$$= \iint [u_p(x, y) f(x)\, dx]\, g(y)\, dy,$$

so

$$\hat{U}_p f(y) = \int u_p(x, y) f(x)\, dx \quad \text{for } \xi\text{-a.e. } y. \tag{13.5}$$

Once more, we claim that this is true for all y. Apply $q\hat{U}_{p+q}$ to both sides of (13.5):

$$q\hat{U}_{p+q}[\hat{U}_p f](y) = \iint q\hat{U}_{p+q}(y, dz) u_p(x, z) f(x)\, dx$$
$$= \int q(\hat{U}_{p+q}u_p)(x, y) f(x)\, dx,$$

which holds for *all* y, even though (13.5) only holds a.e. Now let $q \to \infty$ and note that both sides increase to a limit since $\hat{U}_p f$ and $u_p(x, \cdot)$ are both

p-coexcessive. This leads to

$$\hat{U}_p f(y) = \int u_p(x, y) f(x)\, dx,$$

proving (i). To see that $u_p(\cdot, y)$ is p-excessive, apply $q\hat{U}_{p+q}$ to $u_p(x, \cdot)$:

$$q\hat{U}_{p+q} u_p(x, y) = q \int u_p(x, z) u_{p+q}(z, y)\, dz = U_p g_q(x),$$

where $g_q(z) \stackrel{\text{def}}{=} q u_{p+q}(z, y)$. Note that the right-hand side is a p-potential and therefore p-excessive. Now let $q \to \infty$. The left-hand side increases to $u_p(x, y)$ since $u_p(x, \cdot)$ is p-coexcessive. Therefore the right-hand side also increases, and the limit is p-excessive, being the increasing limit of p-excessive functions. □

13.2. Dual Processes

We will add an isolated point ∂ to $\boldsymbol{E}$ to act as a cemetery; let $\boldsymbol{E}_\partial = \boldsymbol{E} \cup \{\partial\}$. If (U_p) and $(\hat{U}_p)$ are sub-Markovian resolvents on $\boldsymbol{E}$, we extend them to Markovian resolvents on $\boldsymbol{E}_\partial$ as usual. In addition, we introduce:

(P_t), $(\hat{P}_t)$: Borel measurable strong Markov semigroups on $\boldsymbol{E}_\partial$.

(X_t), $(\hat{X}_t)$: right-continuous strong Markov processes on $\boldsymbol{E}$, having left limits in $\boldsymbol{E}$ except perhaps at their lifetimes ζ and $\hat{\zeta}$ respectively, and admitting (P_t) and $(\hat{P}_t)$ as semigroups.

(U_p), $(\hat{U}_p)$: resolvents of the semigroups (P_t) and $(\hat{P}_t)$:

$$U_p f(x) = \int_0^\infty e^{-pt} P_t f(X_t)\, dt, \quad \hat{U}_p f(x) = \int_0^\infty e^{-pt} \hat{P}_t f(\hat{X}_t)\, dt.$$

Our basic hypothesis is

Hypothesis 13.4. (Duality) *(U_p) and $(\hat{U}_p)$ are in duality with respect to a sigma-finite Radon measure ξ.*

Remark 13.5. 1° The assumption that both (X_t) and $(\hat{X}_t)$ are right-continuous actually implies the existence of left limits, except at the lifetime (see Theorem 13.18 with Y the indicator function of the set where left limits exist for all $t < \zeta$).

One can often establish that certain properties hold for $t < \zeta$, but it can be delicate to establish them for $t = \zeta$. This will be a recurring theme. The existence of left limits at ζ is an excellent example. From one point of view the whole machinery of the Martin boundary was created to explain just what one really means by $X_{\zeta-}$.

2° Since U_p and $\hat{U}_p$ are in duality, they both satisfy Meyer's Hypothesis (L) of absolute continuity relative to the duality measure ξ. Consequently, all excessive and co-excessive functions are Borel measurable, and an excessive or co-excessive function which vanishes ξ-a.e. vanishes identically. (See §3.5 Proposition 9.)

3° X and $\hat{X}$ may have branching points. A point x is a **branching point** if $P_0(x, \{x\}) < 1$ and it is a **co-branching point** if $\hat{P}_0(x, \{x\}) < 1$.

We denote the sets of branching and co-branching points by B and $\hat{B}$ respectively. $B_\partial \stackrel{\text{def}}{=} \{x : P_0(x, \{\partial\}) = 1\}$ and $\hat{B}_\partial \stackrel{\text{def}}{=} \{x : \hat{P}_0(x, \{\partial\}) = 1\}$ denote the sets of points branching and co-branching directly to the cemetery. As with Ray processes, (see Proposition 8.8) $P_t(x, B) = 0$ for all $t > 0$ and in fact, B is a polar set for (X_t) and, for the same reason, $\hat{B}$ is polar for $(\hat{X}_t)$, i.e. co-polar.

The sets B_∂ and $\hat{B}_\partial$ are a necessary evil. While the processes don't hit them, their left limits may. Moreover, we also want to treat h-transforms, which often introduce points branching directly to ∂, even if the original processes have none.

Proposition 13.6. *(P_t) and $(\hat{P}_t)$ are in weak duality relative to ξ: if $f, g \in \mathcal{E}^+$ and $t \geq 0$,*

$$\langle f, P_t g\rangle = \langle \hat{P}_t f, g\rangle. \tag{13.6}$$

Proof. Suppose first that f and g are continuous and have compact support in $\boldsymbol{E}$, and thus are ξ-integrable. Then for all $p > 0$

$$\begin{aligned}\int_0^\infty e^{-pt}\langle f, P_t g\rangle\, dt &= \langle f, U_p g\rangle \\ &= \langle \hat{U}_p f, g\rangle = \int_0^\infty e^{-pt}\langle \hat{P}_t f, g\rangle\, dt.\end{aligned}$$

By the uniqueness of the Laplace transform, $\langle f, P_t g\rangle = \langle \hat{P}_t f, g\rangle$ for a.e. $t \geq 0$. We claim they are equal for all $t \geq 0$. Indeed, $t \mapsto f(X_t)$ and $t \mapsto g(\hat{X}_t)$ are right-continuous, hence so are $P_t f$ and $\hat{P}_t g$, and, by bounded convergence, so are $\langle f, P_t g\rangle$ and $\langle \hat{P}_t f, g\rangle$, so (13.6) holds for all $t \geq 0$.

The usual monotone class argument shows that it holds for $f, g \in \mathcal{E}^+$. □

The reason we needed $g = 0$ on B_∂ was that $\xi(B_\partial)$ might not be zero.

Lemma 13.7. *ξ puts no mass on $B \cup \hat{B} - B_\partial \cap \hat{B}_\partial$. Consequently $\xi P_0 = \xi \hat{P}_0$, and $\xi \hat{P}_0(B \cup \hat{B}) = 0$.*

Proof. P_0 puts no mass on the branching points, so $P_0 I_B \equiv 0$, hence

$$0 = \langle 1, P_0 I_B\rangle = \langle \hat{P}_0 1, I_B\rangle.$$

But $\hat{P}_0 1(x) > 0$ if $x \notin \hat{B}_\partial$, so this implies that $\xi((B - \hat{B}_\partial)) = 0$. By symmetry, $\xi(\hat{B} - B_\partial) = 0$ as well, hence $\xi(B \cup \hat{B} - B_\partial \cap \hat{B}_\partial) = \xi(B - \hat{B}_\partial) \cup (\hat{B} - B_\partial) = 0$. The second statement follows because for $x \notin B \cup \hat{B}$, $\hat{P}_0(x, B \cup \hat{B}) = P_0(x, B \cup \hat{B}) = 0$ – indeed, $P_0(x, \cdot)$ is the unit mass at x. The second statement follows easily. □

Remark 13.8. At the price of replacing ξ by $\xi P_0 = \xi \hat{P}_0$ we can assume that ξ puts no mass on the branching or co-branching points. The reader can check that this preserves the duality relation. Thus, we will assume without loss of generality that

$$\xi(B \cup \hat{B}) = 0. \tag{13.7}$$

13.3. Excessive Measures

Definition 13.9. *A σ-finite measure μ on $\boldsymbol{E}$ is* **p-excessive** *if*

(i) $e^{-pt} \mu P_t(A) \leq \mu(A)$, *for all* $A \in \mathcal{E}$, $t \geq 0$.
(ii) $\lim_{t \to 0} \mu P_t(A) = \mu(A)$, *for all* $A \in \mathcal{E}$.

If only (i) *holds, μ is* **p-supermedian** .

Co-excessive and **co-supermedian** measures are defined with $\hat{P}_t$ in place of P_t. Note that an excessive measure sits on $\boldsymbol{E}$, so it doesn't charge ∂. As the following proposition shows, (ii) is almost unnecessary.

Proposition 13.10. *Let μ be a σ-finite measure on $(\boldsymbol{E}, \mathcal{E})$. If μ is p-excessive, then $\mu(B) = 0$. If μ is p-supermedian, then μP_0 is p-excessive. If μ is supermedian and has the property that there is a μ-integrable p-supermedian function which is strictly positive on $\boldsymbol{E} - B_\partial$, then $\mu(B - B_\partial) = 0$, and μ is p-excessive iff $\mu(B_\partial) = 0$.*

Proof. $e^{-pt} \mu P_t$ increases setwise as $t \downarrow 0$ to a limit ν, and ν is dominated setwise by the measure μ, hence ν is a σ-finite measure. Let h be a bounded μ-integrable excessive function which is strictly positive on $\boldsymbol{E} - B_\partial$, and $h(\partial) = 0$. Then

$$\infty > \int h(x)\, \mu(dx) \geq \int h(x)\, \nu(dx) \geq e^{-pt} \int P_t h(x)\, \mu(dx).$$

Since $e^{-pt} P_t h \uparrow h$ as $t \downarrow 0$, the right-hand side increases to $\int h(x)\, \mu(dx) = \int P_0 h(x)\, \mu(dx) = \int h(x)\, \mu P_0(dx)$. Thus $\mu = \nu = \mu P_0$ on $\{h > 0\} = \boldsymbol{E} - B_\partial$. But we know $\mu P_0(B) = 0$ so certainly $\mu(B - B_\partial) = 0$. It follows that $\mu P_0 = \lim_{t \to 0} \mu P_t$ on $\boldsymbol{E}$; thus μP_0 is p-excessive, and equals μ iff $\mu(B_\partial) = 0$. □

Remark 13.11. The relation between supermedian and excessive measures is simpler than the relation between supermedian and excessive functions.

If μ is a supermedian measure, then $\mu|_{E-B_\partial}$ is excessive. Thus we shall only consider excessive measures in what follows.

EXAMPLE 13.12. Let f be a.e. finite and p-coexcessive and let μ be the measure $\mu(dx) = f(x)\,dx$. Then μ is p-excessive. Indeed, if $t > 0$ and $A \in \mathcal{E}$,

$$\begin{aligned} e^{-pt}\mu P_t(A) &= e^{-pt}\int f(x)P_t(x, A)\,dx = e^{-pt}\langle f, P_t I_A\rangle \\ &= e^{-pt}\langle \hat{P}_t f, I_A\rangle \le \langle f, I_A\rangle \\ &= \mu(A) \end{aligned}$$

and as $e^{-pt}\hat{P}_t f \uparrow f$ as $t \downarrow 0$, μ is p-excessive.

Surprisingly, this is not a special case, for all excessive measures are of this type.

Theorem 13.13. *A σ-finite measure μ is p-excessive if and only if there exists an a.e. finite p-co-excessive function g such that*

$$\mu(dy) = g(y)\,dy.$$

Before proving this, let us state a corollary. By (13.7) the duality measure ξ doesn't charge $B \cup \hat{B}$ and $P_0 1 = \hat{P}_0 1 = 1$ on $\boldsymbol{E} - B \cup \hat{B}$. Since $P_0 1$ is excessive and $\hat{P}_0 1$ is co-excessive, $\xi(dy) = \hat{P}_0 1(y)\xi(dy) = P_0 1(y)\xi(dy)$ and we see that:

Corollary 13.14. *The duality measure ξ is both excessive and co-excessive.*

Proof. (of Theorem 13.13) First note that an excessive measure is absolutely continuous with respect to ξ, since $\mu(A) = \lim_{q\to\infty} qU_{p+q}(A)$, and this is zero if $\xi(A) = 0$. Thus $\mu(dx) = g(x)\,\xi(dx)$ for some version g of the Radon-Nikodym derivative $d\mu/d\xi$. To see that g is (almost) p-coexcessive, let $k \in \mathcal{E}^+$ and consider

$$\int k(x)\,\mu(dx) = \int k(x)g(x)\,dx = \langle k, g\rangle,$$

and

$$\int k(x)\,\mu P_t(dx) = \int k(x)\int g(y)P_t(y, dx)\,dy = \langle P_t k, g\rangle.$$

Since $\mu \ge e^{-pt}\mu P_t$ setwise,

$$\langle k, g\rangle \ge e^{-pt}\langle P_t k, g\rangle = e^{-pt}\langle k, \hat{P}_t g\rangle.$$

As k is arbitrary,

$$g \ge e^{-pt}\hat{P}_t g \;\text{ a.e.,} \tag{13.8}$$

so that g is (almost) p-co-excessive. We will modify g to remove the "almost." Let

$$g_q = q\hat{U}_{p+q}\, g.$$

By (13.8) and a judicious use of Fubini's Theorem, $g_q \le g$ ξ-a.s. Now g_q is clearly $(p+q)$-coexcessive, but even better, it is p-coexcessive, since

$$e^{-pt}\hat{P}_t g_q = q\hat{U}_{p+q}(e^{-pt}\hat{P}_t g) \le q\hat{U}_{p+q}\, g,$$

where we use (13.8) and the fact that $\hat{U}_{p+q}(x,\cdot) \ll \xi$. As g_q is $p+q$-coexcessive, $\hat{P}_t g_q \to g_q$ as $t \to 0$. Therefore g_q is p-coexcessive. Moreover, g_q increases with q. Now notice that

$$q\mu U_{p+q}(dy) = dy \int g(x)\xi(dx)\, u(x,y) = g_q(y)\, dy;$$

and, μ being p-excessive, $q\mu U_{p+q}(dy) \uparrow \mu(dy)$. Thus, if $\hat{g} = \lim_{q\to\infty} g_q$, the monotone convergence theorem tells us that

$$\mu(dy) = \hat{g}(y)\, \xi(dy);$$

but $\hat{g}$ is the increasing limit of p-co-excessive functions, so it is itself p-co-excessive. □

13.4. Simple Time Reversal

We continue with the same hypotheses and notation. We assume, in addition, that the processes are defined on the canonical space Ω of right-continuous functions from $\boldsymbol{R}^+$ to $\boldsymbol{E}_\partial$ which admit a lifetime and have left limits in $\boldsymbol{E}$, except possibly at the lifetime. Denote the lifetimes of (X_t) and $(\hat{X}_t)$ by ζ and $\hat{\zeta}$ respectively.

There are two somewhat different connections between duality and time reversal. The first, and simplest, concerns reversal from a fixed time. The second concerns reversal from a random time, as in Chapter 10. We will examine that in the next section, but meanwhile we will look at simple reversal. This is already enough to make sense of the statement that "The dual process is the original process reversed in time."

If f and g are positive Borel functions, then $\langle f, P_t g\rangle = \langle \hat{P}_t f, g\rangle$ is shorthand for $\int f(x)P_t g(x)\, dx = \int g(x)\hat{P}_t f(x)\, dx$. Now ξ is not a probability measure in general, since it is only σ-finite. However, it is not hard to make sense of expectations when (X_t) has "initial measure" ξ. These are just integrals: by definition, $E^\xi\{f(X_t)\} = \int P_t f(x)\, dx$. In fact, if we extend f and g to $\boldsymbol{E}_\partial$ by setting them equal to zero at ∂, then (13.6) can be rewritten

$$E^\xi\{f(X_0)g(X_t)\} = \hat{E}^\xi\{g(\hat{X}_0)f(\hat{X}_t)\}. \tag{13.9}$$

Let us see what happens with three functions. Let f, g, and h be positive, $\mathcal{E}$-measurable functions which vanish at ∂. Using (13.9) and the Markov property repeatedly:

$$
\begin{aligned}
\hat{E}^{\xi}\{f(\hat{X}_0)g(\hat{X}_s)h(\hat{X}_{s+t})\} &= \hat{E}^{\xi}\{f(\hat{X}_0)\,(g(\hat{X}_s)\hat{P}_t h(\hat{X}_s))\}\\
&= E^{\xi}\{g(X_0)\,\hat{P}_t h(X_0) f(X_s)\}\\
&= E^{\xi}\{(g(X_0)\hat{P}_t h(X_0))\,P_s f(X_0)\}\\
&= \hat{E}^{\xi}\{g(\hat{X}_0)P_s f(\hat{X}_0)\hat{P}_t h(\hat{X}_0)\}\\
&= \hat{E}^{\xi}\{(g(\hat{X}_0)P_s f(\hat{X}_0))\,h(\hat{X}_t)\}\\
&= E^{\xi}\{h(X_0)\,\,(g(X_t)P_s f(X_t))\}\\
&= E^{\xi}\{h(X_0)g(X_t)f(X_{t+s})\}.
\end{aligned}
$$

Bring the lifetimes explicitly into the picture: if $f_0, \dots, f_n$ are positive $\mathcal{E}$-measurable functions, we can show by induction that

$$
\begin{aligned}
&\hat{E}^{\xi}\{f_0(\hat{X}_0)f_1(\hat{X}_{t_1})\dots f_n(\hat{X}_{t_n});\,\hat{\zeta} > t_n\}\\
&\qquad = E^{\xi}\{f_n(X_0)f_{n-1}(X_{t_n-t_{n-1}})\dots f_0(X_{t_n});\,\zeta > t_n\}. \quad (13.10)
\end{aligned}
$$

Lemma 13.15. *For each* $t > 0$, $P^{\xi}\{X_{t-}$ doesn't exist or $X_{t-} \neq X_t\} = 0$.

Proof. Let $\Lambda_t = \{X_{t-}$ doesn't exist or $X_{t-} \neq X_t\}$. The paths are rcll (except possibly at ζ) so there are at most countably many t for which $P^{\xi}\{\Lambda_t\} > 0$. Choose a t for which it is zero. Then

$$
P^{\xi}\{\Lambda_{t+s}\} = \int P^x\{\Lambda_{t+s}\}\,dx = \int P_s(P^{\cdot}\{\Lambda_t\})(x)\,dx.
$$

But ξ is an excessive measure, so for all $f \geq 0$, $\xi P_s f \leq \xi f$, and the above is

$$
\leq \int P^x\{\Lambda_t\}\,dx = 0. \qquad \square
$$

Note in particular that $P^{\xi}\{\zeta = t\} = 0$ for all t. Let $t > t_n$ and rewrite (13.10) in the form

$$
\hat{E}^{\xi}\left\{\prod_{j=0}^{n} f_j(\hat{X}_{t_j});\ \hat{\zeta} \geq t\right\} = E^{\xi}\left\{\prod_{j=0}^{n} f_j(X_{t-t_j-})\dots f_0(X_{t-});\ \zeta \geq t\right\}.
\tag{13.11}
$$

We can put (13.11) in a more compact form by defining the **reversal operator** $r_t : \Omega \mapsto \Omega$, which reverses the path from a fixed time t:

$$\omega' = r_t\omega \Longleftrightarrow \omega'(s) = \begin{cases} \omega(t-s-), & \text{if } 0 \le s < t < \zeta(\omega) \\ \partial & \text{if } s \ge t \text{ or } \zeta(\omega) \le t. \end{cases}$$

Remark 13.16. We have to take left limits above since as $r_t\omega \in \Omega$, it must be a right-continuous function.

Similarly, we define the **killing operator** k_t by

$$k_t\omega(x) = \begin{cases} \omega(s) & \text{if } 0 \le s < t, \\ \partial & \text{if } s \ge t. \end{cases}$$

The killing operator simply kills the path at time t, and the reversal operator reverses it from t, and makes the reversed path right-continuous. We say that ω and ω' are ***t*-equivalent** if $\omega(s) = \omega'(s)$, $0 \le s < t$, or equivalently, if $k_t\omega = k_t\omega'$.

If Y is an $\mathcal{F}^0_{t-}$-measurable random variable – i.e. measurable with respect to the *uncompleted* σ-fields – it is constant on t-equivalence classes, and there is no ambiguity in the meaning of $Y \circ r_t$. If Y is measurable with respect to the augmented σ-field $\mathcal{F}_{t-}$, there exists an $\mathcal{F}^0_{t-}$-measurable random variable Y' which is P^ξ-a.s. equal to Y, and $Y' \circ r_t$ is defined unambiguously. Thus we can define $Y \circ r_t$ up to sets of P^ξ-measure zero.

EXAMPLE 13.17. If $Y = f_0(X_0) \dots f_n(X_{t_n})$, where $0 < t_1 < \dots < t_n$, and if $t > t_n$, then (13.11) can be written

$$\hat{E}^\xi\{Y, \hat{\zeta} \ge t\} = E^\xi\{Y \circ r_t; \zeta \ge t\}. \tag{13.12}$$

Theorem 13.18. *If Y is measurable with respect to the P^ξ-completion of $\mathcal{F}^0_{t-}$ for some $t > 0$, then $Y \circ r_t$ is measurable with respect to the $\hat{P}^\xi$-completion of $\mathcal{F}^0_{t-}$ and*

$$E^\xi\{Y, \zeta \ge t\} = \hat{E}^\xi\{Y \circ r_t, \hat{\zeta} \ge t\}. \tag{13.13}$$

Proof. Let $\mathcal{G}$ be the class of $\mathcal{F}^0_{t-}$-measurable random variables Y for which (13.13) holds. Then $\mathcal{G}$ contains Y of the form $f_0(X_{t_0}) \dots f_n(X_{t_n})$. Since $\mathcal{G}$ is clearly a cone which is closed under monotone increasing convergence and dominated decreasing convergence, it contains all positive $\mathcal{F}^0_{t-}$-measurable functions. If Y is measurable with respect to the P^ξ-completion of $\mathcal{F}^0_{t-}$, there are $\mathcal{F}^0_{t-}$-measurable random variables $Y_1 \le Y \le Y_2$ such that $P^\xi\{Y_1 \ne Y_2\} = 0$. Therefore $Y_1 \circ r_t \le Y \circ r_t \le Y_2 \circ r_t$. Since (13.12) holds for Y_1 and Y_2, it holds for Y. □

13.5. The Moderate Markov Property

The import of Theorem 13.18 is that the dual process is essentially the right-continuous reverse of the original strong Markov process (X_t). It is not the reverse in the sense of Theorem 10.1 – there is no reversal time – but the relationship is close enough to be called intimate. Most of the general properties of reverse processes carry over. In particular, we expect (X_t) to be moderately Markov. Indeed, it is. The proof of this is far simpler than it was for the reverse, for we do not have to construct any new transition semigroups. The main idea is to construct a version of the resolvent which is left-continuous along the paths of the left-continuous process (X_{t-}). Recall the moderate Markov property for a rcll process defined in §2.4.

Theorem 13.19. *The processes (X_t) and $(\hat{X}_t)$ are strongly Markov, and (X_{t-}) and $(\hat{X}_{t-})$ are moderately Markov in the following sense, which we state for (X_{t-}): there exists a transition function $\bar{P}_t$ on $\boldsymbol{E}_\partial$ such that for each predictable time $0 < T < \infty$, $f \in b\mathcal{E}$, and initial measure μ, $E^\mu\{f(X_{T+t}) \mid \mathcal{F}_{T-}\} = \bar{P}_t f(X_{T-})$ almost everywhere on the set $\big\{P^\mu\{T = \zeta \mid \mathcal{F}_{T-}\} < 1\big\}$.*

Proof. We know both processes are strongly Markov. Only the moderate Markov property is in question. Notice that $X_{\zeta-}$ may not be in $\boldsymbol{E}$, so we can not expect the moderate Markov property to hold at the lifetime.

By symmetry, it is enough to prove it for (X_t), and by Hypothesis (L), it is enough to prove it for $\mu = \xi$. Let $f \in C(\bar{\boldsymbol{E}})$. Then $t \mapsto U_p f(X_t)$ is rcll, so $t \mapsto U_p f(X_{t-})$ has essential left limits at all $t > 0$. In terms of the dual process, Theorem 13.18 tells us that $t \mapsto U_p f(\hat{X}_t)$ has essential *right* limits at all $t \in (0, \zeta)$. Let us regularize $U_p f$ as in (10.19). Set

$$\bar{U}_p f(x) = \begin{cases} U_p f(x) & \text{if } x \in \hat{B} \\ \liminf_{r\to\infty} r\hat{U}_r\left(U_p f\right)(x) & \text{if } x \in \boldsymbol{E}_\partial - \hat{B}. \end{cases} \tag{13.14}$$

Now $\hat{B}$ is co-polar, and by Lemma 10.7 the limit exists except perhaps on a co-polar set, so $\bar{U}_p f(\hat{X}_t) \equiv \operatorname{ess\,lim}_{s\downarrow t} U_p f(\hat{X}_s)$ a.s. on $(0, \hat{\zeta})$. (By Hypothesis (L) the exceptional co-polar set of the lemma is independent of the initial distribution.) By Theorem 13.18 again, $t \mapsto \bar{U}_p f(X_{t-}) = \operatorname{ess\,lim}_{s\uparrow t} U_p f(X_s)$ a.s. on $(0, \zeta)$, so it is a.s. left-continuous on $(0, \zeta)$. Let T be predictable and let $\Gamma = \{\omega : X_{T-}(\omega) \in \boldsymbol{E},\ \bar{U}_p f(X_{T-}\omega) = \operatorname{ess\,lim}_{s\uparrow T} U_p f(X_s\omega)\}$. Notice that $\Gamma \in \mathcal{F}_{T-}$ and $T = \zeta$ a.s. off Γ.

Now $\Gamma \in \mathcal{F}_{T-}$, so let $T' = T_\Gamma$ (see §1.3 (4)). Then T' is predictable (§1.3 Exercise 7) $T' \geq T$ and $\{T' = T\}$ on $\{T \neq \zeta\}$. (There is no problem if $T' > \zeta$ since then $X_{T'} = \partial$. We need only worry about the set $\{T' \leq \zeta\}$.)

Let (T_n) announce T'. Then

$$\begin{aligned} p\bar{U}_p f(X_{T-}) &= p \lim_{n\to\infty} U_p f(X_{T_n}) \\ &= \lim_{n\to\infty} E\left\{ \int_0^\infty pe^{-pt} f(X_{T_n+t})\,dt \mid \mathcal{F}_{T_n} \right\} \qquad (13.15) \\ &= E\left\{ \int_0^\infty pe^{-pt} f(X_{T'+t})\,dt \mid \mathcal{F}_{T'-} \right\} \quad \text{a.s.} \end{aligned}$$

This holds simultaneously for all f in a countable dense subset of $C(\bar{\boldsymbol{E}})$ and for all rational $p > 0$. Therefore, as in the proof of Lemma 10.13, it extends by continuity to all $f \in C(\bar{\boldsymbol{E}})$ and all $p > 0$.

But this implies the moderate Markov property. Indeed, define

$$\bar{P}_0 f(x) = \liminf_{p\to\infty} p\bar{U}_p f(x), \qquad (13.16)$$

Let T be a predictable time, and let $p \to \infty$ in (13.15). Notice that the limit of the right-hand side exists and equals $E\{f(X_T) \mid \mathcal{F}_{T-}\}$ a.s. on the set $\{T \neq \zeta\}$. It follows that the set Λ of x for which the limit in (13.16) fails to exist for some f in a countable dense subset of $C(\bar{\boldsymbol{E}})$ is copolar or, equivalently, left-polar for (X_t). (Otherwise by Theorem 6.9 there would be a predictable T for which $P\{X_{T-} \in \Lambda\} > 0$.) Thus $\bar{P}_0$ extends to a measure $\bar{P}_0(x, \cdot)$, and $E\{f(X_T) \mid \mathcal{F}_{T-}\} = \bar{P}_0 f(X_{T-})$. Set $\bar{P}_t = \bar{P}_0 P_t$ and apply the strong Markov property at T to see that $E\{f(X_{T+t}) \mid X_{T-}\} = \bar{P}_t f(X_{T-})$. □

Remark 13.20. (X_t) and $(\hat{X}_t)$ are not Ray processes, and their branching behavior is less transparent than that of Ray processes. $\bar{P}_0(x, \cdot)$ is the **branching measure**. Let $B^- \stackrel{\text{def}}{=} \{x : \bar{P}_0(x, \{x\}) < 1\}$. A point in B^- is a **left branching point**. Note that B^- contains the set B of branching points, but it may contain other points, too: if $x \in B$, $\bar{P}_0(x, \{x\}) = 0$, but a priori $\bar{P}_0(x, \{x\})$ can take any value between zero and one.

Lemma 13.21. *The set B^- of left branching points is semi-polar.*

Proof. Let $\delta > 0$ and let $B_\delta^- = \{x : \bar{P}_0(x, D(x, \delta)) < 1 - \delta\}$, where $D(x, \delta) = \{y : d(x, y) < \delta\}$, and d is the metric on $\boldsymbol{E}$. It is not hard to see $B_\delta^- \in \mathcal{E}$, and $B^- = \cup_n B_{1/n}^-$, so that it is enough to show that B_δ^- is semipolar.

Let τ be the infimum of t such that the set $\{s \le t : X_s \in B_\delta^-\}$ is uncountable. (This is the "other penetration time," and it is a stopping time (Dellacherie [2] T22, p. 135.) Let the initial distribution of X be a reference measure. Then B^- is semipolar $\Longleftrightarrow \{t : X_t \in B^-\}$ is a.s. countable $\Longleftrightarrow \tau = \infty$ a.s..

By repeated applications of Theorem 6.9 we can find a sequence (T_n) of predictable times such that $T_n \downarrow \tau$ and $P\{X_{T_n-} \in B_\delta^-\} \to P\{\tau < \infty\}$. As (X_t)

is right-continuous, there is $t > 0$ such that

$$P\{\tau < \infty, \sup_{0\le s\le t} d(X_\tau, X_{\tau+s}) > \delta/2\} < (\delta/2)P\{\tau < \infty\}.$$

On the other hand

$$P\Big\{\sup_{0\le s\le t} d(X_\tau, X_{\tau+s}) > \frac{\delta}{2}\Big\} \ge P\{\tau < T_n < \tau + t,\ d(X_{T_n-}, X_{T_n}) > \delta\}$$

$$\ge \delta P\{\tau < T_n < \tau + t,\ X_{T_n-} \in B_\delta^-\}.$$

where the last inequality comes from the moderate Markov property of (X_{t-}) at T_n. But this is greater than $(\delta/2)P\{\tau < \infty\}$ for large n, which is a contradiction. Thus $\tau = \infty$ a.s., hence B_δ^- is semipolar. □

This leads us to a characterization of quasi-left continuity, which should be compared with Corollary 8.14. It depends only upon the moderate Markov property of (X_{t-}). Duality only serves to establish the moderate Markov property.

Proposition 13.22. *(X_t) is quasi-left continuous on $(0, \zeta)$ iff for all semipolar sets A,*

$$P^\xi\{\exists\, t < \zeta : X_{t-} \in A,\ X_{t-} \ne X_t\} = 0. \tag{13.17}$$

Proof. If (X_t) is quasi-left-continuous and A is semi-polar, the set $\{t : X_t \in A\}$ is countable hence, as there are only countably many discontinuities, so is $\{t : X_{t-} \in A\}$. By Theorem 6.13 there exists a sequence (T_n) of predictable times such that the sets $\{t : \exists\, t < \zeta,\ X_{t-} \in A\}$ and $\{t : \exists\, n \ni t = T_n < \zeta\}$ are a.s. identical. Since $X_{T_n-} = X_{T_n}$ a.s. by quasi-left-continuity, the set in (13.17) is a.s. empty.

Conversely, suppose (13.17) holds for all semi-polar sets. Then it holds for $A = B^-$, which is semi-polar by Lemma 13.21. If T is predictable, then $X_T = X_{T-}$ if $X_{T-} \in B^-$ and $T < \zeta$ by (13.17), while by the moderate Markov property, $P\{X_T = X_{T-} \mid \mathcal{F}_{T-}\} = \bar{P}(X_{T-}, \{X_T\})$, which equals one a.s. if $X_{T-} \in \mathbf{E} - B^-$. Thus $X_T = X_{T-}$ a.s. on $\{T < \zeta\}$, which implies quasi-left continuity. □

Remark 13.23. The proposition says "All semipolar sets" but according to the proof, there is one key semi-polar, namely the set B^- of left-branching points. Indeed, quasi-left continuity fails at T iff $X_{T-} \in B^-$, hence, as G. Mokobodzki pointed out, we can say informally that quasi-left continuity holds off the set B^- of left-branching points.

13.6. Dual Quantities

We want to look at some inner symmetries – one might almost say "inner harmonies" – of the two dual processes. We will use the idea of reversal to identify the "duals" of some familiar quantities and qualities. For an example of the latter, consider Hunt's Hypothesis (B), which is important in the theory of balayage introduced in §3.8. We mentioned then that its meaning was not immediately obvious, but that it would become clearer under duality theory. In fact, it turns out to be the mirror image of something we know quite well.

Theorem 13.24. *One of the processes* (X_t) *and* $(\hat{X}_t)$ *satisfies Hypothesis (B) if and only if the other is quasi-left-continuous on* $(0, \zeta)$.

Proof. Let $A \subset \boldsymbol{E}$ be semipolar for (X_t). Consider the two subsets of the canonical probability space

$$Z_A^t = \{\omega : \exists\, s, 0 < s < t < \zeta(\omega) : \omega(s-) \in A,\ \omega(s-) \neq \omega(s)\}; \quad (13.18)$$

$$\hat{Z}_A^t = \{\omega : \exists\, s, 0 < s < t < \zeta(\omega) : \omega(s) \in A,\ \omega(s) \neq \omega(s-)\}. \quad (13.19)$$

Note that the set in (13.17) is a union of Z_A^t and that for each t, $\hat{Z}_A^t = r_t^{-1} Z_A^t$ and vice-versa, so by Theorem 13.18, $P^\xi\{Z_A^t\} = 0 \Longleftrightarrow \hat{P}^\xi\{\hat{Z}_A^t\} = 0$. By Proposition 13.22, $P^\xi\{Z_A^t\} = 0$ for all semipolar A iff (X_t) is quasi-left-continuous; by Theorem 1 of §3.8 (which is stated for a Hunt process, but which holds for general rcll strong Markov processes) $(\hat{X}_t)$ satisfies Hypothesis (B) iff $\hat{P}^\xi\{\hat{Z}_A^t\} = 0$ for every co-semipolar set A. All that remains to finish the proof is to remark that a set is co-semipolar if and only if it is semipolar. This would be easy to establish here, but we will put it off to the next section in company with some allied results, and simply say that the proof follows from Proposition 13.27. □

Let us look at some other dual quantities. For a first example, if $A \in \mathcal{E}$, let $T_A = \inf\{t > 0 : X_t \in A\}$, and let $S_A = \inf\{t > 0 : X_{t-} \in A\}$ be the **left-entrance time** S_A. Then on $\{\zeta \geq t\}$,

$$r_t^{-1}\{T_A < t\} = \{\omega : \exists\, 0 < s < t \ni \omega(t - s-) \in A\} = \{S_A < t\}. \quad (13.20)$$

Thus, T_A and S_A are duals in a certain sense. It is not yet clear exactly what this sense is, over and above the appealing relation (13.20), but we will come back to it shortly. Note that the left-entrance time was introduced in §3.3 (23) where it was called T_A^-. It will be important in the sequel.

A second example is the first penetration time $\pi_A = \inf\{t : m\{s \leq t : X_s \in A\} > 0\}$. Then, as $X_s = X_{s-}$ for all but countably many s, $r_t^{-1}\{\pi_A < t\} = \{\pi_A < t\}$, so the first penetration time is self-dual.

For a third example, consider the time τ_A, which is the infimum of all t with the property that the set $\{s \leq t : X_s \in A\}$ is uncountable, i.e. "the other penetration time." We can apply exactly the same argument we did with π_A to see that τ_A is self-dual.

These times all have the shift property: if T is any one of them, then $T(\omega) > t \Longrightarrow T(\omega) = t + T(\theta_t \omega)$, or

$$T = t + T \circ \theta_t \quad \text{on} \quad \{T > t\}.$$

Stopping times which satisfy this are called **terminal times**. (See §3.4 Exercise 8.)

One can also find duals of more complicated objects, such as additive and multiplicative functionals. An adapted process $\{A_t,\ t \geq 0\}$ is a **perfect positive additive functional** if for all $\omega \in \Omega$:

(i) $t \mapsto A_t(\omega)$ is positive and right-continuous;
(ii) for each $s \geq 0$, and $t \geq 0$,

$$A_{s+t}(\omega) = A_s(\omega) + A_t(\theta_s \omega). \tag{13.21}$$

An adapted process $\{M_t, t \geq 0\}$ is a **perfect multiplicative functional** if for all $\omega \in \Omega$:

(i) $t \mapsto M_t(\omega)$ is right-continuous and takes values in $[0, 1]$;
(ii) for each $s \geq 0$, and $t \geq 0$,

$$M_{s+t}(\omega) = M_s(\omega) M_t(\theta_s \omega). \tag{13.22}$$

A typical example of an additive functional is given by $A_t \equiv \int_0^t f(X_s)\, ds$, where $f \in \mathcal{E}_b^+$. Then $M_t = e^{-A_t}$ is a multiplicative functional. Neither multiplicative nor additive functionals need be continuous, as one can see by considering the multiplicative functional $M_t = I_{\{t < T_A\}}$.

The word "perfect" above means that (13.21) and (13.22) hold with no exceptional sets. One commonly requires that they only hold a.s. for each pair s and t. However, it is known (see Walsh [4]) that one can perfect such functionals, i.e. there are versions of the functionals which satisfy (13.21) and (13.22) identically. Consequently, we will always assume that the additive and multiplicative functionals we deal with are perfect.

In general, if A_t is an additive functional, e^{-A_t} is a multiplicative functional, but the correspondence doesn't go both ways. Multiplicative functionals can take on the value zero so one may not be able to define an additive functional corresponding to a given multiplicative functional. (Unless, of course, one is willing to accept additive functionals with extended-real values; with the convention that $e^{-\infty} = 0$ and $\log 0 = -\infty$, one can then set $M_t = e^{-A_t}$ and $A_t = \log M_t$ to get a complete correspondence between the two.)

If (A_t) is a continuous additive functional, define

$$\hat{A}_t = A_t \circ r_t, \tag{13.23}$$

i.e. $\hat{A}_t(\omega) = A_t(r_t \omega)$. Then $(\hat{A}_t)$ is adapted and from the additivity of (A_t), if $t + s < \zeta$,

$$\hat{A}_{t+s}(\omega) = A_{t+s}(r_{t+s}\omega) = A_s(r_{t+s}\omega) + A_t(\theta_s r_{t+s}\omega).$$

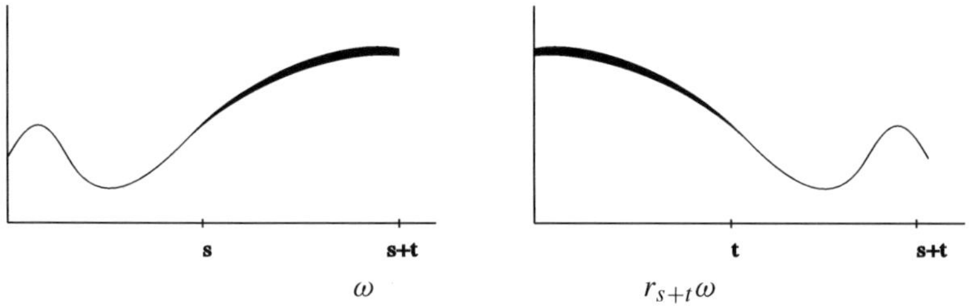

Fig. 13.1. The graphs of ω and $r_{s+t}\omega$.

Now $r_{t+s}\omega$ is s-equivalent to $r_s\theta_t\omega$ and $\theta_s r_{s+t}\omega$ is t-equivalent to $r_t\omega$, which can be seen from the diagram. The heavy portion of the second graph is part of the path $r_{s+t}\omega$, but it is also $r_t(\theta_s\omega)$. The light portion can be written either $\theta_t r_{s+t}\omega$ or $r_s\omega$.

Thus this is

$$= A_s(r_s\theta_t\omega) + A_t(r_t\omega) = \hat{A}_s(\theta_t\omega) + \hat{A}_t(\omega),$$

which shows the additivity of $\hat{A}$. Note that $(A_1 - A_{1-t})(\omega) = A_t(\theta_{1-t}\omega) = \hat{A}_t(r_1\omega)$, since $r_t\theta_{1-t}$ is t-equivalent to $r_1\omega$. This means that the graph of $\{\hat{A}_t : 0 \le t < 1\}$, is just the graph of A_t, rotated by 180°! See Fig. 13.2.

The above formula does not quite work for discontinuous functionals. If M_t is an arbitrary multiplicative functional, define

$$\bar{M}_t(\omega) = \begin{cases} M_{t-}(r_t\omega) & \text{if } t \le \zeta(\omega) \text{ or } \zeta(\omega) = 0, \\ \lim_{s\uparrow\zeta(\omega)} \bar{M}_s(\omega) & \text{if } t > \zeta(\omega) > 0. \end{cases}$$

Then the dual multiplicative functional is

$$\hat{M}_t = \lim_{s\downarrow t} \bar{M}_s. \tag{13.24}$$

It is easy to see that if (M_t) is continuous, that $\hat{M}_t = M_t \circ r_t$, in which case the same calculation we made for additive functionals shows that $\hat{M}$ is

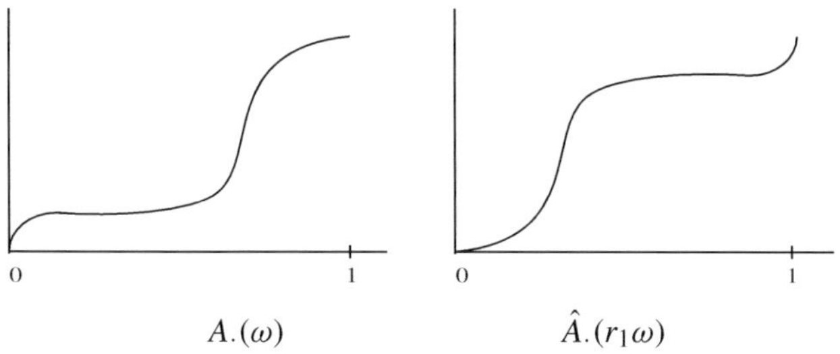

Fig. 13.2. The graphs of $A.(\omega)$ and $\hat{A}.(\omega) = A.(r_1\omega)$.

a multiplicative functional. In general, one has to follow the limits carefully to check (13.22). This also serves to define dual additive functionals, since $M_t \equiv e^{-A_t}$ is then a multiplicative functional. We then have

Theorem 13.25. *If* (A_t) *and* (M_t) *are an additive and a multiplicative functional respectively, then* $(\hat{A}_t)$ *and* $(\hat{M}_t)$ *defined by (13.24) are also additive and multiplicative functionals respectively, and* $(\hat{M}_t)$ *is dual to* (M_t) *in the sense that for* $f, g \in b\mathcal{E}^+$, $t \geq 0$:

$$E^{\xi}\{f(X_0)g(X_t)M_t; \zeta \geq t\} = \hat{E}^{\xi}\{g(\hat{X}_0)f(\hat{X}_t)\hat{M}_t; \hat{\zeta} \geq t\}. \tag{13.25}$$

Proof. We will just prove (13.25). Choose t such that $f(\hat{X}_t)$, $g(X_t)$, M_t and $\hat{M}_t$ are all a.s. continuous at t, which is true of a.e. t. Then $M_t \circ r_t = \hat{M}_t$ so $f(X_0)g(X_t)M_t \circ r_t = f(\hat{X}_0)g(\hat{X}_t)\hat{M}_t$ and (13.25) follows from Theorem 13.18, for a.e. t. But if we let f and g have compact support, so that they are ξ-integrable (for ξ is a Radon measure) then both sides are right-continuous in t and therefore (13.25) holds for all t. □

Remark 13.26. A multiplicative functional can be used to kill a process to get a subprocess: if M_t is a multiplicative functional, then there exists a subprocess (Y_t) of (X_t) – that is, (Y_t) is (X_t) killed at some (randomized) stopping time – and the transition function Q_t of (Y_t) is given by

$$Q_t(x, A) = E^x\{X_t \in A; M_t\}.$$

If $\hat{Q}_t$ is the transition function of the subprocess $(\hat{Y}_t)$ of $(\hat{X}_t)$ induced by $(\hat{M}_t)$, then an application of the Markov property on both sides of (13.25) gives $E^{\xi}\{f(X_0)Q_t g(X_0)\} = \hat{E}^{\xi}\{g(\hat{X}_0)\hat{Q}_t f(\hat{X}_0)\}$, or

$$\langle f, Q_t g\rangle = \langle \hat{Q}_t f, g\rangle.$$

In other words, the subprocesses induced by the dual multiplicative functionals are themselves dual with respect to ξ. Additive functionals also act on dual processes in a dual way. As this takes longer to explain and is harder to prove, we will not state it here. But the above shows that "dual multiplicative functionals act on dual processes in a dual way."

To return to the example which introduced this section, let T be a terminal time, and let $M_t = I_{\{T>t\}}$ be the corresponding multiplicative functional. The corresponding sub-process is just the original process killed at T. If we unwind the definition (13.24) of the dual multiplicative functional $\hat{M}_t$, we see that $\hat{M}_t = I_{\{t<\hat{T}\}}$, where

$$\hat{T} \stackrel{\text{def}}{=} \inf\{t : T \circ r_t < t\} \tag{13.26}$$

is the **co-terminal time dual to** $\boldsymbol{T}$. It follows that the process (X_t) killed at T is dual – with the same duality measure – to the process $(\hat{X}_t)$ killed at $\hat{T}$.

In particular, the process (X_t) killed at T_A and the process $(\hat{X}_t)$ killed at S_A are dual.

13.7. Small Sets and Regular Points

Polar and semi-polar sets were defined in §3.5. We say a set A is **co-polar** if it is polar for the dual process $(\hat{X}_t)$, and **co-semi-polar** if it is semi-polar for $(\hat{X}_t)$. It is **left polar** if, for all $x \in \boldsymbol{E}$, $P^x\{S_A < \zeta\} = 0$, and it is **left co-polar** if it is left polar for $(\hat{X}_t)$.

If $A \in \mathcal{E}$, we say x is **left-regular for A** if $P^x\{S_A = 0\} = 1$, and x is **left-co-regular for A** if $\hat{P}^x\{S_A = 0\} = 1$. We denote the left regular points of A by A^ℓ. The regular points of A are denoted A^r, and the co-regular and left-co-regular points by $\hat{A}^r$ and $\hat{A}^\ell$ respectively.

We have to be careful with S_A, since it is possible that $S_A = \zeta$, whereas $T_A \neq \zeta$ – unless $\partial \in A$, of course, which we have excluded above by specifying that $A \subset \boldsymbol{E}$. Note that as S_A is a terminal time (§3.4 Exercise 8), $x \mapsto P^x\{S_A < \infty\}$ is excessive and $x \mapsto E^x\{e^{-pS_A}\}$ is p-excessive for $p > 0$.

Fortunately, several of the above ideas coincide, so we will not need to keep close track of them all.

Proposition 13.27.

(i) *A set $A \in \mathcal{E}$ is left polar iff it is co-polar and it is left-co-polar iff it is polar.*

(ii) *A set is semipolar iff it is co-semipolar. In particular, $B \cup \hat{B}$ is semipolar, and hence co-semipolar.*

(iii) *Suppose (X_t) and $(\hat{X}_t)$ are quasi-left-continuous on $(0, \zeta)$ and $(0, \hat{\zeta})$ respectively. Then a set is co-polar if and only if it is polar.*

Proof. (i) Let $p_A(x) = P^x\{S_A < \infty\}$ and let $\varrho_t(x) = P^x\{S_A < t;\ \zeta \geq t\}$. Certainly $p_A(x) \leq \sum_{t \in \mathbb{Q}} \varrho_t(x)$. Now

$$\int \varrho_t(x)\, dx = P^\xi\{S_A < t \leq \zeta\} = \hat{P}^\xi\{T_A < t \leq \hat{\zeta}\}$$

since S_A and T_A are dual terminal times. If A is co-polar, this is zero, hence $\varrho_t(x) = 0$ for ξ-a.e. x, and $p_A(x) = 0$ for ξ-a.e. x. But by Hypothesis (L), an excessive function which vanishes almost everywhere vanishes identically, so $p_A \equiv 0$. The converse follows by a similar argument.

The second conclusion follows from consideration of the stopping time τ_A, the "other penetration time." By Theorem 8 of §3.5, A is semipolar iff $P^x\{\tau_A < \infty\} = 0$, and since τ_A is self-dual, $P^\xi\{\tau_A < t \leq \zeta\} = \hat{P}^\xi\{\tau_A < t \leq \hat{\zeta}\}$, so the above argument applies.

For (iii), let $A \in \mathcal{E}$. If A is not left-polar, the Section Theorem (Theorem 6.9) applied to (X_{t-}) implies that there exists a predictable time T such that

$X_{T-} \in A$ and $T < \zeta$ with strictly positive probability. By quasi-left-continuity, $X_T = X_{T-} \in A$ with the same probability, so A is not polar. Thus, by (i), it is not left-co-polar. As $(\hat{X}_t)$ is also quasi-left continuous on $(0, \hat{\zeta})$, A is not co-polar either. By (i) this implies that A is not left-polar, which closes the circle. □

There is a nice characterization of quasi-left continuity, which is essentially a corollary of the proof of Proposition 13.22. Recall that B^- denotes the set of left branching points, and that $B \subset B^-$.

Proposition 13.28. *(X_t) is quasi-left-continuous on $(0, \zeta)$ iff the set B^- is co-polar.*

Proof. Let $T_n \uparrow T$ be bounded stopping times. Define

$$T_n' = \begin{cases} T_n & \text{if } T_n < T \\ \infty & \text{otherwise.} \end{cases}$$

Let $T' = \lim T_n'$. Then T' is predictable, as (T_n') announces it. (We can replace T_n' by $T_n' \wedge n$ in order to have $T_n < T$ on $\{T = \infty\}$.) Thus for any $f \in C(\bar{E})$, we have by the moderate Markov property that

$$E\{f(X_{T'}) - f(X_{T'-}) \mid \mathcal{F}_{T'-}\} = \bar{P}_0 f(X_{T'-}) - f(X_{T'-}) \tag{13.27}$$

a.s. on $\{T' \neq \zeta\}$. If B^- is co-polar, it is left-polar, hence $P\{X_{T'-} \in B^-\} = 0$ and the right-hand side vanishes a.s. since $\bar{P}_0$ is the identity on $\boldsymbol{E} - B^-$. This is true for all f in a countable dense subset of $C(\bar{\boldsymbol{E}})$ so that $X_{T'-} = X_{T'}$ a.s. and in particular on the set $\{T_n < T, \ \forall n\}$. But obviously $\lim_n X_{T_n} = X_T$ on the set $\{\exists n : T_n = T\}$, so (X_t) is quasi-left-continuous on $(0, \zeta)$.

Conversely, suppose (X_t) is quasi-left-continuous on $(0, \zeta)$. As B^- is semi-polar, there exist predictable times T_n with the property that $\{t : X_{t-} \in B^-\} = \{t : \exists n \ni t = T_n\}$ a.s. (Theorem 6.13.) But (13.27) holds for each T_n, and the left-hand side is zero on $\{T_n < \zeta\}$ by quasi-left continuity. Thus the right-hand side vanishes too: $\bar{P}_0 f(X_{T_n-}) = f(X_{T_n-})$. It follows that $X_{T_n-} \notin B^-$ a.s. on $\{T_n < \zeta\}$, which means that (X_{t-}) does not hit B^- in $(0, \zeta)$, hence B^- is left-polar, and therefore co-polar. □

We should keep in mind that time reversal interchanges right and left limits, and right and left continuity: if $s < t$, then $X_s(\omega) = \hat{X}_{t-s-}(r_t\omega)$, so that a statement about X_t will translate into a statement about $\hat{X}_{t-}$. For example, the dual of T_A is not T_A, but S_A. Another easy consequence of reversal is the following:

Proposition 13.29. *Let $f \in \mathcal{E}$. Then*

(i) *$t \mapsto f(X_t)$ is P^x-a.s. right continuous on $(0, \zeta)$, $\forall x$ iff $t \mapsto f(\hat{X}_{t-})$ is $\hat{P}^x$-a.s. left-continuous on $(0, \hat{\zeta})$, $\forall x$;*

(ii) *$t \mapsto f(X_t)$ has left limits on $(0, \zeta)$, P^x-a.s. $\forall x$ iff $t \mapsto f(\hat{X}_{t-})$ has right limits on $(0, \hat{\zeta})$, P^x-a.s. $\forall x$.*

We leave the proof as an exercise. Similarly, we can detail the behavior of co-excessive functions along the paths of (X_t).

Proposition 13.30. *Let f be co-excessive and let μ be a probability measure on $\mathbf{E}$. Then with P^μ-probability one*

(i) *there exist stopping times $0 \le S_1 \le S_2 \le \infty$ such that*

$$\begin{cases} 0 < t \le S_1 & \Longrightarrow f(X_{t-}) = 0, \\ S_1 < t < S_2 & \Longrightarrow 0 < f(X_{t-}) < \infty \text{ and } 0 < f(X_{t-})_+ < \infty, \\ t \ge S_2 & \Longrightarrow f(X_{t-}) = \infty; \end{cases}$$

(ii) *$t \mapsto f(X_{t-})$ is left-continuous and has right limits on $(0, \zeta)$ in the extended reals.*

Proof. Since f is co-excessive, $f(\hat{X}_t)$ is a positive rcll local supermartingale as soon as it is finite. Moreover, if $f(\hat{X}_t)$ vanishes for some t, $f(\hat{X}_s)$ vanishes identically for all $s \ge t$. Let $A = \{x : f(x) < \infty\}$, $Z = \{x : f(x) = 0\}$, and for any t let Λ_t be the set of ω on which there exist extended reals $0 \le \tau_1 \le \tau_2 \le \infty$ (depending on ω) such that for all $s < t \le \zeta$, $\hat{X}_s(\omega) \in A$ if $s \in (0, \tau_1)$, $\hat{X}_s \in \mathbf{E}_f$ if $s \in (\tau_1, \tau_2)$, and $\hat{X}_s(\omega) \in Z$ if $s \ge \tau_2$. (In short, Λ is the set on which $\hat{X}_s$ is successively in A, $\mathbf{E}_f$, and Z for $s < t$, though it need not enter in all three.) Then $\Lambda_t \in \mathcal{F}_t$ and Λ_t^c has zero probability under any initial measure, so

$$0 = \hat{P}^\xi\{\Lambda_t^c;\ \hat{\zeta} \ge t\} = P^\xi\{r_t(\Lambda_t^c);\ \zeta \ge t\}$$

by Theorem 13.18. Since $r_t(\Lambda_t)$ is just the set on which X_{s-} is successively in Z, $\mathbf{E}_f$ and A for $s < t$, (i) follows on taking $S_1 = S_{A \cup \mathbf{E}_f}$ and $S_2 = S_A$. Then (ii) follows by applying the same reasoning to $\Lambda_t' \stackrel{\text{def}}{=} \{\omega : s \mapsto \hat{X}_s(\omega) \text{ is rcll on } 0 < s < t \le \zeta(\omega)\}$. $\square$

Proposition 13.30 tells us about the values of (X_{t-}). What about the values of (X_t)? If f is excessive, let $\mathbf{E}_f^\ell$ be the set of points left regular for $\mathbf{E}_f$. If S_1 and S_2 are the stopping times of Proposition 13.30, then

Corollary 13.31. *Let f be co-excessive and let μ be a probability measure on $\mathbf{E}$. Then $X_t \in \mathbf{E}_f^\ell$ for all $t \in (S_1, S_2)$, and $X_t \in (\mathbf{E} - \mathbf{E}_f^\ell) \cup \{\partial\}$ for all $t < S_1$ or $t > S_2$, P^μ-a.s.*

Proof. By Proposition 13.30 $X_{t-} \in \boldsymbol{E}_f$ on (S_1, S_2). If T is a stopping time, $X_T \notin B$, so $P^{X_T}\{S_{\boldsymbol{E}_f} = 0\}$ equals either zero or one. $S_{\boldsymbol{E}_f} \circ \theta_T > 0 \Longleftrightarrow f(X_{T+t-}) = 0$ or ∞ for all small enough $t > 0$. On $\{T < S_1\} \cup \{T > S_2\}$, $f(X_{T+t-}) = 0$ or ∞ for all small enough t, hence $S_{\boldsymbol{E}_f} \circ \theta_T > 0$ and, by the strong Markov property, $P^{X_T}\{S_{\boldsymbol{E}_f} = 0\} = 0$, so $X_T \notin \boldsymbol{E}^{\ell}_f$. On the other hand, on $\{S_1 < T < S_2\}$, $X_{T+t-} \in \boldsymbol{E}_f$ for sufficiently small t, hence $S_A \circ \theta_T = 0$, and evidently $X_T \in \boldsymbol{E}^{\ell}_f$. Thus by Theorem 6.9, the two processes $\{I_{\boldsymbol{E}^{\ell}_f}(X_t),\ t \geq 0\}$ and $\{I_{\boldsymbol{E}_f}(X_{t-}),\ t \geq 0\}$ are equal, except possibly at S_1 and S_2. □

We showed in Theorem 6 of §3.5 that $A - A^r$ is semi-polar. That proof was for Hunt processes, but it applies to the present case. We leave the proof of the following as an exercise.

Proposition 13.32. *Let* $A \in \mathcal{E}$. *The following sets are semipolar:*

$$A - A^r,\quad A - A^{\ell},\quad A - \hat{A}^r,\quad A - \hat{A}^{\ell}.$$

13.8. Duality and h-Transforms

Let us determine the dual of an h-transform of (X_t). Let h be excessive and let $\hat{\boldsymbol{E}}^{\ell}_h$ be the set of points co-left-regular for $\boldsymbol{E}_h$. By Corollary 13.31, the process $(\hat{X}_t)$ can never return to $\hat{\boldsymbol{E}}^{\ell}_h$ once it leaves it. Let $(\hat{V}_p)$ be the resolvent of the process $(\hat{X}_t)$ killed when it first hits $\boldsymbol{E} - \hat{\boldsymbol{E}}^{\ell}_h$. Then $(\hat{V}_p)$ is just the restriction of $(\hat{U}_p)$ to $\hat{\boldsymbol{E}}^{\ell}_h$, i.e. if $x \in \boldsymbol{E} - \hat{\boldsymbol{E}}^{\ell}_h$, then $\hat{V}_p(x, E) = 0$, while if $x \in \hat{\boldsymbol{E}}^{\ell}_h$, $\hat{V}_p(x, A) = \hat{U}_p(x, A \cap \hat{\boldsymbol{E}}^{\ell}_h)$.

Theorem 13.33. *Let h be excessive and let $(\hat{V}_p)$ be the resolvent of the process $(\hat{X}_t)$ killed upon first hitting $\boldsymbol{E} - \hat{\boldsymbol{E}}^{\ell}_h$. Then $({}_hU_p)$ and $(\hat{V}_p)$ are in duality relative to ξ_h, where*

$$\xi_h(dx) = I_{\boldsymbol{E}_h}(x) h(x)\, \xi(dx). \tag{13.28}$$

Proof. Let $f, g \in \mathcal{E}^+$ vanish on $\boldsymbol{E} - \boldsymbol{E}_h$. Then

$$\langle f, {}_hU_p g\rangle_{\xi_h} = \langle f, I_{\boldsymbol{E}_h} h \frac{1}{h} U_p(hg)\rangle_{\xi} = \langle f, U_p(hg)\rangle_{\xi},$$

since $f = 0$ on $\boldsymbol{E}^c_h$. By duality this is

$$= \langle \hat{U}_p f, hg\rangle_{\xi} = \langle \hat{U}_p f, g\rangle_{\xi_h}.$$

But now $\hat{\boldsymbol{E}}^{\ell}_h - \boldsymbol{E}_h$ and $\boldsymbol{E}_h - \hat{\boldsymbol{E}}^{\ell}_h$ are both semi-polar. Thus they are of potential zero, hence of ξ-measure zero. From the remarks preceding the theorem,

$$\langle f, {}_hU_p g\rangle_{\xi_h} = \langle \hat{U}_p f, g\rangle_{\xi_h} = \langle \hat{V}_p f, g\rangle_{\xi_h}. \tag{13.29}$$

Notice that (13.29) remains true even if f and g don't vanish on $\boldsymbol{E}_h$ since ${}_hU_p g = 0$ on $\boldsymbol{E} - \boldsymbol{E}_h$ and $\hat{V}_p f = \hat{V}_p(f I_{\hat{\boldsymbol{E}}_h^\ell}) = 0$ on $\boldsymbol{E} - \hat{\boldsymbol{E}}_h^\ell$, which differs from $\boldsymbol{E}_h$ by a set of potential zero. Finally, $\hat{V}_p(x, \cdot) \le \hat{U}_p(x, \cdot) \ll \xi(\cdot)$, so $\hat{V}_p(x, \cdot) \ll \xi_h(\cdot)$. Since $(\hat{V}_p)$ is the resolvent of a strong Markov process, the proof is complete. □

The processes (X_t^h) and "$(\hat{X}_t)$ killed at its first exit from $\hat{\boldsymbol{E}}_h^\ell$" have $\boldsymbol{E}_h \cup \hat{\boldsymbol{E}}_h^\ell$ as their natural state space. All other points of $\boldsymbol{E}$ are both branching and co-branching points which branch directly to ∂. In addition, $\boldsymbol{E}_h - \hat{\boldsymbol{E}}_h^\ell$ co-branches to ∂ and $\hat{\boldsymbol{E}}_h^\ell - \boldsymbol{E}_h$ branches to ∂.

Exercises

1. Show that under the hypotheses of §13.2 there exists a bounded p-excessive ξ-integrable function h such that $\{h > 0\} = \boldsymbol{E} - B_\partial$.
2. Let (X_t) be uniform motion to right on line, but when $X_{t-} = 0$, it jumps to one with probability 1/2 and otherwise goes on to hit zero. If $X_0 = 0$, then (X_t) is uniform motion to right. Show that the origin is in B^- but not in B. Find the dual process with respect to Lebesgue measure.
3. Verify that additive functionals of the form $A_t = \int_0^t f(X_s)\, ds$ are self-dual.
4. Prove Proposition 13.29.
5. Show that $r_t \circ r_t = k_t$, $\theta_s \circ r_{s+t} = r_t$ and $k_t \circ r_{s+t} = r_t \circ \theta_s$.
6. Prove Proposition 13.32.
7. Show that $\boldsymbol{E}_f^\ell - \boldsymbol{E}_f$ is semi-polar.
8. Find an example in which $A^r - A$ is not semi-polar.

13.9. Reversal From a Random Time

We saw in Chapter 10 that the reverse of a right-continuous strong Markov process is a left-continuous moderate Markov process. It may not be strongly Markov. However, if there is a right-continuous strong Markov dual, then the right-continuous reverse is also strongly Markov, at least for strictly positive t. (The question of whether the reverse can be extended to include time $t = 0$ is delicate. See §13.10.)

The reverse process and the dual process are connected through h-transforms. If we consider the reversal of h-transforms of both the original process and the dual, we get a symmetric theory. The reverse of an h-transform of (X_t) is a transform of the dual $(\hat{X}_t)$ by a coexcessive function – the function depends on both h and the distribution of X_0 – and the reverse of a transform of $(\hat{X}_t)$ is a transform of (X_t) by an excessive function. This is given in Theorem 13.34. There is one complication: we may have to transform a sub-process of the dual, rather than the dual itself.

Let h be excessive, let (X_t^h) be the h-transform of (X_t), and let μ be the initial distribution of (X_t^h). Assume that the lifetime of (X_t^h) is finite with positive probability. Let $(\tilde{X}_t)$ be the reverse of (X_t^h) from its lifetime, *made right-continuous*:

$$\tilde{X}_t = \begin{cases} X_{(\zeta - t+)} & \text{if } 0 < t < \zeta < \infty \\ \partial & \text{if } t \geq \zeta \text{ or } \zeta = \infty. \end{cases}$$

We know from Theorem 10.1 that $\{\tilde{X}_{t-}, t > 0\}$ is a left-continuous moderate Markov process. Let $({}_hU_p)$ be the resolvent of (X_t^h), and let $(\hat{V}_p)$ be the resolvent of $(\hat{X}_t)$ killed the first time it leaves $\hat{\boldsymbol{E}}_h^\ell$, the set of points left-co-regular for $\boldsymbol{E}_h$. (Remember that $(\tilde{X}_t)$ is the right-continuous reverse process, and $(\tilde{X}_{t-})$ is the unmodified left-continuous reverse.) We want to compare $(\tilde{X}_t)$ with an h-transform of the dual $(\hat{X}_t)$.

Theorem 13.34. *If (X_t^h) has initial distribution μ, then there exists a function v which is excessive with respect to $(\hat{V}_p)$, such that $\{\tilde{X}_t,\ t > 0\}$ is a right-continuous strong Markov process having resolvent $({}_v\hat{V}_p)$. The function v is determined by*

$$v(x)\xi_h(dx) = \mu({}_hU_0)(dx), \tag{13.30}$$

where the measure ξ_h was defined in (13.28).

Proof. If ${}_hP^\mu\{0 < \zeta < \infty\} = 0$, there is nothing to prove, so suppose the probability is strictly positive. Let $k(x) = {}_hE^x\{e^{-\zeta}; \zeta < \infty\}$. This is 1-$h$-excessive and $\mu\,{}_hU_0$-integrable by (10.5), and $\mu\,{}_hU_0$ is σ-finite on $\{x : k(x) > 0\}$.

Note that $\mu\,{}_hU_0$ is an h-excessive measure, and by Theorem 13.33, $({}_hU_p)$ and $(\hat{V}_p)$ are in duality relative to $\xi_h(dx) = I_{\boldsymbol{E}_h}(x)h(x)\,dx$. By Theorem 13.13 there exists a ξ_h-a.e. finite $\hat{V}$-excessive function v such that

$$\mu\,{}_hU_0(dx) = v(x)\xi_h(dx). \tag{13.31}$$

If $(\tilde{U}_p)$ is the resolvent of $(\tilde{X}_t)$, then by Theorem 10.1 $(\tilde{U}_p)$ is in duality with $({}_hU_p)$ relative to the occupation-time measure of (X_t^h), which is $\mu\,{}_hU_0 = v\xi_h$.

Thus for $f, g \in b\mathcal{E}^+$,

$$\langle {}_v\hat{V}_p f, g\rangle_{v\xi_h} = \langle \tilde{U}_p f, g\rangle_{v\xi_h}$$

since both equal $\langle f, {}_hU_p g\rangle_{v\xi_h}$. It follows that ${}_v\hat{V}_p(x,\cdot) = \tilde{U}_p(x,\cdot)$ for $v\xi_h$-a.e. x. Let $A = \{x : \tilde{U}_p(x,\cdot) = {}_v\hat{V}_p(x,\cdot)\}$. Then

$$\begin{aligned} 0 = v\xi_h(A^c) = \mu({}_hU_0)(A^c) &= \int_0^\infty \mu({}_hP_t)(A^c)\,dt \\ &= E\left\{\int_0^\infty I_{A^c}(X_t^h)\,dt\right\} \\ &= E\left\{\int_0^\infty I_{A^c}(\tilde{X}_t)\,dt\right\} \\ &= \int_0^\infty P\{\tilde{X}_t \in A^c\}\,dt, \end{aligned}$$

so for $f \in b\mathcal{E}^+$ and a.e. t,

$$E\left\{\int_0^\infty e^{-ps} f(\tilde{X}_{t+s})\,ds \mid \mathcal{F}_t\right\} = \tilde{U}_p f(\tilde{X}_t) = {}_v\hat{V}_p f(\tilde{X}_t).$$

Now ${}_v\hat{V}_p f$ is p-co-excessive, hence $t \mapsto {}_v\hat{V}_p f(\hat{X}^h_{t-})$ is left-continuous on $(0,\zeta)$, so by time reversal, $t \mapsto {}_v\hat{V}_p f(\tilde{X}_t)$ is right-continuous on $(0,\hat{\zeta})$. This implies that $\{\tilde{X}_y, t > 0\}$ is strongly Markov. □

Remark 13.35. 1° Theorem 13.34 says that $(\tilde{X}_t)$ has the same distribution as the v-transform of the dual killed when it first hits $\boldsymbol{E} - \hat{\boldsymbol{E}}^\ell_h$.

The reason $(\tilde{X}_t)$ is a transform of the killed dual and not the dual itself is easy to see once we remember that an h-transform never enters the set where $h = 0$ or ∞. That means that the reverse process cannot take values there either – so it must be connected with the dual process killed upon entering that set. If h is strictly positive and finite, there is no need for killing, and the reverse is indeed a transform of the dual.

Notice that we only consider the reverse process for strictly positive t. We do not claim that $\{\tilde{X}_t,\ t \geq 0\}$ is a strong Markov process. Indeed, it is not in general without further hypotheses. This is a delicate point which we shall take up first in §13.10, and in even greater depth, in Chapter 14.

2° Suppose the resolvent density $u_p(x,y)$ exists for $p = 0$ as well as for $p > 0$. Then

$$\mu\,{}_hU_0(dy) = \left(\int_{\boldsymbol{E}_h} \mu(dx)\,u_0(x,y)\frac{h(y)}{h(x)}\right)\xi(dy).$$

Compare this with (13.30) to see that

$$v(y) = \int_{\boldsymbol{E}_h} u_0(x,y)\frac{1}{h(x)}\mu(dx).$$

If μ sits on $\boldsymbol{E}_h$, we can write this more compactly as

$$v(y) = \hat{U}_0\left(\frac{1}{h}\mu\right)(y), \quad y \in \hat{\boldsymbol{E}}_h^{\ell} \tag{13.32}$$

i.e. v is the co-potential of the measure $\frac{1}{h}\mu$.

It is worthwhile to spell out three special cases. The first follows from Corollary 11.30, which says that the reverse of (X_t^h) from a finite co-optional time has the same semigroup as the reverse from its lifetime. The second two concern particular choices of μ. We leave their verification to the reader.

Corollary 13.36. *Suppose (X_t^h) has initial measure μ and U_0 has a density $u_0(x, y)$. Let L be an a.e. finite co-optional time. Then*

(i) *the reverse of (X_t^h) from L is a v-transform of $(\hat{X}_t)$, where v is given by*

$$v(y) = \hat{U}_0(\frac{1}{h}\mu)(y);$$

(ii) *if $\mu(dx) = g(x)\xi(dx)$, then $(\tilde{X}_t)$ has resolvent $({}_v\hat{V}_p)$, where v is given by*

$$v(y) = \hat{U}_0\left(\frac{g}{h}\right)(y) = \int_{E_h} u_0(x, y)\frac{g(x)}{h(x)}\,dx.$$

(iii) *If $\mu(dx)$ is the unit mass at x, then $v(y) = u_0(x, y)$, independent of h.*

The following is a consequence of (iii).

Corollary 13.37. *Fix $y \in \boldsymbol{E} - \hat{B}$. Consider the h-transform of (X_t) for $h(x) = u_0(x, y)$; denote it by (X_t^y) instead of (X_t^h). Then for any initial distribution μ,*

$$P^{\mu}\{\zeta < \infty,\ X_{\zeta-}^y = y\} = 1.$$

Proof. If k is a co-excessive function such that $y \in \boldsymbol{E}_k$ and $(\hat{X}_t^k)$ has a finite lifetime with positive probability, start $(\hat{X}_t^k)$ from y, and reverse it from its lifetime. By (iii), the result is a $u_0(\cdot\,, y)$-transform of (X_t); it has a finite lifetime (we only reverse on the set where the lifetime is finite) and its limit at its own lifetime is just $\hat{X}_0^h = y$. □

Remark 13.38. 1° Note that in (iii), (13.32) tells us that $v(y) = u_0(x, y)/h(x)$, but as x is fixed, $1/h(x)$ is just a constant of proportionality which doesn't affect the transform.

2° By symmetry, the $u_0(x, \cdot)$-transform of $(\hat{X}_t)$ converges to x at its lifetime. If we denote it by $(\hat{X}_t^x)$, we conclude that if $X_0^y = x$ and $\hat{X}_0^x = y$, then (X_t^y)

and $(\hat{X}_t^x)$ are reverses of each other. The first is born at x and dies at y; the second is born at y and dies at x. (Notice that the function k was only used to get a finite lifetime, and dropped out of the final results.)

The function v in Corollary 13.36 (ii) is interesting. It is called a co-potential. Let us define this formally. Let μ be a σ-finite measure on $\boldsymbol{E}$. For $p \geq 0$ the **p-potential of μ** is the function $U_p\mu(x) = \int u_p(x, y)\mu(dy)$ and the **p-co-potential of μ** is $\hat{U}_p\mu(y) = \int u_p(x, y)\,\mu(dx)$. The functions $U_p\mu$ and $\hat{U}_p\mu$ are p-excessive and p-co-excessive respectively. (We will write U and $\hat{U}$ in place of U_0 and $\hat{U}_0$, and $u(x, y)$ in place of $u_0(x, y)$. Note that μU is a measure, but $U\mu$ is a function.)

Recall that (U_p) is the potential kernel of the process "$(X_t^p) \equiv (X_t)$ killed at an independent exponential (p) time." There is no guarantee that U_0 is finite, so there may not be non-trivial zero potentials. However (U_p) is bounded for $p > 0$, and there is no lack of non-trivial p-potentials. We state the following theorem for $p = 0$, but it is valid for $p > 0$ as long as we replace (X_t) by (X_t^p) and u_0 by u_p.

Theorem 13.39. *Let μ be a σ-finite measure on $\mathbf{E} - \hat{B}$ and let $h = U\mu$. If $x \in \mathbf{E}_h$, the distribution of $X_{\zeta-}^h$ is given by*

$$_hP^x\{\zeta < \infty,\ X_{\zeta-} \in dy\} = \frac{1}{h(y)} u_0(x, y)\mu(dy). \tag{13.33}$$

Proof. Write $_yP^x$ instead of $_{u(\cdot, y)}P^x$. If $A \in \mathcal{E}$, consider the event $\Lambda = \{X_{\zeta-} \in A\}$. By Proposition 11.10 if $x \in \boldsymbol{E}_h$,

$$_hP^x\{\Lambda\} = \frac{1}{h(x)} \int_A u_0(x, y)\, _yP^x\{\Lambda\}\, \mu(dy). \tag{13.34}$$

If $0 < u(x, y) < \infty$ and if $y \in \mathbf{E} - \hat{B}$, then $_yP^x\{X_{\zeta-} = y\} = 1$ by Corollary 13.37. Since $h(x) < \infty$, $\mu\{y : u_0(x, y) = \infty\} = 0$, so (13.34) is

$$= \frac{1}{h(x)} \int_A u_0(x, y)\, \mu(dy).$$

□

The following proposition tells us that an excessive function is determined up to a multiplicative constant by the way it acts on processes. See the exercises below for some information on fine components.

Proposition 13.40. *Let f and g be excessive and $p \geq 0$, and suppose that for ξ-a.e. $x \in \mathbf{E}_f$, the measures $_fU_p(x, \cdot)$ and $_gU_p(x, \cdot)$ agree on $\mathbf{E}_f$. Then for any f-fine component F of $\mathbf{E}_f$, there exists a constant c such that $f = cg$ on F.*

Proof. Let $h \in b\mathcal{E}^+$. If $x \in \boldsymbol{E}_f \cap \boldsymbol{E}_g$ is non-branching, then the functions $1/f, 1/g, U_p(fh)$, and $U_p(gh)$ are all fine continuous at x, hence so are ${}_fU_ph$ and ${}_gU_ph$. If $h = 0$ on $\boldsymbol{E}_f^c$, then ${}_fU_ph(x) = {}_fU_ph(x)$ for ξ-a.e. $x \in \boldsymbol{E}_f$, hence for every $x \in \boldsymbol{E}_f \cap \boldsymbol{E}_g$ by fine continuity. Note that ${}_fU_p1 \neq 0$ on $\boldsymbol{E}_f$, so that, as ${}_fU_p1 = {}_gU_p1$ a.e. on $\boldsymbol{E}_f$, we conclude $g > 0$ ξ-a.e. on $\boldsymbol{E}_f$.

But as g is excessive, $\{x : g(x) = 0\}$ is fine open as well as fine closed, so $\{g = 0\} \cap \boldsymbol{E}_f$ is a fine open set of ξ-measure zero, and therefore empty. Thus ${}_fU_p = {}_gU_p$ on $E_f \cap \{g < \infty\}$.

Let $x \in \boldsymbol{E}_f \cap \{g < \infty\}$. Then

$$\frac{1}{f(x)} \int u_p(x, y) f(y) h(y)\, dy = {}_fU_ph(x) = {}_gU_ph(x)$$

$$= \frac{1}{g(x)} \int u_p(x, y) g(y) h(y)\, dy.$$

As h is arbitrary, $f(y)/f(x) = g(y)/g(x)$ for ξ-a.e. y such that $0 < u_p(x, y) < \infty$. Therefore if $\alpha = f(x)/g(x)$, then $f(y) = \alpha g(y)$ for ξ-a.e. y for which $u_p(x, y) > 0$. Set $A_\alpha = \{x : f(x) = \alpha g(x)\}$, $N = \boldsymbol{E}_h \cap \{g = \infty\}$ and apply Exercise 18 below to conclude that A_α is an f-fine component of $\boldsymbol{E}_f$. □

Exercises

Definition 13.41. *A Borel set $A \subset \boldsymbol{E} - B$ is* **totally closed** *if*

(i) $x \in A \Longrightarrow P^x\{T_{A^c} < \zeta\} = 0$,
(ii) $x \in A^c - B \Longrightarrow P^x\{T_A < \zeta\} = 0$.

It is a **fine component** *if it is totally closed and has no totally closed proper subset.*

Cofine components (i.e. relative to $\hat{P}^x$) and ***h*-fine components** (relative to ${}_hP^x$) are defined similarly. We could define "totally closed" without excepting the branching points, but that would change the definition significantly.

Prove the following:

9. If $A \subset \boldsymbol{E} - B$ is Borel, A is totally closed iff $A^c - B$ is totally closed.
10. If $A_1, A_2, \ldots$ are totally closed, so are $\cup_n A_n$ and $\cap_n A_n$.
11. If $A \neq \emptyset$ is totally closed, $\xi(A) \neq 0$.
12. Two fine components are either disjoint or identical.
13. For each $x \in \boldsymbol{E} - B$ there exists a unique fine component containing x. (Hint: the main problem is measurability.)
14. $\boldsymbol{E} - B$ is a countable disjoint union of fine components.
15. If h is excessive and A is totally closed, hI_A is excessive on $\boldsymbol{E} - B$.

16. If h is a minimal excessive function, $\boldsymbol{E}_h - B$ is contained in a single fine component.

The foregoing did not involve duality or time reversal, but the following do.

17. If h is excessive and a.e. finite, and if $y \in \boldsymbol{E} - \hat{B}$ is left co-regular for $\boldsymbol{E}_h$, then the set $\{x : u(x, y) > 0\}$ is contained in a single h-fine component of $\boldsymbol{E}_h$.

18. Let F be a fine component, and let $\{A_\alpha,\ \alpha \in I\}$ be a collection of disjoint fine closed Borel subsets of $\boldsymbol{E} - B$ such that

(i) $x \in A_\alpha \Longrightarrow U_p(x, F - A_\alpha) = 0,\ p \geq 0$;

(ii) $\cup_{\alpha \in I} A_\alpha = F - N$, where N is a polar set.

Then there exists α_0 such that $F = A_{\alpha_0}$.

13.10. $X_{\zeta-}$: Limits at the Lifetime

If we reverse from the lifetime, is the (right-continuous) reverse process strongly Markov *with $t = 0$ included in the parameter set?* The answer is "not always." A typical counter-example would be Brownian motion (X_t) in a bounded domain D in the plane, killed as it reaches the boundary. Let the state space be D, rather than $\boldsymbol{R}^2$. Then the reverse $\tilde{X}_t$ is only defined for $t > 0$, since at $t = 0$ it is in the boundary of D, which is not in the state space. Of course, $\tilde{X}_0$ is defined as an element of the plane, and it is fair to ask if $\{\tilde{X}_t, t \geq 0\}$ is then strongly Markov. The answer is the same: "Not always." It depends on the domain. If the domain is sufficiently regular then the reverse is strongly Markov. Otherwise, it may not be. See Exercise 8 of §14. A close examination of this problem leads deeply into boundary theory, and in particular, into the theory of the Martin boundary. We will treat this in Chapter 14. Meanwhile, we will see what we can say without it.

We will look at a situation in which the limit at the lifetime may not exist – as above – but in which the process will be well-behaved when it does exist. This requires some smoothness conditions on the dual resolvent. With these smoothness conditions, the reverse process is indeed strongly Markov, even when the origin is included in the parameter set. The import of this fact will become evident in the succeeding sections, where we show some of its rich consequences in potential theory.

Thus we impose the following hypothesis on the dual resolvent, which will be in force throughout this section.

Hypothesis 13.42. **$(\widehat{\text{SF}})$**

(i) *The resolvent density $u_p(x, y)$ exists for $p \geq 0$;*

(ii) *if $f \in \mathcal{E}_b^+$, then $\hat{U}_p f$ is continuous on $\boldsymbol{E}$ for all $p \geq 0$;*

(iii) *$\hat{U}_p(x, \boldsymbol{E})$ is bounded in x for all $p \geq 0$.*

Under Hypothesis ($\widehat{\mathbf{SF}}$) the dual resolvent has the *strong Feller property,* i.e. it takes bounded Borel functions to continuous functions. There may still be co-branching points, however. Let $\hat{B}_\partial$ denote the co-branching points which branch directly to the cemetery.

Fix an a.e. finite excessive function h, and consider the process (X_t^h). Define a co-optional time L by

$$L = \begin{cases} \zeta & \text{if } X_{\zeta-}^h \in \boldsymbol{E} - \hat{B}_\partial \text{ and } \zeta < \infty \\ 0 & \text{otherwise.} \end{cases}$$

Let $(\hat{Y}_t)$ be the right-continuous process

$$\hat{Y}_t = \begin{cases} X_{L-t-}^h & \text{if } t < L, \\ \partial & \text{if } t \geq L. \end{cases}$$

Let g be a bounded Borel function which is strictly positive on $\boldsymbol{E}_h$, zero off it, such that g/h is bounded and $\int g(x)\,\xi(dx) = 1$. Let $\mu_0(dx) = g(x)\,\xi(dx)$ and set

$$v_0(x) = \hat{U}\left(h^{-1}\mu_0\right)(x). \tag{13.35}$$

Notice that v_0 is bounded, continuous and strictly positive on the cofine closure of $\boldsymbol{E}_h - \hat{B}_\partial$. We will use the process with initial measure μ_0 as a reference.

Proposition 13.43. *Under hypothesis ($\widehat{SF}$), if (X_t^h) has initial measure μ_0, then $\{\hat{Y}_t, t \geq 0\}$ is a right-continuous strong Markov process, and is a v_0-transform of $(\hat{X}_t)$.*

Proof. Let $c_L(x) = {}_hP^x\{L > 0\}$. The process (X_t^h) killed at L is a c_L-transform of (X_t^h), hence it is an hc_L-transform of (X_t) by Theorem 11.28. By replacing h by hc_L if necessary, we may assume that $X_{\zeta-}^h \in \boldsymbol{E} - \hat{B}_\partial$ and that $L = \zeta < \infty$.

By Theorem 13.34, $(\hat{Y}_t)$ is a v_0-transform of $(\hat{X}_t)$ (killed upon leaving $\hat{\boldsymbol{E}}_h^\ell$, but that won't bother us here).

Now v_0 is bounded and continuous by ($\widehat{SF}$) so that $v_0(\hat{Y}_0) = \lim_{t\to 0} v_0(\hat{Y}_t)$. Let $K = \boldsymbol{E} - \boldsymbol{E}_h$. Then for any $p > 0$, $\langle h, p\hat{U}_p I_K\rangle = \langle pU_p h, I_K\rangle = 0$, so $\hat{U}_p(x, K) = 0$ for a.e. x, and hence (Hypothesis (L)) for all $x \in \boldsymbol{E}_h$. Now $g/h > 0$ on $\boldsymbol{E}_h - \hat{B}_\partial$. But now if $\bar{\boldsymbol{E}}_h$ is the closure of $\boldsymbol{E}_h$ in $\boldsymbol{E}$, then if $x \in \bar{\boldsymbol{E}} - \hat{B}_\partial$, $0 < \hat{U}_p(x, \boldsymbol{E}) = \hat{U}_p(x, \boldsymbol{E}_h)$. It follows that $v_0(x) > 0$. But $X_{\zeta-}^h \in \boldsymbol{E}_h - \hat{B}_\partial$, so by the definition of L, $v_0(X_{\zeta-}) > 0$ a.s..

In particular $1/v_0(\hat{Y}_t)$ is right-continuous at zero, therefore right continuous for $t \geq 0$. It follows by Hypothesis ($\widehat{SF}$) that for each bounded Borel f, ${}_{v_0}\hat{U} f(\hat{Y}_t)$ is a.s. right-continuous on $[0, \infty)$, which implies the strong Markov property as claimed. □

We want to extend this to arbitrary initial distributions on $\boldsymbol{E}_h$. We will use the process with initial distribution μ_0 as a reference, and see how the distribution of $X_{\zeta-}$ varies with the initial distribution. If ν is a probability measure on E_h, let K_t^ν be the distribution of $\hat{Y}_t$ given that (X_t^h) has initial measure ν. In particular, $K_0^\nu(A) = {}_hP^\nu\{0 < \zeta < \infty,\ X_{\zeta-} \in A\}$.

Lemma 13.44. *Let μ be an initial probability measure on $\boldsymbol{E}_h$, and let $v(y) = \hat{U}\left(h^{-1}\mu\right)(y)$. Then $K_0^\mu \ll K_0^{\mu_0}$, K_0^μ does not charge $\{x : v(x) = 0 \text{ or } \infty\}$, and*

$$K_0^\mu(dy) = \frac{v(y)}{v_0(y)} K_0^{\mu_0}(dy). \tag{13.36}$$

Moreover, $\{K_t^\mu, t \geq 0\}$ is an entrance law for $({}_vP_t)$.

Proof. Let $k(t)$ be a positive continuous function such that $\int_0^\infty k(t)\,dt = 1$. Set $g_k(x) = {}_hE^x\{k(\zeta)\}$ and denote $\mu_h G$ by ${}_hG^\mu$. From Theorem 10.1 applied to (X_t^h),

$${}_hE^\mu\left\{\int_0^\infty k(t) f(\hat{Y}_t)\,dt\right\} = \int_E f(x) g_k(x)\,{}_hG^\mu(dx).$$

Now ${}_hG^\mu(dx) = (\int_{E_h} \mu(dz) h(z)^{-1} u_0(z,x) h(x))\xi(dx) = v(x)h(x)\xi(dx)$. Similarly, ${}_hG^{\mu_0}(dx) = v_0(x)h(x)\xi(dx)$, so the above is

$$= \int_E f(x) g_k(x) \frac{v(x)}{v_0(x)}\,{}_hG^{\mu_0}(dx) = \int_0^\infty k(t) K_t^{\mu_0}(v_0^{-1} v f)\,dt. \tag{13.37}$$

If f is bounded and continuous, $t \mapsto K_t^\mu f$ is right-continuous, so that $\lim_{t\to 0} K_t^\mu f = K_0^\mu f = {}_hE^\mu\{f(X_{\zeta-})\}$. Similarly, if μ_0 is the initial measure, then $(\hat{Y}_t)$ is a v_0-transform of $(\hat{X}_t)$. But v/v_0 is v_0-co-excessive, so $\{v(\hat{Y}_t)/v_0(\hat{Y}_t),\ t > 0\}$ is a supermartingale and

$$\lim_{t\to 0} K_t^{\mu_0}(v_0^{-1} v f) = \lim_{t\to 0} {}_hE^{\mu_0}\{v(\hat{Y}_t) v_0^{-1}(\hat{Y}_t) f(\hat{Y}_t)\} = K_0^{\mu_0}(v_0^{-1} v f).$$

Thus, if we let (k_n) be a sequence of positive continuous functions with integral one and support in $[0, 1/n)$ and take the limit in (13.37), we see

$$K_0^\mu f = K_0^{\mu_0}(v_0^{-1} v f)$$

for each bounded continuous f. This implies (13.36) .

It follows that K_0^μ does not charge $\{x : v(x) = 0\}$, since $K_0^{\mu_0}$ doesn't. But it can't charge $\{x : v(x) = \infty\}$ either, since if it did, so would K^{μ_0}, and $K_0^{\mu_0}(v_0^{-1} v f)$ would be infinite for $f \equiv 1$, which contradicts the fact that K_0^μ is a sub-probability measure.

Finally, to see that K_t^μ is an entrance law for $({}_vP_t)$, note that $(K_t^{\mu_0})$ is an entrance law for ${}_{v_0}P_t$ by Proposition 13.43. Let k be a positive continuous function of integral one, and let f be bounded and continuous. The laws K_t^μ satisfy

$$\begin{aligned}\int_0^\infty k(t)K_t^\mu f\,dt &= \int_0^\infty k(t)K_0^{\mu_0}(v_0^{-1}vf)\,dt \\ &= \int_0^\infty k(t)K_0^{\mu_0}{}_{v_0}\hat{P}_t(v_0^{-1}vf)\,dt \\ &= \int_0^\infty dt\,k(t)\int_{\mathbf{E}} K_0^{\mu_0}(dx)\frac{v(x)}{v_0(x)}\int_E \hat{P}_t(x,dy)\frac{v(y)}{v(x)}f(y) \\ &= \int_0^\infty k(t)K_0^\mu{}_v\hat{P}_t f\,dt,\end{aligned}$$

where we have used the fact that K_0^μ does not charge the set $\{v = 0 \text{ or } \infty\}$. This holds for all such k, hence $K_t^\mu f = K_0^\mu{}_v\hat{P}_t f$ for a.e. t, and therefore for all t since both sides are right continuous in t. □

Our main theorem now follows easily.

Theorem 13.45. *Under hypothesis ($\widehat{SF}$), for any excessive function h and initial distribution μ on $\mathbf{E}_h$ of (X_t^h), $\{\hat{Y}_t, t \geq 0\}$ is a strong Markov process.*

Proof. We know that $\{\hat{Y}_t, t > 0\}$ is a v-transform of $(\hat{X}_t)$, and that its initial law is K_0^μ, which does not charge $\{v = 0 \text{ or } \infty\}$. Let us compare $(\hat{Y}_t)$ with the v-transform of $(\hat{X}_t)$ which has initial distribution K_0^μ. Call this $(\hat{Y}_t^1)$. Now $\{\hat{Y}_t^1, t \geq 0\}$ is a right-continuous strong Markov process. It has the same transition function as does $(\hat{Y}_t)$, and it has laws $K_t^1 \equiv K_0^\mu{}_vP_t = K_t^\mu$ by Lemma 13.44. But two Markov processes with the same transition function and the same absolute laws have the same distribution. □

The above theorem implicitly assumed the existence of $X_{\zeta-}$ for $(\hat{Y}_t)$ is only the reverse of $(\hat{X}_t^h)$ on the set on which $X_{\zeta-}^h$ exists. Here is one condition which assures this.

Corollary 13.46. *Under Hypothesis ($\widehat{SF}$), if h is excessive and K is a compact subset of $\mathbf{E} - \hat{B}$, then $X_{\zeta-}^h$ exists a.e. on the set $\{T_{K^c} > \zeta\}$.*

Proof. Let v_0 and μ_0 be as in (13.35). Then v_0 is continuous and strictly positive on $\bar{\mathbf{E}}_h - \hat{B}$, so it is bounded away from zero on the compact $K' = \bar{\mathbf{E}}_h \cap K$.

If $\zeta(\omega) < T_{K^c}(\omega)$, $\{X_t^h(\omega),\ 0 \leq t < \zeta\}$ takes values in K', so all its limit points as $t \uparrow \zeta(\omega)$ are also there. We claim there is exactly one such.

Points of K' are not co-branching, so the functions $U_0 f,\ f \in C(\bar{\mathbf{E}})$ separate points of K', and are continuous thanks to Hypothesis ($\widehat{SF}$). Give (X_t^h) the initial measure μ_0, and let its right-continuous reverse from ζ be $\{\tilde{X}_t,\ t > 0\}$. (We don't know about $\tilde{X}_0$ since we haven't assumed that $X_{\zeta-}$ exists.)

However, $v_0^{-1}\hat{U}_0 f$ is v_0-coexcessive and continuous on K', so that the process $\{v_0^{-1}(\tilde{X}_t)\hat{U}_0 f(\tilde{X}_t),\ t > 0\}$ is a supermartingale and thus has limits at $t = 0$ a.s. This is true for a countable family of such functions which separate points of K'. As K' is compact, this implies the existence of $\lim_{t\to 0}\tilde{X}_t = \lim_{t\uparrow\zeta} X_t^h$, a.s. on the set $\{\zeta < T_{K^c}\}$. □

13.11. Balayage and Potentials of Measures

We will now look at some potential theoretic results. We will make heavy use of h-transforms and time-reversal. For comparison, some of the same results are proved under different hypotheses in Chapter 5, using more conventional methods.

As before, (P_t) and $(\hat{P}_t)$ are strong Markov semigroups whose respective resolvents $(U_p)_{p\geq 0}$ and $(\hat{U}_p)_{p\geq 0}$ are in duality with respect to a Radon measure ξ; and $\xi(B \cup \hat{B}) = 0$, where B and $\hat{B}$ are the branching and co-branching points respectively. We suppose that the two semigroups have right-continuous strong Markov realizations (X_t) and $(\hat{X}_t)$ with values in $\boldsymbol{E}_\partial$. Their left limits exist in $\boldsymbol{E} \cup \{\partial\}$ as well, except perhaps at the lifetime. We assume as before that $(\hat{U}_p)_{p\geq 0}$ satisfies Hypothesis $(\widehat{SF})$.

We should note that the hypothesis that $\hat{U}$ is uniformly bounded is stronger than we need. On the other hand, we would automatically have it if we confined ourselves to p-potentials for some $p > 0$, and we do need some type of transience to be sure there are non-constant excessive functions. In the end, we will use Hypothesis $(\widehat{SF})$ mainly to assure that Theorem 13.45 holds. In most cases we could replace Hypothesis $(\widehat{SF})$ by the hypothesis that the conclusion of Theorem 13.45 holds.

Definition 13.47. *A potential $U\mu$ has* **support K** *if K is the closed support of μ, and $U\mu$* **sits on A** *if $\mu(A^c) = 0$; in this case, we also say that A* **supports** *$U\mu$.*

Remark 13.48. If μ is σ-finite on $\boldsymbol{E}$, then μ has the same potential as $\mu\hat{P}_0$, and $\mu\hat{P}_0$ sits on $(\boldsymbol{E} - \hat{B}) \cup \partial$. Indeed, $U(\mu\hat{P}_0)(x) = \int u(x, y)\,\mu\hat{P}_0(dy)$, which equals $\iint u(x, y)\,\mu(dz)\,\hat{P}_0(z, dy) = \int u(x, z)\,\mu(dz) = U\mu(x)$, where we have used the fact that $u(x, \cdot)$ is co-excessive. Thus a potential doesn't always determine its measure, but it does if the measure sits on $\boldsymbol{E} - \hat{B}$. Indeed

Proposition 13.49. *Let μ and ν be σ-finite measures on $\boldsymbol{E} - \hat{B}$ having a.e. finite potentials. If $U\mu = U\nu$, then $\mu = \nu$.*

Proof. Let $h = U\mu = U\nu$ and let $\eta(dx) = g(x)\xi(dx)$ be a probability measure, where $g > 0$ on $\boldsymbol{E}_h$, $g = 0$ on $\boldsymbol{E} - \boldsymbol{E}_h$, and $\hat{U}(g/h)$ is bounded.

(This last is true if, for example, g/h is bounded.) Apply Theorem 13.39 to $U\mu$ and $U\nu$:

$$
\begin{aligned}
{}_hP^\eta\{X_{\zeta-} \in dy\} &= \left(\int \eta(dx)\frac{1}{h(x)}u(x,y)\right)\mu(dy)\\
&= \left(\int \eta(dx)\frac{1}{h(x)}u(x,y)\right)\nu(dy),
\end{aligned}
$$

or

$$\hat{U}\left(\frac{g}{h}\right)(y)\,\mu(dy) = \hat{U}\left(\frac{g}{h}\right)(y)\,\nu(dy).$$

Thus $\mu = \nu$ on the set $\{y : \hat{U}(g/h)(y) > 0\}$. But both μ and ν live on this set, for if $A \in \mathcal{E}$ is such that $\mu(A) > 0$, then, since $g/h > 0$ a.e. on $\boldsymbol{E}_h$,

$$0 < \int dx\,\frac{g(x)}{h(x)}\int_A u(x,y)\,\mu(dy) = \int_A \hat{U}(g/h)\,\mu(dy)$$

and the same is true of ν, so we are done. □

According to Proposition 13.40 an excessive function is determined by the way it acts on processes. For example we can tell when an excessive function is the potential of a measure by the behavior of the h-process at its lifetime.

Proposition 13.50. *An a.e. finite excessive function h is the potential of a measure if and only if*

$$_hP^x\{\zeta < \infty,\ X_{\zeta-} \in \boldsymbol{E} - \hat{B}\} = 1, \quad x \in \boldsymbol{E}_h. \tag{13.38}$$

Proof. If h is the potential of a measure, then (13.38) holds by Theorem 13.39. Conversely, suppose (13.38) holds. Give (X_t^h) the initial measure $d\nu = g\,dx$, where g is a Borel function such that $g > 0$ on $\boldsymbol{E}_h$, $g = 0$ on $\boldsymbol{E} - \boldsymbol{E}_h$ and $\hat{U}(g/h)$ is bounded. Let $(\hat{Y}_t)$ be the reverse of (X_t^h) from its lifetime and let $v = \hat{U}(g/h)$. Then $\{\hat{Y}_t,\ t \geq 0\}$ is a v-transform of $(\hat{X}_t)$ by Corollary 13.36 and Theorem 13.45. Moreover, the initial distribution of $(\hat{Y}_t)$ is the same as the distribution of $X_{\zeta-}$. Call this distribution η. Now $(\hat{Y}_t)$ also has a finite lifetime, so we can reverse it. Let (Z_t) be the reverse of $(\hat{Y}_t)$. By (13.32) $\{X_t,\ t > 0\}$ is a w-transform of (X_t), where $w = U(\eta/v)$. But $Z_t \equiv X_t^h$, so that it is at the same time a w-transform and an h-transform of (X_t). Thus $_wU(x,\cdot) = {}_hU(x,\cdot)$ ξ-a.e. on $\boldsymbol{E}_h$.

Let $F_1, F_2, \ldots$ be the h-fine components of $\boldsymbol{E}_h$. By Proposition 13.40, for each n there exists α_n such that $h = \alpha_n w$ on F_n. Let $A_n = \{y : y \text{ is left co-regular for } F_n\}$. The A_n are disjoint and if $y \in A_n$ then

$\{x : u(x, y) > 0\}$ is contained in F_n by Exercise 17. Thus on F_n,

$$h(x) = \alpha_n w(x) = \alpha_n \int u(x, y)\frac{1}{v(y)}\,\eta(dy) = \alpha_n \int_{A_n} u(x, y)\frac{1}{v(y)}\,\eta(dy).$$

Therefore if we let $\mu_n(dy) = \alpha_n \frac{I_{A_n}(y)}{v(y)}\eta(dy)$ and set $\mu = \sum_n \mu_n$, then $h = U\mu$. □

Remark 13.51. If ${}_hP^x\{\zeta < \infty,\ X_{\zeta-} \in A\} = 1$ for some Borel set $A \subset \boldsymbol{E} - \hat{B}$, we can even affirm that h is the potential of a measure sitting on A. Proposition 13.50 is an important part of the story, but it is not all of it. For the rest of the tale, see Theorem 14.12.

13.12. The Interior Reduite of a Function

We introduced balayage in §3.4 and discussed it further in §5.1, where we characterized the equilibrium measure. We want to examine these results using entirely different methods. Duality allows us to be more precise.

Consider the following problem in electrostatics: put a fixed charge on a conducting region A, let the system go to equilibrium, and ask what the electrical potential is. Physically, we know that the electrical charge will migrate to the boundary of the region, and the equilibrium potential is the electrical potential of the resulting charge distribution. It is constant on the region A and is harmonic off A.

This problem has been well-studied and oft-generalized; it has led to rich mathematics, for it is in the intersection of the theories of classical physics, partial differential equations, complex variables and, more recently, Markov processes.

Let $S_A = \inf\{t > 0 : X_{t-} \in A\}$ and $L_A^- = \sup\{t > 0 : X_{t-} \in A\}$, with the convention that the inf of the empty set is infinity, and the sup of the empty set is zero. Note that $L_A^- = 0 \iff S_A = \infty$; otherwise $L_A^- \ge S_A$. Recall that $T_A = \inf\{t > 0 : X_t \in A\}$. It is possible that $S_A = \zeta$, whereas the only way that $T_A = \zeta$ is if $\partial \in A$.

If h is an a.e. finite excessive function, and if $A \in \mathcal{E}$, let

$$c_A(x) = {}_hP^x\{S_{A-\hat{B}} < \infty\} = {}_hP^x\{L^-_{A-\hat{B}} > 0\}. \tag{13.39}$$

Then $c_A(x)$ is h-excessive, so by Proposition 11.7 there exists an excessive function which equals $c_A(x)h(x)$ on $\{h(x) < \infty\}$. Since h is a.e. finite, the extension is unique. (It is possible to define $p_A h$ for an h which is not a.e. finite by simply taking the smallest extension. It is not clear that this is interesting, however, and we will not pursue it.)

Definition 13.52. *Let h be an a.e. finite excessive function. The* **interior reduite** *$p_A h$ of h on A is the (unique) excessive function which equals $c_A h$ on $\boldsymbol{E}_h$.*

Thus

$$p_A h(x) = h(x)_h P^x\{S_{A-\hat{B}} < \infty\},\ x \in \boldsymbol{E}_h.$$

Remark 13.53. It is possible to define $p_A h$ without using the h-transform; it is connected with the expectation of $h(X_{S_A})$, but the exact connection is not simple, (See the exercises.) This is one reason we gave the somewhat indirect definition above. Another, more important, reason is that this definition works even when A is a polar set, which is one of the most important cases.

Here are some elementary facts about $p_A h$:

Proposition 13.54. *Let g and h be a.e. finite excessive functions and let A, $C \in \mathcal{E}$. Then*

(i) $p_A h \le h$ *and* $p_A h = h$ *on* A^ℓ;
(ii) $g \le h \Longrightarrow p_A g \le p_A h$; $p_A(g + h) = p_A g + p_A h$;
(iii) *if* $A \subset C$ *then* $p_A(p_C h) = p_A h$;
(iv) $p_A h(x) = \sup\{p_K h(x) : K \subset A \text{ compact}\}$.

Proof. It is enough to prove these for x such that $g(x) + h(x) < \infty$. (i) is clear from (13.39) since $c_A(x) \le 1$ and $c_A = 1$ on A^ℓ. (ii) is almost clear from Exercise 22, but there is one delicate point. The sequence (τ_n) of predictable stopping times in the hint to the exercise might depend on the measure $_h P^x$. However, if we do the decomposition for $_{(g+h)}P^x$, we see from Proposition 11.10 that $_{(g+h)}P^x = (g(x) + h(x))^{-1}(g(x)_g P^x + h(x)_h P^x)$, so that if (τ_n) announces S_1 for $_{(g+h)}P^x$, it announces S_1 for both $_g P^x$ and $_h P^x$. To prove (iii), note that $A \subset C \Longrightarrow L_C^- \ge L_A^-$, while the $p_C h$-transform of X has the same law as the h-transform killed at L_C, given that $L_C > 0$. (See Theorem 11.26.) Thus, if $0 < p_C h(x) \le h(x) < \infty$,

$$_{(p_C h)}P^x\{S_A < \infty\} = {}_h P^x\{S_A < \infty \mid L_C > 0\}$$

$$= \frac{_h P^x\{S_A < \infty\}}{_h P^x\{L_C > 0\}} = \frac{\frac{p_A h(x)}{h(x)}}{\frac{p_C h(x)}{h(x)}} = \frac{p_A h(x)}{p_C h(x)}$$

where we used the fact that $A \subset C$ so $S_A < \infty \Longrightarrow L_C > 0$. Multiply both sides by $p_c h$ to get (iii).

Finally, (iv) follows from the fact that for any initial distribution μ on $\boldsymbol{E}_h$, there are $K_n \subset A$ such that $_h P^\mu\{S_{K_n} \downarrow S_A\} = 1$. $\square$

The next proposition is not elementary.

Proposition 13.55. *Assume Hypothesis $(\widehat{SF})$ and that $(\hat{X}_t)$ is quasi-left continuous on $(0, \hat{\zeta})$ (or equivalently, that (X_t) satisfies Hypothesis (B)). Let h be an a.e. finite excessive function. Let (K_n) be a decreasing sequence of* *compacts with intersection K. Then $p_{K_n}h(x) \to p_K h(x)$ for each $x \in \{h < \infty\} \cap (K - K^\ell)$.*

Proof. Since the K_n decrease, the times S_{K_n} increase. Let $S = \lim_n S_{K_n}$. Note that if $S(\omega) < \infty$, then $S(\omega) \le \zeta(\omega)$. Moreover, $p_K h = p_K(p_{K_1}h)$ by Proposition 13.54, so at the price of replacing h by $p_{K_1}h$, we may assume that h is the potential of a measure, and hence that $t \mapsto X^h_{t-}$ is left continuous and has a finite lifetime.

Let $x \in E_h \cap (K - K^\ell)$ and note that $p_{K_n}h(x) = h(x)_h P^x\{S_{K_n} < \infty\}$, so the proposition will be proved if we show that $S_K = S$ with ${}_hP^x$-probability one. There are several possibilities.

(i) If $S(\omega) = 0$, then for each n, $S_{K_n}(\omega) = 0$, so there exists $t \downarrow 0$ such that $X_{t-} \in K_n$. By right continuity, $X_0 \in \cap_n K_n = K$. But $X_0 = x \notin (K - K^\ell)$, so $x \in K^\ell$, hence ${}_hP^x\{S_K = 0\} = h(x)^{-1}E^x\{h(X_0); S_K = 0\} = 1$, so that $S_K = S$, ${}_hP^x$-a.e. on $\{S = 0\}$.

(ii) If $S(\omega) > 0$ and $S_{K_n}(\omega) < S(\omega)$ for all n, then, as one of $X_{S_{K_n}-}(\omega)$ or $X_{S_{K_n}}(\omega)$ is in K_n, left continuity implies that $X_{S-}(\omega) \in \cap_n K_n = K$, hence $S_K(\omega) = S(\omega)$.

(iii) In the remaining case, there is an n such that $S_{K_n}(\omega) = S(\omega) > 0$. Then one of X_{S-} or X_S is in K. If $X_{S-} \in K$, then $S_K = S$ again. If $X_{S-}(\omega) \notin K$ then $X_S \in K$, so either $X_S \in K^\ell$ or $X_S \in K - K^\ell$. On $\{X_S \in K^\ell\}$ the strong Markov property implies that $S_K \circ \theta_S = 0$, and once again $S_K = S$, ${}_hP^x$-a.s.

Finally, on $\{X_S \in K - K^\ell\}$, $X_S \ne X_{S-}$ so by the quasi-left continuity of $(\hat{X}_t)$,

$$
\begin{aligned}
&{}_hP^x\{X_S \in K - K^\ell,\ X_{S-} \ne X_S\} \\
&\qquad\qquad \le {}_hP^x\{\exists t : 0 < t < \zeta,\ X_t \in K - K^\ell,\ X_{t-} \ne X_t\}.
\end{aligned}
$$

The reverse of (X_t) from its (finite) lifetime is a $u(x, \cdot)$-transform of $(\hat{X}_t)$ with initial measure ν, say, so this is

$$
= {}_x\hat{P}^\nu\{\exists t : 0 < t < \hat{\zeta},\ \hat{X}_{t-} \in K - K^\ell,\ \hat{X}_{t-} \ne \hat{X}_t\} = 0
$$

by Proposition 13.22 since $K - K^\ell$ is semipolar while $(\hat{X}_t)$, and therefore its transform, is quasi-left-continuous on $(0, \hat{\zeta})$. □

Definition 13.56. *An excessive function h is* **harmonic** *if, for each compact $K \subset E - \hat{B}$, $p_{K^c}h = h$.*

Here is a characterization of harmonic functions in the same spirit as the characterization of potentials in Proposition 13.50.

Proposition 13.57. (i) *An excessive function h is harmonic if and only if*

$$ {}_hP^x\{X_{\zeta-} \text{ exists in } \boldsymbol{E} - \hat{B},\ 0 < \zeta < \infty\} = 0, \quad \forall\, x \in \boldsymbol{E}_h. \tag{13.40} $$

(ii) *If h is the potential of a measure and $h < \infty$ a.e., then there exists an increasing sequence of compacts $K_n \subset \boldsymbol{E} - \hat{B}$ such that $p_{K_n^c}h \to 0$ a.e. as $n \to \infty$.*

Proof. We prove (ii) first. Let μ be a reference probability measure on $\boldsymbol{E}_h$, i.e. a measure which charges all fine open subsets of $\boldsymbol{E}_h$. Then $X^h_{\zeta-}$ exists in $\boldsymbol{E} - \hat{B}$, ${}_hP^\mu$-a.e.. As $t \mapsto X^h_{t-}$ is lcrl, the set $\{X^h_{t-}(\omega),\ 0 < t < \zeta\}$ has compact closure in $\boldsymbol{E} - \hat{B}$. Thus there exists an increasing sequence of compacts K_n with the property that ${}_hP^\mu\{S_{K_n^c} < \infty\} \to 0$. (If $\boldsymbol{E} - \hat{B}$ is open, we can let $K_n = \{x : d(x, (\bar{\boldsymbol{E}} - \boldsymbol{E}) \cup \hat{B}) \geq 1/n\}$. In general, we can use a capacitability theorem as we did in §3.3 to show the existence of the K_n.)

Thus, for a.e. x, $p_{K_n^c}h(x) = h(x)\,{}_hP^x\{S_{K_n^c} < \infty\} \to 0$.

To prove (i), let K be compact. By Corollary 13.46 $X_{\zeta-}$ exists in $E - \hat{B}$ a.s. on $\{S_{K^c} > \zeta\}$. Thus if h satisfies (13.40), ${}_hP^x\{S_{K^c} \leq \zeta\} = 1 \Longrightarrow p_{K^c}h(x) = h(x)$. Conversely, the argument proving (ii) shows that if the probability in (13.40) is strictly positive, we can find a compact $K \subset \boldsymbol{E} - \hat{B}$ such that $p_{K^c}h(x) < h(x)$ for some $x \in \boldsymbol{E}_h$. □

This brings us to the Riesz decomposition theorem.

Theorem 13.58. *(Riesz Decomposition) An a.e. finite excessive function h can be written uniquely as the sum of a harmonic function and a potential.*

Proof. Let

$$ L = \begin{cases} \zeta & \text{if } X^h_{\zeta-} \in \boldsymbol{E} - \hat{B} \text{ and } \zeta < \infty \\ 0 & \text{otherwise.} \end{cases} $$

and let $M = \zeta - L$ (with the convention that $\infty - \infty = 0$!). Both L and M are co-optional. Let (Y_t) be (X^h_t) killed at L. Then by Theorem 11.26 (Y_t) is a c_L-transform of (X^h_t), where $c_L(x) = {}_hP^x\{L > 0\}$, and it is therefore an hc_L-transform of (X_t) by Proposition 11.28. Since $Y_{\zeta-} = X^h_{L-} \in \boldsymbol{E} - \hat{B}$ and $\zeta < \infty$, $u \overset{\text{def}}{=} hc_L$ is the potential of a measure by Proposition 13.50.

Similarly, the process (Z_t) which is (X^h_t) killed at M is an hc_M-transform of (X_t). Evidently $Z_{\zeta-}$ is not in $\boldsymbol{E} - \hat{B}$. Thus $v = hc_M$ is harmonic by Proposition 13.57.

Now $x \in \boldsymbol{E}_h - \hat{B} \Longrightarrow 1 = {}_hP^x\{\zeta > 0\} = c_L(x) + c_M(x)$ hence $h(x) = u(x) + v(x)$. But now $h \equiv u + v$ by hypothesis (L).

Uniqueness is immediate, for if $h = u + v = u' + v'$ are two such, there exist compacts K_n such that $p_{K_n^c}u \to 0$ and $p_{K_n^c}u' \to 0$, so that $v = v' = \lim_n p_{K_n^c}h$. □

Proposition 13.59. *Let h be an a.e. finite excessive function and let K be compact in $\mathbf{E} - \hat{B}$. Then $p_K h$ is the potential of a measure supported by K.*

Proof. Notice that $X^h_{(L_K^-)-} \in K$ by left continuity, as long as $L_K^- > 0$. (Corollary 13.46 assures us there is no trouble at ζ.) Since the process killed at L_K^- is a $p_K h$-transform of X^h by Theorem 11.26, we conclude by Proposition 13.50 that $p_K h$ is the potential of a measure, and the remark following it assures us that this measure is supported by K. □

Remark 13.60. The proof also shows that the balayage operator $P_K h(x) = E^x\{h(X_{T_K})\}$ which we introduced in §3.4 is the potential of a measure, but this measure need not be supported by K. See Exercise 19.

Here is an easy consequence of Proposition 13.59.

Corollary 13.61. *If $y \in \mathbf{E} - \hat{B}$, there is a unique (up to proportionality) minimal excessive function h with pole y, such that for $x \in \mathbf{E}_h$, $P_t h(x) \to 0$ as $t \to \infty$, namely $u(\cdot, y)$.*

Proof. If h is excessive, $(1/h(x))P_t h(x) \to 0 \Longrightarrow {}_hP^x\{\zeta < \infty\} = 1$. If h is a minimal excessive function with pole y, then ${}_hP^x\{X^h_{\zeta-} = y,\ \zeta < \infty\} = 1$, all $x \in \mathbf{E}_h$ by Theorem 11.21. Therefore by Proposition 13.59 h is a potential supported by $\{y\}$, i.e. $h = au(\cdot, y)$ for some $a > 0$. This shows uniqueness.

On the other hand, if $u(\cdot, y) = f(\cdot) + g(\cdot)$, where f and g are excessive, then by Corollary 13.37,

$$u(\cdot, y) = p_{\{y\}}u(\cdot, y) = p_{\{y\}}f + p_{\{y\}}g = au(\cdot, y) + bu(\cdot, y),$$

since both $p_{\{y\}}f$ and $p_{\{y\}}g$ are potentials supported by $\{y\}$. □

The special case where h itself is the potential of a measure is interesting, for then we can identify p_A exactly. Recall from §1.5 that the *debut* of a set A is the stopping time $D_A = \inf\{t \geq 0 : X_t \in A\}$, which differs from T_A in that $t = 0$ is included in the infimum. Define the kernel $\hat{D}_A(x, C) = \hat{P}^x\{\hat{X}_{D_A} \in C\}$. (Notice that this involves the dual process.)

Proposition 13.62. *Suppose $h = U\mu$ is an a.e. finite potential. If $A \in \mathcal{E}$ has compact closure in $E - \hat{B}$, then $p_A h = U\hat{\mu}$, where $\hat{\mu} = \mu\hat{D}_A$.*

Proof. We will calculate the distribution of $X^h_{(L_A^-)-}$ two ways. We first reverse (X^h_t) from ζ. Let $(\hat{Y}_t)$ be the right continuous reverse of (X^h_t) from ζ. Note that $\hat{Y}_{D_A} = X^h_{(L_A^-)-}$ on the set $\{L_A^- > 0\}$. Suppose that $X^h_0 = x$. Then $(\hat{Y}_t)$

is a $u(x, \cdot)$ transform of $(\hat{X}_t)$ by Corollary 13.36, and by Theorem 13.34, $\hat{Y}_0 = X^h_{\zeta-}$ has the distribution

$$\nu(dy) = \frac{1}{h(x)} u(x, y)\, \mu(dy).$$

Thus $X^h_{(L_A^-)-} = \hat{Y}_{D_A}$ has the distribution

$$\int \nu(dz) \frac{1}{u(x, z)} \hat{D}_A(z, dy)\, u(x, y) = \frac{1}{h(x)} u(x, y)\, \mu\hat{D}_A(dy).$$

Now let us calculate this directly. The process (X^h_t) killed at L_A^- is a $p_A h$-transform of (X_t), and $X^h_{(L_A^-)-} \in \boldsymbol{E} - \hat{B}$, so $p_A h = U\hat{\mu}$ for some measure $\hat{\mu}$. But then, by Theorem 13.39, the distribution of $X^h_{(L_A^-)-}$ must be

$$\frac{1}{h(x)} u(x, y) \hat{\mu}(dy).$$

Compare this with the above calculation. Since x is arbitrary, we see that $\hat{\mu} = \mu\hat{D}_A$. □

Corollary 13.63. *If h is excessive and a.e. finite and if $A \in \mathcal{E}$ has compact closure in $E - \hat{B}$, then $p_A h$ is a potential supported by A together with the points co-regular for A.*

Proof. By Proposition 13.54 (iii), $p_A h = p_A(p_{\bar{A}} h)$. Since $\bar{A}$ is compact, $p_{\bar{A}} h$ is the potential of a measure, say $p_{\bar{A}} h = U\mu$. Thus $p_A h = U(\mu\hat{D}_A)$. But for each x, $\hat{D}_A(x, \cdot)$ sits on the union of A with the points co-regular for A. □

Remark 13.64. The appearance of the dual process in the balayage formulas may initially appear puzzling, but the example of particles, holes, and heat conduction that we discussed in the introduction to this chapter may provide some insight. Suppose there is a source of heat located in the complement of a set A. We can think of the heat source as an emitter of energetic particles which transfer the heat. The temperature at a given point is proportional to the number of particles which are emitted by the source and pass by the point in unit time. These particles originate in A^c, so in order to pass by a point $x \in A$, they must first reach A – which they do at the time D_A – and then continue on to reach x. As far as an observer at x is concerned, one could replace the original distribution of particles, say μ, by their distribution at time D_A. But remember that the particles are governed by the dual semigroup $(\hat{P}_t)$ (it is the "holes" which are governed by (P_t)!) so this distribution is $\mu\hat{D}_A$, not μD_A.

Here is an important characterization of p_A.

Theorem 13.65. *If h is an a.e. finite excessive function and if $A \in \mathcal{E}$, then*

$$p_A h = \sup\{U\mu : U\mu \leq h,\ \operatorname{supp}(\mu) \text{ is compact in } A - \hat{B}\}.$$

Proof. By Proposition 13.54 (iv), $p_A h = \sup\{p_K h : K \subset A \text{ compact}\}$. Since $p_K h$ is a potential with compact support in $A - \hat{B}$, $p_A h$ is certainly dominated by the right-hand side. But if $U\mu \leq h$, $p_A h \geq p_A U\mu = U(\mu \hat{D}_A)$; if μ has compact support in $A - \hat{B}$, then $\mu \hat{D}_A = \mu$, so this equals $U\mu$, and we are done. □

Corollary 13.66. *If h is an a.e. finite excessive function and if $A \subset C \in \mathcal{E}$, then*

$$p_A p_C h = p_C p_A h = p_A h.$$

Proof. From Proposition 13.54 we need only prove the second equality. Since $p_A h \geq p_C p_A h$, we need only show "$\leq$". But

$$\begin{aligned} p_A h &= \sup\{U\mu : U\mu \leq h,\ \operatorname{supp}(\mu) \text{ is compact in } A - \hat{B}\} \\ &\leq \sup\{U\mu : U\mu \leq p_A h,\ \operatorname{supp}(\mu) \subset C - \hat{B}\} \\ &= p_C p_A h. \end{aligned}$$

□

Here are two results on domination of potentials. The first is nearly trivial. The second will be sharpened later in Corollary 13.83.

Proposition 13.67. *An excessive function dominated by a potential is itself a potential.*

Proof. If h is excessive and $h \leq f$, where f is a potential, then for any compact K, $p_K f \geq p_K h = p_K u + p_K v$, where u is a potential and v is harmonic. Choose K_n so that $p_{K_n^c} f \to 0$, which we can do by Proposition 13.57 (ii). Then $p_{K_n^c} v \to 0$. As v is harmonic, $v = p_{K_n^c} v = 0$. □

Proposition 13.68. *Let h be excessive and let f be an a.e. finite potential. If h dominates f on a neighborhood of its support, $h \geq f$ everywhere.*

Proof. Let $\Gamma = \{y : h(y) \geq (n-1) f(y)/n\}$. Fix $x \in E_f$ and write $f = U\mu$. Then by hypothesis, Γ contains a neighborhood of $\operatorname{supp}(\mu)$. By Theorem 13.39, the distribution of $X^f_{\zeta-}$ is $(1/f(x)) u(x, y)\, \mu(dy)$, which sits on Γ. Thus we have ${}_f P^x\{T_\Gamma < \zeta\} = 1$. But now ${}_h X^f_t / f(X^f_t)$ is a supermartingale, so

$$\frac{h(x)}{f(x)} \geq {}_f E^x \left\{ \frac{h}{f}(X^f_{T_\Gamma}) \right\} \geq 1 - \frac{1}{n}.$$

Therefore $h \geq f$ on E_f. But clearly $h \geq f$ on $\{f = 0\}$, and $\{h = \infty\}$ has ξ-measure zero by hypothesis. Thus $h \geq f$ a.e., and hence, $h \geq f$ everywhere by Hypothesis (L). $\square$

The case $h \equiv 1$ is important. We discussed it in §5.1, and we will look at it again in our present setting. We will only consider it briefly; we will return to it later when we discuss capacity.

Define $C_A(x) = p_A 1(x)$. Then C_A is called the **equilibrium potential** of A. By Corollary 13.63 it is the potential of a measure, μ_A, called the **equilibrium measure**. This sits on $A \cup \hat{A}^r$. It ties in with the value of the process at its last exit from A, or more precisely, with the time $L_A^- = \sup\{t > 0 : X_t- \in A\}$. The exact relation is given by the following.

Proposition 13.69. *Let $A \in \mathcal{E}$ have compact closure in **E**. The equilibrium potential of A satisfies:*

(i) $C_A = U\mu_A$, *where μ_A is a measure which sits on $A \cup \hat{A}^r$;*
(ii) *for $x \in \mathbf{E}$, the distribution of $X_{(L_A^-)-}$ is given by*

$$P^x\{X_{(L_A^-)-} \in dy\} = C_A(x)^{-1} u(x, y)\, \mu_A(dy). \tag{13.41}$$

Proof. (i) is just a re-statement of the discussion above. To see (ii), notice that L_A^- is co-optional and $L_A^- > 0 \iff S_A < \infty$, so that $P^x\{S_A < \infty\} = p_A 1(x) = C_A(x)$. By Theorem 11.26, the process killed at L_A^- is a $c_{L_A^-}$-transform of (X_t), but $c_{L_A^-} = C_A$. Thus the left-hand limit of the transformed process at its lifetime is just the left-hand limit of the original process at L_A^-. Then (13.41) follows from Theorem 13.39. $\square$

13.13. Quasi-left-continuity, Hypothesis (B), and Reduites

The interior reduite $p_K h$ is the largest potential supported by K and dominated on K by h. Under more stringent hypotheses it equals $P_A h$, and both equal the exterior reduite, which we will define below. The fact that these three are equal is one of the big theorems of the subject, and one of Hunt's triumphs was to prove it for Hunt processes.

It may not hold for more general processes, however. Exercise 19 gives an example satisfying our present assumptions in which all three are different. Question: what extra hypotheses do we need to prove this equality? We will see that in a certain sense, we need additional time-symmetry. This is a theme: to get nicer and nicer results, we need more and more symmetry.

The dual properties of quasi-left continuity and Hypothesis (B) are keys. (See Theorem 13.24.) They affect the deeper relations between hitting probabilities and potentials. It is important to understand how.

Let h be an a.e. finite excessive function and $A \subset \boldsymbol{E}$. Let $\mathcal{R}(h, A)$ be the class of all excessive functions which are greater than or equal to h on some open neighborhood of A, where the neighborhood may depend on the function. Define

$$R_A^0 h(x) \stackrel{\text{def}}{=} \inf\{g(x) : g \in \mathcal{R}(h, A)\}.$$

Thanks to Hypothesis (L), $R_A^0 h$ is supermedian. Its excessive regularization is

$$R_A h(x) \stackrel{\text{def}}{=} \lim_{t \downarrow 0} P_t R_A^0 h(x).$$

Then $R_A h$ is excessive, and the set $\{x : R_A h(x) \neq R_A^0 h(x)\}$ is semipolar and therefore of ξ-measure zero. (See §3.6 Theorem 1.)

Definition 13.70. *$R_A h$ is called the* **exterior reduite** *(or reduced function)* **of h over A**.

The exterior reduite is defined for all sets A. Let us compare it with $p_A h$ and $P_A h$, both defined for Borel sets or, slightly more generally, analytic sets. Let us start with a slight refinement of Theorem 3 of §3.8. That stated that T_A and S_A (called T_A^- there) were the same for a Hunt process satisfying Hypothesis (B). A Hunt process is quasi-left-continuous, so quasi-left continuity and Hypothesis (B) together imply that $S_A = T_A$. We want to see what each implies separately. We will shorten the proof somewhat by unashamed use of the Section Theorem.

Lemma 13.71. *Let $A \in \mathcal{E}$. Then*

(i) *if (X_t) is quasi-left-continuous, $T_A \leq S_A$ a.s.;*
(ii) *if (X_t) satisfies Hypothesis (B), $S_A \leq T_A$.*

Proof. (i) First note that there exists a sequence (τ_n) of predictable times converging to S_A on the set $\{S_A < \infty\}$. To see this, note that it is enough to prove it on the set $\{S_A < M\}$ for an arbitrary M, hence we might as well assume $S_A < 1$ whenever it is finite. Let J_{kn} be the half-open interval $J_{kn} = (k/n, (k+1)/n]$. For each n and $k < n$ apply the Section Theorem to the left-continuous process $\{X_{t-}, \ t \in J_{kn}\}$: there exists a predictable time τ_{kn} such that $P\{X_{\tau_{kn}-} \in A\} \geq P\{\exists t \in J_{nk} : X_{t-} \in A\} - 2^{-n}$. Set $\tau_{kn} = \infty$ on the set $\{\tau_{nk} \notin J_{kn}\} \cap \{X_{\tau_{kn}-} \notin A\}$. The result will still be predictable by Exercise 5 of §1.3. Then $\tau_n \stackrel{\text{def}}{=} \min_{k \leq n} \tau_{kn}$, being a finite minimum of predictable times, is predictable, and $X_{\tau_n} \in A$ on $\{\tau_n < \infty\}$. Moreover, $S_A \leq \tau_n \leq S_A + 1/n$ except for a set of probability at most $n 2^{-n}$. Clearly $\tau_n \to S_A$, as claimed.

Now by quasi-left-continuity, $X_{\tau_n} = X_{\tau_n -} \in A$, so that $T_A \le \tau_n$. Let $n \to \infty$ to see that $T_A \le S_A$.

(ii) Suppose that (X_t) satisfies Hypothesis (B). As in (i) the Section Theorem will produce stopping times – not necessarily predictable – (σ_n) such that $\sigma_n \to T_A$ and $X_{\sigma_n} \in A$ a.s. on $\{\sigma_n < \infty\}$. We claim $S_A \le \sigma_n$ for all n, which will prove $S_A \le T_A$. Clearly $S_A \le \sigma_n$ on the set $\{X_{\sigma_n -} = X_{\sigma_n} \in A\}$. On the set $\{X_{\sigma_n} \in A^\ell\}$, the strong Markov property implies that $S_A \circ \theta_{\sigma_n} = 0$, so once again $S_A \le \sigma_n$. The remaining case, $X_{\sigma_n} \in A - A^\ell$ and $X_{\sigma_n -} \ne X_{\sigma_n}$, has probability zero, since

$$P\{X_{\sigma_n} \in A - A^\ell,\ X_{\sigma_n -} \ne X_{\sigma_n}\} \le P\{\exists t > 0 : X_t \in A - A^\ell,\ X_{t-} \ne X_t\};$$

but this vanishes by Hypothesis (B) since $A - A^\ell$ is semipolar (§3.8 (7).) Thus $S_A \le \sigma_n$ a.s., and the conclusion follows. □

This brings us to the first of two main results of this section. Recall that an excessive function is **regular** if $t \mapsto f(X_t)$ is continuous wherever (X_t) is.

Theorem 13.72. *Assume Hypothesis* $(\widehat{SF})$, *let h be an a.e. finite excessive function, let $G \subset E$ be open and let $K \subset E$ be compact. Then*

(i) $P_G h = p_G h$;
(ii) *if* $G \cap (B - B_\partial) = \emptyset$, *then* $p_G h = P_G h = R_G h$;
(iii) *suppose that (X_t) satisfies Hypothesis (B). If there exists an open neighborhood G of K such that $G \cap (B - B_\partial) = \emptyset$ then $p_K h = R_K h$;*
(iv) *if (X_t) is quasi-left continuous and also satisfies Hypothesis (B), and if h is bounded and regular, then $p_K h = P_K h$; if in addition there exists an open neighborhood G of K such that $G \cap (B - B_\partial) = \emptyset$, then $p_K h = R_K h = P_K h$.*

Remark 13.73. The hypotheses become stronger as we go down the list, and by (iv) we essentially have Hunt processes in duality, since if (X_t) satisfies Hypothesis (B), $(\hat{X}_t)$ must be quasi-left continuous, at least on $(0, \zeta)$. By the same token, we could replace Hypothesis (B) in (iii) by quasi-left-continuity of $(\hat{X}_t)$ on $(0, \zeta)$. In (iv), (X_t) is quasi-left-continuous on $(0, \infty)$, not just on $(0, \zeta)$.

Proof. (i) $X_t \in G \Longrightarrow \exists s \downarrow t \ni X_{s-} \in G \Longrightarrow S_G \le t$;
$X_{t-} \in G \Longrightarrow \exists s \uparrow t \ni X_s \in G \Longrightarrow T_G \le t$;
thus $S_G = T_G$. Since neither T_G nor S_G can equal ζ, we have for $x \in \boldsymbol{E}_h$ that

$$p_G h(x) = h(x)_h P^x\{S_G < \zeta\} = h(x)_h P^x\{T_G < \zeta\} = E^x\{h(T_G)\} = P_G h(x).$$

Since both sides are excessive and vanish on $\{h = 0\}$, they are equal everywhere.

(ii) If $x \in G - B$ and if $h(x) < \infty$, then $P^x\{T_G = S_G = 0\} = 1$, so $p_G h(x) = h(x)$. If $x \in B_\partial$, $P_G h(x) = h(x) = 0$, so $p_G h = h$ on G. Thus $p_G h \in \mathcal{R}(h, G)$, so it is included in the inf, which implies that $R^0_G h \leq p_G h$.

Conversely, if $g \in \mathcal{R}(h, G)$, then

$$p_G h = \sup_{K \subset G \text{ cpct}} p_K h \leq \sup_{K \subset G \text{ cpct}} p_K g \leq g,$$

by Propositions 13.54 and 13.68. Thus $p_G h \leq R^0_G h$. But $p_G h$ is excessive, so after regularization, we see that $p_G h \leq R_G h$, hence there is equality.

(iii) If $g \in \mathcal{R}(h, K)$, then $g \geq h$ on some open $G \supset K$. Then $p_K h \leq p_G h \leq p_G g \leq g$. We may assume $G \cap (B - B_\partial) = \emptyset$. Then by (ii), $p_K h \leq p_G h = R_G h$.

Now let G_n be a decreasing sequence of open relatively compact neighborhoods of K, all contained in G, such that $G \supset G_n \supset \bar{G}_{n+1} \supset \dots \downarrow K$. Then for each n, $p_{\bar{G}_n} h \geq p_{G_n} h = R_{G_n} h \geq R^0_K h \geq p_K h$. Let $n \to \infty$ and notice that by Proposition 13.55 that if $x \in K - K^\ell$, $\lim_n p_{\bar{G}_n} h(x) = p_K h(x)$. Thus $R^0_K h = p_K h$ except possibly on the semipolar set $K - K^\ell$. It follows by regularization that $R_K h = p_K h$.

(iv) Let G_n be as above. Equality is clear on $\{h = 0\}$, so suppose $h(x) > 0$. Under the given hypotheses, $S_{G_n} = T_{G_n} \leq S_K$, and $T_{G_n} < S_K$ on $\{S_K = \zeta\}$ so

$$p_K h(x) = h(x)_h P^x\{S_K < \infty\} = \lim_n h(x)_h P^x\{T_{G_n} < \zeta\} = \lim_n E\{h(X_{T_{G_n}})\}.$$

Let $S = \lim_n T_{G_n}$ and note that $S \leq \zeta$ on $\{S < \infty\}$. By quasi-left continuity, $X_S = \lim X_{T_{G_n}}$, so $X_S \in \cap_n \bar{G}_n = K$. Thus $S = T_K$. Since h is regular, $h(X_t)$ is continuous whenever (X_t) is, and in particular, $h(X_{T_K}) = \lim h(X_{T_{G_n}})$. Since h is bounded, we can go to the limit in the expectation to see this is

$$= E^x\{h(X_{T_K})\} = P_K h(x).$$

The second statement now follows from (iii). □

This brings us to the second main result of this section.

Theorem 13.74. *Assume Hypothesis* $(\widehat{SF})$ *and let* h *be an a.e. finite excessive function. Assume that* $B - B_\partial = \emptyset$.

(i) *If* (X_t) *satisfies Hypothesis (B), then* $p_A h = R_A h$ *for all Borel (even analytic) sets* $A \subset \mathbf{E}$.

(ii) *If* (X_t) *is both quasi-left-continuous and satisfies Hypothesis (B), and if* h *is bounded and regular, then* $p_A h = R_A h = P_A h$ *for all Borel (even analytic) sets* $A \subset E$.

Proof. There is a big idea in this proof: Choquet's Theorem. It will give us both (i) and (ii) by the same argument.

By (ii) and (iii) of Theorem 13.72, $p_A h = R_A h$ for both open and compact A. Let η be a probability measure on E which is equivalent to ξ, such that $\int h \, d\eta < \infty$. Let $C(A) = \int R_A h(x)\, \eta(dx)$. We claim that $C(A)$ is a capacity. We must check the conditions of Theorem 5, §3.3. First, $C(A)$ is defined for all $A \subset E$. Next, if $K_n \downarrow K$ are compacts, $p_{K_n} h \downarrow p_K h$ a.e. by Proposition 13.55 (In fact there is convergence outside of a semipolar.) Therefore $C(K_n) \downarrow C(K)$.

Noting that $C(A) = \int h(x)_h P^x\{S_A < \infty\}\, \eta(dx)$, we can copy the proof of Proposition 3 in §3.3, simply changing X_s to X_{s-}, to conclude that C is strongly subadditive. Now the proof of Lemma 4 of §3.3 applies word-for-word – it only depends on strong sub-additivity – to see that $A_n \uparrow A \Longrightarrow C(A_n) \uparrow C(A)$, proving that C is a capacity. But now Choquet's Theorem (§3.3) implies that each analytic (and therefore, each Borel) set is capacitable. That is, if A is analytic,

$$\begin{aligned} C(A) &= \sup\{C(K) : A \supset K \text{ compact}\} \\ &= \inf\{C(G) : A \subset G \text{ open}\}. \end{aligned}$$

It follows that for each analytic set A,

$$\begin{aligned} \int p_A h(x)\, \eta(dx) &= \sup\left\{ \int p_K h(x)\, \eta(dx) : \ K \subset A \right\} \\ &= \inf\left\{ \int R_G h(x)\, \eta(dx) : A \subset G \right\} \\ &= \int R_A h(x)\, \eta(dx). \end{aligned}$$

But $p_A h \le R_A h$, so this implies equality η-a.e., therefore ξ-a.e., therefore everywhere.

This proves (i), but note that all we used was that $p_A h$ agreed with $R_A h$ on both open and compact sets. Thus the same argument holds when we replace $p_A h$ by $P_A h$ and use (ii) and (iv) of Theorem 13.72. □

13.14. Fine Symmetry

Time is not symmetric: entropy and age increase with it. Nor is duality: duality implies not symmetry, but something more like a mirror reflection. However, the laws of physics appear to be symmetric in time, and many of the stochastic processes used to model physical situations – such as Brownian motion and stable processes – are self-dual. It seems to be a fact – physical, mathematical, or metaphysical – that nice things become nicer as they approach symmetry, and they reach an apogee when there is just the right amount of tension between symmetry and asymmetry. So we will move close to symmetry, but we shall take care not to reach it.

First, let us put the two processes on the same footing. We assume that both resolvents, (U_p) and $(\hat{U}_p)$ satisfy Hypothesis $(\widehat{SF})$, so that both are strong

Feller. This will hold throughout this section, but we will re-emphasize it by saying that "(SF) and $(\widehat{SF})$ hold."

This has several consequences: for example, both excessive and co-excessive functions must be lower-semi-continuous. We will not use this fact as such; what we will use are two other results, first, that the conclusion of Theorem 13.45 holds for both (X_t) and $(\hat{X}_t)$, so that we can connect the limits at the lifetime with both co-potentials and potentials; and the following, which is little more than a remark:

Lemma 13.75. *Suppose (SF) and $(\widehat{SF})$ hold. Then $B = B^-$ and $\hat{B} = \hat{B}^-$.*

Proof. Let $f \in b\mathcal{E}$. Since $U_p f$ is continuous, $t \mapsto Uf(\hat{X}_t)$ is right-continuous, so that the function $\bar{U}_p f$ defined in (13.14) equals $U_p f$ itself. It follows that the transition function $\bar{P}_t$ of Theorem 13.19 is identical to P_t. Thus the definitions of branching and left-branching points coincide. The same holds for co-branching points by symmetry. □

Proposition 13.76. *Suppose (SF) and $(\widehat{SF})$ hold. Then (X_t) and $(\hat{X}_t)$ are both quasi-left-continuous on $(0, \zeta)$ (and therefore both satisfy Hypothesis (B)) if and only if B is co-polar and $\hat{B}$ is polar.*

Proof. This is immediate from Proposition 13.28 and the fact that branching and left-branching points coincide for both (X_t) and $(\hat{X}_t)$. □

This allows us to express our hypotheses purely in terms of B and $\hat{B}$. Let us denote the symmetric difference of sets A and B by $A \vartriangle C = (A - C) \cup (C - A)$.

Proposition 13.77. *Suppose (SF) and $(\widehat{SF})$ hold and that B is copolar and $\hat{B}$ is polar. If $A \in \mathcal{E}$, then*

(i) *A is polar $\Longleftrightarrow$ A is co-polar;*
(ii) *$B \cup \hat{B}$ is polar and co-polar;*
(iii) *$A^\ell = A^r$ and $\hat{A}^\ell = \hat{A}^r$;*
(iv) *$A^r \vartriangle \hat{A}^r$ is semi-polar.*

Proof. Suppose A is polar. If it is not co-polar, it is not left-polar either, for the two are the same under these hypotheses (Proposition 13.27) so by the Section Theorem (Theorem 6.9) there exists a predictable time T such that $P\{X_{T-} \in A, T < \zeta\} > 0$. As (X_t) is quasi-left-continuous on $(0, \zeta)$, $P\{X_T \in A\} = P\{X_{T-} \in A, T < \zeta\} > 0$, which contradicts the polarity of A. Thus polar implies co-polar. The reverse implication follows by symmetry.

This proves (i), (ii) is an immediate corollary, and, as (iii) follows from the same considerations, we leave it as an exercise. This brings us to the proof of (iv), which is the hard part of the proposition. By intersecting A with large compacts, we can assume without loss of generality that the closure of A is compact.

Let $F = \hat{A}^r - A^r$ and $p > 0$. Set

$$f(x) = E^x\{e^{-pS_F}\}.$$

Then f is p-excessive and in terms of the p-process, (X_t^p) (i.e. (X_t) killed at an independent $\exp(p)$ time) this is

$$= P_p^x\{S_F < \infty\} = p_F^p 1(x),$$

where P_p^x is the distribution of (X_t^p) and p_F^p is the corresponding interior reduite operator. By Corollary 13.63 f is the p-potential of a measure μ supported by $F \cup \hat{F}^r$, so

$$f(x) = U_p\mu(x).$$

Now let us calculate $p_A^p f$. By Proposition 13.62, if $\nu = \mu\hat{D}_A^p$, $p_A^p f = U_p\nu$. But μ is supported by $F \cup \hat{F}^r \subset \hat{A}^r$ and $\hat{D}_A(x, \{x\}) = 1$ for $x \in \hat{A}^r$, so $\nu = \mu$. Thus

$$p_A^p f = f.$$

Thus $f(x) = p_A^p f(x) \le p_A^p 1(x) = E^x\{e^{-S_A}\}$. But if $x \in F$ then $x \notin A^r$. Now $A^r = A^\ell$. If $x \notin A^\ell$, $P^x\{S_A > 0\} = 1$ and we conclude that $f(x) = E^x\{e^{-pS_A}\} < 1$. Thus $f(x) < 1$ on F, which implies that F is thin, and therefore semi-polar. By symmetry (interchange (X_t) and $(\hat{X}_t)$) $A^r - \hat{A}^r$ is semi-polar. □

A strong Markov process generates an intrinsic topology, called the fine topology (see §3.5). It describes the local behavior of the process, albeit in a limited way. For instance, if $X_0 = x$, and A is a set not containing x, then (X_t) hits A immediately iff x is in the fine closure of A. If f is a function, $f(X_t)$ is continuous at $t = 0$ iff f is fine continuous at x. So the fine topology dictates where the process goes – or does not go – right out of the starting gate. It does not do so well for the global behavior: for instance, all regular diffusions on the line share the same fine topology. (That is trivial, for in that case the fine topology is the Euclidean topology.) It is also true – and far less trivial – on $\boldsymbol{R}^n$: if $n \ge 2$, all smooth non-singular diffusions on $\boldsymbol{R}^n$ share the same fine topology. In this case the fine topology is neither first nor second countable, so it is non-metrizable. (Of course, one has to define "smooth" and "non-singular" carefully!)

The upshot of this is that, heuristically at least, processes having the same fine topology are locally similar. In the present case, there are two processes, (X_t) and $(\hat{X}_t)$. We call the intrinsic topology of (X_t) the **fine topology**, and that of $(\hat{X}_t)$, the **cofine topology.**

These need not be the same. The standard counter-example, as so often happens, is uniform motion to the right on $\boldsymbol{R}$, whose dual (with respect to Lebesgue measure) is uniform motion to the left. Then $[x, x+\varepsilon)$ is a fine neighborhood of x and $(x-\varepsilon, x]$ is a cofine neighborhood. Their intersection is $\{x\}$, which is neither a fine nor cofine open set. Furthermore, the only polar set is the empty set, so a singleton is non-polar, but has no regular points. The set $\{1/n : n = 1, 2, \dots\}$ is semi-polar and cothin but not thin; the function $f(x) = I_{(-\infty,0)}(x)$ is excessive but not regular, and the fine and cofine closures of $(0, 1)$ are $[0, 1)$ and $(0, 1]$ respectively, which differ by a non-polar set.

We will investigate the case where the fine and cofine topologies are (nearly) the same. The consequences of this are far-reaching. One should keep the above example in mind to appreciate them.

Definition 13.78. *We say that the fine and cofine topologies are* **nearly equivalent** *if for each set Λ which is open in one of the two, there exists a set Γ which is open in the other for which $\Lambda \vartriangle \Gamma$ is polar.*

Notice that if Λ and Γ are fine open and cofine open sets respectively, and if they differ by a polar set, then $\Lambda \cap \Gamma$ is both fine and cofine open. (See the exercises.)

Definition 13.79. *A p-excessive function f is* **regular** *if with probability one, $t \mapsto f(X_t)$ is continuous at each point of continuity of $t \mapsto X_t$.*

This brings us to the main result of this section.

Theorem 13.80. *Suppose (SF) and $(\widehat{SF})$ hold and that B is co-polar and $\hat{B}$ is polar. Then the following are equivalent:*

(i) *the cofine topology and fine topologies are nearly equivalent;*
(ii) *all p-excessive functions are regular;*
(iii) *every semi-polar set is polar;*
(iv) *if $A \in \mathcal{E}$, $X_{T_A} \in A^r$ a.s. on $\{T_A < \infty\}$;*
(v) *every non-polar set contains a regular point;*
(vi) *for every $A \in \mathcal{E}$, $A^r \vartriangle \hat{A}^r$ is polar.*

Proof. (i) $\Longrightarrow$ (ii): The idea is simple: an excessive function f is fine continuous; if the fine and cofine topologies are equal, it is also cofine continuous. Thus both $t \mapsto f(X_t)$ and $t \mapsto f(\hat{X}_t)$ are continuous. By time-reversal, the latter implies that $t \mapsto f(X_{t-})$ is left-continuous. Then $f(X_t)$ must be continuous at each point of continuity of (X_t). However, some care is needed to make this idea rigorous since there are exceptional polar sets involved.

Let f be excessive. By considering $f \wedge n$ if necessary, we may assume it is bounded. As f is fine continuous, for any $a < b$ the set $\Lambda_{ab} \stackrel{\text{def}}{=} \{x : a < f(x) < b\}$ is fine open. Thus by (i) there exists a cofine open set Γ_{ab} such that $\Lambda_{ab} \vartriangle \Gamma_{ab}$ is polar. Let

$$D = B \cup \hat{B} \cup \bigcup_{a<b\in \boldsymbol{Q}} \Lambda_{ab} \vartriangle \Gamma_{ab}.$$

Then D is a countable union of polar sets, so it is polar and therefore copolar. If $x \notin D$ and $f(x) \in (a, b)$, then $x \in \Lambda_{ab} \cap \Gamma_{ab}$. Now (see Exercise 26) $\Lambda_{ab} \cap \Gamma_{ab}$ is both fine and cofine open, hence Γ_{ab} is a cofine neighborhood of x. This is true for all rational $a < f(x) < b$, so it follows that f is cofine continuous at each $x \in \boldsymbol{E} - D$. Thus, f is cofine continuous off a polar (and therefore copolar) set. Now suppose that we have shown that $t \mapsto f(\hat{X}_t)$ is right continuous. By time-reversal, $t \mapsto f(X_{t-})$ is left-continuous on $(0, \zeta)$. Bearing in mind that $f(X_t)$ has left limits, we see:

$$\lim_{s\downarrow t} f(X_t) = f(X_t), \qquad \lim_{s\uparrow t} f(X_s) = \lim_{s\uparrow t} f(X_{s-}) = f(X_{t-}).$$

Thus if $X_{t-} = X_t$, $f(X_s)$ is continuous at t, hence f is regular.

It remains to show that $f(\hat{X}_t)$ is right continuous if f is cofine continuous off a copolar set. (This may seem obvious, until one sees examples in which such a function is everywhere-discontinuous in the usual topology. But given all the machinery at hand, it is not hard.) Since we know $f(\hat{X}_t)$ has *essential* left and right limits at all points, we can define a process by $Z_t = \operatorname{ess\,lim}_{s\downarrow t} f(\hat{X}_s)$. This is right continuous (Proposition 6.2). To show that $f(\hat{X}_t)$ is right-continuous, we need only show it equals Z_t for all t. For this it is sufficient to show the two are equal at an arbitrary stopping time T (the Section Theorem again.) Let $g(x) = \hat{P}^x\{\operatorname{ess\,lim}_{s\downarrow t} f(\hat{X}_s) = f(x)\}$. Then $g(x) = 1$ if f is cofine continuous at x, and therefore $g(x) = 1$ at all x outside of a co-polar set. Apply the strong Markov property at T:

$$\hat{P}\{f(X_T) = Z_T\} = \hat{P}\{f(X_T) = \operatorname*{ess\,lim}_{s\downarrow 0} f(\hat{X}_{T+s})\} = \hat{E}\{g(\hat{X}_T)\} = 1,$$

since the set $V \equiv \{x : g(x) \neq 1\}$ is co-polar, so that $\hat{P}\{\hat{X}_T \in V\} = 0$. Thus $t \mapsto f(\hat{X}_t)$ is right-continuous as claimed.

(ii) $\Longrightarrow$ (iii): A semipolar set is a countable union of thin sets, so it is enough to show a thin set is polar, and, for this, it is enough to show a compact thin set is polar. Thus, let K be a compact thin set and let $f(x) = E^x\{e^{-T_K}\}$, which is a bounded 1-excessive function, and therefore regular by hypothesis. Let G_n be open relatively compact sets whose closures decrease to K.

Under our hypotheses, (X_t) satisfies Hypothesis (B) and K is semi-polar, so $X_{T_K-} = X_{T_K}$ if $T_K < \infty$ (Theorem 1 of §3.8), and $T_{G_n} < T_K$ on $\{0 < T_K < \infty\}$. Suppose T_K is finite with positive probability.

On $\{T_K < \infty\}$,

$$f(X_{T_{G_n}}) = E\{e^{-(T_K - T_{G_n})} \mid \mathcal{F}_{T_G}\} \longrightarrow 1$$

as $n \to \infty$. As f is regular,

$$1 = \lim_n f(X_{T_{G_n}}) = f(\lim_n X_{T_{G_n}}) = f(X_{T_K})$$

as $n \to \infty$. But $f(x) = 1 \Longrightarrow x \in K^r$; K is thin, so $K^r = \emptyset$, a contradiction. Thus $T_K = \infty$, so K is polar.

(iii) $\Longrightarrow$ (iv): This is clear if A is polar, since then $A^r = \emptyset$ and T_A is a.s. infinite. If A is not polar, then $X_{T_A} \in A \cup A^r = A^r \cup (A - A^r)$. Since $A - A^r$ is semi-polar, and hence polar by (iii), the process never hits it, so $X_{T_A} \notin A - A^r$, which implies (iv).

(iv) $\Longrightarrow$ (v): If A is non-polar, it contains a non-polar compact subset K. Then T_K is finite with positive probability. By (iv), $X_{T_K} \in K^r$ with positive probability, hence K^r is non-empty. But $K^r \subset K$ so $K^r \subset A \cap A^r$, which implies (v).

(v) $\Longrightarrow$ (vi): $\hat{A}^r \vartriangle A^r$ is semipolar by Proposition 13.77. Thus it is a countable union of thin sets. If C is one of those thin sets, it can contain no regular points, hence by (v) it is polar. A countable union of polars is polar, hence $\hat{A}^r \vartriangle A^r$ is polar.

(vi) $\Longrightarrow$ (i): This is almost immediate by complementation. If A is fine open, then $C \equiv A^c$ is fine closed, so $C = C \cup C^r$. By (v), $C^r \vartriangle \hat{C}^r$ is polar, so $C \vartriangle (C \cup \hat{C}^r)$ is polar. But $C \cup \hat{C}^r$ is cofine closed, hence $\hat{A} \equiv (C \cup \hat{C}^r)^c$ is cofine open and $A \vartriangle \hat{A}$ is polar. □

If we now add the hypothesis that any one of the above equivalent conditions holds, we get a potential theory which gives us some of the most refined results. Accordingly, we make the following definition.

Definition 13.81. *We say that there is* **near symmetry** *if* $(\widehat{SF})$ *and* (SF) *hold,* $(B - B_\partial) \cup (\hat{B} - \hat{B}_\partial) = \emptyset$, *if* (X_t) *and* $(\hat{X}_t)$ *are quasi-left continuous, and if the fine and cofine topologies are nearly equal.*

Remark 13.82. To all intents and purposes, (X_t) and $(\hat{X}_t)$ are two Hunt processes in duality, both having strongly Feller resolvents, and sharing (nearly) the same fine topology. We could of course replace the latter by any of the equivalent conditions of Theorem 13.80. ("Semi-polar $\Longrightarrow$ polar" is a popular choice. We prefer "fine = cofine" in order to emphasize symmetry.) The hypotheses $(\widehat{SF})$ and (SF), imply that the processes are transient. This is automatically true if we are discussing p-excessive functions for strictly positive p, but it is a restriction when $p = 0$. The problem of dealing with recurrent processes is out of the scope of this book – this part of it, at least – so we will leave it aside.

Corollary 13.83. *Suppose there is near symmetry. Let $h = U\mu$ be a potential which sits on a Borel set A. Then*

(i) *(Maximum Principle)* $\sup_{x \in \boldsymbol{E}} h(x) = \sup_{x \in A} h(x)$;
(ii) *Let f a bounded excessive function. If $f \geq h$ on A, then $f \geq h$ everywhere.*
(iii) *If A is polar and h is not identically zero, then h is unbounded.*

Proof. Notice that (i) follows from (ii), (f = constant). For (ii), note that for each x, $h = \sup_K \int_K u(x, y)\,\mu(dy) = \sup_K U\mu_K(x)$, where the sup is over all compacts $K \subset A$ and μ_K is the restriction of μ to K. Together with Proposition 13.54 this shows $h = p_A h$. If f is excessive, it is regular by Theorem 13.80 (ii). Thus, by Theorem 13.74,

$$h = p_A h = P_A h.$$

Now h and f are fine continuous, so if $h \leq f$ on A, then $h \leq f$ on $A \cup A^r$. As the distribution of X_{T_A} is concentrated on $A \cup A^r$, it follows that

$$h \leq P_A f \leq f,$$

which proves (ii).

(iii) If μ has support in a compact set K and is bounded, say $U\mu \leq M$, then $U\mu = p_K U\mu \leq M p_K 1 = M P_K 1$ by Theorem 13.74. But this equals $MP\{T_A < \infty\}$, which is zero if K is polar. □

Remark 13.84. Note that (i) and (ii) refine Proposition 13.68, which required domination on a neighborhood of the support.

13.15. Capacities and Last Exit Times

Let $A \in \mathcal{E}$ have compact closure in E. We showed (Proposition 13.69) that the equilibrium potential of A was $C_A(x) = p_A 1(x) = U\mu_A(x)$, where μ_A is a measure on $A \cup \hat{A}^r$. Under near-symmetry, (X_t) and $(\hat{X}_t)$ both satisfy (SF), so we have the same results for the **equilibrium co-potential** of A, $\hat{C}_A = \hat{P}_A 1 = \hat{U}\hat{\mu}_A$, where $\hat{\mu}_A$ is the equilibrium measure for $(\hat{X}_t)$; $\hat{\mu}_A$ sits on $A \cup A^r$. We define the **capacity** $C(A)$ **of** $\boldsymbol{A}$ by $C(A) = \mu_A(\boldsymbol{E})$ and the **co-capacity** $\hat{C}(A)$ **of** $\boldsymbol{A}$ by $\hat{C}(A) = \hat{\mu}_A(\boldsymbol{E})$.

Proposition 13.85. *Under near symmetry, if $A \in \boldsymbol{E} - B \cup \hat{B}$ has compact closure in $\boldsymbol{E}$,*

(i) *μ_A and $\hat{\mu}_A$ do not charge polar sets;*
(ii) *μ_A and $\hat{\mu}_A$ sit on $\hat{A}^r \cap A^r$;*
(iii) *$C_A(x) = 1$ if $x \in A^r$; $\hat{C}_A(x) = 1$ if $x \in \hat{A}^r$;*

(iv) *C_A is the largest potential which sits on $A \cup \hat{A}^r$ and which is dominated by 1; and it is the smallest excessive function which dominates 1 on A, up to a polar set;*

(v) *the capacity and the co-capacity of A are equal.*

Proof. (i) follows from Corollary 13.83. To see (ii), note that μ_A sits on $A \cup \hat{A}^r$, but $A - \hat{A}^r$ and $A^r \vartriangle \hat{A}^r$ are polar, and μ_A doesn't charge polars, so it also sits on $A^r \cap \hat{A}^r$. The same is true of $\hat{\mu}_A$ by symmetry.

(iii) follows from the definitions of A^r and $\hat{A}^r$: if $x \in A^r$, T_A is P^x-a.s. zero, hence it is finite.

(iv) follows from Theorem 13.74 (ii).

To see (v), note that by (ii), $\mu_A(\boldsymbol{E}) = \mu_A(A^r \cap \hat{A}^r)$ so that by (iii), $\mu_A(\boldsymbol{E}) = \int_{A^r \cap \hat{A}^r} \hat{C}_A(y)\, \mu_A(dy) = \int_{A^r \cap \hat{A}^r} \mu(dy) \int_{\boldsymbol{E}} u(x, y) \hat{\mu}_A(dx)$. Since $\hat{\mu}$ sits on $A^r \cap A^r$, we have

$$\mu_A(\boldsymbol{E}) = \int_{\boldsymbol{E}} \int_{\boldsymbol{E}} \hat{\mu}_A(dx)\, u(x, y)\, \mu_A(dy).$$

The same calculation for $\hat{\mu}_A$ shows that

$$\hat{\mu}_A(\boldsymbol{E}) = \int_{\boldsymbol{E}} \int_{\boldsymbol{E}} \mu_A(dy)\, u(x, y)\, \hat{\mu}(dx),$$

and (iv) follows by Fubini's theorem. □

Exercises

19. Consider $\boldsymbol{E} = [-1, 0] \cup \{1\}$ and let (X_t) be the right-continuous process which is reflected Brownian motion on $[-1, 0)$, until it first hits the origin. It then jumps to 1, stays there an exponential time, and is killed.
    (i) Show the origin is a degenerate branching point with $P_0(0, \{1\}) = 1$.
    (ii) Show that $p_{\{0\}} 1(x) = I_{(-\infty,0)}(x)$ while $P_{\{0\}} 1(x) \equiv 0$.
    (iii) Show that the smallest excessive function which dominates 1 at 0 is identically one on $\boldsymbol{E}$.
    (iv) Show that this process satisfies Hypothesis $(\widehat{SF})$.

    Conclude that all three reduites, $p_{\{0\}}1$, $P_{\{0\}}1$ and $R_{\{0\}}1$ are different.
20. Show that under the hypotheses of §13.2, if $p > 0$ there exists a bounded p-excessive ξ-integrable function h such that $\{x : h(x) > 0\} = E - B_\partial$.
21. Let ξ' be an excessive reference measure. Show that there is a co-excessive function h such that (X_t) is in duality with the h-transform of $(\hat{X}_t)$.
22. Show that there exists a sequence (T_n) of stopping times such that $T_n < \zeta$ on $S_A < \infty$, such that $T_n \uparrow S_A$ and $\{T_n < \infty\} \downarrow \{S_A < \infty\}$ ${}_hP^x$-a.s., and that $p_A h(x) = \lim_n E^x\{h(X_{T_n})\}$.

Hint: Decompose S_A into

$$S_1 = \begin{cases} S_A & \text{if } X_{S_A-} \in A \\ \infty & \text{otherwise;} \end{cases} \qquad S_2 = \begin{cases} S_A & \text{if } X_{S_A-} \notin A \\ \infty & \text{otherwise.} \end{cases}$$

Then $S_2 < \infty \Longrightarrow S_2 < \zeta$ (why?) and S_1 is predictable (why? It's not immediately obvious!) so there is a sequence (τ_n) announcing it ${}_hP^X$-a.s. Thus put

$$T_n = \begin{cases} \tau_n & \text{if } \tau_n < \infty \\ S_2 & \text{otherwise} \end{cases}$$

Then if $h(x) < \infty$,

$$E^x\{h(X_{T_n})\} = h(x)\,{}_hE^x\{T_n < \infty\} \to h(x)_hP^x\{S_A < \infty\} = p_Ah(x).$$

23. Let (X_t) be uniform translation to the right on $\boldsymbol{R}$. Let $h(x) = 1$ if $x < 0$, and $h(x) = 1/2$ if $x \geq 0$. Then $p_{\{0\}}h \neq P_{\{0\}}h$. ($h$ is excessive but not regular.) Describe the lifetime of the h-transform.
24. Let $\{x_0, x_1\} \subset \boldsymbol{E}$, where x_0 is a branching point which branches directly to a trap, x_1. If G is open and contains x_1 but not x_2, show $R_G1 \neq P_G1$.
25. Show that quasi-left-continuity on $(0, \zeta)$ and Hypothesis (B) are preserved under h-transforms.
26. Let Λ and Γ be fine and cofine open sets respectively and suppose $\Lambda \vartriangle \Gamma$ is polar. If polar $\Longleftrightarrow$ co-polar, show that $\Lambda \cap \Gamma$ is both fine and cofine open.
27. Assume $(\widehat{SF})$ and that X is quasi-left-continuous on $(0, \zeta)$ and satisfies Hypothesis (B). If $A \in \mathcal{E}$ and $h(x) = u(x, y)$, $k(y) = u(x, y)$, show that $P_Ah(x) = \hat{P}_Ak(y)$ (in other notation, $P_Au(\cdot, y)(x) = \hat{P}_Au(x, \cdot)(y)$). (Hint: $P_Ah(x) = E^x\{h(X_{T_A})\} = h(x)_hE^x\{T_A < \zeta\}$ and use the fact that the reverse of X_h from its lifetime is a k-transform of $\hat{X}$ with initial point y to conclude this equals $k(y)_k\hat{P}^y\{T_A < \zeta\}$.)
28. Show that $C(A)$ is a Choquet capacity. (This, of course, was the seminal example of a capacity.)

Notes on Chapter 13

Physicists will recognize the electrons and holes as a much-simplified version of the solid-state theory (see Kittel [1]) in which we have ignored, for instance, energy bands and the exclusion principle. (See for instance Kittel [1].)

To date, it appears that no two accounts of duality agree completely on the details. We have continued that tradition. In contrast to most treatments, we allow both branching and left-branching points, and make no assumptions about quasi-left continuity. This allows us to show the interesting relations between branching points, quasi-left continuity and Hypothesis (B), but in most cases branching points don't affect the fundamental ideas in the proofs. Therefore, we will not dwell on the exact hypotheses in the historical notes below; we will just follow the ideas. Much of the material – with completely different proofs – can be found in Blumenthal and Getoor [1], which remains the standard reference on duality. Our approach was pioneered by Kunita and Watanabe in [2].

The duality hypothesis in its present form was introduced in Hunt's third memoir [5] as Hypothesis (F). Hunt showed that it was the key to the extension of the more delicate parts of classical potential theory. Theorem 13.2 is due to Kunita and Watanabe in [1] and [2].

§13.4 and §13.6 are taken from Walsh [5], except for Theorem 13.24 which was proved in Smythe and Walsh [1]. This sharpens Meyer's earlier result that Hypothesis (B) is satisfied for standard processes in duality. Proposition 13.28 is from the same paper.

Theorem 13.33, which gives the duality between h-transforms, is implicit in Hunt's paper [5] and explicit in the important paper by Kunita and Watanabe [1]. The relations between h-transforms and reversal in §13.8–13.10 are from the same paper. In particular, the key Theorem 13.45 was proved by Kunita and Watanabe. Our exposition follows Kunita and Watanabe in spirit, if not in detail.

§13.12. The probabilistic definition of the interior reduite and Theorem 13.65 are from Walsh [1]. Proposition 13.69 is due to Chung. Theorem 13.74 is due to Hunt. Theorem 13.80 applies to Brownian motion and contains several classical results, including the Evans-Kellogg theorem. In particular, it implies – and gives a different proof of – Theorems 9, 10 and the Corollary of §4.5. Hunt [5] saw the importance of the condition "semipolar implies polar" and proved its equivalence to (ii), (v) and (vi).

Corollary 13.83 (ii) is actually true under the weaker condition that f is μ-a.e. finite; in that case it implies the complete maximum principle.

Chapter 14

The Martin Boundary

Boundary behavior is an important part of potential theory at all levels. We introduced it briefly in §4.4 and §4.5, where we discussed the Dirichlet problem, the Poisson kernel, and some related questions for the (important!) special case of Brownian motion on the ball. Now it is time to discuss it in greater depth.

The difficulty with boundary theory in the present setting is that, quite simply, there is no boundary. The state space is locally compact, but it is not necessarily embedded in a larger space in which to define the boundary. We will have to create one.

Given our experience with Ray processes, we might guess that the key is a compactification. It is, and it leads to the Martin boundary. This is closely related to the Ray-Knight compactification: indeed, it is essentially a Ray-Knight compactification relative to the dual process.

The idea of the Martin boundary was originally purely analytic, with no reference to probability: R. S. Martin constructed it to extend the Poisson representation of harmonic functions from the ball to arbitrary domains in Euclidean space. However, J. L. Doob adapted it to Brownian motion and Markov chains, and it is now a staple of probabilistic potential theory.

We will study it under our duality hypotheses. This will tie up some loose ends, as it were, and complete the answers to some partially-answered questions we have discussed earlier.

For example, the Riesz representation theorem tells us that an excessive function is the sum of a potential and a harmonic function. We have an integral representation for the potential, but not for the harmonic function. We also have a relation between h-transforms and $X_{\zeta-}$, but only when ζ is finite and $X_{\zeta-}$ is in E.

14.1. Hypotheses

We still assume the duality hypotheses of §13.2. The main additional hypotheses we need is that the original process is transient, and that the dual potential kernel is smooth. To be specific, we assume that:

(i) for every compact $K \subset E$, and for all x, $P^x\{L_K^- < \infty\} = 1$, where L_K^- is the last exit time of (X_{t-}) from K;

(ii) Hypothesis $(\widehat{SF})$ holds;

(iii) $(\hat{U}_p)$ separates points, and if x and y are distinct co-branching points, $P_0(x, \cdot)|_E$ and $P_0(y, \cdot)|_E$ are not proportional;

(iv) $\hat{B}_\partial = \emptyset$.

We make no smoothness assumptions on the resolvent (U_p). The last two hypotheses are somewhat technical: (iii) assures us that not only $\hat{U}$, but $\hat{\kappa}$ defined below separates points. (iv) is just for convenience. The set $\hat{B}_\partial$ of points which co-branch directly to the cemetery could be treated as a special case – it is a closed ξ-null set after all – but as we would have to do this several times, we simply assume that it is empty. In fact, apart from the separation property, branching and co-branching points play no rôle in the Martin boundary. The reader should simply assume at the first reading that there are no co-branching points, in which case the entire difficulty will disappear.

Let $f_0 \geq 0$ be a bounded Borel function such that

(i) $\hat{U} f_0$ is a continuous strictly positive function on E;

(ii) the measure $\gamma(dx) \stackrel{\text{def}}{=} f_0(x)\xi(dx)$ is a probability measure;

(iii) $\gamma U_1(dx)$ is a reference measure.

Remark 14.1. f_0 and $\hat{U} f_0$ are important for they determine the class of excessive functions represented: it is exactly the class of excessive functions which are γ-integrable. We can always satisfy (i) – (iii) by choosing f_0 to be strictly positive and ξ-integrable. Then $\hat{U} f_0$ is bounded, continuous, and strictly positive, thanks to $(\widehat{SF})$ and the fact that $\hat{B}_\partial$ is empty. Then γ itself will be a reference measure. However, condition (iii) doesn't require f_0 to be strictly positive; this is useful in applications.

We write $u(x, y) = u_0(x, y)$ for the potential density, which is, as before, excessive in x and co-excessive in y. We denote the potential kernels by U and $\hat{U}$ instead of U_0 and $\hat{U}_0$.

Let f_0 be a strictly positive, bounded Borel function on E, such that $\int f_0(x)\xi(dx) = 1$. Then $\hat{U} f_0$ is continuous by $(\widehat{SF})$, and it is strictly positive since $\hat{B}_\partial = \emptyset$. Define the probability measure $\gamma(dx) = f_0(x)\xi(dx)$ and, if h is a γ-integrable excessive function, let $\gamma^h(dx) = C_h h(x)\gamma(dx)$, where $C_h = \left(\int h(x)\gamma(dx)\right)^{-1}$ is chosen to make γ^h a probability measure.

14.2. The Martin Kernel and the Martin Space

The following notation will be used throughout this chapter. Let $E_1 \subset E_2 \subset \dots$ be a sequence of open, relatively compact sets such that for all n, $\bar{E}_n \subset E_{n+1}$ and $\cup_n E_n = \boldsymbol{E}$. Let L_n be the last exit time of E_n for (X_{t-}):

$$L_n = \sup\{t : X_{t-} \in E_n\}.$$

Then $L_1 \le L_2 \le \ldots < \infty$ a.s. Clearly $L_n \le \zeta$, and $\lim_{n\to\infty} L_n = \zeta$. It may happen that $L_n = \zeta$ for some n, in which case $L_m = \zeta$ for all $m \ge n$. In any case, if $X_{\zeta-}$ exists, it equals $\lim_{n\to\infty} X_{L_n-}$.

Define a kernel κ by

$$\kappa(x, y) = \frac{u(x, y)}{\hat{U} f_0(y)}, \; x, y \in \boldsymbol{E},$$

where f_0 is defined above. This is well-defined since $\hat{U} f_0$ is strictly positive. We write $\hat{\kappa} f(y) = \int \kappa(x, y) f(x)\xi(dx)$ and $\kappa\mu(x) = \int \kappa(x, y)\mu(dy)$.

Here are some of its elementary properties.

Proposition 14.2.

(i) *For any* $y \in \boldsymbol{E}$, $x \mapsto \kappa(x, y)$ *is excessive;*
(ii) $\int \kappa(x, y) f_0(x)\xi(dx) = 1$, $y \in \boldsymbol{E}$;
(iii) *if* f *is a bounded Borel function with support in an open, relatively compact set* $A \subset \boldsymbol{E}$, *then* $\hat{\kappa} f$ *is continuous and bounded, and there is a constant* C_A *such that* $\|\hat{\kappa} f\|_\infty \le C_A \|f\|_\infty$.

Proof. (i) follows since $x \mapsto u(x, y)$ is excessive, and (ii) is a direct calculation: $\int f_0(x)\kappa(x, y)\xi(dx) = \int f_0(x)u(x, y)\xi(dx)/\hat{U} f_0(y) = 1$.

To see (iii), note that $\hat{\kappa} f(y) = \hat{U} f(y)/\hat{U} f_0(y)$. Since both numerator and denominator are continuous by $(\widehat{SF})$ and the denominator never vanishes, the ratio is continuous on $\boldsymbol{E}$. We must show it is bounded.

Let A be an open relatively compact set such that f vanishes identically on A^c. As $\kappa(\cdot, y)$ is excessive, $\kappa(\cdot, y) \ge p_A\kappa(\cdot, y)$ and $p_A\kappa(\cdot, y)$ is the potential of a measure μ_y, say, which is supported by $\bar{A}$ (Corollary 13.63), so $\kappa(\cdot, y) \ge \int_{\bar{A}} u(\cdot, z)\mu_y(dz) = \int_{\bar{A}} \kappa(\cdot, z)\hat{U} f_0(z)\mu_y(dz)$. There is equality at non-branching $x \in A$, so $\kappa(\cdot, y) = \int_{\bar{A}} \kappa(\cdot, z)\hat{U} f_0(z)\mu_y(dz)$ on $A - B$. Thus

$$\begin{aligned}
\hat{\kappa} f(y) &= \int_A \xi(dx) f(x) \int_{\bar{A}} \kappa(x, z)\hat{U} f_0(z)\mu_y(dz) \\
&= \int_{\bar{A}} \hat{\kappa} f(z)\hat{U} f_0(z)\mu_y(dz) \\
&\le \sup\{|\hat{\kappa} f(z)|, z \in \bar{A}\} \int_{\bar{A}} \hat{U} f_0(z)\mu_y(dz).
\end{aligned}$$

But now

$$\begin{aligned}
\int_{\bar{A}} \hat{U} f_0(z)\mu_y(dz) &\le \int_{\bar{A}} \int_{\boldsymbol{E}} f_0(w)u(w, z)\xi(dw)\mu_y(dz) \\
&\le \int_{\boldsymbol{E}} f_0(w)\kappa(w, y)\xi(dw) = 1,
\end{aligned}$$

and (iii) follows with $C_A = \sup\{\hat{\kappa} I_A(y), y \in A\}$. ☐

We now construct the Martin space. (In fact, as Meyer points out in [5], there is a whole class of spaces which can be given the name "Martin space," since there are different possible choices for the metric d below. We will construct one of the simplest of them.) We define a metric d as follows. Let (f_n) be a set of functions which is dense in $\mathcal{C}_0$. Then let

$$d(x, y) = \sum_{n=1}^{\infty} 2^{-n} \frac{|\hat{\kappa} f_n(x) - \hat{\kappa} f_n(y)|}{1 + |\hat{\kappa} f_n(x) - \hat{\kappa} f_n(y)|}.$$

By the resolvent equation, if $p > 0$, then $\hat{U}_p f = \hat{U}(f - p\hat{U}_p f)$, so that the range of $\hat{U}$ is the same as the range of the co-resolvent. The kernel $\hat{\kappa}$ separates points by hypothesis and the f_n are dense in $\mathcal{C}_0$, so the $\hat{\kappa} f_n$ also separate points of $\boldsymbol{E}$. Thus $x \neq y \Longrightarrow d(x, y) > 0$. Then d is a bounded metric, and, as the $\hat{\kappa} f_n$ are continuous, it is equivalent to the original metric on each E_n.

Let $\boldsymbol{E}^*$ be the completion of $\boldsymbol{E}$ in the metric d. Then $\boldsymbol{E}^*$ is a compact space and $\boldsymbol{E}$ is a dense subset of it. Since the f_n are dense in $\mathcal{C}_0$, it follows that for any $f \in \mathcal{C}_0$, the function $\hat{\kappa} f$ can be extended to $\boldsymbol{E}^*$ by continuity. Let $\mathfrak{B} = \boldsymbol{E}^* - \boldsymbol{E}$. We call $\boldsymbol{E}^*$ the **Martin space** and $\mathfrak{B}$ the **Martin boundary**. Note that a sequence (y_j) which has compact support in $\boldsymbol{E}$ converges to a point of $\boldsymbol{E}$ iff it converges to the point in the original topology. In general, $y_j \to y \in \boldsymbol{E}^*$ iff for all $f \in \mathcal{C}_0$, $\hat{\kappa} f(y_j) \to \hat{\kappa} f(y)$. (Unfortunately this does not imply that $\kappa(x, y_j)$ itself converges for a given x.)

This shows that the Martin space does not depend on the particular set of functions (f_n); however it may depend on f_0, since the kernel κ does.

Proposition 14.3. *The kernel $\kappa(x, y)$ on $\boldsymbol{E} \times \boldsymbol{E}$ can be extended to a Borel measurable kernel on $\boldsymbol{E} \times \boldsymbol{E}^*$, called the* **Martin kernel**, *which we also denote by κ, in such a way that*

(i) $\hat{\kappa} f(\eta) = \int \kappa(x, \eta) f(x) \xi(dx)$, $\eta \in \mathfrak{B}$;
(ii) $y \mapsto \kappa(x, y)$ *is lower semi-continuous on $\boldsymbol{E}^*$ for fixed $x \in \boldsymbol{E}$*;
(iii) $\int \kappa(x, y) f_0(x) \xi(dx) \leq 1$, $y \in \mathfrak{B}$;
(iv) *if $y \in \boldsymbol{E}^*$, $x \mapsto \kappa(x, y)$ is excessive*;
(v) *if $y \neq z \in \mathfrak{B}$, $\kappa(\cdot, y) \neq \kappa(\cdot, z)$*;
(vi) *if $y \to z$ in the topology of $\boldsymbol{E}^*$ and if $f \in \mathcal{C}_0$, then $\hat{\kappa} f(y) \to \hat{\kappa} f(z)$.*

Proof. If $f \in \mathcal{C}_0$, $\hat{\kappa} f$ can be extended by continuity to $\boldsymbol{E}^*$. Thus for each point $\eta \in \mathfrak{B}$, $f \mapsto \hat{\kappa} f(\eta)$ is a positive linear functional on $\mathcal{C}_0$, which therefore corresponds to a Radon measure on $\boldsymbol{E}$, which we denote $k(\eta, dx)$. Then for each $f \in \mathcal{C}_0$, $\int k(\eta, dx) f(x) = \hat{\kappa} f(\eta)$. Let us define a function $\kappa(x, \eta)$ by

$$\kappa(x, \eta) = \sup_{\alpha > 0} \left\{ \int \alpha u_\alpha(x, z) k(\eta, dz) \right\}.$$

(Notice that this formula holds if $\eta \in \boldsymbol{E}$, for then $k(\eta, dz) = \kappa(z, \eta)\xi(dz)$ and $x \mapsto \kappa(x, \eta)$ is excessive.) Now the resolvent is continuous in α, so it is enough to take the supremum over, say, rational α. It follows that κ is Borel. Let us now verify its properties. We start with (ii).

(ii) Note that $y \mapsto u_\alpha(x, y)$ is co-excessive, and therefore lower-semi-continuous on $\boldsymbol{E}$. Thus there exists an increasing sequence $g_{n\alpha} \in \mathcal{C}_0$ such that $u_\alpha(x, \cdot) = \lim_n g_{n\alpha}(\cdot)$. It follows that for all $\eta \in \boldsymbol{E}^*$, $\kappa(x, \eta) = \sup\{\alpha \hat{\kappa} g_{n\alpha}(\eta) : n, \alpha > 0, \alpha \in \boldsymbol{Q}\}$. Thus $\kappa(x, \cdot)$ is the supremum of continuous functions, and hence lower-semi-continuous on $\boldsymbol{E}^*$.

(iii) Let $f \in \mathcal{C}_0$. For $y \in \boldsymbol{E}$, $\hat{\kappa} f(y) = \int f(x)\kappa(x, y)\,\xi(dx)$. Let $y \to \eta \in \mathfrak{B}$, and use Fatou's lemma and the lower-semi-continuity of κ to see that

$$\hat{\kappa} f(\eta) = \lim_{y \to \eta} \hat{\kappa} f(y) = \lim_{y \to \eta} \int f(x)\kappa(x, y)\xi(dx)$$
$$\geq \int f(x)\kappa(x, \eta)\xi(dx). \tag{14.1}$$

This holds for all positive Borel f and in particular for f_0, which proves (iii).

(i) Let $f \in \mathcal{C}_0$ and note that by the definition of κ,

$$\int f(x)\kappa(x, \eta)\xi(dx) \geq \limsup_{\alpha \to \infty} \int f(x) \int \alpha u_\alpha(x, z) k(\eta, dz)\xi(dx).$$

Interchange the order of integration and use Fatou's lemma: this is

$$\geq \int \lim_{\alpha \to \infty} (\alpha \hat{U}_\alpha f(z)) k(x, dz) = \int \hat{P}_0 f(z) k(x, dz),$$

so that, combining this with (14.1),

$$\hat{\kappa} f(\eta) \geq \int f(x)\kappa(x, \eta)\xi(dx) \geq \int \hat{P}_0 f(z) k(x, dz). \tag{14.2}$$

By the usual monotone class argument, (14.2) extends to all positive measurable f, not just continuous f. Let $g_n = I_{E_n - \hat{B}}$. Let $\hat{B}_n = \hat{B} \cap \{x : \hat{P}_0(x, E_n) > 0\}$. Since by assumption no points co-branch exclusively to ∂, $\cup_n \hat{B}_n = \hat{B}$. But $\hat{P}_0 g_n(x) \geq g_n(x) = 0$. Since $\hat{\kappa} g_n(\eta) \geq \hat{\kappa}(\hat{P}_0 g_n)(\eta)$, evidently the two are equal, so that $k(\eta, \hat{B}_n) = 0$, hence $k(\eta, \hat{B}) = 0$. Thus there is equality in (14.2), and (i) follows easily.

(iv) To see that $\kappa(\cdot, \eta)$ is excessive, note that by definition, for $\alpha > 0$,

$$\kappa(x, \eta) \geq \int \alpha u_\alpha(x, z) k(\eta, dz)$$
$$= \int \alpha u_\alpha(x, z)\kappa(z, \eta)\xi(dz) = \alpha U_\alpha \kappa(\cdot, \eta),$$

which implies that $x \mapsto \kappa(x, \eta)$ is supermedian. Thus $\alpha U_\alpha \kappa(\cdot, \eta)$ is increasing, so that by the definition of κ,

$$\kappa(x, \eta) = \lim_{\alpha\to\infty} \int \alpha U_\alpha(x, z)\kappa(z, \eta)\xi(dz),$$

hence κ is excessive in the first variable.

(v) This is almost immediate: if $\kappa(\cdot, \eta) = \kappa(\cdot, \eta')$, then $\hat{\kappa} f(\eta) = \hat{\kappa} f(\eta')$ for all f_n, so any Cauchy sequence in $\boldsymbol{E}$ which converges to η also converges to η'. But the $\hat{\kappa} f_n$ separate points of $\boldsymbol{E}$, and also of $\boldsymbol{E}^*$ by the remarks preceding the theorem, so $\eta = \eta'$.

Finally, (vi) is true by construction. □

14.3. Minimal Points and Boundary Limits

We say that η is a **minimal point** of $\boldsymbol{E}^*$ if $\kappa(\cdot, \eta)$ is a minimal excessive function. (See §11.4.) Let $\mathfrak{B}_{\min}$ be the set of minimal points $\eta \in \mathfrak{B}$. First, we prove a lemma.

Lemma 14.4. *Let h be a γ-integrable excessive function, and let (X_t^h) have initial measure γ^h. Let g be co-excessive and let $m(y) = g(y)/\hat{U} f_0(y)$, $Z_{-n} = m(X^h_{L_n-})$, and $\mathcal{F}_{-n} = \sigma\{X^h_{L_j-}, j \geq n\}$. Then*

(i) *for $x \in \boldsymbol{E}_h$, $\lim_{t\to\zeta-} m(X_{t-}^h)$ exists in the extended reals ${}_hP^x$-a.s.*
(ii) *For $b > 0$, $\{b \wedge Z_n, \mathcal{F}_n, n = \ldots, -2, -1\}$ is a positive supermartingale.*

Proof.

(i) Let (X_t^h) have initial measure γ^h, and let (Y_t^n) be its reverse from L_n. Then by Theorem 11.29 and Corollary 11.30, (Y_t^n) is a $(\hat{U} f_0)$-transform of $(\hat{X}_t)$. Thus m is $(\hat{U} f_0)$-co-excessive, so $m(Y_t^n)$ is a positive supermartingale (if it is integrable) and, in any case, if $0 < a < b$, $b \wedge m(Y_t^n)$ is a bounded supermartingale. Let β_n be the number of upcrossings of $[a, b]$ by $b \wedge \kappa\nu(Y_t^n)$; then by the upcrossing inequality,

$$E\{\beta_n\} \leq \frac{b}{b-a}.$$

But by reversal, β_n is also the number of upcrossings of $[a, b]$ by $\{m(X_{t-}), 0 < t < L_n\}$. As $n \to \infty$, β_n increases to β_∞, the number of upcrossings by $\{m(X_{t-}), 0 < t < \zeta\}$. Thus, by monotone convergence, $E\{\beta_\infty\} \leq b/(b-a)$. Therefore, $\beta_\infty < \infty$ a.s. This is true simultaneously for all rational pairs $a < b$ a.s. and we conclude that $m(X_{t-})$ has no oscillatory discontinuities, and that, in particular, $\lim_{t\to\zeta-} \kappa\nu(X_{t-})$ exists in the extended reals.

This is true for the initial measure γ^h, which is equivalent to ξ on $\boldsymbol{E}^h$, so that for ξ-a.e. $x \in \boldsymbol{E}_h$, $P^x\{\lim_{t\to\zeta-} g(X_{t-}) \text{ exists}\} = 1$. But this probability defines an excessive function of x; it equals one ξ-a.e., so by Hypothesis (L) it must be identically one.

(ii) Consider Y_t^{n+k}, the process reversed from L_{n+k}. The last exit from E_n by (X_{t-}^h) corresponds to the first entrance of E_n by (Y_t^{n+k}); call that time τ_n^{n+k}. Then

$$X_{L_n-}^h = Y_{\tau_n^{n+k}}^{n+k}. \tag{14.3}$$

Since $b \wedge m(Y_t^{n+k})$ is a bounded supermartingale, we have $m(b \wedge Y_{\tau_{n+1}^{n+k}}^{n+k}) \le E\{b \wedge m(Y_{\tau_n^{n+k}}^{n+k}) \mid Y_{\tau_1^{n+k}}^{n+k}, \dots, Y_{\tau_n^{n+k}}^{n+k}\}$. Now let $k \to \infty$ and use (14.3) to get the conclusion. □

Lemma 14.5. *Let h be a γ-integrable excessive function. For ξ-a.e. x and all $y \in \boldsymbol{E}$,*

$$ {}_hP^y\{\kappa(x, X_{\zeta-}) = \lim_{t\to\zeta-} \kappa(x, X_{t-})\} = 1.$$

Proof. The limit in question exists by Lemma 14.4. Let us call it $Y(x)$, being careful to choose a version which is measurable in the pair (ω, x). Since $\kappa(x, \cdot)$ is lower-semi-continuous, $Y(x) \ge \kappa(x, X_{\zeta-}^h)$. On the other hand, if $f \in \mathcal{C}_0$, $\hat{\kappa} f$ is continuous on $\boldsymbol{E}^*$, so

$$\int \kappa(x, X_{\zeta-}^h) f(x)\xi(dx) = \hat{\kappa} f(X_{\zeta-}^h) = \lim_{t\to\zeta-} \hat{\kappa} f(X_{t-}^h)$$

$$= \lim_{t\to\zeta-} \int \kappa(x, X_{t-}^h) f(x)\xi(dx) \ge \int Y(x) f(x)\xi(dx)$$

by Fatou. Since $Y(x) \ge \kappa(x, X_{\zeta-}^h)$, there must be equality a.s. for ξ-a.e. x as claimed. □

Proposition 14.6. *Let h be a γ-integrable excessive function. Then for each $x \in \boldsymbol{E}_h$, ${}_hP^x\{X_{\zeta-}$ exists in $\boldsymbol{E}^*\} = 1$.*

Proof. Apply Lemma 14.4 to $\hat{\kappa} f_n = \hat{U} f_n / \hat{U} f_0$ to see that $\lim_{t\to\zeta-} \kappa f_n(X_{t-}^h)$ exists for all n. This implies the existence of $X_{\zeta-}^h$. □

14.4. The Martin Representation

A potential $U\mu$ can be written in the form $\kappa\nu$, with $d\nu = \hat{U} f_0 d\mu$. By the Riesz representation theorem, any excessive function is the sum of a potential and a harmonic function. The Martin representation theorem tells us that a harmonic function can be written in the same form, but for a measure sitting on the Martin boundary. Combining the two, it follows that any γ-integrable excessive function can be written in the form $\kappa\mu$, where μ is a finite measure on $\boldsymbol{E}^*$. The aim of this section is to prove that. At the same time, we will sharpen the representation by identifying the measure μ in terms of the distribution of $X_{\zeta-}$, and showing that this measure always sits on the minimal points of $\boldsymbol{E}^*$.

Before going further, we should alert the reader to two subtle points. The first is the rôle of the measures γ and γ^h. It is natural to use γ as an initial measure of (X_t) because it is a reference measure, but more importantly, because its reverse from any finite co-optional time has potential kernel $\kappa(x, y)$. Then γ^h is a natural initial measure for (X_t^h) for exactly the same reasons. Indeed, by Proposition 11.29, the reverse of (X_t^h) with initial measure γ^h has exactly the same semigroup as the reverse of (X_t) with initial measure γ.

The second point is that, from a certain point of view, we are extending the results and consequences of §13.10 from $\boldsymbol{E}$ to the Martin space $\boldsymbol{E}^*$. We proved these results by time reversal from the lifetime. We could do the same here if the lifetime were finite. However, it may not be, and even if $P\{\zeta < \infty\} = 1$, it may not be true that ${}_hP\{\zeta < \infty\} = 1$ for all excessive h. (That question is generally non-trivial.) Therefore we can not reverse directly. Instead, we reverse from the L_n, which *are* finite and tend to ζ, and then take a limit. We begin with a lemma.

Lemma 14.7. *Let h be a γ-integrable excessive function. Let μ_n be the distribution of $X^h_{L_n-}$ under the initial measure $\gamma^h(dx) = h(x)\gamma(dx)$. Then*

$$h(x) = \int_{\bar{E}_n} \kappa(x, y)\mu_n(dy), \; x \in \boldsymbol{E}_n - B. \tag{14.4}$$

Proof. By Proposition 13.69 applied to ${}_hP^x$,

$${}_hP^x\{X_{L_n-} \in dy\} = \frac{1}{h(x)}\hat{U} f_0(y)h(y)\nu_{E_n}(dy).$$

It remains to express this in terms of μ_n. But

$$\mu_n(dy) = \int f_0(x)h(x)_hP^x\{X_{L_n-} \in dy\}\xi(dx) = \hat{U} f_0(y)h(y)\nu_{E_n}(dy)$$

since $\int f_0(x)\kappa(x, y)\xi(dx) = 1$ for $y \in \boldsymbol{E}$. Thus, for a Borel set A

$${}_hP^x\{X_{L_n-} \in A\} = \frac{1}{h(x)}\int_A \kappa(x, y)\mu_n(dy).$$

Multiply by $h(x)$, take $A = \boldsymbol{E}$, and note that $L_n > 0$ if $x \in E_n - B$ to get (14.4). □

This brings us to the main theorem of this section.

Theorem 14.8. *(Martin Representation) Let h be a γ-integrable excessive function. Then there exists a unique finite measure μ on $(\boldsymbol{E} - \hat{B}) \cup \mathfrak{B}_{\min}$ such that*

$$h(x) = \int \kappa(x, y)\mu(dy). \tag{14.5}$$

Moreover, for any finite measure μ on $\boldsymbol{E} \cup \mathfrak{B}_{\min}$, (14.5) defines an excessive function on $\boldsymbol{E}$. The measure μ is related to the distribution of $X_{\zeta-}$ by ${}_hP^{\gamma^h}\{X_{\zeta-} \in dx\} = C_h\mu(dx)$ where $(C_h)^{-1} = \int h(x) f_0(x)\xi(dx)$. Moreover, h is a potential if μ sits on $\boldsymbol{E} - \hat{B}$, and h is harmonic if μ sits on $\mathfrak{B}_{\min}$.

Proof. Let $n \to \infty$ in (14.4). Note that $X^h_{L_n-} \to X^h_{\zeta-}$, so that μ_n converges weakly to μ. Thus for each $f \in \mathcal{C}_0$,

$$\begin{aligned}
\int f(x)h(x)\xi(dx) &= \iint \kappa(x, y) f(x)\xi(dx)\mu_n(dy) \\
&= \int \hat{\kappa} f(y)\mu_n(dy) \\
&\longrightarrow \int \hat{\kappa} f(y)\mu(dy) \\
&= \int f(x) \left(\int \kappa(x, y)\mu(dy) \right) \xi(dx).
\end{aligned}$$

This is true for all $f \in \mathcal{C}_0$, so it follows that $h(x) = \int \kappa(x, y)\mu(dy)$ for ξ-a.e. x. Both sides are excessive, so there is equality for all x.

Next, we show that μ sits on $\mathfrak{B}_{\min}$. The crux of the proof is to show that $X_{\zeta-}$ is a.s. a minimal point. For this we prove a lemma. We write ${}_\eta P$ and X^η_t in place of ${}_{\kappa(\cdot,\eta)}P$ and $X^{\kappa(\cdot,\eta)}$ respectively.

Lemma 14.9. *A point $\eta \in \boldsymbol{E}^*$ is minimal iff ${}_\eta P^{\gamma^h}\{X_{\zeta-} = \eta\} = 1$.*

Proof. This is true for $\eta \in \boldsymbol{E} - \hat{B}$ by Corollary 13.61, so we need only consider $\eta \in \mathfrak{B}$. Since $X_{\zeta-}$ is measurable with respect to the invariant field $\mathcal{I}$, its distribution must be a point mass by Proposition 11.17. Suppose the point mass is at a point η'. By (14.5) and the above we have $\kappa(\cdot, \eta) = \kappa(\cdot, \eta')$, which means that $\eta' = \eta$ by Proposition 14.3. Conversely, suppose that $X^\eta_{\zeta-} = \eta$ a.s.. If $\kappa(\cdot, \eta) = g_1 + g_2$, where the g_i are excessive, then by Proposition 11.10, for all $x \in \boldsymbol{E}$,

$$\begin{aligned}
1 &= P^x\{X^\eta_{\zeta-} = \eta\} \\
&= \frac{g_1(x)}{g_1(x) + g_2(x)} P^x\{X^{g_1}_{\zeta-} = \eta\} + \frac{g_2(x)}{g_1(x) + g_2(x)} P^x\{X^{g_1}_{\zeta-} = \eta\},
\end{aligned}$$

which implies that both probabilities on the right-hand side equal one. It follows from (14.5) that both g_1 and g_2 are proportional to $\kappa(\cdot, \eta)$, and hence the latter is minimal. This proves the lemma. □

Returning to the proof of the theorem, let us show that μ sits on $\mathfrak{B}_{\min}$. We will do this in a round-about way, by calculating the conditional distribution of the process (X_t^h) conditioned on $X_{\zeta-}^h$. (Warning: this argument has some sharp bends.) We assume we have defined the process on a nice probability space, such as the canonical space. We can then define a regular conditional probability given $X_{\zeta-}^h$, which allows us to speak rigorously of the process conditioned on $\{X_{\zeta-}^h = y\}$. We claim that it is a $\kappa(\cdot, X_{\zeta-}^h)$-transform of (X_t).

To see this, we first find the conditional distribution of $X^n \stackrel{\text{def}}{=} \{X_t, 0 \le t < L_n\}$ given X_{L_n-}. Suppose $x \in E_n$. Then L_n is a strictly positive, a.s.-finite co-optional time, and we can compute the distribution of (X_t^n) directly: we claim it has the distribution of a $\kappa(\cdot, X_{L_n-}^h)$-transform of (X_t). This follows from a double-reversal argument. Let $(\tilde{X}_t)$ be the reverse of (X_t) from L_n. By Corollary 13.36 (ii), $(\tilde{X}_t)$ is a $u(x, \cdot)$-transform of $(\hat{X}_t)$; by Proposition 13.43 its initial value is in $\boldsymbol{E}$, so we can condition on it. But note that conditioning on $\{X_{L_n-} = y\}$ is equivalent to conditioning on $\{\tilde{X}_0 = y\}$, for the two sets are equal. Thus, the process $\{X_t, t < L_n\}$ conditioned on $\{X_{L_n-} = y\}$ is the reverse of a transform of $(\hat{X}_t)$ which *starts* at y, and this is, again by Corollary 13.36 (iii), a $u(\cdot, y)$-transform of (X_t). But $u(\cdot, y)$ and $\kappa(\cdot, y)$ are proportional, so (X_t^n) is also a $\kappa(\cdot, y)$-transform, as claimed!

Now we will show that the finite-dimensional distributions of (X_t^n) converge as $n \to \infty$ to those of a $\kappa(\cdot, y)$-transform. Let $Y = \prod_{j=1}^n f_j(X_{t_j}^h)$, where $f_j \in \mathcal{C}_0$ and $t_1 < t_2 < .. < t_n \le t$. We note that by the strong Markov property of the reversed process, $E^x\{Y \mid X_{L_n-}^h, X_{L_{n+1}-}^h, \dots\} = E^x\{Y \mid X_{L_n-}^h\} = \kappa(\cdot, X_{L_n-})E\{Y\}$. On the set $\{X_{L_n-} = y\}$ this is:

$$\frac{1}{\kappa(x, y)} \int \dots \int P_{t_1}(x, dx_1) f_1(x_1) \dots$$

$$P_{t_n - t_{n-1}}(x_{n-1}, dx_n) f_n(x_n) \kappa(x_n, X_{L_n-}) = \frac{\hat{\kappa}\nu(X_{L_n-})}{\kappa(x, X_{L_n-})},$$

where $\nu(dz) = \int \dots \int P_{t_1}(x, dx_1) f_1(x_1) \dots P_{t_n - t_{n-1}}(x_{n-1}, dz) f(z)$. By Lemma 14.4 (i), both numerator and denominator have limits with probability one as $n \to \infty$; and by part (ii) of the same lemma, the sequence is a positive reverse supermartingale, so the limit can only be zero if the supermartingale vanishes. Thus the ratio itself has a limit, and Lemma 14.5 identifies the limit as $\hat{\kappa}\nu(X_{\zeta-})/\kappa(x, X_{\zeta-})$. As this is measurable with respect to $\sigma\{X_{\zeta-}\}$, we have proved that the process conditioned on $\cup_n \mathcal{F}_{-n}$ is the same as the process conditioned on $X_{\zeta-}$, and that the process conditioned on $\{X_{\zeta-} = y\}$ is a $\kappa(\cdot, y)$-transform of (X_t).

We are just about to pull the proverbial rabbit out of the proverbial hat. Does the reader see – *without reading the following sentence* – why $X_{\zeta-}$ is a.s. a minimal point?

Indeed it is, for by definition, the terminal value of the process X_t^h conditioned on $\{X_{\zeta-} = y\}$ (i.e. of the y-transform) must be y itself; the conditioned process is (X_t^y), and therefore $X_{\zeta-}^y = y$ a.s. ... so y is minimal by Lemma 14.9!

We have finished the hard work. The characterization of potentials and harmonic functions follows easily from Propositions 13.50 and 13.57 by the following argument. We have just seen that $C_h\mu(\boldsymbol{E} - \hat{B}) = {}_hP^{\gamma^h}\{X_{\zeta-} \in \boldsymbol{E} - \hat{B}\}$. This equals one if μ sits on $\boldsymbol{E} - \hat{B}$ and zero if μ sits on $\mathfrak{B}_{\text{min}}$. In the former case, we conclude that ${}_hP^x\{X_{\zeta-} \in \boldsymbol{E} - \hat{B}\} = 1$ for ξ-a.e. x, and hence for all $x \in \boldsymbol{E}_h$, since $x \mapsto {}_hP^x\{X_{\zeta-} \in \boldsymbol{E} - \hat{B}\}$ is h-excessive. Thus h is a potential. Similarly, in the latter case, the probability is identically zero, and therefore h is harmonic.

Finally, to see uniqueness, suppose h has the representation (14.5) and a second representation, say $h = \int \kappa(\cdot, y)\nu(dy)$. Then for any Borel set $A \subset \boldsymbol{E}^*$,

$$
\begin{aligned}
C_h\mu(A) &= {}_hP^{\gamma^h}\{X_{\zeta-} \in A\} \\
&= C_h \int \xi(dx) f_0(x) h(x) \int_{\boldsymbol{E}^*} \kappa(x, \eta)_\eta P^x\{X_{\zeta-} \in A\}\nu(dy).
\end{aligned}
$$

But ${}_\eta P^x\{X_{\zeta-} \in A\} = I_A(\eta)$ if η is minimal, so this is

$$
= C_h \int_A \int \kappa(x, \eta) f_0(x) \xi(dx)\nu(dy) = C_h\nu(A).
$$

Thus $\nu = \mu$, proving uniqueness. □

14.5. Applications

If we apply Proposition 11.10 to the Martin representation, we see that if h is given by (14.5) and $\Lambda \in \mathcal{F}$, $x \in \boldsymbol{E}$, that

$$
{}_hP^x(\Lambda) = \int_{(E - \hat{B})\cup\mathfrak{B}_{\text{min}}} \kappa(x, y)_y P^x(\Lambda)\mu(dy). \tag{14.6}
$$

This brings us to our characterization of the invariant field $\mathcal{I}$. This can be thought of as the dual to Blumenthal's zero-one law.

Theorem 14.10. *Let h be a γ-integrable excessive function and let $x \in \boldsymbol{E}$. Then, up to ${}_hP^x$-null sets, the invariant field equals $\sigma\{X_{\zeta-}\}$.*

Proof. Let $\Lambda \in \mathcal{I}$, $x \in \boldsymbol{E}$. For each minimal point y, ${}_yP^x\{\Lambda\} = 0$ or 1 (Proposition 11.17). Let $A = \{y : {}_yP^x(\Lambda) = 1\}$ and let $\check{A} = \{X_{\zeta-} \in A\}$. Then

${}_yP^x\{X_{\zeta-} = y\} = 1$, so that

$$y \in A \Longrightarrow {}_yP^x\{\check{\Lambda}\} = 1 = {}_yP^x\{\Lambda\};$$

$$y \in A^c \Longrightarrow {}_yP^x\{\check{\Lambda}\} = 0 = {}_yP^x\{\Lambda\}.$$

In both cases, ${}_yP^x\{\Lambda - \check{\Lambda}\} = {}_yP^x\{\check{\Lambda} - \Lambda\} = 0$. □

Remark 14.11. We mentioned that the Martin compactification is essentially the same as the dual Ray-Knight compactification. We say "essentially" because the lifetime ζ may not be finite. If the lifetime were finite for (X_t) and for all h-transforms of it, then the Martin and Ray-Knight compactifications would be the same. (See Theorem 14.12 below.) But if the lifetime can be infinite, there is a chance that the Ray-Knight compactification might be strictly smaller.

However, it is profitable to think of the Martin compactification as the dual Ray-Knight compactification. This helps explain the rôle of the non-minimal points. They are simply the co-branching points of the Martin boundary. In the same vein, the fact that $X_{\zeta-}$ is never in the set of non-minimal points is equivalent to the fact that that the reverse process never hits the co-branching points. Theorem 14.10 can be seen as a reflection of the Blumenthal zero-one law for the reversed process, for the invariant field of (X_t) is essentially $\tilde{\mathcal{F}}_{0+}$ for $(\tilde{X}_t)$, the process reversed from ζ. Once again, we are making heavy use of the word "essentially" since, as we remarked, we can't really reverse time on the set $\{\zeta = \infty\}$, but intuitively, that is what is happening.

Define $\{\kappa_p, p > 0\}$ by $\kappa_p(x, y) = u_p(x, y)/\hat{U} f_0(y)$, and define $\hat{\kappa}_p$ by $\hat{\kappa}_p f(y) = \int \kappa_p(x, y) f(x)\xi(dx)$. Note that this is the resolvent of the Uf_0-transform of $(\hat{X}_t)$. We leave it to the reader to show that $\hat{\kappa}_p$ as well as $\hat{\kappa}$ can be extended continuously to $\boldsymbol{E}^*$. (See Exercise 1.) Then we have the following extension of Theorem 13.45.

Theorem 14.12. *Let h be a γ-integrable excessive function with the property that ${}_hP^\gamma\{\zeta < \infty\} = 1$. Let (X_t^h) be the h-transform of (X_t) with initial distribution γ^h. If $(\hat{Y}_t)$ is the right-continuous reverse of (X_t^h) from its lifetime, where finite, then $\{\hat{Y}_t, t \geq 0\}$ is a strong Markov process.*

Proof. We know that $\{\hat{Y}_t, t > 0\}$ is a $\hat{U} f_0$-transform of $\hat{X}_t$, so that its resolvent is $(\hat{\kappa}_p)$. But for each $f \in \mathcal{C}_0$, $\hat{\kappa}_p f$ is continuous on $\boldsymbol{E}^*$, and $\lim_{t\to 0} \hat{Y}_t = \lim_{t\to\zeta-} X_{t-}^h$ exists in $\boldsymbol{E}^*$. It is then easy to conclude that $E\{\int_0^\infty e^{-ps} f(\hat{Y}_{s+t})ds \mid \mathcal{F}_t\} = \hat{\kappa} f(\hat{Y}_t)$ and to see that as $t \mapsto 0$ this tends to $E\{\int_0^\infty e^{-ps} f(\hat{Y}_s)ds \mid \hat{\mathcal{F}}_0\} = \hat{\kappa}_p f(\hat{Y}_0)$ so that $\hat{\kappa}_p$ is the resolvent of $(\hat{Y}_t)$ on all of $\boldsymbol{E}^*$. But then $t \to \hat{\kappa}_p f(\hat{Y}_t)$ is right continuous for all $t \geq 0$, which implies the strong Markov property. □

Notice that this implies that it is possible to start a transform of $(\hat{X}_t)$ from the Martin boundary. It is natural to ask what happens to κ_p at points η of the Martin boundary for which ${}_\eta P\{\zeta = \infty\} = 1$. In that case, if $p > 0$, $\hat{\kappa}_p \equiv 0$. This is one reason we used the potential kernel, not the resolvent, to define the Martin boundary. Had we used the resolvent, we would have only picked up the minimal points η for which (X_t^η) has a finite lifetime.

14.6. The Martin Boundary for Brownian Motion

Let us now consider the situation we investigated in §4.5. We adopt the notation of that section. Let D be a bounded domain in $\boldsymbol{R}^n$, for some $n \geq 1$, and consider a Brownian motion $\{X_t, t \geq 0\}$ in D, killed when it first hits the boundary of D. Denote its transition function by $P_t(x, dy)$, which has a density $p_t(x, y)$ with respect to Lebesgue measure. We denote its potential kernel by $g_D(x, y)$. (This was denoted $u(x, y)$ above, but we will use the "g_D" instead for "Green's function," or "Green's potential.") The basic fact about g_D is that it is symmetric in x and y: $g_D(x, y) = g_D(y, x)$, and therefore, it is self-dual with respect to Lebesgue measure. Since $G_D f(x) = \hat{G}_D f(x) \stackrel{\text{def}}{=} \int_D g_D(x, y) f(y) dy$ is bounded and continuous for any bounded Borel f, G_D satisfies the strong Feller property $(SF) = (\widehat{SF})$. There are no branching or co-branching points. The lifetime ζ is the first hit of the boundary of D, and it is a.s. finite.

It was shown in §4.5 Theorem 3 that "excessive" = "positive superharmonic." It is an interesting exercise to show that the notions of "harmonic" in §4.3 and §13.11 are the same.

Thus, hypotheses §14.1 are satisfied, and the Martin boundary $\mathfrak{B}$ exists. The kernel κ extends to a kernel $\kappa(x, y)$ on $\boldsymbol{E} \times \boldsymbol{E}^*$. It is superharmonic in x, $G_D f_0$-superharmonic in y, and lower-semi-continuous in y.

A priori, $\boldsymbol{E}^*$ might depend on the choice of function f_0. However, it does not, providing we let f_0 have compact support.

Indeed, let $f_0 \geq 0$ be a bounded Borel function with compact support in D, such that $\int f_0(x)dx = 1$. Then $G_D f_0$ is bounded, continuous and strictly positive on D, and the measure $(\int f_0(x) g_D(x, y) dx) dy$ is equivalent to Lebesgue measure, and is hence a reference measure. Thus all our conditions are satisfied, and we can define the Martin kernel by

$$\kappa(x, y) = \frac{g_D(x, y)}{G_D f_0(y)}.$$

The Martin space can then be described as the smallest compact space D^* containing D as a dense subspace, such that all functions of the form κf, $f \in \mathcal{C}_0$, have continuous extensions to D^*. The Martin boundary is $\mathfrak{B} = D^* - D$, and $\mathfrak{B}_{\text{min}}$ denotes the set of minimal points of $\mathfrak{B}$.

Let $\gamma(dx) = f_0(x)dx$, and note that, since positive superharmonic functions are locally integrable (why?), *every* positive superharmonic function is

γ-integrable. Thus the representation theorem applies to *all* superharmonic functions.

Theorem 14.13. *Let h be a positive superharmonic function on D. Then there exists a unique finite measure μ on $D \cup \mathfrak{B}_{\min}$ such that*

$$h(x) = \int_{D\cup\mathfrak{B}_{\min}} \kappa(x, y)\mu(dy). \tag{14.7}$$

If μ sits on D, h is a potential, and if μ sits on $\mathfrak{B}_{\min}$, h is harmonic.

While the process (X_t) is killed upon first hitting the boundary, h-transforms may be killed in D, before they reach the boundary. The behavior of arbitrary h-transforms can be deduced from (14.7), Proposition 14.6, and the following.

Proposition 14.14. *If y is a minimal point of $\boldsymbol{E}$, then $\kappa(\cdot, y)$ is a minimal positive superharmonic function with pole y and the conditioned process (X_t^y) converges to y at its lifetime. If $y \in D$, $\kappa(\cdot, y)$ is a potential, and the lifetime of (X_t^y) is finite. If $y \in \mathfrak{B}_{\min}$, $\kappa(\cdot, y)$ is harmonic and (X_t^y) converges to the boundary of D, and its lifetime equals the first exit time of D.*

Theorem 14.10 translates directly into:

Theorem 14.15. *For each $x \in D$ and positive superharmonic function h, the invariant sigma field is equal to $\sigma(X_{\zeta-})$ up to sets of ${}_hP^x$-measure zero.*

Note that the region D now has two boundaries: its geometrical boundary $\bar{D} - D$ and its Martin boundary $D^* - D$. If the geometrical boundary is smooth, the two are the same; however, they are generally different. The following example is a good illustration.

Example 14.16. Let D be the unit disk in $\boldsymbol{R}^2$ with the line segment from the origin to $(1, 0)$ removed. The geometrical boundary is the unit circle plus the ray. The Martin boundary is larger. Notice that there are two ways for a Brownian motion to approach a boundary point on the ray: it can either approach it from above or below. Since the (minimal) Martin boundary is in one-to-one correspondence with the modes of Brownian convergence to the boundary (this is another way of phrasing Proposition 14.15) it follows that each point of the ray splits into two points of the Martin boundary, so that the Martin compactification is isomorphic to the "pac-man" in Fig. 14.1

14.7. The Dirichlet Problem in the Martin Space

Let us return to the Dirichlet problem of §4.4, posed this time for Brownian motion on the Martin space. Let f be a Borel measurable function on $\mathfrak{B}$.

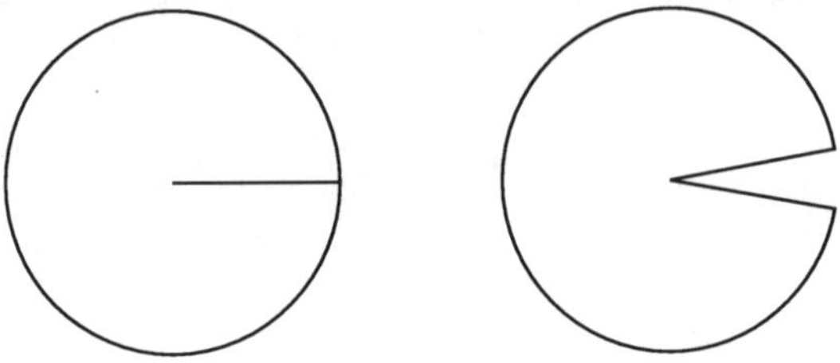

Fig. 14.1. The disk minus a ray and its Martin compactification.

To avoid inessential difficulties, let us suppose it is bounded. Problem: find a function h which is harmonic in D and which satisfies, in some sense, $\lim_{x\to\eta} h(x) = f(\eta)$ for $\eta \in \mathfrak{B}$. We don't assume f is continuous, so we can't expect the limit to hold at every point, and, in fact, part of the problem is to find the sense in which the limit exists. This will take us back to the fine topology of §3.5, but in a new setting.

Strangely enough, the solution of this problem is easy, given what we know. It is

$$h_f(x) \stackrel{\text{def}}{=} E^x\{f(X_{\zeta-})\}.$$

The problem is not in finding the solution, it is in making the problem precise! Let us see in what sense h_f has the boundary value f.

First, as in §4.4, this implies the existence of **harmonic measure** $H_D(x, A)$ on the Martin boundary, defined by

$$H_D(x, A) = P^x\{X_{\zeta-} \in A\}.$$

This is harmonic in x for fixed A, and it defines a probability measure on $\mathfrak{B}$ for fixed x. Notice that by the harmonicity in x, if $H_D(x, A)$ vanishes for one $x \in E$, it vanishes for all. Thus we can say that a set has harmonic measure zero without ambiguity.

We saw in §4.4 that h_f was harmonic. (This also follows from Theorem 14.10: $f(X_{\zeta-})$ is invariant, i.e. $\mathcal{I}$-measurable, so its expectation has to be harmonic.) By the Markov property,

$$h_f(X_{t\wedge\zeta}) = E\{f(X_{\zeta-}) \mid \mathcal{F}_{t\wedge\zeta}\}.$$

Thanks to the martingale convergence theorem, this converges a.s. to

$$E\{f(X_{\zeta-}) \mid \mathcal{F}_{\zeta}\} = f(X_{\zeta-})$$

since the latter is $\mathcal{F}_{\zeta}$-measurable. It follows that

$$\lim_{t\to\zeta-} h_f(X_t) = f(X_{\zeta-}) \text{ a.s.} \tag{14.8}$$

Thus f is the boundary value of h_f in the sense that it is the limit of h_f along Brownian paths. We would like to state this in a more conventional way, by defining a topology at the boundary. By Theorem 14.13, if $\eta \in \mathfrak{B}_{\text{min}}$, $\kappa(\cdot, \eta)$ is a minimal harmonic function with pole at η. We say a set A is **thin** at η if for $x \in D$, ${}_\eta P^x\{L_A < \zeta\} = 1$. We say that $C \subset D$ is a **fine neighborhood of** η if C^c is thin at η, and we say that a function f on D has the **fine limit** L at η if for every $\varepsilon > 0$, $\{x : |f(x) - L| < \varepsilon\}$ is a fine neighborhood of η, and we write fine $\lim_{x\to\eta} f(x) = L$.

If A is a fine neighborhood of η, then its complement is thin, so that the process (X_t^η) must leave A^c before ζ. Thus, A is a fine neighborhood of η iff $X_t^y \in A$ for t in some interval $(\zeta - \epsilon, \zeta)$.

Here is the connection between fine limits and limits along Brownian paths.

Lemma 14.17. *Let f be Borel on D, and $L \in \boldsymbol{R}$. Then* fine $\lim_{x\to\eta} f(x) = L$ *iff*

$$ {}_\eta P^x\{\lim_{t\to\zeta-} f(X_t) = L\} = 1. $$

Proof. Let $\varepsilon > 0$. If the fine limit equals L, then with probability one, X_t^η is in the set $A_\varepsilon = \{x : |f(x) - L| < \varepsilon\}$ for t in some interval $(\zeta - \delta, \zeta)$ (δ is random, of course); this is true simultaneously for a set of $\varepsilon \downarrow 0$, implying that $\lim_{t\to\zeta-} f(X_t^\eta) = L$. Conversely, if $f(X_t^\eta)$ converges to L at ζ, it is almost surely in the set A_ε for some (random) time interval $(\zeta - \delta, \zeta)$, so that A_ε is a fine neighborhood of η. It follows that the fine limit equals L. □

Theorem 14.18. *Let f be a bounded Borel function on the Martin boundary. Then for a.e. (harmonic measure) $\eta \in \mathfrak{B}$,* fine $\lim_{x\to\eta} h_f(x) = f(\eta)$.

Proof. $1 = P^x\{\lim_{t\to\zeta-} h_f(X_t) = f(X_{\zeta-})\}$. Condition on $X_{\zeta-}$ whose distribution is $H_D(x, d\eta)$. Then

$$ \begin{aligned} 1 &= \int {}_\eta P^x\{\lim_{t\to\zeta-} h_f(X_t) = f(X_{\zeta-})\} H_D(x, d\eta) \\ &= \int {}_\eta P^x\{\lim_{t\to\zeta-} h_f(X_t) = f(\eta)\} H_D(x, d\eta). \end{aligned} $$

But $H_D(x, \cdot)$ is a probability measure, so the probability in the integrand must equal one $H_D(x, \cdot)$-a.e. □

As might be expected, the idea of fine limits at the boundary is quite useful. See the exercises for some further examples of fine-limit theorems.

Exercises

1. Show that Proposition 14.2 (i), (iii) and Proposition 14.3 hold with κ replaced by $p\kappa_p$ for any $p > 0$.

2. Prove that the conclusion of Theorem 14.12 holds for any initial measure on $\boldsymbol{E}_h$.
3. Show that the Martin boundary and the geometrical boundary of the ball coincide.
4. Show that in $\boldsymbol{R}^2$, the Martin boundary of a region is conformally invariant. That is, if D is the conformal image of the unit disk, there is a one-to-one correspondence between the points of the unit circle and the minimal points of the Martin boundary. (Hint: Brownian motion itself is conformally invariant modulo a time change.)
5. Let f be a measurable function on $\mathfrak{B}$. Show that if f is $H_D(x, \cdot)$-integrable for one $x \in D$, it is $H_D(x, \cdot)$-integrable for all $x \in D$. (If so, we say that f is integrable with respect to harmonic measure.)
6. Generalize the Dirichlet problem and its solution to functions which are integrable with respect to harmonic measure.
7. State and prove Theorem 14.18 in the general case, i.e. under the hypotheses of §14.4.
8. Let D be the disk-minus-a-ray of Fig. 14.1. Show that if (X_t) is Brownian motion on $\boldsymbol{R}^2$ killed when it hits the boundary of D, and that if $(\tilde{X}_t)$ is the right-continuous reverse from the lifetime, that $\{X_t, t \geq 0\}$ is not a strong Markov process.

Notes on Chapter 14

The Martin boundary was created by R. S. Martin in his pivotal paper [1]. Although it took some time to realize it, this paper links classical potential theory, convex analysis and Markov processes, and for Markov processes, it links conditioning, h-transforms, and reversal. In addition, it may still be the best fundamental account of the Martin boundary, for it constructs it in its purest setting (harmonic functions on a Euclidean domain) and it sets out the ideas – at least the analytic ideas – quite clearly.

Our treatment is a synthesis of the approaches of H. Kunita and T. Watanabe [2] and G. Hunt [4]. We use Kunita and Watanabe's hypotheses (slightly adapted to our situation) and their construction of the kernel $\kappa(x, y)$. Hunt's ideas are used to derive the existence and particularly, the uniqueness, of the representation and its connection with minimal excessive functions. Let us say something of the history of the Martin boundary and mention some of the ideas sparked by Martin's work.

From the analytic point of view, the Martin boundary has to do with the representation of harmonic functions in a region D of Euclidean space by an integral over a boundary kernel. The prototype of these is the well-known Poisson representation. This had long been known for the ball, and had been generalized to smooth or nearly-smooth domains which one could think of as perturbations of the ball. Martin showed how to define an intrinsic boundary for any region, no matter how irregular its boundary, and how to construct the representation from that.

Doob [5] discovered the deep connection with Brownian motion, including the relations between the h-transform, conditioning, and the Martin boundary. He used this to prove Fatou-type theorems for harmonic functions and their ratios at the Martin boundary. He then showed that one could do all of this for Markov chains as well [7]. This strongly suggested the possibility of further generalization. G. Hunt, in the seminal paper [4] established the connection between the Martin boundary and time reversal, again for Markov chains. Hunt used what he called "approximate Markov Chains" to handle the fact that one might have to reverse from an infinite time. K. L. Chung [9] showed that for a continuous-time Markov chain killed at its first infinity, the reversal time would be finite and one could construct the boundary directly by time-reversal.

Martin was motivated by a proof involving the normal derivative of the Green's function (i.e. the potential kernel) at the boundary of a smooth region. He remarked that the normal derivative could be written as the boundary limit of the ratio of two Green's functions with different poles. This led to his definition of the kernel, which would be, in our notation, $K(x, y) = \frac{u(x,y)}{u(x_0,y)}$. He defined the Martin boundary as the set of all limit points of $K(x, y)$ as y tends to the geometric boundary. The probabilistic interpretation of Martin's kernel – that it is the potential kernel of the reverse of a Brownian motion from its first hit of the boundary – is due to Hunt (in the Markov chain setting), although Doob gave a strong hint of it earlier in [5].

We should point out that we replaced $u(x_0, y)$ by the function $\hat{U} f_0(y)$. The latter is bounded, continous, and strictly positive, while the former may be infinite when $y = x_0$, and in our general setting, it may also vanish on other sets.

It was clear after Doob's work that these ideas must hold in great generality, but there were formidable technical obstacles to proving it. Kunita and Watanabe [2] showed the existence of the Martin boundary and the representation theorem under essentially the same hypotheses used here, and they showed the critical importance of the dual resolvent. P. A. Meyer [5], in his account of the Kunita-Watanabe paper, proved the representation by a theorem of Choquet. This introduced the connection with convex analysis: the set of potentials, the set of positive harmonic functions, and the set of excessive functions are all convex cones, and Martin's representation theorem is a representation of the elements of the cone by an integral over its extremal rays, which are made up of minimal excessive functions. This aligned Martin's theorem with Choquet's and the Krein-Milman theorems. Mokobodzki [1, 6] used Choquet's theorem to prove the most general representation theorem to date for excessive functions.

Chapter 15

The Basis of Duality: A Fireside Chat

We've seen that duality hypotheses let one extend some of the more delicate results of classical potential theory. This raises the question, "What does the duality hypothesis itself really add?" In other words, given a strong Markov process satisfying Hypothesis (L), what more is needed to assure it has a strong Markov dual?

Let (X_t) be such a process. Let (U_p) be its resolvent, and let ν be a reference probability measure.

Once again, we want to chat about a question without going into technical details, so we'll just give the definitions and results, with the merest indication of proof, concentrating on the intuition. The key idea in what follows is that the dual process is closely connected to the right-continuous reverse of (X_t).

We need at least two things for duality: first, a duality measure ξ: ξ is an excessive reference measure, so at least one excessive reference measure must exist. Next, a strong Markov dual has a fine topology, which we call the cofine topology. So there must be a cofine topology. (Of course, this begs the question: without a dual, how does one define the cofine topology? We'll see below.) This turns out to be enough: together, they imply the existence of a strong Markov dual.

15.1. Duality Measures

Let's take them in order. First comes the question of a duality measure. The answer is quite satisfying. It turns out to be a transience/recurrence question: if the process is either transient or recurrent, there is such a measure. The only case in which there is none is when the state space is a non-trivial melange of transient and recurrent states.

Definition 15.1. *A point x is* **finely transient** *if there is a fine neighborhood V_x of x such that $U_0(x, V_x) < \infty$. It is* **finely recurrent** *otherwise. The set of finely transient points is denoted $\mathbf{E}^T$ and the set of finely recurrent points is denoted $\mathbf{E}^R$.*

Theorem 15.2. *There exists a non-trivial σ-finite excessive reference measure on $\boldsymbol{E}$ if and only if for each $x \in \boldsymbol{E}^T$*

$$U_p(x, \boldsymbol{E}^R - \{\partial\}) = 0, \quad \forall p > 0. \tag{15.1}$$

The process can have all recurrent or all transient states or half-and-half, as long as it cannot get from $\boldsymbol{E}^T$ to $\boldsymbol{E}^R$. (Consider the trivial example in which $\boldsymbol{E}$ consists of two states, say 0 and 1; if $X_0 = 0$, then (X_t) stays at 0 for an exponential length of time, then jumps to 1 and stays there forever after. Why is there no non-trivial reference measure?)

Remark 15.3. There are p-excessive reference measures for any $p > 0$. Indeed, the corresponding process, (X_t^p), has a finite lifetime, so there are no finely recurrent points. Thus (15.1) is always satisfied.

It turns out that any σ-finite excessive reference measure will do as a duality measure. This might seem surprising, but in fact, one can see from Theorem 13.33 that a change of duality measure just entails replacing the dual process by an h-transform.

15.2. The Cofine Topology

This brings us to the question of a cofine topology. We must define it in terms of (X_t). Rather than defining its open sets or its closure properties, we characterize it by its actions on functions. If f is a function on $\boldsymbol{E}$, and if $\mathcal{T}$ is a topology on $\boldsymbol{E}$, let

$$f^{\mathcal{T}}(x) = \mathcal{T} - \limsup_{y \to x, y \neq x} f(y).$$

Definition 15.4. *Let $\mathcal{T}$ be a topology on* **E** *which is finer than the metric topology. Then $\mathcal{T}$ is said to be a* **cofine topology** *if*

(i) *for all $\mathcal{T}$-continuous f, $t \mapsto f(X_{t-})$ is a.s. left-continuous;*
(ii) *for all $t > 0$ and $f \in b\mathcal{E}$, the set*

$$\{\omega : s \mapsto f(X_{s-}(\omega)) \text{ has essential left limits at all } s \in (0, t)\}$$

is a.s. contained in

$$\{\omega : s \mapsto f^{\mathcal{T}}(X_{s-}(\omega)) \text{ is left-continuous on } (0, t)\}.$$

To understand this definition, consider the right-continuous reverse $\tilde{X}_s \equiv X_{(t-s)-}$. Then (i) says that a cofine continuous function is right-continuous along the paths of $(\tilde{X}_t)$. This alone isn't enough to determine the topology – the metric topology satisfies it, for instance. But (ii) can be thought of as a description of the fine closure of a set: if f is the indicator function of a fine open set, $f^{\mathcal{T}}$ is the indicator of its fine closure, and (ii) says that it behaves as it should along the paths of $(\tilde{X}_t)$.

These conditions do not define a unique topology in general, for there may not be enough functions satisfying (ii) to determine it. But excessive and supermedian functions satisfy (ii), and, of course, co-excessive functions do too, once we have a dual to define them. This brings us to the theorem.

Theorem 15.5. *Suppose there exists a σ-finite excessive reference measure ξ. Then*

(i) *there exists a left-continuous moderate Markov process $(\hat{X}_{t-})$ such that (X_t) and $(\hat{X}_{t-})$ are in duality with respect to ξ;*

(ii) *there exists a rcll strong Markov process $(\hat{X}_t)$ such that (X_t) and $(\hat{X}_t)$ are in duality with respect to ξ if and only if there is a cofine topology.*

Remark 15.6. 1° Note that there is always a left-continuous moderate Markov dual, whether or not there is a cofine topology. Thus the mere existence of a moderate Markov dual adds nothing new, nothing not already present in the initial process. The entire question hinges on whether the right-continuous modification of the left-continuous dual is strongly Markov. The existence of a cofine topology is necessary and sufficient for this.

2° The fine topology of $(\hat{X}_t)$ now becomes **the** cofine topology; it is also **a** cofine topology in the sense of the definition, although there may be others.

Outline of Proof. The idea is straightforward: reverse the process; the reverse will be (up to an h-transform) the dual. There will be a strong Markov dual iff – up to the same h-transform – the *right-continuous* reverse is a strong Markov process.

First kill (X_t) at an exponential time to be sure that there is a finite lifetime to reverse from, then reverse it to get a left-continuous moderate Markov process $(\tilde{X}_t)$ with resolvent $(\tilde{U}_p)$. By Theorem 10.1 (X_t) and $(\tilde{X}_t)$ are in duality with respect to a certain measure. This may not be the given excessive measure ξ, but this can be taken care of by an h-transform, proving (i).

If there is a strong Markov dual, there will be a cofine topology. Conversely, suppose there is a cofine topology. For $g \in b\mathcal{E}$, $t \mapsto \tilde{U}_p g(\tilde{X}_t)$ is lcrl. By re-reversing, we see that $t \mapsto \tilde{U}_p(X_t)$ is rcll, and in particular, has left limits, so $t \mapsto \tilde{U}_p g(X_{t-})$ has essential left limits at all times. (As usual, we can interchange X_t and X_{t-} without affecting the essential limits.) Apply the definition of cofine topology: $(\tilde{U}_p g)^{\mathcal{T}}(X_{t-})$ is left-continuous by (i) hence, by reversing yet again, $(\tilde{U}_p g)^{\mathcal{T}}(\tilde{X}_{t+})$ is right continuous. But $(\tilde{U}_p g)^{\mathcal{T}}$ is another version of the resolvent, and right continuity of the resolvent along the paths implies the strong Markov property. This means that $(\tilde{X}_{t+})$ is a strong Markov process.

This is the key. The remainder of the proof of (ii) requires some surgery on the semigroup to assure that the reverse satisfies Hypothesis (L) with respect to ξ, and then one must undo the original exponential killing, and h-transform the result to get the proper dual, $(\hat{X}_t)$. Note that the cofine topology is only used to show that the right-continuous reverse is strongly Markov; the left-continuous reverse was already moderately Markov, without any help from the cofine topology.

Remark 15.7. If (X_t) has a finite lifetime, or if we are only interested in problems which do not involve the long-term behavior of (X_t), and can therefore consider the process killed at an exponential time, then the duality measure poses no problem: it always exists. There will be a right continuous strong Markov dual if and only if there is a cofine topology.

There are other conditions which are equivalent to the existence of a cofine topology, some of them unexpected. We will give one of them, a representation of terminal times which is interesting in its own right.

Every probability student must wonder at some point if all stopping times are first-hitting times. But it doesn't take long to realize that fixed times and second-hitting times are counter-examples. A more sophisticated conjecture would be that all *terminal* times are first-hitting times. But the first jump time is a counter-example to this. However, if we consider the pair (X_{t-}, X_t), we get a surprisingly nice answer. If (X_t) is a Ray process, then terminal times are first-hitting times of the pair (X_{t-}, X_t) if and only if there is a cofine topology. Moreover, we have:

Theorem 15.8. *Suppose (X_t) is a Ray process which satisfies Hypothesis (L). Let T be an exact terminal time, and let Δ be the diagonal of $\mathbf{E} \times \mathbf{E}$.*

(i) *If T is totally inaccessible, there exists a Borel set $K \subset (\mathbf{E} - B) \times (\mathbf{E} - B) - \Delta$ such that*

$$T = \inf\{t > 0 : (X_{t-}, X_t) \in K \quad a.s..$$

(ii) *If T is predictable, and if there exists a cofine topology, then there is a Borel set $C \subset \mathbf{E}$ such that*

$$T = \inf\{t > 0 : X_{t-} \in C\} \quad a.s..$$

(iii) *If T is accessible, and if there exists a cofine topology, there exists a Borel set $K \subset (B \times \mathbf{E}) \cup \Delta$ such that*

$$T = \inf\{t > 0 : (X_{t-}, X_t) \in K\} \quad a.s..$$

(iv) *Conversely, if (ii) holds for all predictable exact terminal times, then there exists a cofine topology.*

Remark 15.9. Note that the representation of totally inaccessible times does not require the cofine topology. Similar results hold without Hypothesis (L), but then the sets K and C may depend on the initial measure, and they may be nearly Borel rather than Borel.

Our definition of the cofine topology is somewhat unwieldy, for it is non-constructive and it involves all of the paths of the process. Here is a potential-theoretic characterization which leads to a construction of the topology. Consider the following condition.

There exists a left-polar set N and $p \geq 0$ such that if $y \in \boldsymbol{E} - N$, there exists at most one minimal excessive function with pole y. (15.2)

Theorem 15.10. *Assume hypothesis* (L). *If* (15.2) *holds, there exists a cofine topology.*

To see this, one first shows that there is a left-polar Borel set $N_1 \supset N$ such that for each $y \in \boldsymbol{E} - N$, there exists a minimal p-excessive function g_y with pole y. Then a set $C \subset \boldsymbol{E}$ is a cofine neighborhood of y if $y \in C$ and if $p^x_{g_y}\{L^-_{C^c} < j\} = 1$, $X \in \boldsymbol{E}_{g_y}$.

Note the similarity to the definition in the last chapter of a fine neighborhood of a Martin boundary point: the only difference is that $L^-_{C^c}$ replaces L_{C^c}.

Notes on Chapter 15

Theorems 15.2 and 15.5 are due to Smythe and Walsh [1]. The definition 15.4 of the cofine topology and Theorem 15.8 (i)–(iii) are from Walsh and Weil [1], and (iv) is from in Walsh and Weil [2]. The cofine topology was first defined – without using the dual process – in [1]. Theorem 15.10 is from Walsh [1]. The condition (15.2) is related to the Green's function hypothesis in axiomatic potential theory. Indeed, a measurable choice of the functions g_y gives a Green's function.

Bibliography

L. V. Ahlfors
[1] *Complex Analysis*, Second Edition, McGraw Hill Co., New York, 1966.
J. Azéma
[1] Théorie générale des processus et retournement du temps, *Ann. École Norm. Sup. Ser. 4*, **6** (1973), 459–519.
R. M. Blumenthal and R. K. Getoor
[1] *Markov Processes and Potential Theory*, Academic Press, New York, 1968.
M. Brelot
[1] *Eléments de la Théorie Classique du Potentiel*, Fourth Edition, Centre de Documentation Universitaire, Paris, 1969.
[2] Le problème de Dirichlet. Axiomatique et frontière de Martin, *J. Math. Pures et Appl.* 9e série, t. 35 (1956), 297–335.
K. L. Chung
[1] *A Course in Probability Theory*, Third Edition, Academic Press, New York, 2001
[2] *Markov Chains with Stationary Transition Probabilities*, Second Edition, Springer-Verlag 1967.
[3] *Boundary Theory for Markov Chains*, Princeton University Press, Princeton NJ, 1970.
[4] A simple proof of Doob's theorem. Sém. de Prob. V, (1969/70), p. 76. *Lecture Notes in Mathematics No. 191*, Springer-Verlag, Berlin Heidelberg New York.
[5] On the fundamental hypotheses of Hunt Processes, Instituto di alta matematica, *Symposia Mathematica* **IX** (1972), 43–52.
[6] Probabilistic approach to the equilibrium problem in potential theory, *Ann. Inst. Fourier* **23** (1973), 313–322.
[7] Remarks on equilibrium potential and energy, *Ann. Inst. Fourier* **25** (1975), 131–138.
[8] Excursions in Brownian motion, *Arkiv für Mat.* **14** (1976), 155–177.
[9] On the Martin Boundary for Markov chains, *Proc. Nat. Acad. Sci.* **48** (1962), 963–968.
[10] On the boundary theory for Markov chains, *Acta Math.* **110** (1963), 19–77; **115** (1966), 111–163.
K. L. Chung and J. L. Doob
[1] Fields, optionality and measurability, *Amer. J. Math.* **87** (1965), 397–424.
K. L. Chung and J. Glover
[1] Left continuous moderate Markov processes, *Z. Wahrscheinlichkeitstheorie* **49** (1979), 237–248.
K. L. Chung and P. Li
[1] Comparison of probability and eigenvalue methods for the Schrödinger equation, *Advances in Math.* (to appear).
K. L. Chung and K. M. Rao
[1] A new setting for potential theory (Part I), *Ann. Inst. Fourier* **30** (1980), 167–198.
[2] Equilibrium and energy, *Probab. Math. Stat.* **1** (1981).
[3] Feynman-Kac functional and the Schrödinger equation, *Seminar in Stochastic Processes* **1** (1981), 1–29.
K. L. Chung and S. R. S. Varadhan
[1] Kac functional and Schrödinger equation, *Studia Math.* **68** (1979), 249–260.
K. L. Chung and J. B. Walsh
[1] To reverse a Markov process, *Acta Math.* **123** (1969), 225–251.

[2] Meyer's theorem on predictability, *Z. Wahrscheinlichkeitstheorie und Verw. Gebiete* **29** (1974), 253–256.
[3] To reverse a Markov process, *Acta Math.* **123** (1970), 225–251.

R. Courant
[1] *Differential and Integral Calculus, Interscience* 1964.

M. Cranston and T. McConnell
[1] The lifetime of conditioned Brownian motion, *Z. Wahrscheinlichkeitstheorie und Verw. Gebiete* **65** (1983), 1–11.

C. Dellacherie
[1] Ensembles aléatoires I, II, Sém. de Prob. III (1967/68), pp. 97–136, *Lecture Notes in Mathematics No. 88*, Springer-Verlag.
[2] Capacités et Processus Stochastiques, Springer-Verlag Berlin Heidelberg New York 1972.
[3] Mesurabilité des debuts et théorème de section: le lot a la portée de toutes les bourses, Sém. de Prob. XV (1979/80), pp. 351–370, *Lecture Notes in Mathematics No.* Springer-Verlag, Berlin Heidelberg New York.

C. Dellacherie and P. A. Meyer
[1] *Probabilités et Potentiel*, Hermann, Paris 1975.
[2] *Probabilités et Potentiel: Théorie des Martingales*, Hermann, Paris 1980.

J. L. Doob
[1] *Stochastic Processes*, Wiley and Sons, New York, 1953.
[2] Semimartingales and subharmonic functions, *Trans. Amer. Math. Soc.* **77** (1954), 86–121.
[3] A probability approach to the heat equation, *Trans. Amer. Math. Soc.* **80** (1955), 216–280.
[4] Separability and measurable processes, *J. Fac. Sci. U. of Tokyo* **17** (1970), 297–304.
[5] Conditional Brownian motion and the boundary limits of harmonic functions, *Bull. Soc. Math. France* **85** (1957) 431–458.
[6] Probability and the first boundary value problem, *Ill. J. Math.* **2** (1958), 19–36.
[7] Compactification of the discrete state space of a Markov process, *Z. Wahrscheinlichkeitstheorie und Verw. Gebiete* **10** (1968), 236–251.

E. B. Dynkin
[1] *Markov Processes*, Springer-Verlag, Berlin, Heidelberg, New York, 1965.

E. B. Dynkin and A. A. Yushekevich
[1] *Theorems and Problems in Markov Processes*, (in Russian), Nauk, Moscow, 1967.

W. Feller and H. P. McKean
[1] A Diffusion equivalent to a countable Markov chain, *Proc. Nat'l Acad. Sci. USA* **42** (1956), 351–354.

O. Frostman
[1] Potential d'équilibre et capacités des ensembles avec quelques applications à la théorie des fonctions, *Medd. Lunds Univ. Math. Sem.* **3** (1935), 1–118.

R. K. Getoor
[1] Markov processes: Ray processes and right processes, *Lecture Notes in Mathematics No. 440*, Springer-Verlag, Berlin Heidelberg New York 1975.
[2] Transience and recurrence of Markov Processes, Sém. de Prob. XIV (1978/79), pp. 397–409, *Lecture Notes in Mathematics No. 784*, Springer-Verlag, Berlin Heidelberg New York.
[3] The Brownian escape process, *Ann. Probability* **7** (1979), 864–867.

I. I. Gihman and A. V. Skorohod
[1] *The Theory of Stochastic Processes II* (translated from the Russian), Springer-Verlag Berlin Heidelberg New York 1975.

L. L. Helms
[1] *Introduction to Potential Theory*, Wiley and Sons, New York, 1969.

G. A. Hunt
[1] Some theorems concerning Brownian motion, *Trans. Amer. Math. Soc.* **81** (1956), 294–319.
[2] Markoff processes and potentials, I, *Illinois J. Math.* **1** (1957), 44–93.
[3] *Martingales et Processus de Markov*, Dunod, Paris, 1966.
[4] Markoff chains and Martin boundaries, *Ill. J. Math.* **4** (1960), 316–340.
[5] Markoff processes and potentials III, *Ill. J. Math.* **2** (1958), 151–213.

N. Ikeda, M. Nagasawa, and K. Sato

[1] A time reversion of Markov processes with killing, *Kodai Math. Sem. Rep.* **16** (1964), 88–97.

K. Ito

[1] *Lectures on Stochastic Processes*, Tata Institute, Bombay 1961.

M. Jacobsen and J. W. Pitman

[1] Birth, death and conditioning of Markov Chains, *Ann. Prob.* **5** (1977), 430–450.

M. Kac

[1] On some connections between probability theory and differential and integral equations. *Proc. Second Berkeley Symposium on Math. Stat. and Probability*, 189–215. University of California Press, Berkeley 1951.

S. Kakutani

[1] Two-dimensional Brownian motion and harmonic functions, *Proc. Imp. Acad. Tokyo* **22** (1944), 706–714.

J. L. Kelley

[1] *General Topology*, D. Van Nostrand Co., New York, 1955.

O. D. Kellogg

[1] *Foundations of Potential Theory*, Springer-Verlag Berlin, 1929.

R. Z. Khas'minskii

[1] On positive solutions of the equation $\mathcal{A}u + Vu = 0$. *Theory of Probability and its Applications* (translated from the Russian) **4** (1959), 309–318.

C. Kittel

[1] *Introduction to Solid State Physics*, 3rd edition, John Wiley and Sons, New York, London, and Sydney, 1968.

F. Knight

[1] Note on regularisation of Markov processes, *Ill. J. Math.* **9** (1965), 548–552.

A. Kolmogoroff

[1] Zur theorie der Markoffschen Ketten, *Math. Ann.* **112** (1936), 155–160.

[2] Zur Unkehrbarkeit der statistischen Naturgesetze, *Math. Ann.* **113** (1937), 766–772.

H. Kunita and T. Watanabe

[1] On certain reversed processes and their applications to potential theory and boundary theory, *J. Math. Mech.* **15** (1966), 393–434.

[2] Markov processes and Martin boundaries I, *Ill. J. Math.* **9** (1965), 485–526.

[3] Some theorems concerning resolvents over locally compact spaces, *Proc. 5th Berkeley Symp.* **II** (1967), 131–164.

P. Lévy

[1] *Théorie de l'Addition des Variables Aléatoires*, Second edition, Gauthier-Villars, Paris, 1954.

R. S. Martin

[1] Minimal positive harmonic functions, *Trans. Amer. Math. Soc.* **49** (1941), 137–172.

P. A. Meyer

[1] *Probability and Potentials*, Blaisdell Publishing Co. 1966.

[2] Processus de Markov, *Lecture Notes in Mathematics No. 26*, Springer-Verlag, Berlin Heidelberg New York 1967.

[3] Processus de Markov: la frontière de Martin, *Lecture Notes in Mathematics No. 77*, Springer-Verlag, Berlin, Heidelberg, New York, 1968.

[4] Le retournement du temps, d'apres Chung et Walsh, Séminaire de Probabilités V, Univ. de Strasbourg, *Lecture Notes in Mathematics No. 191*, Springer-Verlag 1971.

[5] Processus de Markov: la frontière de Martin, *Lecture Notes in Mathematics, 77*, Springer-Verlag, 1967.

[6] Représentation intégrale des fonctions excessives: resultats de Mokobodzki, Séminaire de Probabilités V, Univ. de Strasbourg, *Lecture Notes in Math. No. 191*, Springer-Verlag, 1971.

P. A. Meyer, R. T. Smythe and J. B. Walsh

[1] Birth and death of Markov processes, *Proc. Sixth Berkeley Symposium on Math. Stat. and Probability*, Vol. III, pp. 295–305, University of California Press, 1972.

P.-A. Meyer and J. B. Walsh

[1] Quelques applications des resolvants de Ray, *Invent. Math.* **14** (1971), 143–166.

[2] Un résultat sur les résolvantes de Ray, Séminaire de Probabilités, VI (Univ. Strasbourg, année universitaire 1970–1971), 168–172, *Lecture Notes in Math. No. 258*, Springer-Verlag, 1972.

G. Mokobodzki
[1] Dualité formelle et représentation intégrale des fonctions excessives, Actes du Congrés International des mathématiciens, Nice, 1970, tome 2, 531-535, Gauthier-Villars, Paris, 1971.
M. Nagasawa
[1] Time reversions of Markov processes, *Nagoya Math. J.* **24** (1964), 177–204.
E. Nelson
[1] The adjoint Markov process, *Duke Math. J.* **25** (1958), 671–690.
A. O. Pittenger and M. J. Sharpe
[1] Regular Birth and Death Times, *Z. Wahrsch. verw. Gebiete* **58** (1981), 231–246.
S. Port and C. Stone
[1] *Brownian Motion and Classical Potential Theory*, Academic Press, New York 1978.
K. M. Rao
[1] Brownian motion and classical potential theory, *Aarhus University Lecture Notes* No. 47, 1977.
D. B. Ray
[1] Resolvents, transition functions, and strongly Markovian processes, *Ann. Math.* **70** (1959), 43–72.
H. L. Royden
[1] *Real Analysis*, Second Edition, Macmillan, New York, 1968.
M. Sharpe
[1] *General Theory of Markov Processes; Pure and Applied Mathematics 133*, Academic Press, 1988.
R. T. Smythe and J. B. Walsh
[1] The existence of dual processes, *Invensiones Math.* **19** (1973), 113–148.
F. Spitzer
[1] Electrostatic capacity, heat flow, and Brownian motion, *Z. Wahrscheinlichkeitstheorie* **3** (1964), 110–121.
J. M. Steele
[1] A counterexample related to a criterion for a function to be continuous, *Proc. Amer. Math. Soc.* **79** (1980), 107–109.
E. C. Titchmarsh
[1] *The Theory of Functions*, Second Edition, Oxford University Press, 1939.
J. B. Walsh
[1] The cofine topology revisited, *Proceedings of Symposia in Pure mathematics XXX1, Probability, Amer. Math. Soc.* (1977), pp. 131–152.
[2] Two footnotes to a theorem of Ray, Séminaire de Probabilités, V (Univ. Strasbourg, année universitaire 1969–1970), 283–289, Lecture Notes in Math., Vol. 191, Springer, Berlin, 1971.
[3] Transition functions of Markov processes, Séminaire de Probabilités, VI (Univ. Strasbourg, année universitaire 1970–1971), 215–232, Lecture Notes in Math., Vol. 258, Springer, Berlin, 1972.
[4] The perfection of multiplicative functionals, Séminairede, Probabilités VI (Univ. Strasbourg, année universitaire 1970–1971), Lecture Notes in Math. No. 258, 233–242, Berlin-Heidelberg-New York: Springer 1972.
[5] Markov processes and their functionals in duality, *Z. Wahrscheinlichkeitstheorie verw. Geb.* **24** (1972), 229–246.
[6] Some topologies connected with Lebesgue measure, Séminaire de Probabilités V, 290–311, Univ. de Strasbourg, *Lecture Notes in Math. 191*, Springer-Verlag, 1971.
[7] Time reversal and the completion of Markov processes, *Invent. Math.* **10** (1970), 57–81.
J. B. Walsh and M. Weil
[1] Représentation de temps terminaux et applications aux fonctionelles additives et aux systémes de Lévy, *Ann. Sci. de l'École Normale Sup.* 4e serie, t. 5 (1972), 121–155.
[2] *Terminal times of Ray processes II, Symposia Mathematica*, Vol. IX (Convegno di Calcolo delle Probabilità, INDAM, Rome, 1971), pp. 107–120, Academic Press, London, 1972.

T. Watanabe
[1] On the theory of Martin boundaries induced by countable Markov processes, *Mem. Coll. Sci. Univ. Kyoto Ser. A Math.* **33** (1960/1961), 39–108.
J. G. Wendel
[1] Hitting spheres with Brownian motion, *Ann. of Probability* **8** (1980), 164–169.
J. Wermer
[1] Potential Theory, *Lecture Notes in Mathematics No. 408*, Springer-Verlag, Berlin Heidelberg New York 1974.
D. Williams
[1] *Diffusions, Markov Processes, and Martingales*, Vol. 1, Wiley and Sons, New York 1979.
K. Yosida
[1] *Functional Analysis*, Springer-Verlag, Berlin Heidelberg New York 1965.

Index

Grundlehren der mathematischen Wissenschaften

A Series of Comprehensive Studies in Mathematics

A Selection

307. Schwarz: Quantum Field Theory and Topology
308. Schwarz: Topology for Physicists
309. Adem/Milgram: Cohomology of Finite Groups
310. Giaquinta/Hildebrandt: Calculus of Variations I—The Lagrangian Formalism
311. Giaquinta/Hildebrandt: Calculus of Variations II—The Hamiltonian Formalism
312. Chung/Zhao: From Brownian Motion to Schrödinger's Equation
313. Malliavin: Stochastic Analysis
314. Adams/Hedberg: Function Spaces and Potential Theory
315. Bürgisser/Clausen/Shokrollahi: Algebraic Complexity Theory
316. Saff/Totik: Logarithmic Potentials with External Fields
317. Rockafellar/Wets: Variational Analysis
318. Kobayashi: Hyperbolic Complex Space
319. Bridson/Haefliger: Metric Spaces of Non-Positive Curvature
320. Kipnis/Landim: Scaling Limits of Interacting Particle Systems
321. Grimmett: Percolation
322. Neukirch: Algebraic Number Theory
323. Neukirch/Schmidt/Wingberg: Cohomology of Number Fields
324. Liggett: Stochastic Interacting Systems—Contact, Voter and Exclusion Processes
325. Dafermos: Hyberbolic Conservation Laws in Continuum Physics
326. Waldschmidt: Diophantine Approximation on Linear Algebraic Groups
327. Martiner: Perfect Lattices in Euclidean Spaces
328. Van der Put/Singer: Galois Theory of Linear Differential Equations
329. Korevaar: Tauberian Theory—A Century of Developments